ADVA

MATHEMATICS

BOOK 1

Martin and Patricia Perkins

Bell & Hyman

First published 1983 by
BELL & HYMAN
An imprint of Unwin Hyman Limited
Denmark House
37–39 Queen Elizabeth Street
London SEI 2QB

British Library Cataloguing in Publication Data

Perkins, Martin
    Advanced maths.
    Book 1
    1. Mathematics–1961–
    I. Title   II. Perkins, Patricia
    510   QA39.2
ISBN 0 7135 1272 5

Printed in Great Britain at the
University Press, Cambridge

Contents

Preface

This is the first of two books designed to cover most Advanced Level single subject Mathematics syllabuses. The present volume provides a complete first year course for two-year candidates and should also be suitable for subsidiary level mathematics courses of various kinds. Students in schools and colleges embarking on advanced work come from a variety of backgrounds. We have therefore tried to introduce each new topic gradually to cater for those to whom it is unfamiliar.

We believe that the split between 'modern' and 'traditional' mathematics has been damaging and consider that the subject should be seen as a whole: we have attempted in this book to use what we see as the best in these two approaches to create a coherent forward-looking course. SI units are used throughout and we assume that students will have access to an adequate 'scientific' calculator.

To give students the opportunity to apply their knowledge of pure mathematics at an early stage, the chapters on applied mathematics occur at intervals throughout the two books. Many of the new combined syllabuses contain some elementary numerical methods, probability and statistics, as well as a vectorial approach to mechanics, so these topics naturally find their place in the course.

Topics are presented in a logical sequence, but the text is arranged in short chapters to allow individual students or teachers to vary the order of study. The chapters are divided into numbered sections each followed by an exercise. The early questions in these exercises are intended to provide plenty of routine practice in newly acquired skills, whereas the aim of later questions is to develop a greater understanding of the ideas involved. There is a selection of further questions in the miscellaneous exercise at the end of each chapter.

We are most grateful for permission to use questions from past G.C.E. Advanced Level examinations. These are acknowledged as follows:

University of London University Entrance and Schools Examination Council   (L)

The Associated Examining Board   (AEB)

Joint Matriculation Board   (JMB)

University of Cambridge Local Examinations Syndicate   (C)
Oxford and Cambridge Schools Examination Board   (O & C)
Oxford Delegacy of Local Examinations   (O)
Southern Universities Joint Board   (SU)
Welsh Joint Education Committee   (W)

Our thanks are also due to the staff of Messrs. Bell and Hyman, especially Peter Kaner and Anne Forsyth for their helpful criticism, and to many friends and colleagues for their encouragement and patience.

M.L.P.
P.P.

List of symbols

s.f.	significant figures
d.p.	decimal places
$\approx$	is approximately equal to
$\equiv$	is identically equal to
$<$	is less than
$\leqslant$	is less than or equal to
$>$	is greater than
$\geqslant$	is greater than or equal to
$\Rightarrow$	implies
$\Leftrightarrow$	is equivalent to
$\in$	is an element of
$:$	such that
$\subseteq$	is a subset of
$\subset$	is a proper subset of
$\varnothing$	empty set
$\mathscr{E}$	universal set
A'	complement of set A
$n(A)$	number of elements in the set A
$A \times B$	Cartesian product of sets A and B
$\mathbb{N}$	set of natural numbers $\{0, 1, 2, 3, \ldots\}$
$\mathbb{Z}$	set of integers
$\mathbb{Z}^+$	set of positive integers
$\mathbb{Q}$	set of rational numbers
$\mathbb{R}$	set of real numbers
$\mathbb{R}^2$	set of ordered pairs of real numbers (x, y)
$\cup$	union
$\cap$	intersection
$f : x \rightarrow y$	the function f maps x to y
$f(x)$	image of x under f
f^{-1}	inverse of function f
gf	function f followed by function g

$\sqrt{x}$	non-negative square root of x
$\lvert x \rvert$	modulus (or magnitude) of x
$[x]$	integral part of x
$\rightarrow$	tends to
$\lim\limits_{x \to a} f(x)$	limit of $f(x)$ as x tends to a
∞	infinity
δ	small increment
$f'(x)$ or $\dfrac{dy}{dx}$	first derivative of $y = f(x)$
$f''(x)$ or $\dfrac{d^2 y}{dx^2}$	second derivative of $y = f(x)$
$\displaystyle\int f(x)\,dx$	indefinite integral
$\displaystyle\int_a^b f(x)\,dx$	definite integral
$\Big[f(x) \Big]_a^b$	$f(b) - f(a)$
$n!$	n factorial
$_nP_r$	number of permutations of r objects chosen from n
$_nC_r$	number of combinations of r objects chosen from n
$P(A)$	probability of event A
$P(A')$	probability that A does not occur
$P(A \cup B)$	probability of event A or event B (or both)
$P(A \cap B)$	probability of event A and event B
$P(A\lvert B)$	probability of event A given B
Σ	the sum of
u_n	nth term of a sequence or series
S_n	sum to n terms of a series
$\triangle$	determinant
$\det \mathbf{A}$ or $\lvert \mathbf{A} \rvert$	determinant of matrix $\mathbf{A}$
$\begin{vmatrix} a & b \\ c & d \end{vmatrix}$	determinant of matrix $\begin{pmatrix} a & b \\ c & d \end{pmatrix}$
$\mathbf{PQ}$	vector represented by line segment $\overrightarrow{PQ}$
$\lvert \mathbf{PQ} \rvert$ or PQ	magnitude of vector $\mathbf{PQ}$
$\lvert \mathbf{a} \rvert$ or a	magnitude of vector $\mathbf{a}$
$\mathbf{i, j, k}$	unit vectors in the directions of the x-, y- and z-axes
$\dot{x}, \dot{y}, \dot{\theta}, \dots$	first derivative of $x, y, \theta, \dots$ with respect to time
$\ddot{x}, \ddot{y}, \ddot{\theta}, \dots$	second derivative of $x, y, \theta, \dots$ with respect to time

g	acceleration due to gravity, taken as $9{\cdot}8\,\mathrm{m\,s^{-2}}$ unless otherwise stated
μ	coefficient of friction
λ	angle of friction
	or modulus of elasticity
ω	angular velocity
$\bar{x}$	mean of observations $x_1, x_2, \ldots$
var (x)	variance of observations $x_1, x_2, \ldots$
s.d.	standard deviation

The use of the symbols $=$ and $\approx$

In numerical work it is important to distinguish between exact values and approximate values. When giving exact answers we use the symbol $=$. For approximate results we use either the symbol $=$ (together with an appropriate phrase, such as 'by calculator', 'to 3 s.f.'), or the symbol $\approx$. The use of the symbol $\approx$ may show that a result has been obtained by an approximate method, but often indicates that a more accurate result obtained by calculator has been rounded to the given number of significant figures (or, in the case of angles, to the nearest tenth of a degree).

1 Real numbers and their properties

1.1 The language of sets

A *set* is a collection of objects. These objects are the *elements* of the set. A set may be defined either by listing the elements or by describing their properties, e.g. $A = \{2, 4, 7\}$, B is the set of all triangles, $C = \{\ldots, -5, -3, -1, 1, 3, \ldots\}$ or $\{$odd integers$\}$. Using the colon : to mean 'such that', we may also write $P = \{p : p$ is a prime number$\}$, which is read 'P is the set of all elements p such that p is a prime number'. Whatever form is chosen it must always be possible to decide from the definition whether or not an object belongs to the set.

The statement $x \in S$ means 'x is an element of S' or 'x belongs to S'. Similarly $x \notin S$ means 'x is not an element of the set S'. Thus, for the sets defined above, $2 \in A$ but $2 \notin C$ and $9 \in C$ but $9 \notin P$.

Two sets are *equal* if they contain exactly the same elements, e.g. $\{16, 25, 36, 49, 64, 81\} = \{36, 81, 16, 25, 64, 49\} = \{$two digit squares$\}$. The way in which the elements are listed or described is not important.

Sometimes all the elements of one set are contained in another. If every element of a set A also belongs to a set B, then A is a *subset* of B, written $A \subseteq B$. If A is a subset of B and there is at least one element of B which is not an element of A, then A is a *proper subset* of B and we write $A \subset B$. For instance, if $X = \{a, c\}$, $Y = \{a, c, e\}$ and $Z = \{a, b, c, d\}$, then $X \subset Y$ and $X \subset Z$ but $Y \not\subset Z$. We may also write $Y \supset X$ and $Z \supset X$ using the symbol $\supset$ to mean 'contains as a proper subset'.

Three important properties of the set inclusion relation follow from the definition.

(1) Any set A is a subset of itself, i.e. $A \subseteq A$.
(2) $A \subseteq B$ and $B \subseteq C \Rightarrow A \subseteq C$.
(3) $A \subseteq B$ and $B \subseteq A \Leftrightarrow A = B$.

[Note that if p and q represent statements, then $p \Rightarrow q$ means 'if p then q' or 'p implies q'. The symbol $\Leftarrow$ means 'is implied by'. We write $p \Leftrightarrow q$ meaning 'p implies and is implied by q' when the statements p and q are equivalent.]

All the objects under consideration in any mathematical discussion form a set called the *universal set* (or *universe of discourse*), often denoted by $\mathscr{E}$. For instance, in a geometrical problem the universal set may be a set of points, a set of lines or a set of plane figures. Once a universal set $\mathscr{E}$ has been defined any other set considered must be one of its subsets. Thus for any set A, $A \subseteq \mathscr{E}$.

The set with no elements is called the *null* or *empty set*, written $\varnothing$ or sometimes $\{\ \}$. The empty set $\varnothing$ is considered to be a subset of any set A, i.e. $\varnothing \subseteq A$.

The *complement* of a set A is the set containing all the elements of $\mathscr{E}$ which are not elements of A and is written A'. For instance, if $\mathscr{E} = \{1,2,3,4,5\}$ and $A = \{1,3,5\}$ then $A' = \{2,4\}$. It follows from the definition that for any set A, $(A')' = A$. We also find that $\varnothing' = \mathscr{E}$ and $\mathscr{E}' = \varnothing$.

Consider now two sets X and Y. If $x \in X$ and $y \in Y$, then (x,y) is called an *ordered pair*. Although the set $\{3,4\}$ is equal to the set $\{4,3\}$, the ordered pairs $(3,4)$ and $(4,3)$ are not the same. The set of all ordered pairs (x,y) where $x \in X$ and $y \in Y$ is called the *product set* or *Cartesian product* of X and Y, written $X \times Y$. For example, if $X = \{1,2,3\}$ and $Y = \{4,5\}$,

then $\quad X \times Y = \{(1,4),(1,5),(2,4),(2,5),(3,4),(3,5)\}$
and $\quad Y \times X = \{(4,1),(5,1),(4,2),(5,2),(4,3),(5,3)\}$.

This example demonstrates the fact that in general $X \times Y \neq Y \times X$.

Exercise 1.1

1. State whether the following statements are true or false:
(a) $4 \in \{\text{even integers}\}$ (b) $\{a\} \in \{a,b,c\}$
(c) $51 \notin \{\text{prime numbers}\}$ (d) $\varnothing \subset \{3,7,11,12\}$
(e) $\{x,y\} = \{\{x\},\{y\}\}$ (f) $0 \subset \{0,1,2\}$
(g) $\{p : p \text{ is prime and } p \neq 2\} \subset \{x : x \text{ is an odd integer}\}$
(h) $\{n : n \text{ is divisible by 2, 3 and 4}\} = \{k : k \text{ is a multiple of 24}\}$.

2. Use the symbols $=$ and $\subset$ to write down any relationships between the following:
$A = \{1,2,0\}, \quad B = \{2,1\}, \quad C = \{0,1,2\}, \quad D = \{2,3\}, \quad E = \{3\}$.

3. Use the symbols $=$, $\subset$ and $\in$ to write down any relationships between the following:
$\varnothing, \quad X = \{0\}, \quad Y = \{0,1\}, \quad 0, \quad P = \{\{0\}\}, \quad Q = \{\varnothing\}$.

4. In each of the following cases state the connection between the statements p and q by writing $p \Leftrightarrow q$, $p \Rightarrow q$ or $p \neq q$.
(a) p: $x \in A$ and $A \subset B$, q: $x \in B$
(b) p: $1 \in X$ and $2 \in X$, q: $\{1,2\} \subseteq X$
(c) p: $a = x$ and $b = y$, q: $\{a,b\} = \{x,y\}$
(d) p: $x \notin A$ and $x \in B$, q: A is a proper subset of B.

5. Write down the connection between the sets X and Z if
(a) $W \subset Y$, $X \subset W$ and $Y \subset Z$,   (b) $X \subseteq Y$, $Y \subseteq Z$ and $Z \subseteq X$.

6.   Find the sets A, B, C and D which are distinct subsets of the set $\{1,2,3\}$ given that:
$2 \in C$, $3 \notin D$, $A \subset B$, $B \subset C$, $B \subset D$ and $D \not\subset C$.

7.   List the elements of the following product sets, where $A = \{x\}$, $B = \{p,q\}$ and $C = \{f,g,h\}$
(a) $A \times C$,   (b) $A \times B$,   (c) $B \times A$,   (d) $A \times A$.

8.   If $A \times B = \{(1,1), (2,1), (1,2), (3,2), (3,1), (2,2)\}$, list the elements of the sets A and B.

9.   If the universal set $\mathscr{E} = \{0,1,2\}$, write down the complements of the following sets
(a) $\{1\}$,   (b) $\varnothing$,   (c) $\{0,1\}$,   (d) $\{0,1,2\}$.

10.   Discuss the following statements
(a) $x \in A' \Leftrightarrow x \notin A$
(b) $A = \{\text{even integers}\} \Leftrightarrow A' = \{\text{odd integers}\}$
(c) $A \subset B \Leftrightarrow B' \subset A'$
(d) $A \subset B$ and $B' \subset C' \Rightarrow A \not\subset C$.

1.2 Rational and irrational numbers

Most people are familiar with the counting numbers 0,1,2,3,4,... The set $\{0,1,2,3,4,...\}$ containing all such numbers is called the set of *natural numbers* $\mathbb{N}$. Any two natural numbers, a and b, may be added to produce a third, $a+b$. However, the solution of the equation $a+x = b$, where a, $b \in \mathbb{N}$, is a natural number only if $a \leqslant b$, 'a is less than or equal to b'.

To solve the equation $a+x = b$ for all elements a, $b \in \mathbb{N}$, it is necessary to extend the set of natural numbers to create a new set containing all numbers of the form $b-a$. For example, the set must include as elements $3-8$, i.e. -5, and $4-6$, i.e. -2. All such numbers form the set of *integers*, $\mathbb{Z} = \{..., -2, -1, 0, 1, 2,...\}$. Clearly the set $\mathbb{N}$ is a subset of $\mathbb{Z}$ and we may write $\mathbb{N} \subset \mathbb{Z}$. The set of *positive integers* $\{1,2,3,4,...\}$ is denoted by $\mathbb{Z}^+$. Since the set $\mathbb{N}$ contains all non-negative integers including zero, $\mathbb{Z}^+ \subset \mathbb{N}$, i.e. $\mathbb{Z}^+$ is a proper subset of $\mathbb{N}$.

The addition, subtraction or multiplication of two integers always produces another integer. However, to solve all equations of the type $bx = a$, where a, $b \in \mathbb{Z}$ and $b \neq 0$, a further extension of the number system is needed. The enlarged set must include elements such as $3/4$, $-7/11$, $10/3$. The set containing all numbers of the form a/b, where a is any integer and b is any non-zero integer, is the set of *rational numbers* $\mathbb{Q}$. Since any integer k can be expressed in the form a/b by writing $a = k$ and $b = 1$, we may write $\mathbb{Z} \subset \mathbb{Q}$.

The natural numbers, $\mathbb{N}$, the integers, $\mathbb{Z}$, and the rationals, $\mathbb{Q}$, are all subsets of the set of *real numbers* $\mathbb{R}$. The real numbers may be represented as the points on a straight line called the *real number line*. As shown in the diagram, the integers are represented by points evenly spaced along the line. Every point on the line represents some real number.

Real numbers which are not rational are called *irrational numbers*. To prove that such numbers exist, we consider $\sqrt{2}$, the positive root of the equation $x^2 = 2$. If $\sqrt{2}$ is rational, then it can be expressed as a fraction in its lowest terms i.e. as a/b, where a and b are integers with no common factor greater than 1.

$\sqrt{2} = \dfrac{a}{b} \Rightarrow 2 = \dfrac{a^2}{b^2} \Rightarrow a^2 = 2b^2 \Rightarrow a^2$ is an even number. Since no odd number has

an even square, a itself must be an even number. If we let $a = 2c$, then $(2c)^2 = 2b^2$ so that $b^2 = 2c^2$. This implies that b is an even number. Thus a and b are both divisible by 2, which contradicts the assumption that a and b have no common factor greater than 1. Hence the assumption that $\sqrt{2}$ is rational must be false. We deduce that $\sqrt{2}$ is an irrational number.

All rational numbers and many irrational numbers, such as $\sqrt{2}$, which are roots of equations of the form $a_0 x^n + a_1 x^{n-1} + \ldots + a_{n-1} x + a_n = 0$, where $a_0, a_1, \ldots, a_n \in \mathbb{Z}$, are known as *algebraic numbers*. Real numbers which are not algebraic are called *transcendental numbers*, the most familiar of these being π.

We now consider the expression of rational and irrational numbers as decimals. A proper fraction a/b is converted into decimal form by the process of long division. Each division by b leaves a remainder, which may be $0, 1, 2, \ldots$, $b-2$ or $b-1$. If at any stage this remainder is zero, then the process is complete and a/b has been expressed as a finite decimal. If at every stage a non-zero remainder is obtained, then there are $(b-1)$ different possible remainders. This means that in the first b divisions at least one remainder, R, must be repeated. After the repetition of R, the sequence of remainders following R must also be repeated. Thus a recurring decimal will be produced. Hence a rational number may be expressed either as a finite decimal or as a recurring decimal.

Conversely every finite decimal represents a rational number, e.g. $0 \cdot 45$ $= 45/100 = 9/20$. Any recurring decimal may be written in rational form by a method similar to the following:

let $x = 0 \cdot 4\dot{5} = 0 \cdot 454545\ldots$
then $100x = 45 \cdot 454545\ldots$
subtracting $99x = 45$
giving $x = 45/99 = 5/11$.

These results together imply that in decimal form rational numbers are either finite or recurring and irrational numbers are infinite and non-recurring.

Exercise 1.2

1. If $X = \{2, 0, -2, 0\cdot\dot{7}, \sqrt[3]{2}, \pi, 7\cdot3\}$, write down all the elements of X which are (a) integers, (b) rational, (c) irrational, (d) algebraic, (e) transcendental.

2.   Repeat question 1 for $X = \{7, -1, 4/3, \ \pi + 2, 3 \cdot 142, \sqrt{\pi}, \sqrt{3} + 2\}$.

3.   List the elements of the set A when (i) $x \in \mathbb{N}$, (ii) $x \in \mathbb{Z}$, (iii) $x \in \mathbb{Q}$, (iv) $x \in \mathbb{R}$, given that

(a) $A = \{x : x^2 - 3x - 4 = 0\}$ (b) $A = \{x : x^2 - 5 = 0\}$
(c) $A = \{x : 2x^2 - 3x + 1 = 0\}$ (d) $A = \{x : x^2 + 5 = 0\}$

4.   State which of the following rational numbers may be expressed as a finite decimal. (You need not write the numbers in decimal form.)
$$\frac{1}{2}, \frac{1}{3}, \frac{7}{16}, \frac{4}{9}, \frac{1}{25}, \frac{8}{11}, \frac{1}{625}, \frac{5}{64}, \frac{2}{7}, \frac{4}{45}, \frac{1}{150}, \frac{51}{75}.$$

5.   Express the following recurring decimals in rational form:
(a) $0 \cdot \dot{5}$,   (b) $0 \cdot 4 \dot{9}$,   (c) $0 \cdot \dot{2} \dot{7}$,   (d) $0 \cdot 0 \dot{3} \dot{7}$,   (e) $0 \cdot \dot{1} 4285 \dot{7}$.

1.3 Binary operations on real numbers

A *binary operation*, $*$, is a rule for combining two objects, a and b, to give a third object c, written $c = a * b$. In later chapters it will be shown that binary operations can be performed on objects such as vectors, matrices and sets. However, we will consider here only binary operations in which two real numbers are combined to form another real number. The most important of these are addition, subtraction, multiplication and division.

When adding or multiplying two numbers the order in which the numbers are written is not important, e.g. $3 + 5 = 5 + 3$ and $4 \times 6 = 6 \times 4$. For this reason addition and multiplication are said to be *commutative*. However, $8 - 3 \neq 3 - 8$ and $15 \div 5 \neq 5 \div 15$. Thus subtraction and division are not commutative. More formally, the *commutative laws* for addition and multiplication state that

$$a + b = b + a, \quad ab = ba \quad \text{for all} \quad a, b \in \mathbb{R}.$$

We next examine the addition or multiplication of three or more real numbers. The order in which the operations are performed does not affect the result, e.g. $5 + (4 + 3) = (5 + 4) + 3$ and $5 \times (4 \times 3) = (5 \times 4) \times 3$. This property of addition and multiplication is called *associativity*. Subtraction and division are not associative, e.g. $5 - (4 - 3) = 4$ but $(5 - 4) - 3 = -2$; $24 \div (4 \div 2) = 12$ but $(24 \div 4) \div 2 = 3$. In general, the *associative laws* for addition and multiplication state that

$$a + (b + c) = (a + b) + c, \quad a(bc) = (ab)c \quad \text{for all} \quad a, b, c \in \mathbb{R}.$$

Another important property which involves both addition and multiplication is the *distributive law*. This is the rule used when 'removing brackets' or 'factorising' in elementary algebra. It states that

$$a(b+c) = ab+ac \quad \text{for all} \quad a, b, c \in \mathbb{R}.$$

Multiplication is said to be *distributive over addition*.

Within the set of real numbers there are two numbers with special properties, namely 0 and 1. We find that

$$a+0 = 0+a = a, \quad a \times 1 = 1 \times a = a \quad \text{for all} \quad a \in \mathbb{R}.$$

Both addition of 0 and multiplication by 1 leave a real number a unchanged (i.e. identical in value). Hence 0 and 1 are called *identity elements*. The number 0 is the identity for addition and 1 is the identity for multiplication.

This leads to the notion of an *inverse* under a binary operation. If an inverse exists then the combination of an element with its inverse gives the identity element. Thus, since $a+(-a) = (-a)+a = 0$, the inverse of a real number a with respect to addition is $(-a)$. Similarly, since $a \times \dfrac{1}{a} = \dfrac{1}{a} \times a = 1$ (assuming $a \neq 0$), the inverse of any non-zero real number a with respect to multiplication is $1/a$.

In further study of the structure of the real number system, it is often convenient to regard the subtraction of a number a as the addition of $(-a)$ and division by a, where $a \neq 0$, as multiplication by $1/a$.

We now consider the use made of inverses and the properties of binary operations in the solution of a simple equation.

$$3(x+2) = 8$$
$\Rightarrow \qquad\qquad 3x+6 = 8$ distributive law
$\Rightarrow \{3x+6\}+(-6) = 8+(-6)$ using the inverse of 6 for $+$
$\Rightarrow 3x+\{6+(-6)\} = 2$ associative law for $+$
$\Rightarrow \qquad\qquad 3x+0 = 2$
$\Rightarrow \qquad\qquad\quad 3x = 2$ property of 0
$\Rightarrow \qquad \frac{1}{3} \times (3x) = \frac{1}{3} \times 2$ using the inverse of 3 for $\times$
$\Rightarrow \qquad (\frac{1}{3} \times 3)x = \frac{2}{3}$ associative law for $\times$
$\Rightarrow \qquad\qquad 1 \times x = \frac{2}{3}$
$\Rightarrow \qquad\qquad\qquad x = \frac{2}{3}$ property of 1.

In work with real numbers we usually take their properties for granted. However, the terms introduced here will be useful when comparing the behaviour of real numbers with that of other mathematical objects.

Exercise 1.3

1. Write down the inverses with respect to addition of the real numbers 5, $\frac{1}{4}$, -2, k and $-n$.

2. Write down the inverses with respect to multiplication of the real numbers 2, $-\frac{1}{3}$, -1, 4/5 and $1/x$.

3.   Given that $x*y = x^y$, where x and y are positive integers, state whether the binary operation $*$ is (a) commutative, (b) associative.

4. Repeat question 3 given that $x*y$ is defined as the highest common factor of x and y.

5.   Repeat question 3 given that $x*y = \frac{1}{2}(x+y)$.

6.   Solve the following equations, stating which properties of real numbers are used.
(a) $3x = 12$ (b) $x-4 = 5$ (c) $2x+5 = 9$
(d) $2x = x-2$ (e) $3(x+2) = x$ (f) $x^2 - 2x = 0$.

1.4 Introduction to inequalities

An inequality is a statement involving one of the symbols $<$ 'is less than', $>$ 'is greater than', $\leqslant$ 'is less than or equal to', $\geqslant$ 'is greater than or equal to'. Inequalities such as $2x+3 < 7$, can be 'solved' by methods similar to those used to solve equations.

$$2x+3 < 7$$
Subtract 3 from both sides $$2x < 4$$
Divide both sides by 2 $$x < 2$$

The result can be represented as a set of points on the real number line. An open circle is used to indicate that $x = 2$ is not a solution of the inequality.

In general we find that when simplifying inequalities, any number may be added to or subtracted from both sides. However, when both sides are multiplied or divided by a number, the inequality remains valid only if the number is positive. If both sides of an inequality are multiplied or divided by a negative number, then the inequality sign must be reversed. For instance, $3 < 4$ but $-3 > -4$. Similarly, if $-x > 2$ then $x < -2$.

Example 1   Simplify the inequality $2x-5 \leqslant 5x+4$.

$$2x-5 \leqslant 5x+4 \Leftrightarrow \quad 2x \leqslant 5x+9$$
$$\Leftrightarrow -3x \leqslant 9$$
$$\Leftrightarrow \quad x \geqslant -3$$

This result is illustrated in the diagram below.

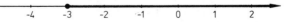

The set of numbers satisfying an equation or inequality is called the *solution set*. For instance, the solution set of the equation $2x+3 = 7$ is the set $\{2\}$ containing one element. The solution set of the inequality $2x+3 < 7$ is the set $\{x : x < 2\}$, which has infinitely many elements. [To avoid ambiguity we may describe this set more fully by writing $\{x \in \mathbb{R} : x < 2\}$.]

Example 2   Find, and represent on the real number line, the solution set of the inequalities $0 < 4x - 3 \leqslant 9$.

$$0 < 4x - 3 \leqslant 9 \Leftrightarrow 3 < 4x \leqslant 12$$
$$\Leftrightarrow \tfrac{3}{4} < x \leqslant 3$$

∴ the solution set is $\{x : \tfrac{3}{4} < x \leqslant 3\}$.

The *modulus* of a real number x, written $|x|$, is the magnitude of x. For instance, $|3| = 3$ and $|-5| = 5$. Using this notation inequalities such as $-2 < x < 2$ can be written in the form $|x| < 2$.

Example 3   Find the range of values of x for which $|3x| > 12$.

$$|3x| > 12 \Rightarrow 3x < -12 \quad \text{or} \quad 3x > 12$$
$$\Rightarrow x < -4 \quad \text{or} \quad x > 4.$$

Exercise 1.4

1.   Represent the following sets on the real number line
(a) $\{x : x > -1\}$ (b) $\{x : x \leqslant 4\}$ (c) $\{x : 0 < x < 3\}$
(d) $\{x : -3 \leqslant x \leqslant 1\}$ (e) $\{x : |x| \leqslant 3\}$ (f) $\{x : |x| > 2\}$.

2.   Find, and represent on the real number line, the solution sets of the following inequalities.
(a) $x + 2 < 9$ (b) $x - 4 \geqslant 1$ (c) $1 - x > 2$
(d) $2x - 5 \leqslant 3$ (e) $1 - 3x > 7$ (f) $5x + 3 < x$

3.   Find the range of values of x satisfying the following inequalities.
(a) $2x - 3 < 4 - 5x$ (b) $2x + 9 \leqslant 4x - 7$
(c) $2(x + 1) < x - 1$ (d) $\tfrac{1}{2}(3x + 1) \geqslant \tfrac{1}{3}(4x + 5)$
(e) $|2x| \geqslant 10$ (f) $|\tfrac{1}{2}x| < 3$

4.   Simplify the following inequalities.
(a) $-1 < 2x < 1$ (b) $0 \leqslant -3x \leqslant 6$
(c) $-2 < 5x - 7 < 3$ (d) $-3 \leqslant 1 - 4x < 3$
(e) $|x - 1| < 2$ (f) $|x + 3| \leqslant 7$

5.   State whether the following statements are true or false.
(a) $x < k \Rightarrow 1 - x < 1 - k$ (b) $-x < k \Rightarrow -k < x$
(c) $x^2 > 1 \Rightarrow x > 1$ (d) $x^2 < 1 \Rightarrow x < 1$
(e) $x < y \Rightarrow x^2 < y^2$ (f) $x < y < 0 \Rightarrow x^2 > y^2$
(g) $x < y \Rightarrow \dfrac{1}{x} > \dfrac{1}{y}$ (h) $0 < x < y \Rightarrow \dfrac{1}{x} > \dfrac{1}{y}$
(i) $x < y \Rightarrow |x| < |y|$ (j) $x < y < 0 \Rightarrow |y| < |x|$
Check your answers by considering some numerical examples.

1.5 Surds and indices

The reader will be familiar with the basic laws of indices which state that, for positive integers m and n,

$$a^m \times a^n = a^{m+n} \quad \text{and} \quad (a^m)^n = a^{mn}.$$

For example, $a^2 \times a^3 = (a \times a) \times (a \times a \times a) = a^5$
and $\quad\quad\quad (a^2)^3 = (a \times a)^3 = (a \times a) \times (a \times a) \times (a \times a) = a^6.$

The rule for division is that, for $m > n$, $a^m \div a^n = a^{m-n}$. By assuming that this rule also holds for $m \leqslant n$, we can give meanings to the expressions a^0 and a^{-n}, (provided that $a \neq 0$).

We write $\quad a^0 \ = a^{n-n} = a^n \div a^n \ = 1$
and $\quad\quad\quad a^{-n} = a^{0-n} = a^0 \div a^n = 1 \div a^n = 1/a^n.$

Thus it is consistent with the laws of indices to let $a^0 = 1$ and $a^{-n} = 1/a^n$.

We extend this process by assuming that the laws of indices also apply to fractional powers of positive real numbers. It then follows that $(a^{1/2})^2 = a^{1/2} \times a^{1/2} = a^1 = a$. Hence $a^{1/2}$ must be a square root of a. It is usual to take $a^{1/2}$ to be the positive square root $\sqrt{a}$. More generally, we can write

$$(a^{1/n})^n = a^{(1/n)\cdot n} = a^1 = a.$$

Hence, if n is a positive integer, $a^{1/n}$ is considered to be the positive nth root of a, written $\sqrt[n]{a}$. It follows that

$$a^{m/n} = (a^{1/n})^m = (\sqrt[n]{a})^m \quad \text{and} \quad a^{m/n} = (a^m)^{1/n} = \sqrt[n]{(a^m)},$$

where m and n are positive integers.

We can now summarise the laws of indices for rational powers of a positive real number a.

$$\boxed{\begin{array}{ll} a^m \times a^n = a^{m+n} & a^0 = 1, \quad a^{-n} = 1/a^n \\ a^m \div a^n = a^{m-n} & \\ (a^m)^n = a^{mn} & a^{m/n} = \sqrt[n]{(a^m)} = (\sqrt[n]{a})^m. \end{array}}$$

Example 1 Evaluate (a) 2^{-2} (b) $16^{1/2}$ (c) $8^{2/3}$ (d) $\left(\dfrac{4}{9}\right)^{-1/2}$.

(a) $2^{-2} = \dfrac{1}{2^2} = \dfrac{1}{4}$,   (b) $16^{1/2} = \sqrt{16} = 4$,   (c) $8^{2/3} = (\sqrt[3]{8})^2 = 2^2 = 4$,

(d) $\left(\dfrac{4}{9}\right)^{-1/2} = \left(\dfrac{9}{4}\right)^{1/2} = 9^{1/2}/4^{1/2} = \dfrac{\sqrt{9}}{\sqrt{4}} = \dfrac{3}{2} = 1\tfrac{1}{2}.$

[We have not considered rational powers of negative numbers at this stage because of the difficulties which arise. One problem is that no negative number has a real nth root when n is even. In the case of odd values of n, contradictory results can be produced. For instance, $(-8)^{1/3} = \sqrt[3]{(-8)} = -2$, but $(-8)^{1/3} = (-8)^{2/6} = \sqrt[6]{\{(-8)^2\}} = \sqrt[6]{64} = 2.$]

Equations involving indices can be solved in a variety of ways.

Example 2   Solve the equation $8^x = 32$.

Expressing both sides of the equation as powers of 2,
$$8^x = 32 \Leftrightarrow (2^3)^x = 2^5 \Leftrightarrow 2^{3x} = 2^5$$
$$\therefore \quad 3x = 5, \text{ i.e. } x = 1\tfrac{2}{3}.$$

Example 3   Solve the equation $2^{2x} - 5 \cdot 2^x + 4 = 0$.

Substituting $y = 2^x$, $2^{2x} - 5 \cdot 2^x + 4 = 0$
$$\Leftrightarrow \quad y^2 - 5y + 4 = 0$$
$$\Leftrightarrow (y-1)(y-4) = 0$$
$$\Leftrightarrow y = 0 \quad \text{or} \quad y = 4$$
$$\Leftrightarrow 2^x = 1 \quad \text{or} \quad 2^x = 4$$
$$\Leftrightarrow x = 0 \quad \text{or} \quad x = 2.$$

Example 4   Solve the equation $9^{x+1} - 3^{x+3} - 3^x + 3 = 0$.

Substituting $y = 3^x$, $9^{x+1} - 3^{x+3} - 3^x + 3 = 0$
$$\Leftrightarrow 9 \cdot 9^x - 3^3 \cdot 3^x - 3^x + 3 = 0$$
$$\Leftrightarrow \quad 9y^2 - 27y - y + 3 = 0$$
$$\Leftrightarrow \quad (9y-1)(y-3) = 0$$
$$\Leftrightarrow y = 1/9 \quad \text{or} \quad y = 3$$
$$\Leftrightarrow 3^x = 1/9 \quad \text{or} \quad 3^x = 3$$
$$\Leftrightarrow x = -2 \quad \text{or} \quad x = 1.$$

Numbers such as $\sqrt{3}$ and $\sqrt[3]{5}$ are called *surds*. It is often helpful to simplify expressions involving surds. For instance, since $(\sqrt{2} \cdot \sqrt{3})^2 = \sqrt{2} \cdot \sqrt{3} \cdot \sqrt{2} \cdot \sqrt{3}$ $= (\sqrt{2})^2(\sqrt{3})^2 = 2 \cdot 3 = 6$, we see that $\sqrt{2} \cdot \sqrt{3} = \sqrt{6}$. This is an illustration of the useful result $(\sqrt{a})(\sqrt{b}) = \sqrt{(ab)}$.

Example 5   Simplify (a) $\sqrt{50}$, (b) $\sqrt{18}/\sqrt{2}$, (c) $\sqrt{12} - \sqrt{27} + \sqrt{3}$.

(a) $\sqrt{50} = \sqrt{(25 \cdot 2)} = \sqrt{25} \cdot \sqrt{2} = 5\sqrt{2}$
(b) $\sqrt{18}/\sqrt{2} = \sqrt{(18/2)} = \sqrt{9} = 3$
(c) $\sqrt{12} - \sqrt{27} + \sqrt{3} = \sqrt{4} \cdot \sqrt{3} - \sqrt{9} \cdot \sqrt{3} + \sqrt{3} = 2\sqrt{3} - 3\sqrt{3} + \sqrt{3} = 0.$

It is usual to *rationalise* fractions which involve surds, so that the denominators are integers. For instance, $1/\sqrt{2}$ may be written $\sqrt{2}/2$. To rationalise a denominator of the form $a + \sqrt{b}$, we use the fact that $(a + \sqrt{b})(a - \sqrt{b})$ reduces to the 'difference of squares' $a^2 - (\sqrt{b})^2$, i.e. $a^2 - b$.

Example 6   Simplify $\dfrac{1+\sqrt{3}}{2+\sqrt{3}}$.

$$\frac{1+\sqrt{3}}{2+\sqrt{3}} = \frac{(1+\sqrt{3})(2-\sqrt{3})}{(2+\sqrt{3})(2-\sqrt{3})} = \frac{2-\sqrt{3}+2\sqrt{3}-3}{4-2\sqrt{3}+2\sqrt{3}-3} = \frac{-1+\sqrt{3}}{1} = -1+\sqrt{3}.$$

The remaining examples concern surds and fractional indices in algebraic expressions.

Example 7   Simplify $x\sqrt{\left(\dfrac{1}{x^2}-1\right)}.$

$$x\sqrt{\left(\frac{1}{x^2}-1\right)} = x\sqrt{\left(\frac{1-x^2}{x^2}\right)} = x \cdot \frac{\sqrt{(1-x^2)}}{\sqrt{(x^2)}} = \sqrt{(1-x^2)}.$$

Alternative method:

$$x\sqrt{\left(\frac{1}{x^2}-1\right)} = \sqrt{(x^2)}\sqrt{\left(\frac{1}{x^2}-1\right)} = \sqrt{x^2\left(\frac{1}{x^2}-1\right)} = \sqrt{(1-x^2)}.$$

Example 8   Simplify $(x+1)^{3/2}-(x+1)^{1/2}$.

$$(x+1)^{3/2}-(x+1)^{1/2} = (x+1)^{1/2}(x+1)^1 - (x+1)^{1/2}$$
$$= (x+1)^{1/2}\{(x+1)-1\} = x(x+1)^{1/2}.$$

Exercise 1.5

In questions 1 to 4 simplify the given expressions.

1. (a) 7^2,   (b) 4^0,   (c) 5^{-1},   (d) $36^{1/2}$,
    (e) $8^{1/3}$,   (f) 2^{-3},   (g) $(12\frac{1}{4})^{1/2}$,   (h) $27^{-1/3}$.

2. (a) $16^{3/4}$,   (b) $(1\frac{1}{2})^{-1}$,   (c) $64^{2/3}$,   (d) $9^{-1/2}$,
    (e) $\left(\dfrac{3}{5}\right)^{-2}$,   (f) $(3\frac{3}{8})^{1/3}$,   (g) $\left(\dfrac{2}{5}\right)^0$,   (h) $\left(\dfrac{1}{4}\right)^{-3/2}$.

3. (a) $\sqrt{54}$,   (b) $\sqrt{98}$,   (c) $\sqrt{72}$,   (d) $\sqrt{68}$,
    (e) $\sqrt{12}\times\sqrt{27}$,   (f) $\sqrt{252}\div\sqrt{63}$,   (g) $\sqrt{75}-\sqrt{48}$,   (h) $\sqrt{32}-\sqrt{18}$.

4. (a) $\sqrt{32}\times\sqrt{15}\div\sqrt{24}$,   (b) $\sqrt{80}-\sqrt{20}+\sqrt{45}$,
    (c) $\sqrt{84}\times\sqrt{140}\div\sqrt{120}$,   (d) $\sqrt{112}-\sqrt{63}-\sqrt{28}$.

5. Express as a single square root
    (a) $3\sqrt{2}$,   (b) $5\sqrt{5}$,   (c) $x\sqrt{y}$,   (d) $2a^2\sqrt{b}$.

6. Rationalise the denominators of the following fractions, simplifying your answers

(a) $\dfrac{10}{\sqrt{5}}$ (b) $\dfrac{3+\sqrt{28}}{\sqrt{7}}$ (c) $\dfrac{3\sqrt{20}-\sqrt{60}}{2\sqrt{3}}$

(d) $\dfrac{\sqrt{5}+2}{\sqrt{5}-1}$ (e) $\dfrac{1+2\sqrt{2}}{5-3\sqrt{2}}$ (f) $\dfrac{\sqrt{3}+\sqrt{2}}{\sqrt{3}-\sqrt{2}}$.

In questions 7 to 9 solve the given equations.

7. (a) $9^x = 27$,   (b) $4^x = 128$,   (c) $5^{x+3} = 1$.

8. (a) $2^{x-3} = 4^{x+1}$ (b) $3^{2x} \cdot 3^{x-1} = 9$
    (c) $2^x \cdot 5^{x+1} = \frac{1}{2}$ (d) $3^x \cdot 2^{2x-3} = 18$.

9. (a) $2^{2x} - 9.2^x + 8 = 0$ (b) $3^{2x} - 10.3^x + 9 = 0$
   (c) $4^x - 3.2^{x+1} + 8 = 0$ (d) $2^{2x+1} + 4 = 2^{x+3} + 2^x$
   (e) $3^{2x-3} - 4.3^{x-2} + 1 = 0$ (f) $16^x - 5.2^{2x-1} + 1 = 0$.

10. Simplify the following expressions

(a) $x^2 \sqrt{\left(1 - \dfrac{1}{x^3}\right)}$ (b) $\dfrac{1}{x}\sqrt{(x^2 + x^4)}$

(c) $(x-2)^{3/2} + 2(x-2)^{1/2}$ (d) $(2x-1)^{-1/2} + (2x-1)^{1/2}$.

1.6 Logarithms

If a and x are positive real numbers and $x = a^p$, then p is the *logarithm* of x to the base a, written $\log_a x$.

Thus

$$x = a^p \Leftrightarrow \log_a x = p.$$

Any statement in index form, such as $25 = 5^2$, has an equivalent logarithmic form, $\log_5 25 = 2$. To determine the logarithm to the base a of a number x, we must find the power of a which is equal to x.

Example 1   Find the logarithms to the base 4 of (a) 16, (b) 2, (c) $\frac{1}{4}$, (d) 4, (e) 1, (f) 8.

(a) $16 = 4^2$ $\therefore \log_4 16 = 2$ (b) $2 = 4^{1/2}$ $\therefore \log_4 2 = \frac{1}{2}$
(c) $\frac{1}{4} = 4^{-1}$ $\therefore \log_4 \frac{1}{4} = -1$ (d) $4 = 4^1$ $\therefore \log_4 4 = 1$
(e) $1 = 4^0$ $\therefore \log_4 1 = 0$ (f) $8 = 4^{3/2}$ $\therefore \log_4 8 = \frac{3}{2}$.

As illustrated in this example, for any base a,

$$\log_a a = 1 \quad \text{and} \quad \log_a 1 = 0.$$

Because logarithms are themselves indices, the laws of logarithms are closely related to the laws of indices.

Let $x = a^p, y = a^q$ then $\log_a x = p, \log_a y = q$.

Using the laws of indices:

$xy = a^p \times a^q = a^{p+q}.$ $\therefore \log_a xy = p + q = \log_a x + \log_a y$

$\dfrac{x}{y} = a^p \div a^q = a^{p-q}$ $\therefore \log_a \dfrac{x}{y} = p - q = \log_a x - \log_a y$

$x^n = (a^p)^n = a^{pn}$ $\therefore \log_a x^n = pn = n \log_a x.$

Summarising:

$$\log_a xy = \log_a x + \log_a y$$
$$\log_a \frac{x}{y} = \log_a x - \log_a y$$
$$\log_a x^n = n \log_a x.$$

Example 2 Express $\log (x^3/\sqrt{y})$ in terms of $\log x$ and $\log y$.

$\log (x^3/\sqrt{y}) = \log (x^3/y^{1/2}) = \log (x^3) - \log (y^{1/2}) = 3 \log x - \frac{1}{2} \log y.$

Example 3 Express $\frac{1}{3} \log 8 - \log \frac{2}{5}$ as a single logarithm.

$\frac{1}{3} \log 8 - \log \frac{2}{5} = \log 8^{1/3} + \log \left(\frac{2}{5}\right)^{-1} = \log 2 + \log \frac{5}{2} = \log \left(2.\frac{5}{2}\right) = \log 5.$

Example 4 Solve the equation $\log_6 x + \log_6 (x+5) = 2$.

$\log_6 x + \log_6 (x+5) = 2 \quad \Rightarrow \quad \log_6 x(x+5) = \log_6 36$
$$\Rightarrow \qquad x(x+5) = 36$$
$$\Rightarrow \quad x^2 + 5x - 36 = 0$$
$$\Rightarrow (x+9)(x-4) = 0.$$

Hence either $\quad x = -9 \quad$ or $\quad x = 4$.

However, since $\log_6 x$ and $\log_6 (x+5)$ are undefined for $x = -9$, the only valid solution of the given equation is $x = 4$.

We now establish some important relationships between logarithms to different bases.

Let $\quad \log_a x = p \quad$ so that $\quad x = a^p$.

Taking logarithms to the base x,

$\log_x x = \log_x (a^p) \Leftrightarrow 1 = p \log_x a \Leftrightarrow \log_x a = 1/p$

$$\therefore \quad \log_x a = 1/\log_a x.$$

Taking logarithms to the base b,

$\log_b x = \log_b (a^p) \Leftrightarrow \log_b x = p \log_b a \Leftrightarrow p = \log_b x / \log_b a$

$$\therefore \quad \log_a x = \log_b x / \log_b a.$$

In particular, when $b = 10$, $\log_a x = \log_{10} x / \log_{10} a$.

Example 5   Given that $\log_2 N = k$, express in terms of k

(a) $\log_2 N^2$,   (b) $\log_2 2N$,   (c) $\log_8 N$,   (d) $\log_N 4$.

(a) $\log_2 N^2 = 2\log_2 N = 2k$,   (b) $\log_2 2N = \log_2 2 + \log_2 N = 1 + k$,
(c) $\log_8 N = \log_2 N/\log_2 8 = \log_2 N/\log_2 2^3 = k/3$,
(d) $\log_N 4 = \log_2 4/\log_2 N = \log_2 2^2/\log_2 N = 2/k$.

Example 6   Solve the equation $2\log_4 x + 1 - \log_x 4 = 0$.

Substituting $\log_4 x = y$, $2\log_4 x + 1 - \log_x 4 = 0$

$$\Leftrightarrow \qquad 2y + 1 - 1/y = 0$$
$$\Leftrightarrow \qquad 2y^2 + y - 1 = 0$$
$$\Leftrightarrow \qquad (2y - 1)(y + 1) = 0$$
$$\Leftrightarrow \qquad y = \tfrac{1}{2} \quad \text{or} \quad y = -1$$
$$\Leftrightarrow \log_4 x = \tfrac{1}{2} \quad \text{or} \quad \log_4 x = -1$$
$$\Leftrightarrow \qquad x = 4^{1/2} \quad \text{or} \quad x = 4^{-1}$$

Hence $x = 2$ or $x = \tfrac{1}{4}$.

For some years the main use of logarithms in elementary work was in performing calculations involving multiplication and division. However, many people now use electronic calculators for such operations in preference to slide rules and sets of mathematical tables. Nevertheless, we now review briefly the use of tables of *common logarithms*, i.e. logarithms to the base 10. We will use the notation $\lg x$ instead of the more cumbersome $\log_{10} x$.

Logarithm tables usually give $\lg x$ for values of x between 1 and 10, e.g. $\lg 5\cdot 3 = 0\cdot 7243$. For other values of x, $\lg x$ is found by expressing x in *standard form*, i.e. in the form $k \times 10^n$, where $1 \leqslant k < 10$ and n is an integer. For example,

$\lg 530 = \lg (5\cdot 3 \times 10^2) \quad = \lg 5\cdot 3 + \lg 10^2 \quad = 0\cdot 7243 + 2 = 2\cdot 7243,$
$\lg 0\cdot 53 = \lg (5\cdot 3 \times 10^{-1}) = \lg 5\cdot 3 + \lg 10^{-1} = 0\cdot 7243 - 1 = -0\cdot 2757,$

which is often written as $\bar{1}\cdot 7243$.

Anti-logarithm tables give values of 10^p for $0 \leqslant p < 1$. They are used to find a number given its logarithm, e.g. if $\lg x = 0\cdot 78$, then $x = 10^{0\cdot 78} = 6\cdot 026$. For other values of p, 10^p is found by using the laws of indices.

If $\lg x = 1\cdot 78$, then $x = 10^{1\cdot 78} = 10^{0\cdot 78} \times 10^1 = 60\cdot 26$.
If $\lg x = \bar{1}\cdot 78$, then $x = 10^{0\cdot 78 - 1} = 10^{0\cdot 78} \times 10^{-1} = 0\cdot 6026$.
If $\lg x = -1\cdot 22$, then $x = 10^{-1\cdot 22} = 10^{0\cdot 78} \times 10^{-2} = 0\cdot 06026$.

[For further numerical work involving logarithms see §4.2.]

Exercise 1.6

1.   Express the following statements in logarithmic form
(a) $16 = 2^4$ (b) $1/9 = 3^{-2}$ (c) $4 = 16^{1/2}$
2.   Express the following statements in index form
(a) $\log_4 64 = 3$ (b) $\log_5 0\cdot 2 = -1$ (c) $\log_9 27 = 1\cdot 5$
3.   Find the logarithms to the base 2 of
(a) 8   (b) $\tfrac{1}{4}$ (c) 1   (d) 2   (e) $\sqrt{2}$

4. Find the logarithms to the base 9 of
(a) 81   (b) 3   (c) 1/9   (d) 27   (e) 1
5. Find the following logarithms
(a) $\log_3 81$ (b) $\log_{27} 3$ (c) $\log_4 0.5$ (d) $\log_{100} 10$
(e) $\log_8 0.25$ (f) $\log_{0.5} 8$ (g) $\log_6 6\sqrt{6}$ (h) $\log_5 0.04$
6. Express in terms of $\log x$ and $\log y$
(a) $\log x^2 y$ (b) $\log \sqrt{(xy)}$ (c) $\log (x^4/y^3)$
7. Express in terms of $\log A$, $\log B$ and $\log C$
(a) $\log (AB^2/C^3)$ (b) $\log (A\sqrt{B}/C^{-2})$ (c) $\log \sqrt{(A^3 B^2 C)}$
8. Express as a single logarithm
(a) $\log 14 - \log 21 + \log 6$ (b) $4 \log 2 + \frac{1}{2} \log 25$
(c) $\frac{3}{2} \log 9 - 2 \log 6$ (d) $2 \log (\frac{2}{3}) - \log (\frac{8}{9})$
9. Find the values of
(a) $\log_a 32/\log_a 2$ (b) $\log_x 125/\log_x 25$ (c) $\log_3 x/\log_9 x$
10. Given that $\log_3 2 = p$ and $\log_3 5 = q$, express in terms of p and q
(a) $\log_3 60$ (b) $\log_3 6.4$ (c) $\log_{10} 2$.
11. Given that $\log_5 x = t$, express in terms of t
(a) $\log_5 5x^2$ (b) $\log_x 5$ (c) $\log_{25} x$ (d) $\log_x 0.2$

In questions 12 to 14 solve the given equations.

12. (a) $\log_2 x^4 + \log_2 4x = 12$ (b) $\log_5 x + \log_5 (1/x^3) = 2$
13. (a) $\log_3 x + \log_3 (x+6) = 3$ (b) $\log_4 2x + \log_4 (x+1) = 1$
14. (a) $\log_3 x = 4 \log_x 3$ (b) $2 \log_4 x + 3 \log_x 4 = 7$
    (c) $3 \log_8 x = 2 \log_x 8 + 5$ (d) $\log_5 x + \log_x 25 = 3$
15. Solve the simultaneous equations: $\log_2 x^2 + \log_2 y^3 = 1$,
$$\log_2 x - \log_2 y^2 = 4.$$

Exercise 1.7   (*miscellaneous*)

In questions 1 to 4, assuming that $A = \{x \in \mathbb{R} : x^2 < 4\}$, $B = \{x \in \mathbb{Z} : 0 \leqslant x \leqslant 2\}$ and $C = \{-2, -1, 0, 1, 2\}$, decide whether the given statements are true or false.

1. (a) $A \subset C$,   (b) $B \subset C$,   (c) $B \subset A$,   (d) $B \subset \mathbb{Z}$.

2. (a) $2 \in A$,   (b) $\sqrt{2} \in A$,   (c) $2 \in B$,   (d) $\sqrt{2} \in B$.

3. (a) $(-1, 1) \in A \times B$,   (b) $(1, -1) \in A \times B$,
    (c) $(2, 2) \in A \times C$,   (d) $(0, 0) \in A \times C$.

4. (a) $n \in A$ and $n \in B \Leftrightarrow n \in \{0, 1\}$,
    (b) $n \notin B$ and $n \in C \Rightarrow n \in \{-2, -1, 0\}$,
    (c) $n \notin A$ and $n \in C \Rightarrow n \in B$,
    (d) $n \in A$ and $n \notin C \Leftrightarrow n \notin \mathbb{Z}$.

5. If the universal set $\mathscr{E}$ is the set of real numbers $\mathbb{R}$, find the complements of the following sets
(a) {rational numbers},   (b) {negative numbers},
(c) $\{x : x \leqslant 4\}$,   (d) $\{x : x < 0 \text{ or } x > 0\}$,
(e) $\{x : x^2 < 0\}$,   (f) $\{x : -1 < x < 1\}$.

6.   Given that $x * y = \sqrt{(xy)}$, where x and y are real numbers, decide whether the binary operation $*$ is (a) commutative (b) associative.

7.   Use the properties of real numbers to prove that
(a) $k + x = k + y \Rightarrow x = y$,
(b) $kx = ky$ and $k \neq 0 \Rightarrow x = y$.

8. Find the range of values of x satisfying the following inequalities.
(a) $3x + 1 < x - 5$,   (b) $2(2x - 3) \geqslant 5x + 2$,
(c) $0 < 2x - 1 \leqslant 3$,   (d) $x(x + 2) > 2x + 9$,
(e) $|x + 2| < 5$,   (f) $|x - 3| \geqslant 2$.

9.   Arrange the following in order of magnitude:
$$7, \quad 2\sqrt{11}, \quad 4\sqrt{3}, \quad 5\sqrt{2}, \quad 3\sqrt{5}.$$

10. Simplify the following expressions
(a) $54^{1/3} \times 2^{-4/3}$,   (b) $25^{1/6} \times 200^{1/3}$,   (c) $54^{1/3} - 16^{1/3}$.

11. Find in the form $p + q\sqrt{2}$, where p and q are integers,
(a) the reciprocal of $3 - 2\sqrt{2}$,   (b) the square of $3 - 2\sqrt{2}$,
(c) the square roots of $3 - 2\sqrt{2}$.

12. Solve the simultaneous equations
$$2^x + 3^y = 5, \quad 2^{x+3} - 3^{y+2} = 23.$$

13. Given that $\log_3 2 \approx 0\cdot 63$, find the values of x for which
(a) $6^x . 2^{3-x} = 1$,   (b) $3^{2x} = 3^x + 2$,   (c) $2 . 9^x + 3^{x-1} + 4 = 2 . 3^{x+1}$.

14. Write down the value of x given that
(a) $\log_x 8 = \frac{3}{4}$,   (b) $\log_x 0\cdot 01 = \log_{0\cdot 1} 100$,
(c) $\log_9 x = -\frac{1}{2}$,   (d) $\log_3 x = \log_9 4$.

15. Simplify the following expressions
(a) $3\log_2 \left(\dfrac{5}{3}\right) - 2\log_2 \left(\dfrac{10}{9}\right) + \log_2 \left(\dfrac{1}{30}\right)$,
(b) $\log_a \sqrt{(a^4 + 1)} - \frac{1}{2}\log_a \left(1 + \dfrac{1}{a^4}\right)$,
(c) $\log_4 \{(a^2 + 1)^2 - (a^2 - 1)^2\} - \log_2 2a$.

16. Solve the following equations
(a) $\log_2 4x = 8\log_x 2$,   (b) $\log_9 x = \log_3 3x$.

17. Prove that, if a, b and c are positive real numbers, then
$$(\log_a b)(\log_b c) = \log_a c.$$

2 Functions and graphs

2.1 Mappings and functions

A *mapping* or *function* f is a rule which assigns to an object x an *image* y. We write $f : x \rightarrow y$ or $x \overset{f}{\rightarrow} y$ and say 'f maps x to y'. The notation $y = f(x)$ is also used.

Consider a function f which maps each day of the week onto its initial letter. This function is represented in the diagram below. If X is the set containing the days of the week and Y is the set containing all letters of the alphabet, then f can be described as a mapping from X into Y, written $f : X \rightarrow Y$. The set X is called the *domain* of f and the set Y is called the *codomain*. By definition every element in the domain of a function has one and only one image in the codomain.

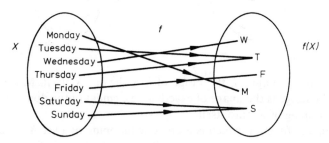

The set $\{W, T, F, M, S\}$ containing all the images under f is the *range* of the function, denoted by $f(X)$. Any element in the range is the image of one or more elements in the domain, e.g. F is the image of Friday, but T is the image of both Tuesday and Thursday. As every element of the range $f(X)$ belongs to the codomain, $f(X) \subset Y$. In this example, as there are letters of the alphabet, such as L and D, which are not images under f, the range $f(X)$ is a proper subset of the codomain Y.

We now examine some functions with special properties.

(1) Let X be the set of natural numbers $\{0,1,2,3,\dots\}$ and let Y be the set of single digits $\{0, 1, 2, \dots, 8, 9\}$. Let $f : X \rightarrow Y$ be the function which maps a natural number to its final digit.

17

As before X is the domain of f and Y is the codomain. However, as every element of Y is the final digit of at least one natural number, the set Y is also the range of f, i.e. $f(X) = Y$.

In general, if every element of the codomain Y of a function f is the image of at least one element of the domain X, then f is said to be a mapping of X *onto* Y. [A function which is 'onto' can also be called *surjective*.]

(2) Let g be the mapping of the set of natural numbers $\mathbb{N}$ into itself defined by $g : x \rightarrow x^2$.

In this case the set $\mathbb{N}$ is both the domain and the codomain of g, but the range of g is a proper subset of $\mathbb{N}$, $\{0, 1, 4, 9, 16, \ldots\}$. Hence g is a mapping 'into' rather than 'onto' $\mathbb{N}$. However, every element of the range is the image of only one element of the domain.

In general, if every element of the range $f(X)$ of a function f is the image of exactly one element of the domain X, then f is said to be *one-one*, sometimes written 1-1. A function which is not one-one may be described as *many-to-one*. Thus the function g defined above is one-one, but both the function f in (1) and the function defined earlier on the set of days of the week are many-to-one. [A one-one function can also be called *injective*.]

(3) Let $X = \{1, 2, 3, 4, 5\}$ and let h be a mapping from X to itself defined by $h : x \rightarrow 6 - x$.

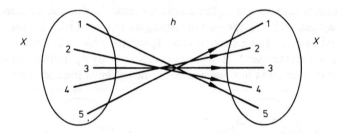

As shown in the diagram, the range of h is the set X itself. Moreover, every element of the range is the image of exactly one element of the domain. Hence h is a one-one mapping of X onto itself.

In general, any function which is a one-one mapping of a set X onto a set Y is said to form a *one-one correspondence* between X and Y. The symbol $\leftrightarrow$ is sometimes used to link object and image under a one-one correspondence. For example, under the mapping h, $1 \leftrightarrow 5$, $2 \leftrightarrow 4$, $3 \leftrightarrow 3$, $4 \leftrightarrow 2$, $5 \leftrightarrow 1$. [A function which is both one-one and 'onto' is said to be *bijective*.].

Exercise 2.1

1.   Decide whether the following are valid definitions of a mapping f from the set $X = \{0, 1, 2, \ldots, 10\}$ into the set of real numbers $\mathbb{R}$.

(a) f maps any element of X onto its cube,

(b) f maps any element of X onto its square root,

(c) f maps any element of X onto its reciprocal.

2.   Decide whether the following are valid definitions of a mapping f from the set of all triangles into the set $\mathbb{R}$.

(a) f maps any triangle onto its area,

(b) f maps any triangle onto its height,

(c) f maps any triangle onto the number of vertices it has.

3.   Let f be the function defined on the set $\mathbb{Z}^+$ of positive integers which maps each integer onto the sum of its digits.

(a) Find $f(4), f(19)$ and $f(301)$.

(b) Decide whether f is one-one or many-to-one.

(c) Are there any elements of $\mathbb{Z}^+$ which do not belong to the range of f? If so, give an example.

(d) Is f a mapping from $\mathbb{Z}^+$ onto itself?

4.   Repeat question 3 for the function f from $\mathbb{Z}^+$ into itself denfined by $f: x \rightarrow 2x$.

5.   Repeat question 3 for the function f from $\mathbb{Z}^+$ into itself defined by $f:x \rightarrow x^2 - x + 1$.

6.   Repeat question 3 for the function f from $\mathbb{Z}^+$ into itself which maps a positive integer onto the sum of its factors, e.g. $f(12) = 1 + 2 + 3 + 4 + 6 + 12 = 28$.

In questions 7 to 15 a mapping f from a set X into a set Y is defined. In each case find the range of the mapping, then decide whether (a) f maps X onto Y, (b) f is one-one, (c) f is a one-one correspondence between X and Y.

7. $X = Y = \mathbb{R}, f : x \rightarrow 2x$.

8. $X = Y = \mathbb{R}, f : x \rightarrow 2$.

9. $X = Y = \mathbb{R}, f : x \rightarrow \frac{1}{2}x$.

10. $X = Y = \mathbb{R}, f : x \rightarrow x^2$.

11. $X = Y = \mathbb{Z}, f : x \rightarrow x - 1$.

12. $X = Y = \mathbb{Z}, f : x \rightarrow 1 - x$.

13. $X = \{\text{all polygons}\}$, $Y = \{x \in \mathbb{R} : x > 0\}$, f maps a polygon onto the length of its perimeter.

14. $X = \{x \in \mathbb{R} : 0 < x < 1\}$, $Y = \mathbb{R}, f : x \rightarrow 1/x$.

15. $X = \{\text{all sets}\}$, $Y = \mathbb{N}$, f maps any set onto the number of subsets it has, e.g. $f : \{a, b, c\} \rightarrow 8$.

2.2 The graph of a function

A function f from a set X to a set Y can be represented as the set of all ordered pairs (x, y) such that $f : x \rightarrow y$, i.e. as a subset of the product set $X \times Y$. By marking off the elements of X along a horizontal axis and the elements of Y along a vertical axis, the function may be represented by a set of points with coordinates (x, y). This set of points is called the *graph* of the function.

Of particular interest are functions which map the set of real numbers $\mathbb{R}$ into itself. Such functions are represented as sets of ordered pairs (x, y) which are subsets of the product set $\mathbb{R} \times \mathbb{R}$, also written $\mathbb{R}^2$. Thus the graph of any function $f : \mathbb{R} \to \mathbb{R}$ is the set of all points (x, y) in the x, y plane such that $f : x \to y$. Points are plotted using horizontal and vertical axes called the x-axis and y-axis respectively. The coordinates (x, y) are often called the *Cartesian coordinates* of the point after René Descartes† who first described the system. The number x is called the x-coordinate or *abscissa* and y the y-coordinate or *ordinate*. Since every point (x, y) on the graph of a function f satisfies the relation $y = f(x)$, this is called the *Cartesian equation* of the graph. For instance, the graph of the function $f : x \to x + 2$ has equation $y = x + 2$.

It is clearly impossible to plot all the points on the graph of a function whose domain is the set of all real numbers. However, we can draw a graph of a function defined in a given interval of the form $\{x : a \leqslant x \leqslant b\}$.

Example 1   Draw the graph of the function defined by the equation $y = 2x^2 - 5x$, taking $-1 \leqslant x \leqslant 4$.

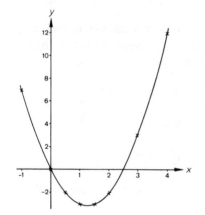

x	-1	0	1	2	3	4
y	7	0	-3	-2	3	12

To improve accuracy we consider two additional values of x.

x	$\frac{1}{2}$	$1\frac{1}{2}$
y	-2	-3

Although drawing a graph can be useful when considering a particular function, important features of the graph may lie beyond the points plotted. In later chapters we will see that there are ways of examining the general behaviour of a function. These can be used to produce a sketch indicating the form of the graph over the whole domain.

An equation $y = f(x)$ describes the relation between x and y as their values vary over the domain and range of the function f. Thus x and y are called *variables*. Since the value of x may be chosen freely from the domain of f, x is called the *independent variable*. However, the value of y depends on the value

---

† *Descartes, René* (1596–1650) French rationalist philosopher and mathematician. In mathematics his greatest achievements were advances in the theory of equations and the invention of coordinate geometry, as described in his essay *La Géométrie* (1637).

chosen for x, so y is called the *dependent variable*. The statement 'y is a function of x' is another way of expressing this relationship.

When defining a function on the set of real numbers, either in the form $f: x \rightarrow y$ or $y = f(x)$, the domain is usually taken to be the set of values of x for which a value of y can be determined. For instance, the domain of the function $f: x \rightarrow 1/(x-1)$ is the set containing all real numbers except 1, written $\mathbb{R} - \{1\}$. There is no value of x such that $f(x) = 0$. Hence the range of the function is the set $\mathbb{R} - \{0\}$. Since a negative number has no real square root, the domain of the function defined by $y = \sqrt{(1-x^2)}$ is the set $\{x \in \mathbb{R}: -1 \leqslant x \leqslant 1\}$. It follows that the range of the function is the set $\{y \in \mathbb{R}: 0 \leqslant y \leqslant 1\}$.

[In general an equation connecting variables x and y defines a *relation* on the set $\mathbb{R}$. Such an equation defines a function only if for every value of x in the domain there is exactly one value of y. For example, the equation $y^2 = x$ defines a relation, but not a function, because for any positive value of x there are two possible values of y.]

Exercise 2.2

In questions 1 to 6 write down the equation of the graph of the given function, then draw the graph taking values of x in the stated interval.

1. $f: x \rightarrow 2 - x$, $-1 \leqslant x \leqslant 3$.

2. $f: x \rightarrow 2x + 1$, $-2 \leqslant x \leqslant 2$.

3. $f: x \rightarrow 4 - x^2$, $-3 \leqslant x \leqslant 1$.

4. $f: x \rightarrow x^3 - 2x^2$, $-2 \leqslant x \leqslant 3$.

5. $f: x \rightarrow 1/x^2$, $\frac{1}{2} \leqslant x \leqslant 4$.

6. $f: x \rightarrow 1/(1-x)$, $-5 \leqslant x \leqslant \frac{1}{2}$.

In questions 7 to 18 decide whether the given equation defines on some subset of $\mathbb{R}$ a function of the form $f: x \rightarrow y$. If so, determine the largest possible domain for f and the corresponding range, and state whether f is one-one or many-to-one.

7. $y = x + 4$.

8. $x + y = 4$.

9. $y = 4x + 1$.

10. $y = x^2 + 4$.

11. $y^2 = x + 4$.

12. $y = 1 - x^2$.

13. $y = \dfrac{1}{x^2}$.

14. $y = \dfrac{1}{x-1}$.

15. $y = \dfrac{1}{4-x}$.

16. $x^2 + y^2 = 4$.

17. $y = \sqrt{x}$.

18. $y^3 = x^2$.

2.3 Points, lines and curves

Any equation connecting variables x and y may be considered as the Cartesian equation of a curve or graph in the x, y plane. The curve consists of the set of points whose coordinates (x, y) satisfy the equation. It is possible to decide whether or not a point lies on a curve by substituting its coordinates into the equation.

Example 1   Which of the points $A(-1,1)$, $B(1,-2)$ and $C(-2,2)$ lie on the curve $x^2+y^2 = 5$?

At A: $x^2+y^2 = (-1)^2+1^2 = 2$ i.e. $x^2+y^2 \neq 5$
At B: $x^2+y^2 = 1^2+(-2)^2 = 5$ i.e. $x^2+y^2 = 5$
At C: $x^2+y^2 = (-2)^2+2^2 = 8$ i.e. $x^2+y^2 \neq 5$
$\therefore$ B lies on the curve $x^2+y^2 = 5$, but A and C do not.

Since the x-coordinate of any point on the y-axis is 0, points of intersection of a curve with the y-axis are found by substituting $x = 0$ in the equation. Similarly points of intersection with the x-axis are found by substituting $y = 0$. The point $(0,0)$, where the axes themselves intersect, is called the *origin*.

Example 2   Find the points of intersection of the curve $y = x^2-1$ with the coordinate axes.

$x = 0 \Rightarrow y = -1$
$\therefore$ the curve cuts the y-axis at the point $(0,-1)$.

$y = 0 \Leftrightarrow x^2-1 = 0 \Leftrightarrow (x+1)(x-1) = 0$
$\Leftrightarrow x+1 = 0$ or $x-1 = 0$
$\Leftrightarrow$ $x = -1$ or $x = 1$

$\therefore$ the curve cuts the x-axis at the points $(-1,0)$ and $(1,0)$.

We now consider pairs of points in the x, y plane and the straight lines joining them.

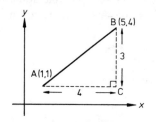

The diagram shows the points $A(1,1)$ and $B(5,4)$. In triangle ABC, AC and BC are parallel to the coordinate axes, so that $\angle ACB = 90°$.
    By Pythagoras' theorem,
$AB^2 = AC^2 + BC^2$

$\therefore$ $AB^2 = (5-1)^2+(4-1)^2 = 4^2+3^2 = 25$

Hence the distance between the points A and B is 5.

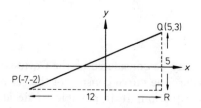

Similarly, if P and Q are the points $(-7,-2)$ and $(5,3)$ respectively, $PQ^2 = PR^2+QR^2$

$\therefore$ $PQ^2 = (5+7)^2+(3+2)^2$
        $= 12^2+5^2 = 169$
$\therefore$ $PQ = 13$.

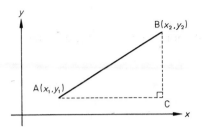

Thus, for a typical pair of points $A(x_1, y_1)$ and $B(x_2, y_2)$,

$$AC = x_2 - x_1, \quad BC = y_2 - y_1$$

$$\therefore \quad \boxed{AB = \sqrt{\{(x_2 - x_1)^2 + (y_2 - y_1)^2\}}.}$$

Example 3   Find the distance between the points $S(-1, 2)$ and $T(3, -1)$.

$$ST = \sqrt{\{(-1-3)^2 + (2-(-1))^2\}} = \sqrt{\{4^2 + 3^2\}} = \sqrt{25} = 5.$$

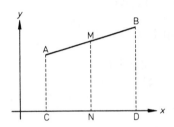

Let M be the mid-point of the line joining the points $A(x_1, y_1)$ and $B(x_2, y_2)$. If AC, MN and BD are drawn parallel to the y-axis then N is the mid-point of CD.

Since $CD = x_2 - x_1$,

$$CN = \tfrac{1}{2}(x_2 - x_1)$$

$$\therefore \quad ON = OC + CN = x_1 + \tfrac{1}{2}(x_2 - x_1) = \tfrac{1}{2}(x_1 + x_2),$$

i.e. the x-coordinate of M is $\tfrac{1}{2}(x_1 + x_2)$.

By a similar argument the y-coordinate of M is $\tfrac{1}{2}(y_1 + y_2)$

$$\therefore \quad \boxed{\text{the mid-point of } AB \text{ is } (\tfrac{1}{2}\{x_1 + x_2\}, \tfrac{1}{2}\{y_1 + y_2\}).}$$

Example 4   Find the mid-points of the lines joining the points (a) $(4, -1)$ and $(0, 3)$ (b) $(-5, 1)$ and $(-2, 5)$.

(a) The mid-point is $(\tfrac{1}{2}\{4+0\}, \tfrac{1}{2}\{-1+3\})$, i.e. $(2, 1)$.
(b) The mid-point is $(\tfrac{1}{2}\{-5-2\}, \tfrac{1}{2}\{1+5\})$, i.e. $(-3\tfrac{1}{2}, 3)$.

Example 5   Given that the points $A(5, -3)$, $B(3, 6)$, $C(-1, 7)$ are the vertices of a parallelogram $ABCD$, find the coordinates of D.

Since the diagonals of a parallelogram bisect each other, the diagonals AC and BD of the parallelogram $ABCD$ have the same mid-point.
The mid-point of AC is $(\tfrac{1}{2}\{5-1\}, \tfrac{1}{2}\{-3+7\})$, i.e. $(2, 2)$.
Letting D be the point (h, k), the mid-point of BD is $(\tfrac{1}{2}\{3+h\}, \tfrac{1}{2}\{6+k\})$.
$\therefore \quad \tfrac{1}{2}\{3+h\} = 2$ and $\tfrac{1}{2}\{6+k\} = 2$ i.e. $h = 1, k = -2.$
Hence the coordinates of D are $(1, -2)$.

Exercise 2.3

1. Find the points of intersection with the coordinate axes of the graphs with the following equations.
(a) $y = x - 1$ (b) $y = 2x + 5$
(c) $y = x^2 + 3x$ (d) $y = 2x^2 + 3x - 2$
(e) $y = \dfrac{2}{x - 2}$ (f) $y = \dfrac{x - 3}{x - 6}$

2. In each of the following cases determine which points lie on the given curve.
(a) $y = x^2 + x$; $(1, 3)$, $(-2, 6)$, $(-3, 6)$, $(0, 2)$.
(b) $y = 1/(x - 2)$; $(0, \frac{1}{2})$, $(3, 1)$, $(-1, -1)$, $(6, \frac{1}{3})$.
(c) $y = 3x^2 - 7x + 5$; $(2, 3)$, $(-5, 115)$, $(-2, 29)$, $(1, 4)$.
(d) $y = x^3 - 5x - 9$; $(-2, -7)$, $(1, -11)$, $(4, 35)$, $(-1, -5)$.
(e) $4x^2 + y^2 = 20$; $(3, -4)$, $(-2, 2)$, $(5, 0)$, $(-1, -4)$.
(f) $x^2 + y^2 - 8x + 4y - 5 = 0$; $(4, -2)$, $(7, -6)$, $(0, 1)$, $(-1, -1)$.

3. Find the distances between the following pairs of points.
(a) $(0, 7)$, $(12, -2)$ (b) $(-3, -5)$, $(-7, -8)$
(c) $(-2, 14)$, $(3, 2)$ (d) $(5, -1)$, $(11, 7)$
(e) $(4, -10)$, $(-4, 5)$ (f) $(-4, -4)$, $(3, 20)$
(g) $(7, -4)$, $(5, -1)$ (h) $(-2, 3)$, $(1, -3)$.

4. Find the mid-points of the lines joining the points given in question 3.

5. Given that B is the mid-point of the line segment AC, find the coordinates of C when
(a) $A(1, 2)$, $B(3, 1)$ (b) $A(6, 2)$, $B(3, -2)$
(c) $A(-5, 4)$, $B(-2, -1)$ (d) $A(4, -3)$, $B(3, -1)$.

6. Find the coordinates of the vertex D of the parallelogram $ABCD$ given
(a) $A(5, 0)$, $B(-3, 2)$, $C(1, -6)$ (b) $A(-2, 7)$, $B(4, 5)$, $C(6, -1)$
(c) $A(3, -4)$, $B(0, -7)$, $C(2, 5)$ (d) $A(1, -5)$, $B(-3, 7)$, $C(-2, 8)$.

7. Determine whether the triangles with vertices given below are (i) isosceles, (ii) right-angled.
(a) $A(4, 7)$, $B(-3, 8)$, $C(9, 2)$ (b) $P(3, -1)$, $Q(-4, 2)$, $R(1, -6)$
(c) $S(-2, 5)$, $T(4, 3)$, $U(-5, -4)$ (d) $X(5, 9)$, $Y(15, 3)$, $Z(-1, -1)$.

8. Given points $P(5, 5)$, $Q(-1, -3)$ and $R(-2, 4)$, show that the mid-point M of the line PQ is the centre of the circle through P, Q and R.

9. Show that the point $P(4, 5)$ lies on the perpendicular bisector of the line joining the points $A(5, -2)$ and $B(-3, 4)$. Find the distance of P from the line AB.

10. Show that the triangle with vertices $D(2, 5)$, $E(3, -7)$ and $F(-6, 1)$ is isosceles and find its area.

11. Given that $A(0, -5)$, $B(-7, 2)$, $C(2, 11)$ are the vertices of a parallelogram $ABCD$, find the coordinates of D. Prove that $ABCD$ is a rectangle and find its area.

12. Given that $A(1, -7)$, $B(6, 1)$ and $C(1, 9)$ are the vertices of a parallelogram $ABCD$, find the coordinates of D. Prove that $ABCD$ is a rhombus and find its area.

13. The points $P(-1, -2)$, $Q(3, 4)$ and $R(11, 6)$ are the vertices of a parallelogram $PQRS$. A point T is taken on RS produced so that S is the mid-point of RT. Find the coordinates of S and T. Prove that $RP = RT$.

14. The coordinates of the points A, B, C are $(t, -1)$, $(-2, 1)$ and $(4, 3)$ respectively. Given that $AB = AC$, calculate the value of t. The line BC is produced to D so that $BC = CD$. Calculate the coordinates of D and the length of AD.

15. The vertices of a quadrilateral $ABCD$ are $A(-4, 3)$, $B(2, 1)$, $C(1, -2)$ and $D(-2, -3)$. Prove that $AB = AD$ and $BC = CD$. Find the area of the quadrilateral.

2.4 Gradient of a straight line

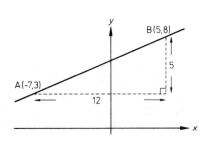

The *gradient* of a straight line is a quantity used for measuring the direction of a line in relation to the co-ordinate axes. Consider the straight line which passes through the points $A(-7, 3)$ and $B(5, 8)$. Moving from A to B results in increases of 12 in the x-coordinate and 5 in the y-coordinate. The gradient of AB is the ratio of these increases,

$$\text{i.e.} \quad \text{gradient of } AB = \frac{\text{increase in } y}{\text{increase in } x} = \frac{5}{12}.$$

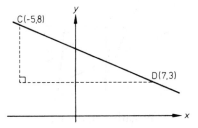

For the points $C(-5, 8)$ and $D(7, 3)$ the increase in x is 12, but there is a decrease in y of 5. This can be considered as an increase of -5

$$\therefore \quad \text{gradient of } CD = -\frac{5}{12}.$$

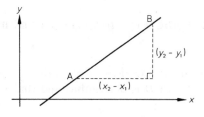

In general, for a line passing through the points $A(x_1, y_1)$ and $B(x_2, y_2)$, we have:

$$\text{gradient} = \frac{y_2 - y_1}{x_2 - x_1} \text{ or } \frac{y_1 - y_2}{x_1 - x_2}.$$

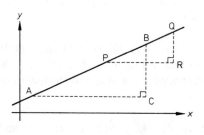

By considering similar triangles ABC and PQR, we see that $\dfrac{BC}{CA} = \dfrac{QR}{RP}$.

Thus the value of the gradient of a line does not depend on the particular points used to calculate it, but only on the angle the line makes with the x-axis. As *parallel* lines make equal angles with the x-axis, they must have equal gradients.

Example 1   Find the gradients of the lines joining
(a) $A(4, -3)$ and $B(2, 1)$ (b) $C(4, 3)$ and $D(-2, 1)$.

(a) Gradient of $AB = \dfrac{-3-1}{4-2} = \dfrac{-4}{2} = -2.$

(b) Gradient of $CD = \dfrac{3-1}{4-(-2)} = \dfrac{2}{6} = \dfrac{1}{3}.$

All the points on a line parallel to the x-axis have the same y-coordinate. Hence lines parallel to the x-axis have *zero gradient*. Since division by zero is undefined, we can give no finite numerical value to the gradient of lines joining points with the same x-coordinate. Thus lines parallel to the y-axis may be said to have *infinite gradient*. The diagrams below show the significance of the sign of the gradient of a line.

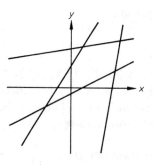

Positive gradients

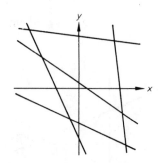

Negative gradients

To find the relationship between the gradients of *perpendicular* lines, consider a pair of perpendicular lines through the origin with gradients m_1 and m_2.

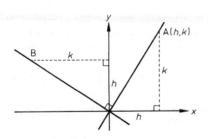

Let $A(h, k)$ be a typical point on the first line and let B be a point on the second line such that $OB = OA$. If OA is rotated anti-clockwise through 90°, the point A moves to B. Hence the coordinates of B are $(-k, h)$. Thus,

gradient of $OA = m_1 = \dfrac{k}{h}$ and

gradient of $OB = m_2 = -\dfrac{h}{k}$

$$\therefore \quad m_1 m_2 = \frac{k}{h} \times \left(-\frac{h}{k}\right), \quad \text{i.e.} \quad m_1 m_2 = -1.$$

For any pair of perpendicular lines there is a corresponding pair through the origin, therefore the product of the gradients of perpendicular lines is -1.

Example 2 Write down the gradients of lines perpendicular to lines with gradients (a) 2, (b) $\frac{1}{2}$, (c) $-5/4$.

The gradients of the perpendicular lines are
(a) $-\frac{1}{2}$, (b) -2, (c) 4/5.

Exercise 2.4

1. Find the gradients of the lines joining the points
(a) $(3, 5)$ and $(-1, 2)$ (b) $(4, 6)$ and $(5, -1)$
(c) $(-2, 6)$ and $(-1, 4)$ (d) $(-2, -3)$ and $(-9, -6)$
(e) $(-4, 0)$ and $(-3, 5)$ (f) $(0, -7)$ and $(2, -5)$

2. Write down the gradients of lines perpendicular to those given in question 1.

3. By comparing the gradients of AB and BC, or otherwise, decide whether or not the points A, B and C are collinear.
(a) $A(1, -6)$, $B(3, -2)$, $C(-7, 6)$ (b) $A(-2, 6)$, $B(5, 7)$, $C(-9, 5)$
(c) $A(3, 4)$, $B(-1, -8)$, $C(5, 10)$ (d) $A(1, 1)$, $B(6, -2)$, $C(-3, 4)$.

4. Given the points $A(8, -1)$, $B(5, 5)$, $C(-5, 5)$ and $D(-3, 1)$, show that
(a) $ABCD$ is a trapezium, (b) AB is perpendicular to BD.

5. Given the points $X(-6, -2)$, $Y(-4, 3)$ and $Z(6, -1)$, show that $\angle XYZ$ is a right angle. If M and N are the mid-points of XY and YZ, verify that MN is parallel to XZ.

6. The coordinates of the points P, Q and R are $(4, -3)$, $(-1, 2)$ and $(2, t)$

respectively. Calculate the values of t given that (a) P, Q and R are collinear, (b) $\angle QPR = 90°$, (c) $\angle PRQ = 90°$.

7. The coordinates of the points A and B are $(1,6)$ and $(5,-2)$ respectively. Find the coordinates of the point C on the x-axis, given that (a) C also lies on AB, (b) $\angle ABC$ is a right angle, (c) C lies on the perpendicular bisector of AB.

8. A quadrilateral $ABCD$ has vertices $A(-1,-2)$, $B(5,0)$, $C(4,3)$ and $D(1,2)$. Show that $\angle ABC$ and $\angle BCD$ are right angles. Calculate the lengths of AB, BC and CD and hence find the area of the quadrilateral.

9. Given the points $P(7,-4)$, $Q(9,8)$, $R(-3,10)$, $S(-12,-7)$, show that (a) QS is perpendicular to PR, (b) QS passes through the mid-point of PR.

10. Given that the triangle with vertices $X(1,2)$, $Y(2,a)$ and $Z(-4,-3)$ is right-angled at Y, find the two possible values of a and the area of the triangle in each case.

11. A quadrilateral $ABCD$ has vertices $A(10,5)$, $B(5,-4)$, $C(-5,-7)$ and $D(-3,6)$. Show that AC is perpendicular to BD and calculate the area of the quadrilateral.

12. Show that the points $D(5,-3)$, $E(2,6)$, $F(-1,10)$ and $G(-4,9)$ are the vertices of a trapezium. Show also that DE is perpendicular to FG and that the area of the trapezium is 30 square units.

2.5 The equation of a line

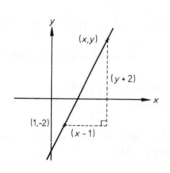

The equation of a straight line is an equation connecting the variables x and y which is satisfied by any point (x,y) on the line. The gradient is used to find a suitable equation. Consider the line with gradient 2, which passes through the point $(1,-2)$. If the point (x,y) lies on the line, then its gradient is given by $\dfrac{y+2}{x-1}$.

Hence $\dfrac{y+2}{x-1} = 2$ for any point (x,y) on the line.

Rearranging, $y+2 = 2(x-1)$,   i.e. $y+2 = 2x-2$

$\therefore$ the equation of the line is $y = 2x-4$.

The same method can be used to find the equation of a line with gradient m passing through the point (x_1, y_1).

The gradient of the line $= \dfrac{y - y_1}{x - x_1} = m$

$\therefore$ the equation of the line is $y - y_1 = m(x - x_1)$.

Example 1   Find the equation of the line with gradient $-\frac{1}{2}$ which passes through the point $(2, -3)$.

The equation of the line is $y + 3 = -\frac{1}{2}(x - 2)$
$$\text{i.e. } 2(y + 3) = -(x - 2)$$
$$2y + 6 = -x + 2$$

$\therefore$ the required equation is $x + 2y + 4 = 0$.

Example 2   Find the equation of the line which passes through the points $(2, -2)$ and $(1, 3)$.

The gradient of the line $= \dfrac{-2 - 3}{2 - 1} = -5$

$\therefore$ the equation of the line is $y + 2 = -5(x - 2)$, i.e. $5x + y - 8 = 0$.

The general form of the straight line equation emerges when we consider a line with gradient m passing through the point $(0, c)$ on the y-axis. Its equation is $y - c = m(x - 0)$, i.e. $y = mx + c$. Hence the equation $y = mx + c$ represents a straight line with gradient m, where c is the *intercept* on the y-axis. In particular, $y = mx$ is a line through the origin $(0, 0)$ and $y = c$ is a line parallel to the x-axis. Lines parallel to the y-axis have equations of the form $x = k$.

Example 3   Find the gradient of the line $3x + 4y = 8$ and the intercept on the y-axis.

The equation may be written $4y = -3x + 8$, i.e. $y = -\frac{3}{4}x + 2$

$\therefore$ the gradient of the line is $-\frac{3}{4}$ and the intercept on the y-axis is 2, i.e. the line cuts the y-axis at the point $(0, 2)$.

A straight line is usually sketched by finding its points of intersection with both x- and y-axes.

Example 4   Sketch the line $2x - 3y = 12$.

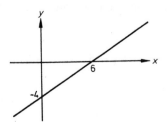

$x = 0 \Rightarrow -3y = 12 \Rightarrow y = -4$

$\therefore$ the line cuts the y-axis at $(0, -4)$.

$y = 0 \Rightarrow 2x = 12 \Rightarrow x = 6$

$\therefore$ the line cuts the x-axis at $(6, 0)$.

The general equation of a straight line can also be written $y - mx = c$. This form is sometimes used when obtaining the equation of a line with given gradient. The constant c is determined by substituting the coordinates of a point on the line.

Example 5   Find the equation of the line through $(-1, 2)$ with gradient 3.

The equation of the line is $y - 3x = 2 - 3(-1)$, i.e. $y - 3x = 5$.

The properties of the gradients of parallel and perpendicular lines are used in the next example.

Example 6   Find the equations of the lines parallel and perpendicular to the line $3x - 2y = 1$, which pass through the point $(2, 3)$.

Method 1   Rearranging the given equation, we obtain $y = \dfrac{3}{2}x - \dfrac{1}{2}$

∴   the gradient of the given line is 3/2.
Hence the gradient of any parallel line is 3/2.
The equation of the parallel line through $(2, 3)$ is

$$y - 3 = \frac{3}{2}(x - 2), \quad \text{i.e.} \quad 2(y - 3) = 3(x - 2)$$

$$2y - 6 = 3x - 6$$

which may be written $3x - 2y = 0$.
The gradient of any perpendicular line is $-\frac{2}{3}$.
∴   the equation of the perpendicular line through $(2, 3)$ is

$$y - 3 = -\tfrac{2}{3}(x - 2), \quad \text{i.e.} \quad 3(y - 3) = -2(x - 2)$$
$$3y - 9 = -2x + 4$$

which may be written $2x + 3y = 13$.

In time, the reader may prefer to adopt a quicker method based on the fact that any line parallel to $3x - 2y = 1$ has an equation of the form $3x - 2y = c$, and any perpendicular line the form $2x + 3y = k$.

Method 2   The equation of the required parallel line is

$$3x - 2y = 3.2 - 2.3, \quad \text{i.e.} \quad 3x - 2y = 0.$$

The equation of the required perpendicular line is

$$2x + 3y = 2.2 + 3.3, \quad \text{i.e.} \quad 2x + 3y = 13.$$

The point of intersection of two straight lines is the point which lies on both lines, therefore its coordinates must satisfy both equations. Hence these are found by solving the equations simultaneously.

Example 7   Find the point of intersection of the lines $3x + y = 5$ and $3x - 2y + 1 = 0$.

At the point of intersection $3x + y = 5$ (1)
and $3x - 2y = -1$ (2)

Subtracting (2) from (1): $3y = 6$ $\therefore y = 2$
Substituting in (1): $3x + 2 = 5$ $\therefore x = 1$
$\therefore$ the point of intersection is $(1, 2)$.

[This method will break down if the two lines are parallel and thus have no point of intersection.]

Exercise 2.5

1.   Find the equations of the straight lines through the given points with the given gradients.
(a) $(-2, 3); 4$ (b) $(2, -3); -\frac{1}{2}$ (c) $(4, -1); -3$
(d) $(-5, -7); \frac{2}{3}$ (e) $(-12, 5); -5/12$ (f) $(t^2, 2t); 1/t$.

2.   Find the equations of the lines joining the following points.
(a) $(0, -7), (2, -5)$ (b) $(-2, 6), (-1, 4)$
(c) $(4, 7), (-5, 7)$ (d) $(4, 3), (-4, -3)$
(e) $(-2, -3), (-9, -6)$ (f) $(6, 2), (6, -3)$
(g) $(p^2, 2p), (q^2, 2q)$ (h) $(p, 1/p), (q, 1/q)$.

3.   Find the gradients and intercepts on the y-axis of the following lines.
(a) $2y = 5x + 9$,   (b) $4y - 7x + 13 = 0$,   (c) $2x + 8y + 19 = 0$,
(d) $3x - 6y - 17 = 0$,   (e) $2y + 5 = 0$,   (f) $5x + 9y - 27 = 0$.

4.   In each of the following cases sketch the given lines and find their points of intersection, if any.
(a) $4x + 3y = 7, 3x - 4y + 1 = 0$,   (b) $y = 4x - 10, 4y = 11 - x$,
(c) $2x - 5y = 15, 3x - 7y = 21$,   (d) $5y = 17 - 4x, 8x + 10y = 19$,
(e) $8x + 7y = 23, 8y = 7x - 6$,   (f) $5y - 2x = 3, 2y - 5x = 4$.

5.   Decide whether the pairs of lines given in question 4 are (i) perpendicular, (ii) parallel.

6.   In each of the following cases find the equations of the lines parallel and perpendicular to the given line through the given point.
(a) $y = 2x - 3, (-1, 3)$ (b) $3x + 2y = 7, (0, 2)$
(c) $3x - 5 = 0, (3, -4)$ (d) $3x - 9y = 4, (-2, -1)$.

7.   Find the coordinates of the foot of the perpendicular from the point $(-2, 4)$ to the line $x - 2y - 5 = 0$.

8.   Find the coordinates of the foot of the perpendicular from the point $(4, 5)$ to the line $2x + 5y = 4$.

9.   Prove that the lines $3x - 2y + 7 = 0$, $5x + 4y = 3$ and $2x + 5y = 8$ are concurrent. Find the equation of the line through the point of intersection which has gradient 3.

10. Find the equation of the perpendicular bisector of the line joining the points $(-2, 3)$ and $(8, -7)$.

11. Given the points $A(9, -4)$, $B(2, 10)$ and $C(-4, 2)$, show that the perpendicular bisector of BC intersects AB on the x-axis.

12. The points $A(7, 3)$, $B(1, -4)$, $C(-5, -1)$ are three vertices of a trapezium $ABCD$. Given that $\angle BCD = 90°$, find (a) the equations of AD and CD, (b) the coordinates of D.

13. The points $P(-3, 1)$, $Q(-1, 5)$, $R(6, 4)$ are three vertices of a trapezium $PQRS$ in which PQ is parallel to SR. Given that the diagonals intersect at right angles, find the coordinates of S.

14. The point $A(7, 5)$ is one vertex of a parallelogram $ABCD$. The equations of the sides BC and CD are $3y = 2x - 9$ and $2y = 3x - 1$ respectively. Find the coordinates of B and D. Show that the diagonals cut at right angles.

15. The equations of two sides of a triangle PQR are $x + 2y + 4 = 0$ and $3x - 4y + 37 = 0$. Given that $\angle PQR = 90°$ and that P is the point $(6, -5)$, find the coordinates of Q.

16. Find the reflection of the point $P(3, 1)$ in the line l with equation $y = 2x$. Use your answer to find the reflection in l of the line $y = x - 2$.

2.6 Linear inequalities

In the previous section it was shown that the equation $y = mx + c$ represents a straight line. We now consider inequalities, such as $y > mx + c$, and the sets of points they represent in the x, y plane.

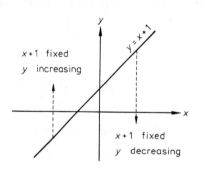

Consider the line $y = x + 1$. At any point on the line, the value of y is equal to the value of $x + 1$. If we move away from the line in a direction parallel to the y-axis, then the value of $x + 1$ remains unchanged. However, the value of y either increases or decreases as shown in the diagram. Thus, for points above the line $y > x + 1$ and for points below the line $y < x + 1$.

By a similar argument, we find that the inequality $y > mx + c$ represents the set of points above the line $y = mx + c$, and $y < mx + c$ the set of points below the line $y = mx + c$.

Example 1   Sketch the region of the x, y plane represented by the inequality $y > 2x + 3$.

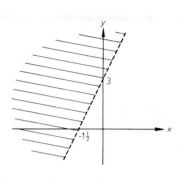

On the boundary of the region $y = 2x + 3$.

$$x = 0 \Rightarrow y = 3$$
$$y = 0 \Rightarrow 2x + 3 = 0 \Rightarrow x = -1\tfrac{1}{2}$$

$\therefore$ the boundary line cuts the axes at $(0, 3)$ and $(-1\tfrac{1}{2}, 0)$. Hence $y > 2x + 3$ represents the shaded region shown in the diagram.
[The broken line is used to indicate that points on the boundary do not belong to the region.]

An alternative approach is used in the next example.

Example 2   Sketch the region of the x, y plane represented by the inequality $x + 3y - 2 \leqslant 0$,

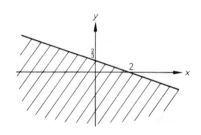

On the boundary of the region $x + 3y - 2 = 0$.

$$x = 0 \Rightarrow 3y - 2 = 0 \Rightarrow y = \tfrac{2}{3}$$
$$y = 0 \Rightarrow x - 2 = 0 \Rightarrow x = 2$$

$\therefore$ the boundary line cuts the axes at $(0, \tfrac{2}{3})$ and $(2, 0)$.
At the origin, $x + 3y - 2 = -2$
$\therefore$ the origin lies in the required region.
Hence $x + 3y - 2 \leqslant 0$ represents the set of points on and below the line $x + 3y - 2 = 0$.

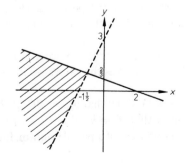

Combining the results of Examples 1 and 2, we obtain the region in which the inequalities $y > 2x + 3$ and $x + 3y - 2 \leqslant 0$ hold simultaneously.

Exercise 2.6

1. Sketch the region of the x, y plane represented by the following inequalities.
(a) $y > x + 1$ (b) $y < 3x - 6$ (c) $y \leqslant 4 - x$ (d) $2y + x \geqslant 7$
(e) $2y < 5x$ (f) $3y - 2x > 8$ (g) $2x - 5 < 3$ (h) $y^2 \leqslant 4$.

2. Sketch the region of the x, y plane represented by the following simultaneous inequalities.

(a) $2y > x + 5$
$\quad x > -1$

(b) $\quad x < y$
$\quad x + y \leqslant 8$

(c) $-2 \leqslant x \leqslant 2$
$\quad -3 \leqslant y \leqslant 3$

(d) $\quad -x < y < x$
$\quad 2 < 2x + y < 4$

(e) $y + 2 > 0$
$\quad y < 2x + 4$
$\quad y + 3x < 6$

(f) $\quad x + 3y \leqslant 9$
$\quad y - 3x \leqslant 9$
$\quad 3x - 4y \leqslant 9$.

3. Given that x and y satisfy the inequalities $2x - y < 8$ and $3x + 5y + 1 > 0$, find the range of possible values for y.

4. Given that x and y satisfy the inequalities $4x + 3y \leqslant 15$ and $3x - 2y < 7$, find the range of possible values for x.

5. Given that $2x + y > 0$, $2y > x$ and $y < 3$, find the range of possible values of (a) x, (b) y, (c) $x + y$.

6. Repeat question 5 given that $x + 4y \leqslant 17$, $y - 3x \leqslant 14$, $x < 0$ and $y > 0$.

Exercise 2.7   *(miscellaneous)*

[Leave answers in surd form where appropriate.]

1. Find the range of the following functions and decide in each case whether the function is (i) one-one, (ii) 'onto'.
(a) f is the function which maps the set of real numbers $\mathbb{R}$ into itself such that
$\quad f : x \to x^2 - 4$.
(b) g is the function which maps the set of positive integers $\mathbb{Z}^+$ into the set of rational numbers $\mathbb{Q}$ defined by $g : x \to 1/x$.
(c) h is the function from $\mathbb{R} \times \mathbb{R}$ into $\mathbb{R}$ defined by $h : (r, s) \to r + s$.

2. Draw the graph of the function f defined by $f : x \to x(x - 3)$ for the domain $-2 \leqslant x \leqslant 5$. Find the range of the function corresponding to this domain. Find also the largest possible domain for which the range is $0 \leqslant f(x) \leqslant 18$.

3. Given that the curve $y = (x + a)/(x + b)$ passes through the points $(2, 1\frac{1}{2})$ and $(-1, 3)$, find the coordinates of the points A and B at which the curve cuts the x- and y-axes and calculate the length of the chord AB.

4. Show that the point $(t^2 - 5, t^3 - 5t)$ lies on the curve $y^2 = x^2(x + 5)$ for all values of t. Show also that if P, Q and R are the points at which t takes the values $-2, 1$ and 3 respectively, then P, Q and R are collinear.

5. Given the points $A(-3, -2)$, $B(4, 2)$ and $C(-4, 1)$, show that $AB = BC$. If the point $P(h, k)$ is also equidistant from A and C, show that $3k = h + 2$.

6. If the points $(2, -1), (6, 7)$ and $(-2, 7)$ are three vertices of a parallelogram, find the three possible positions of the fourth vertex.

7. Find the equations of the following straight lines:

(a) the line through $(1,3)$ with gradient 2,

(b) the line through the points $(3, -5)$ and $(-2,4)$,

(c) the line through $(2, -1)$ parallel to the line $4x - 3y = 2$,

(d) the line through $(-3,6)$ perpendicular to the line $3x - 5y = 7$.

8. The sides AB, BC of the parallelogram $ABCD$ have equations $y + 3x = 1$ and $y = 5x - 7$ respectively. If the coordinates of D are $(5, 10)$, find the coordinates of A, B and C.

9. A triangle has vertices $A(-3,0)$, $B(5,4)$, $C(7, -2)$. If D, E and F are the mid-points of BC, CA and AB respectively, find the point of intersection of the medians AD and BE. Show that this point also lies on the median CF.

10. Show that the points $(5,3)$, $(-1, -5)$ and $(-2,2)$ are three vertices of a square and find the coordinates of the fourth vertex. Find also the area of the square.

11. The points $(2,4)$, $(-1,2)$, $(2, -1)$ are the mid-points of the sides of a triangle. Find the coordinates of the vertices of the triangle.

12. The points A, B, C and D have coordinates $(3, -2)$, $(p,3)$, $(6,2)$ and (q,r) respectively. Given that quadrilateral $ABCD$ is a rhombus, find (a) the values of p, q and r, (b) the area of the rhombus.

13. The line $y = x + 5$ is the perpendicular bisector of the line joining the points $P(3, 10)$ and $Q(h,k)$. Find in terms of h and k (a) the coordinates of the mid-point of PQ and (b) the gradient of PQ. Deduce that $h + 3 = k$ and that $h + k = 13$.

14. The equation of the side QR of a triangle PQR is $3x - 4y = 10$, the coordinates of P are $(3,6)$ and the gradient of PQ is 7. Given that $PQ = PR$, find (a) the coordinates of the foot of the perpendicular from P to QR, (b) the equation of the side PR.

15. Given that X, Y and Z are the feet of the perpendiculars from the origin O to the lines with equations $x = 3$, $x + y = 18$ and $8x - 6y = 25$, show that X, Y and Z lie on a straight line.

16. The line $3x + 4y = 30$ cuts the x-axis at A and the y-axis at B. Find (a) the perpendicular distance from the origin O to the line AB, (b) the equation of the perpendicular bisector of AB.

17. P is the point of intersection of the lines $3x - y = 7$ and $x + 4y + 2 = 0$. Find (a) the equation of the line parallel to $2x + 3y = 40$ which passes through P, (b) the perpendicular distance of P from the line $2x + 3y = 40$.

18. The points $A(3, -1)$, $B(-2,4)$, $C(-1,7)$ are three vertices of the quadrilateral $ABCD$. Given that $\triangle ADC$ is the reflection in the line AC of $\triangle ABC$, find (a) the coordinates of D, (b) the area of quadrilateral $ABCD$.

19. The points $(3, -2)$, $(4,5)$, $(-3,6)$ are the vertices of a triangle. Find the coordinates of the centre and the radius of the circumscribed circle.

20. Draw a diagram illustrating the region S of the x, y plane which is defined by the simultaneous inequalities $x + y \geqslant 7$, $2x + y \leqslant 13$, $2x + 3y \leqslant 19$, and give the coordinates of the vertices of S. Prove that, if the line $y = kx$ intersects S, then $1/6 \leqslant k \leqslant 2\frac{1}{2}$.

The point P lies on $y = kx$ and is in the region S. Prove that, when $1/6 \leqslant k \leqslant 3/5$, the maximum value for the y-coordinate of P is $13k/(2+k)$, and find the corresponding expression when $3/5 \leqslant k \leqslant 2\frac{1}{2}$.   (C)

3 Polynomials and equations

3.1 The quadratic function

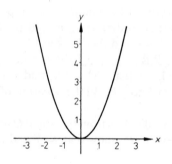

A *quadratic function* is a function of the form $f(x) = ax^2 + bx + c$, i.e.
$f : x \to ax^2 + bx + c$, where $a \neq 0$. Thus the graph of a quadratic function has an equation of the form $y = ax^2 + bx + c$.

As shown in the diagram, the graph of $y = x^2$ is symmetrical about the y-axis. The lowest point of the graph, called the *minimum point*, occurs at the origin $(0,0)$. Hence the *minimum value* of the function $f(x) = x^2$ is zero.

The graphs of some related functions are shown in the diagrams below.

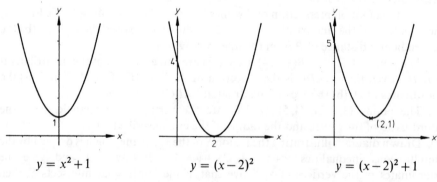

$$y = x^2 + 1$$

$$y = (x-2)^2$$

$$y = (x-2)^2 + 1$$

In each case a curve identical to the curve $y = x^2$ is produced. The term $+1$ in the equation $y = x^2 + 1$ moves the curve up 1 unit. In the graph of $y = (x-2)^2$ the original curve is moved 2 units to the right. The result of moving the curve 2 units to the right and 1 unit up is the graph with equation $y = (x-2)^2 + 1$, i.e. $y = x^2 - 4x + 5$.

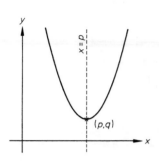

Hence the equation $y = (x-p)^2 + q$ represents a graph of the same shape as $y = x^2$, moved p units to the right and q units up. Thus it is symmetrical about the line $x = p$ and has a minimum point (p, q). This is confirmed by considering the range of the function $f(x) = (x-p)^2 + q$. Since $(x-p)^2$ is never negative, the least possible value of $f(x)$ is given by $(x-p)^2 = 0$. Hence $f(x)$ takes its minimum value q when $x = p$, and its range is the set $\{y \in \mathbb{R} : y \geqslant q\}$.

Example 1   Find the minimum value of the function $f(x) = x^2 - 2x + 5$ and sketch its graph.

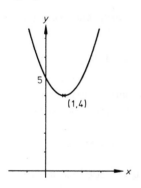

$$f(x) = x^2 - 2x + 5 = (x^2 - 2x + 1) + 4$$
$$= (x-1)^2 + 4$$

$\therefore$ $f(x)$ has a minimum value 4 given by $x = 1$.

Since $f(0) = 5$, the graph cuts the y-axis at the point $(0, 5)$.

[Note that $x^2 - 2x + 5$ is expressed as $(x-1)^2 + 4$ by 'completing the square' of the form $x^2 - 2x + \ldots$. The -1 in the bracket is half the original coefficient of x.]

We now consider the graph of $y = ax^2 + bx + c$ for values of a other than 1.

Example 2   Sketch the graph of $y = 5 - 4x - x^2$.

$$y = 5 - 4x - x^2 = 5 - (x^2 + 4x) = 9 - (x^2 + 4x + 4) = 9 - (x+2)^2$$

$\therefore$ y has a maximum value 9 given by $x = -2$.

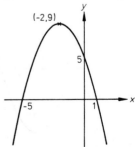

$x = 0 \Rightarrow y = 5$ $\therefore$ the graph cuts the y-axis at $(0, 5)$.
$y = 0 \Leftrightarrow \quad 5 - 4x - x^2 = 0$
$\qquad \Leftrightarrow (5+x)(1-x) = 0$
$\qquad \Leftrightarrow 5 + x = 0 \quad \text{or} \quad 1 - x = 0$
$\qquad \Leftrightarrow \quad\quad x = -5 \quad \text{or} \quad x = 1$

$\therefore$ the graph cuts the x-axis at $(-5, 0)$ and $(1, 0)$.

Example 3   Sketch the curve $y = 2x^2 + 5x$.

$$y = 2x^2 + 5x = 2\left\{x^2 + \frac{5}{2}x\right\}, \quad \text{but} \quad \left(x + \frac{5}{4}\right)^2 = x^2 + \frac{5}{2}x + \frac{25}{16}$$

$$\therefore \quad y = 2\left\{\left(x^2 + \frac{5}{2}x + \frac{25}{16}\right) - \frac{25}{16}\right\} = 2\left(x + \frac{5}{4}\right)^2 - \frac{25}{8}.$$

Hence y has a minimum value $-\dfrac{25}{8}$ given by $x = -\dfrac{5}{4}$.

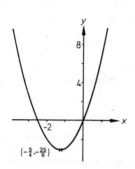

$$x = 0 \Rightarrow y = 0$$
$$y = 0 \Leftrightarrow 2x^2 + 5x = 0$$
$$\Leftrightarrow x(2x + 5) = 0$$
$$\Leftrightarrow x = 0 \quad \text{or} \quad 2x + 5 = 0$$
$$\Leftrightarrow x = 0 \quad \text{or} \quad x = -\frac{5}{2}$$

$\therefore$ the curve cuts the axes at the points

$\left(-\dfrac{5}{2}, 0\right)$ and $(0, 0)$.

In general, the equation $y = ax^2 + bx + c$ represents a curve called a *parabola*. For $a > 0$, the curve has a minimum point and is symmetrical about a vertical axis through this point. However, as shown in Example 2, when $a < 0$ the curve is inverted and has a maximum point on the axis of symmetry.

The last two examples in this section deal with quadratic inequalities.

Example 4   Prove that $2x^2 + 8x + 9 > 0$ for all real values of x.

$$2x^2 + 8x + 9 = 2(x^2 + 4x) + 9 = 2(x^2 + 4x + 4) - 8 + 9 = 2(x + 2)^2 + 1.$$

For all real values of x, $(x + 2)^2 \geqslant 0 \therefore 2x^2 + 8x + 9 > 0$.

Example 5   Find the values of x for which $x^2 - 1 < x + 5$.

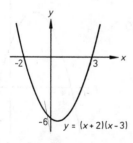

$$x^2 - 1 < x + 5$$
$$\Leftrightarrow \quad x^2 - x - 6 < 0$$
$$\Leftrightarrow (x + 2)(x - 3) < 0.$$

The sketch shows that the curve $y = (x + 2)(x - 3)$ lies below the x-axis for $-2 < x < 3$.
Hence if $x^2 - 1 < x + 5$, then $-2 < x < 3$.

Exercise 3.1

1. Find the maximum or minimum value of each of the quadratic functions given below and sketch the graph of the function.
(a) $x^2 + 4x - 5$ (b) $2x - x^2$ (c) $x^2 - 3x + 4$ (d) $x^2 + 4$
(e) $4x^2 - 12$ (f) $4x^2 - 12x + 9$ (g) $2 + x - 3x^2$ (h) $4x - 6x^2 - 9$.

2. Find the range of each of the following functions.
(a) $f : x \rightarrow x^2 - 5$ (b) $f : x \rightarrow x^2 - 6x$ (c) $f : x \rightarrow 2x^2 + 2x + 1$ (d) $f : x \rightarrow 6 - 5x - x^2$.

3. Express the function $f(x) = x^2 - 4x + 8$ in the form $(x - p)^2 + q$. Deduce that $f(x)$ is positive for all real values of x.

4. Prove that the expression $x - 2 - x^2$ is negative for all real values of x.

5. Prove that the following inequalities hold for all real values of x.
(a) $2x^2 + 5 > 4x$ (b) $2(2x - 1) < 3x^2$ (c) $9x^2 - 30x + 25 \geqslant 0$ (d) $2x < 2x^2 + 1$.

6. Find the values of x for which
(a) $x^2 - 2x < 0$ (b) $x^2 + 8 \leqslant 6x + 3$ (c) $x^2 - 4 \geqslant 0$
(d) $4x + 1 < x^2 + 4$ (e) $3x^2 + 2x < 1$ (f) $2x^2 + 7x + 3 \geqslant 0$.
[You may find sketch graphs helpful.]

7. Find the minimum value of the function $3x^2 - 12x + 5$ and sketch the curve $y = 3x^2 - 12x + 5$. Find the range of values of c for which the function $3x^2 - 12x + c$ is positive for all values of x.

8. Find the range of values of k such that $x^2 + 3x + k \geqslant 0$ for all real values of x.

9. If $y = x^2 + 2px + q$, find the values of p and q in each of the following cases.
(a) $y = 0$ when $x = 2$ and when $x = 4$,
(b) $y = 1$ when $x = 2$ and $y = 7$ when $x = -4$,
(c) y has a minimum value of 2 when $x = -1$.

10. The maximum value of the function $f(x) = ax^2 + bx + c$ is 10. Given that $f(3) = f(-1) = 2$, find $f(2)$.

3.2 Quadratic equations

An equation of the form $ax^2 + bx + c = 0$ is called a *quadratic equation*. Many such equations are solved by factorising. Others may be solved by completing the square as follows:

$$
\begin{aligned}
x^2 - 6x + 7 = 0 &\Leftrightarrow x^2 - 6x + 9 = 2 \\
&\Leftrightarrow (x - 3)^2 = 2 \\
&\Leftrightarrow x - 3 = \pm\sqrt{2} \\
&\Leftrightarrow x = 3 \pm \sqrt{2}.
\end{aligned}
$$

Hence if $x^2 - 6x + 7 = 0$ then $x = 1 \cdot 59$ or $4 \cdot 41$ (to 2 d.p.).

However, in some cases the method breaks down.

$$x^2 - 2x + 5 = 0 \Leftrightarrow x^2 - 2x + 1 = -4$$
$$\Leftrightarrow (x-1)^2 = -4$$

Since $(x-1)^2$ can never be negative, there are no real values of x which satisfy this equation.

Let us now consider the general quadratic equation.

$$ax^2 + bx + c = 0 \Leftrightarrow x^2 + \frac{b}{a}x + \frac{c}{a} = 0$$

$$\Leftrightarrow x^2 + \frac{b}{a}x + \frac{b^2}{4a^2} = \frac{b^2}{4a^2} - \frac{c}{a}$$

$$\left(x + \frac{b}{2a}\right)^2 = \frac{b^2 - 4ac}{4a^2}.$$

Hence the nature of the roots of the equation $ax^2 + bx + c = 0$ depends on the sign of $b^2 - 4ac$.

(i) If $b^2 - 4ac > 0$, then $x + \dfrac{b}{2a} = \pm \dfrac{\sqrt{(b^2 - 4ac)}}{2a}$

$\therefore$ the equation has two distinct real roots given by

$$x = \frac{-b \pm \sqrt{(b^2 - 4ac)}}{2a}.$$

(ii) If $b^2 - 4ac = 0$, then $\left(x + \dfrac{b}{2a}\right)^2 = 0$

$\therefore$ the equation has two equal roots given by $x = -\dfrac{b}{2a}$.

(iii) If $b^2 - 4ac < 0$, then $\left(x + \dfrac{b}{2a}\right)^2 < 0$, which is not possible for real values of x

$\therefore$ the equation has no real roots.

The diagrams opposite show the curve $y = ax^2 + bx + c$ in these three cases, for $a > 0$. Each curve has a minimum point at $x = -\dfrac{b}{2a}$ and is symmetrical about the line $x = -\dfrac{b}{2a}$.

In case (iii) the graph shows that if $a > 0$ and $b^2 - 4ac < 0$ the function $ax^2 + bx + c$ is always positive. Similarly if $a < 0$ and $b^2 - 4ac < 0$ the function $ax^2 + bx + c$ is always negative.

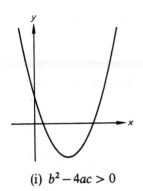

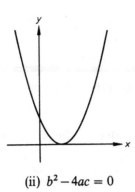

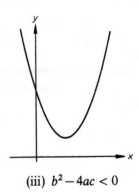

(i) $b^2 - 4ac > 0$ (ii) $b^2 - 4ac = 0$ (iii) $b^2 - 4ac < 0$

Example 1   Solve the equation $3x^2 - 4x - 2 = 0$.

$$3x^2 - 4x - 2 = 0 \Rightarrow x = \frac{-(-4) \pm \sqrt{\{(-4)^2 - 4.3.(-2)\}}}{2.3}$$

$$\Rightarrow x = \frac{4 \pm \sqrt{(16 + 24)}}{6}$$

$$\Rightarrow x = \frac{4 \pm \sqrt{40}}{6}$$

$\therefore$ the roots of the equation are -0.39 and 1.72 (to 2 d.p.).

Example 2   Find the range of values of p for which the equation $x^2 + px + p = 0$ has real roots.

The equation $ax^2 + bx + c = 0$ has real roots if $b^2 - 4ac \geqslant 0$
$\therefore$ the equation $x^2 + px + p = 0$ has real roots if $p^2 - 4p \geqslant 0$
    i.e.   if $p(p-4) \geqslant 0$.
Since p and $(p-4)$ are both positive when $p > 4$ and both negative when $p < 0$, the equation has real roots when $p \leqslant 0$ and when $p \geqslant 4$.

Exercise 3.2

1.   Find the real roots, if any, of the following quadratic equations, giving your answers in surd form.
(a) $x^2 - 3x + 1 = 0$ (b) $2x^2 + 3x - 1 = 0$ (c) $1 - 2x - x^2 = 0$
(d) $3x^2 - 4x + 2 = 0$ (e) $4x^2 - 12x - 27 = 0$ (f) $2x^2 - 6x + 3 = 0$.

2.   Find the values of k for which the following equations have equal roots.
(a) $x^2 + 2kx + k + 6 = 0$,   (b) $(x+1)(x+3) = k$.

3.   Find the range of values of p for which each of the given equations has two distinct real roots.
(a) $x^2 + 2px - 5p = 0$ (b) $3x^2 + 3px + p^2 = 1$
(c) $x(x+3) = p(x-1)$ (d) $p(x^2 - 1) = 3x + 2$.

4.   Given that the following expressions are positive for all real values of x, find in each case the set of possible values of a.
(a) $x^2 + a + a^2$ (b) $x^2 + ax + a^2$ (c) $x^2 - 2ax + 3a$ (d) $ax^2 + 2ax + a^2$.

5.   Find the values of p for which the expression $x^2 + (p+3)x + 2p + 3$ is a perfect square.

6.   Show that there is no real value of k for which the equation $x^2 + (3-k)x + k^2 + 4 = 0$ has real roots.

7.   Find any values of p for which the curve $y = 3px^2 + 2px + 1$ touches the x-axis.

8.   Find the set of values of k for which the equation $kx^2 + x + k - 1 = 0$ has real roots, one positive and one negative.

9.   Find the set of values of k for which the roots of the equation $x^2 + kx - k + 3 = 0$ are real and of the same sign.

10. Given that $y = px^2 + 2qx + r$, show that y will have the same sign for all real values of x if and only if $q^2 < pr$.

3.3 Sum and product of roots

The quadratic equation with roots α and β may be written

$$(x-\alpha)(x-\beta) = 0 \quad \text{i.e.} \quad x^2 - (\alpha+\beta)x + \alpha\beta = 0.$$

The equation $ax^2 + bx + c = 0$ may be written $x^2 + \dfrac{b}{a}x + \dfrac{c}{a} = 0$.

Hence if the equation $ax^2 + bx + c = 0$ has roots α and β,

$$\alpha + \beta = -\frac{b}{a}, \quad \alpha\beta = \frac{c}{a}.$$

Example 1   Write down the sum and product of the roots of the equation $3x^2 - 6x + 2 = 0$.

$$\text{The sum of the roots} = -\frac{(-6)}{3} = 2.$$

$$\text{The product of the roots} = \frac{2}{3}.$$

When the sum and product of the roots of a quadratic equation are known, the equation may be written down in the form

$$x^2 - (\text{sum of roots})x + (\text{product of roots}) = 0.$$

Example 2   Write down the quadratic equation, the sum and product of whose roots are $\frac{3}{4}$ and -7 respectively.

The equation is $x^2 - \frac{3}{4}x + (-7) = 0$ i.e. $4x^2 - 3x - 28 = 0$.

Example 3   Given that one root of the equation $3x^2 + 4x + k = 0$ is three times the other, find k.

Let the roots of the given equation be α and 3α.

The sum of the roots $= \alpha + 3\alpha = -\dfrac{4}{3}$ i.e. $4\alpha = -\dfrac{4}{3}$

The product of the roots $= \alpha \cdot 3\alpha = \dfrac{k}{3}$ i.e. $3\alpha^2 = \dfrac{k}{3}$

Hence $\alpha = -\frac{1}{3}$ and $k = 1$.

If α and β are the roots of a given quadratic equation, it is usually possible to find the value of a symmetrical function of α and β by expressing it in terms of $\alpha + \beta$ and $\alpha\beta$. [A symmetrical function of α and β is one which is unaltered if α and β are interchanged.]

Example 4   If α and β âre the roots of the equation $3x^2 - 5x + 1 = 0$, find the values of (i) $\dfrac{1}{\alpha} + \dfrac{1}{\beta}$ (ii) $\alpha^2 + \beta^2$.

$$\alpha + \beta = \frac{5}{3}, \qquad \alpha\beta = \frac{1}{3}$$

(i) $\dfrac{1}{\alpha} + \dfrac{1}{\beta} = \dfrac{\alpha + \beta}{\alpha\beta} = \dfrac{5}{3} \div \dfrac{1}{3} = \dfrac{5}{3} \cdot \dfrac{3}{1} = 5$

(ii) $\alpha^2 + \beta^2 = (\alpha + \beta)^2 - 2\alpha\beta = \dfrac{25}{9} - 2 \cdot \dfrac{1}{3} = \dfrac{19}{9}$.

Example 5   If α and β are the roots of the equation $x^2 + 7x + 5 = 0$, find an equation whose roots are $\alpha + 1$ and $\beta + 1$.

Method 1: $\alpha + \beta = -7$, $\alpha\beta = 5$
$(\alpha + 1) + (\beta + 1) = \alpha + \beta + 2 = -7 + 2 = -5$
$(\alpha + 1)(\beta + 1) = \alpha\beta + \alpha + \beta + 1 = 5 - 7 + 1 = -1$
$\therefore$ the required equation is $x^2 - (-5)x + (-1) = 0$
i.e. $x^2 + 5x - 1 = 0$.

Method 2: For the equation with roots $x = \alpha + 1$ or $x = \beta + 1$,
either $x - 1 = \alpha$ or $x - 1 = \beta$.

Since the equation with roots α and β is $x^2 + 7x + 5 = 0$, the required equation must take the form

$$(x-1)^2 + 7(x-1) + 5 = 0$$
$$x^2 - 2x + 1 + 7x - 7 + 5 = 0$$
$$\text{i.e.} \quad x^2 + 5x - 1 = 0.$$

Exercise 3.3

1. Find the sums and products of the roots of the following equations.
(a) $2x^2 - 5x - 3 = 0$ (b) $6 + x - x^2 = 0$ (c) $3x^2 - 5 = 0$
(d) $x^2 + px - p = 0$ (e) $x(x-2) = 5(x-1)$ (f) $4x - 1/x = 3$.

2. Find equations, the sums and products of whose roots are, respectively:
(a) $3, 2$ (b) $-\frac{1}{2}, \frac{3}{4}$ (c) $0, -4$
(d) $3/7, -1/14$ (e) $p - q, p + q$ (f) $a/b, 1/ab$.

3. Given that one root of the equation $2x^2 - kx + k = 0$, where $k \neq 0$, is twice the other, find k.

4. Given that the two roots of the equation $x^2 + (7-p)x - p = 0$ differ by 5, find the possible values of p.

5. If α and β are the roots of the equation $ax^2 + bx + c = 0$, prove that
(i) if $\beta = 4\alpha$ then $4b^2 = 25ac$,   (ii) if $\beta = \alpha + 1$ then $a^2 = b^2 - 4ac$.

6. If α and β are the roots of the equation $x^2 - 3x - 2 = 0$, find the values of
(i) $\alpha + \beta$, (ii) $\alpha\beta$, (iii) $\alpha^2 + \beta^2$, (iv) $\alpha^3 + \beta^3$.

7. If α and β, where $\alpha > \beta$, are the roots of the equation $x^2 + 2x - 5 = 0$, find the values of (i) $(\alpha + \beta)^2$, (ii) $(\alpha - \beta)^2$, (iii) $\alpha - \beta$, (iv) $\alpha^2 - \beta^2$.

8. If α and β are the roots of the equation $2x^2 - x - 4 = 0$, find the values of
(a) $\alpha^2 + \beta^2$ (b) $(\alpha - \beta)^2$ (c) $\alpha^3 + \beta^3$
(d) $\dfrac{1}{\alpha} + \dfrac{1}{\beta}$ (e) $\dfrac{1}{\alpha^2} + \dfrac{1}{\beta^2}$ (f) $\dfrac{\alpha}{\beta^2} + \dfrac{\beta}{\alpha^2}$.

9. If $p + q = 5$ and $p^2 + q^2 = 19$, find the value of pq and write down an equation in x whose roots are p and q.

10. If $a - b = 3$ and $a^2 + b^2 = 65$, write down an equation in x whose roots are a and b.

11. If α and β are the roots of the equation $x^2 - 4x + 2 = 0$, find equations whose roots are
(a) $3\alpha, 3\beta$,   (b) $\alpha + 3, \beta + 3$,   (c) $\alpha + 3\beta, \beta + 3\alpha$.

12. If α and β are the roots of the equation $x^2 + x + k = 0$, find equations whose roots are
(a) α^2, β^2,   (b) $-\alpha, -\beta$,   (c) $\alpha - 1, \beta - 1$.

13. If α and β are the roots of the equation $ax^2 + bx + c = 0$, find equations whose roots are
(a) $\dfrac{1}{\alpha}, \dfrac{1}{\beta}$,   (b) $\alpha - \beta, \beta - \alpha$,   (c) $\dfrac{\alpha}{\alpha + \beta}, \dfrac{\beta}{\alpha + \beta}$.

14. Given that the equations $x^2 - 2x + p = 0$ and $2x^2 - 5x + q = 0$ have a common root, by eliminating the terms in x^2, show that this common root is $q - 2p$. Find expressions for the other roots of the equations.

15. Given that the equations $2x^2 + 3x + k = 0$ and $3x^2 + x - 2k = 0$ have a common root, find the possible values of k.

3.4 Non-linear simultaneous equations

In this section we deal with pairs of simultaneous equations in which at least one of the equations is non-linear. It is often necessary to solve such equations when finding the points of intersection of a line and a curve.

Example 1   Find the points of intersection of the line $y = x + 1$ and the curve $2x^2 + y^2 = 6$.

At the points of intersection
$$y = x + 1$$
and $$2x^2 + y^2 = 6$$
$$\therefore \qquad 2x^2 + (x + 1)^2 = 6$$
$$2x^2 + x^2 + 2x + 1 = 6$$
$$3x^2 + 2x - 5 = 0$$
$$(3x + 5)(x - 1) = 0$$

Hence either $3x + 5 = 0$ or $x - 1 = 0$
$$x = -5/3 \qquad\qquad x = 1$$

$\therefore$ the points of intersection are $(-5/3, -2/3)$ and $(1, 2)$.

Example 2   Show that the line $y = 3x - 1$ does not meet the curve $y = x^2 + 3x$.

At any points of intersection $y = 3x - 1$
and $y = x^2 + 3x$
$$\therefore \quad x^2 + 3x = 3x - 1, \text{ i.e. } x^2 = -1.$$

Since this equation has no real roots, there are no real points of intersection between the line and the curve.

Example 3   Find the points of intersection of the line $y = -x$ and the curve $y = x^3 + 6x^2 + 8x$.

At the points of intersection $x^3 + 6x^2 + 8x = -x$
   i.e. $x^3 + 6x^2 + 9x = 0$
$$x(x+3)^2 = 0$$
∴   either $x = 0, y = 0$ or $x = -3, y = 3$.
Hence the points of intersection are $(0,0)$ and $(-3,3)$.

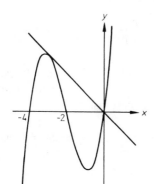

The significance of the fact that $x = -3$ appears as a repeated root in the above example is shown in this sketch graph. The line $y = -x$ is a tangent to the curve at the point $(-3,3)$.

In general, when solving simultaneous equations, one linear and the other quadratic, a method similar to the following is used.

Example 4   Solve the equations $2x - 3y = 4$ (1)
$$x^2 - 3xy + 3y^2 = 7 \qquad (2)$$

Rearranging (1) $3y = 2x - 4$
Equation (2) × 3 $3x^2 - 9xy + 9y^2 = 21$
Substituting for $3y$ $3x^2 - 3x(2x-4) + (2x-4)^2 = 21$
$$3x^2 - 6x^2 + 12x + 4x^2 - 16x + 16 = 21$$
$$x^2 - 4x - 5 = 0$$
$$(x-5)(x+1) = 0$$
∴   either $x = 5$ or $x = -1$.

Substituting in (3) to find the corresponding values of y, we obtain the solutions $x = 5, y = 2$ and $x = -1, y = -2$.

In certain cases there is an alternative to the method of Example 4.

Example 5   Solve the equations $x + 2y = 3$ (1)
$$3x^2 + 7xy + 2y^2 = 12 \qquad (2)$$

Factorising (2) $(x+2y)(3x+y) = 12$
Dividing by (1) $3x + y = 4$
Multiplying (1) by 3 $3x + 6y = 9$.
Subtracting $5y = 5$ ∴ $y = 1$
Substituting in (1) $x + 2 = 3$ ∴ $x = 1$
Hence the equations have one soloution $x = 1, y = 1$.

[Note that when dealing with non-linear simultaneous equations it is always advisable to check that the solutions obtained satisfy the original equations.]

Exercise 3.4

1. Find any points of intersection between the given curves and straight lines.
(a) $y = x^2 + 1$, $y = 5x - 3$
(b) $y = 3 - x^2$, $y = 7 - 4x$
(c) $y = x - 3$, $xy + 2 = 0$
(d) $x^2 + y^2 + 4x = 1$, $y = 2 - x$
(e) $y = 2x - 7$, $2xy + 3y + 5x + 11 = 0$
(f) $4x^2 - 9y^2 = 36$, $2x + 3y + 2 = 0$.

In questions 2 to 14 solve the given pairs of simultaneous equations.

2. $x + y = 1$
   $xy = -12$.

3. $3x = 2y$
   $xy = 24$.

4. $x + y = 1$
   $x^2 - y^2 = 5$.

5. $x - 5y = 8$
   $x^2 + 9y^2 + 3x - xy = 30$.

6. $3x + y = 5$
   $5x^2 - 2xy + y^2 = 5$.

7. $3x + 2y = 10$
   $3x^2 - 2xy + 4y^2 = 25$.

8. $x - 3y = 2$
   $x^2 - 2xy - 3y^2 = 12$.

9. $x + y = -1$, $\dfrac{1}{x} + \dfrac{1}{y} = \dfrac{1}{2}$.

10. $2x + y = 1$, $\dfrac{1}{x} - \dfrac{1}{y} = 2$.

11. $x + y = 1$
    $x^3 + y^3 = 91$.

12. $2x + 3y = 3$
    $8x^3 + 27y^3 = 117$.

13. $x^2 - 2xy + 3y^2 = 3$
    $x^2 + 2xy + 3y^2 = 11$.

14. $4xy = -15$
    $4x^3 + 4y^3 = 49$.

In questions 15 to 18 solve the given pairs of simultaneous equations by eliminating either the terms in x^2 or the terms in y^2.

15. $x^2 + y^2 + 2x - 3y = 3$
    $2x^2 - y^2 + 4x + 18y = 54$.

16. $x^2 + 2y^2 + 4y = 4$
    $x^2 - y^2 + 2x - 2y = 0$.

17. $x^2 + xy + y^2 = 3$
    $x^2 - y^2 = 3$.

18. $x^2 + y^2 - 3x = 9$
    $x^2 - y^2 + x + 4y = 3$.

3.5 Products and factors

A function of the form $P(x) = a_0 x^n + a_1 x^{n-1} + \ldots + a_n$ $(a_0 \neq 0)$, where n is a positive integer, is called a *polynomial* of degree n. The real numbers a_0, $a_1, \ldots$ are called the *coefficients* of x^n, $x^{n-1}, \ldots$. For instance, $x^5 - 4x^2 + 3x$ is a polynomial of degree 5. A *linear* function, such as $ax + b$, is a polynomial of degree 1. A *quadratic* function is a polynomial of degree 2. A polynomial of degree 3 is often called a *cubic*.

The product of two polynomials can be found by either of the methods illustrated overleaf.

Example 1   Multiply $2x^3 - x^2 + 3$ by $3x + 1$.

$$\begin{aligned}
(3x+1)(2x^3 - x^2 + 3) &= 3x(2x^3 - x^2 + 3) + 1(2x^3 - x^2 + 3) \\
&= 6x^4 - 3x^3 + 9x + 2x^3 - x^2 + 3 \\
&= 6x^4 - x^3 - x^2 + 9x + 3.
\end{aligned}$$

OR

$$\begin{array}{r}
2x^3 - x^2 \qquad + 3 \\
3x + 1 \\
\hline
2x^3 - x^2 \quad + 3 \\
6x^4 - 3x^3 \quad + 9x \\
\hline
6x^4 - \ x^3 - x^2 + 9x + 3.
\end{array}$$

When the complete product is not required, the coefficients of particular terms can be picked out.

Example 2   Find the coefficient of x^4 in the product of $x^3 - 3x^2 + x - 2$ and $2x^3 + x^2 - 3x + 1$.

The diagram shows the mental processes involved.

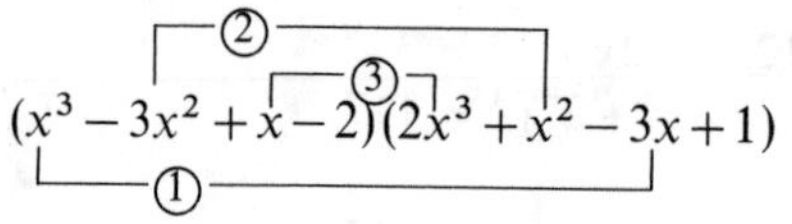

The coefficient of x^4 in the product

$$= 1.(-3) + (-3).1 + 1.2 = -3 - 3 + 2 = -4.$$

Sometimes it is possible to *factorise* a polynomial, i.e. express it as a product. This process is useful when simplifying fractions.

Example 3   Simplify $\dfrac{2x}{x^2 - 4} + \dfrac{7}{2x^2 + x - 6}$.

$$\begin{aligned}
\frac{2x}{x^2-4} + \frac{7}{2x^2+x-6} &= \frac{2x}{(x-2)(x+2)} + \frac{7}{(2x-3)(x+2)} \\[2mm]
&= \frac{2x(2x-3)}{(x-2)(x+2)(2x-3)} + \frac{7(x-2)}{(x-2)(x+2)(2x-3)} \\[2mm]
&= \frac{4x^2 - 6x + 7x - 14}{(x-2)(x+2)(2x-3)} \\[2mm]
&= \frac{4x^2 + x - 14}{(x-2)(x+2)(2x-3)} \\[2mm]
&= \frac{(4x-7)(x+2)}{(x-2)(x+2)(2x-3)}
\end{aligned}$$

$$\therefore \quad \frac{2x}{x^2-4} + \frac{7}{2x^2-x-6} = \frac{4x-7}{(x-2)(2x-3)}.$$

It is often difficult to find factors of a polynomial of degree 3 or more. Testing possible factors by long division is a cumbersome process as the next example shows.

Example 4   Divide $x^4 - 3x + 5$ by $x - 2$.

$$
\require{enclose}
\begin{array}{r}
x^3 + 2x^2 + 4x + 5 \\
x-2 \enclose{longdiv}{x^4 \qquad\qquad\quad -3x + 5} \\
\underline{x^4 - 2x^3} \\
2x^3 \\
\underline{2x^3 - 4x^2} \\
4x^2 - 3x \\
\underline{4x^2 - 8x} \\
5x + 5 \\
\underline{5x - 10} \\
15
\end{array}
$$

$\therefore$ when $x^4 - 3x + 5$ is divided by $x - 2$, the quotient is $x^3 + 2x^2 + 4x + 5$ and the remainder is 15.

This result can be expressed in the form of an *identity*:

$$x^4 - 3x + 5 \equiv (x-2)(x^3 + 2x^2 + 4x + 5) + 15.$$

[The symbol $\equiv$ means 'is identically equal to', i.e. equal for all values of x.]

More generally, we can consider the division of a polynomial $P(x)$ by $(x-a)$. If $Q(x)$ is the quotient and R the remainder, then we may write

$$P(x) \equiv (x-a)Q(x) + R.$$

Since this identity holds for all values of x, we substitute $x = a$ to obtain $P(a) = (a-a)Q(a) + R = R$. Thus, when a polynomial $P(x)$ is divided by $(x-a)$ the remainder is $P(a)$. This result is called the *remainder theorem*.

Example 5   Find the remainder when $P(x) = 2x^3 + 7x^2 - 5x - 4$ is divided by $x + 3$.

$$P(-3) = 2(-3)^3 + 7(-3)^2 - 5(-3) - 4 = -54 + 63 + 15 - 4 = 20$$

$\therefore$ the remainder when $P(x)$ is divided by $x + 3$ is 20.

If for a polynomial $P(x)$ we find that $P(a) = 0$, we deduce from the remainder theorem that there is no remainder when $P(x)$ is divided by $(x-a)$. Thus we have the *factor theorem*, which states that

> if $P(a) = 0$ then $(x-a)$ is a factor of $P(x)$.

We use this result to test for linear factors of polynomials.

Example 6   Factorise $f(x) = 6x^3 + 13x^2 - 4$.
Since the constant term is -4, we first try values of x which are factors of -4, i.e. ± 1, ± 2 or ± 4.

$$f(1) = 6.1^3 + 13.1^2 - 4 = 6 + 13 - 4 \neq 0$$
$$f(-1) = 6(-1)^3 + 13(-1)^2 - 4 = -6 + 13 - 4 \neq 0$$
$$f(2) = 6.2^3 + 13.2^2 - 4 = 48 + 52 - 4 \neq 0$$
$$f(-2) = 6(-2)^3 + 13(-2)^2 - 4 = -48 + 52 - 4 = 0.$$

Hence, by the factor theorem, $x+2$ is a factor of $f(x)$. By inspection or by long division,

$$f(x) = (x+2)(6x^2 + x - 2)$$
$$\therefore \quad f(x) = (x+2)(2x-1)(3x+2).$$

Note that in some cases a cubic polynomial is the product of a linear factor and an irreducible quadratic factor. For example,

$$x^3 - a^3 = (x-a)(x^2 + ax + a^2)$$
$$x^3 + a^3 = (x+a)(x^2 - ax + a^2).$$

Exercise 3.5

1. Find the product of
(a) $2x^2 - x + 7$ and $x+2$
(b) $3x^3 - 2x^2 + 5x - 1$ and $2x - 3$
(c) $x^2 + 3x - 2$ and $4x^2 - x + 1$
(d) $5x^3 - 2x + 3$ and $x^2 - 1$
(e) $3x^4 - 2x^3 + 6x - 4$ and $3x + 2$
(f) $x^5 - x^4 + x^3 - x^2 + x - 1$ and $x + 1$.

2. Find the coefficients of the given terms in the following products.
(a) $(2x - 5)(x^3 - x^2 + 2x - 3)$; x, x^3.
(b) $(x^2 - 2x + 3)(3x^2 + x - 2)$; x, x^2.
(c) $(x^3 - x - 1)(2x^2 + 3x - 1)$; x^2, x^4.
(d) $(3x^3 - x^2 - 2x + 5)(x^3 + 2x^2 - x - 2)$; x^3, x^5.

3. Simplify the following expressions.

(a) $\dfrac{2}{x^2 - 1} - \dfrac{7}{2x^2 - 3x - 5}$

(b) $\dfrac{1}{x+2} + \dfrac{6x}{x^3 + 8}$

(c) $\dfrac{x^4 - x}{2x - 3x^2} \times \dfrac{3x^2 + x - 2}{1 - x^2}$

(d) $\dfrac{8x^3 + 1}{4x^2 - 8x + 3} \div \left(2x - \dfrac{1}{1 - 2x} \right)$.

4. Find the remainder when
(a) $5x^2 + 3x - 7$ is divided by $x - 2$ (b) $x^3 - 2x^2 + 8x - 3$ is divided by $x - 5$
(c) $x^4 - 3x^3 + 4x^2 - x + 6$ is divided by (d) $x^5 + 16$ is divided by $x + 2$
$x + 1$
(e) $3x^2 - 2ax - 4a^2$ is divided by $x + a$ (f) $a^3 - 3a^2b + 2b^3$ is divided by $a - b$.

5. Find by long division the quotient and remainder when
(a) $3x^2 - 2x + 5$ is divided by $x + 1$
(b) $4x^3 - 4x^2 + 5x + 1$ is divided by $2x - 3$
(c) $x^3 - 3$ is divided by $x + 3$
(d) $3x^5 - x^4 - 6x^3 + 11x^2 - 1$ is divided by $3x - 1$
(e) $x^4 - 2x^3 + 6x - 5$ is divided by $x^2 - x - 1$
(f) $2x^4 - x^3 + 3x^2 - 7$ is divided by $x^2 - 2$.

6. Factorise the following polynomials
(a) $2x^3 - 3x^2 + 1$ (b) $3x^3 - 2x^2 - 7x - 2$
(c) $x^4 - x^2 - 72$ (d) $x^5 + x^3 + x$
(e) $4x^3 - 13x + 6$ (f) $4x^4 - 4x^3 - 9x^2 + x + 2$.

7. When the function $x^2 + ax + b$ is divided by $x - 1$ and $x + 2$, the remainders are 4 and 5 respectively. Find the values of a and b.

8. When $x^3 + px^2 + qx + 1$ is divided by $x - 2$ the remainder is 9; when divided by $x + 3$ the remainder is 19. Find the values of p and q.

9. Given that the expression $2x^3 + ax^2 + b$ is divisible by $x + 1$ and that there is a remainder of 16 when it is divided by $x - 3$, find the values of a and b.

10. The expression $ax^3 - 8x^2 + bx + 6$ is exactly divisible by $x^2 - 2x - 3$. Find the values of a and b.

11. The polynomials $x^3 + 4x^2 - 2x + 1$ and $x^3 + 3x^2 - x + 7$ leave the same remainder when divided by $x - p$. Find the possible values of p.

12. Given that the expressions $x^3 - 4x^2 + x + 6$ and $x^3 - 3x^2 + 2x \mp k$ have a common factor, find the possible values of k.

13. If $x - a$ is a factor of the expression $ax^3 - 3x^2 - 5ax - 9$, find the possible values of a and factorise the expression for each of these values.

14. Prove that when a polynomial $P(x)$ is divided by $(ax - b)$ the remainder is $P(b/a)$. Hence find the remainder when (a) $12x^2 - 4x + 9$ is divided by $2x - 1$, (b) $9x^3 - x + 5$ is divided by $3x + 2$.

15. Factorise (a) $8x^3 - 36x^2 + 46x - 15$, (b) $12x^3 + 4x^2 - 5x - 2$,
(c) $3x^3 - x^2 + 4x + 4$.

3.6 Polynomial equations

In practice it is often difficult to solve polynomial equations. However, sometimes solutions can be found by factorisation.

Example 1   Solve the equation $2x^4 - 5x^2 - 12 = 0$.

$$2x^4 - 5x^2 - 12 = 0 \Leftrightarrow \quad (2x^2 + 3)(x^2 - 4) = 0$$
$$\Leftrightarrow (2x^2 + 3)(x - 2)(x + 2) = 0$$
$$\therefore \quad 2x^2 + 3 = 0 \quad \text{or} \quad x - 2 = 0 \quad \text{or} \quad x + 2 = 0.$$

Since $2x^2 + 3 = 0$ has no real roots, either $x = 2$ or $x = -2$.

Example 2   Solve the equation $x^3 + 3x^2 + x - 1 = 0$.

Let $f(x) = x^3 + 3x^2 + x - 1$,   then
$f(1) = 1 + 3 + 1 - 1 \neq 0$, $f(-1) = -1 + 3 - 1 - 1 = 0$
$\therefore \quad x + 1$ is a factor of $f(x)$.
By division $f(x) = (x + 1)(x^2 + 2x - 1)$.

Hence if $x^3 + 3x^2 + x - 1 = 0$,
either $x + 1 = 0$ or $x^2 + 2x - 1 = 0$

$$x = -1 \qquad\qquad x = \frac{-2 \pm \sqrt{(4 + 4)}}{2}$$
$$= -1 \pm \sqrt{2}$$

$\therefore$ the roots of the given equation are -1, $-1 - \sqrt{2}$ and $-1 + \sqrt{2}$.

Equations involving surds can sometimes be expressed in polynomial form. However, since extra roots may be introduced in this process all solutions must be tested in the original equation.

Example 3   Solve the equation $\sqrt{(3 - x)} - \sqrt{(7 + x)} = 2$.

Rearranging $\sqrt{(3 - x)} = \sqrt{(7 + x)} + 2$.
Squaring both sides $3 - x = \{\sqrt{(7 + x)} + 2\}^2$
i.e. $3 - x = (7 + x) + 4\sqrt{(7 + x)} + 4$
$\therefore$ $-2x - 8 = 4\sqrt{(7 + x)}$
$\therefore$ $-(x + 4) = 2\sqrt{(7 + x)}$.
Squaring both sides $x^2 + 8x + 16 = 4(7 + x)$
$\therefore$ $x^2 + 4x - 12 = 0$
                                        $(x + 6)(x - 2) = 0$.
Hence either $x = -6$ or $x = 2$.
If $x = -6$, $\sqrt{(3 - x)} - \sqrt{(7 + x)} = \sqrt{9} - \sqrt{1} = 2$.
If $x = 2$, $\sqrt{(3 - x)} - \sqrt{(7 + x)} = \sqrt{1} - \sqrt{9} = -2$.
Thus the only root of the given equation is $x = -6$.

In Example 1 a polynomial equation of degree 4 was found to have 2 real roots, whereas the cubic equation in Example 2 had 3 real roots. In general, if $x = a$ is a root of a polynomial equation $P(x) = 0$ then $(x - a)$ is a factor of $P(x)$.

Thus, since a polynomial of degree n can have up to n factors of the form $(x-a)$, a polynomial equation of degree n can have up to n real roots.

In §3.2 it was shown that if the roots of the quadratic equation $ax^2 + bx + c = 0$ are α and β, then $\alpha + \beta = -b/a$ and $\alpha\beta = c/a$. Similar expressions can be obtained for equations of higher degree. For example, if the equation $ax^3 + bx^2 + cx + d = 0$ has roots α, β and γ, then it may be written both in the form

$$x^3 + \frac{b}{a}x^2 + \frac{c}{a}x + \frac{d}{a} = 0$$

and as

$$(x-\alpha)(x-\beta)(x-\gamma) = 0$$

i.e.

$$x^3 - (\alpha + \beta + \gamma)x^2 + (\alpha\beta + \beta\gamma + \gamma\alpha)x - \alpha\beta\gamma = 0.$$

Hence

$$\alpha + \beta + \gamma = -\frac{b}{a}, \; \alpha\beta + \beta\gamma + \gamma\alpha = \frac{c}{a}, \; \alpha\beta\gamma = -\frac{d}{a}.$$

Example 4   Given that 1 and -2 are two roots of the equation $3x^3 - x^2 + px + q = 0$, find the third root and the values of p and q.

Since the sum of the roots is $1/3$, the third root is

$$1/3 - \{1 + (-2)\}, \quad \text{i.e.} \quad 4/3.$$

The product of the roots $= -\frac{q}{3} = 1.(-2).\frac{4}{3} \quad \therefore \quad q = 8.$

Since $x = 1$ satisfies the equation, $3 - 1 + p + q = 0$

$$\therefore \quad p = -2 - q = -10.$$

Hence the third root is $4/3$ and the values of p and q are -10 and 8 respectively.

Exercise 3.6

In questions 1 to 14 solve the given equations.

1. $x^3 - 8x^2 + 5x + 14 = 0.$

2. $6 - x - 4x^2 - x^3 = 0.$

3. $4 + 7x + 2x^2 - x^3 = 0.$

4. $6x^3 + 31x^2 + 48x + 20 = 0.$

5. $4x^4 + 23x^2 - 6 = 0.$

6. $x^4 - 11x^2 + 18 = 0.$

7. $x^3 - 6x - 4 = 0.$

8. $x^3 - 6x^2 + 12x - 9 = 0.$

9. $2x^3 + x^2 - 6x - 3 = 0.$

10. $x^4 - 5x^3 + 4x^2 + 6x - 4 = 0.$

11. $\sqrt{(x+6)} - \sqrt{(x+1)} = 1.$

12. $\sqrt{(12+x)} - \sqrt{(13-x)} = 1.$

13. $\sqrt{(3x+1)} - \sqrt{(x-1)} = 2.$

14. $\sqrt{(9-4x)} - \sqrt{(5-x)} = 2.$

15. If $f(x) = x^3 + px + q$ is exactly divisible by $(x+2)$ and $(x-3)$, find the values of p and q. With these values of p and q, find the roots of the equation $f(x) = 0$.

16. Find the value of k if one root of the equation $12x^3 + kx^2 - 17x + 6 = 0$ is $\frac{1}{2}$. When k has this value, find the other roots of the equation.

17. Write down the sums and products of the roots of the following equations.
(a) $2x^3 + 3x^2 - 8x - 12 = 0$ (b) $2x^3 + 5x^2 - 3x = 0$
(c) $3x^3 + 4x^2 - 5x - 2 = 0$ (d) $x^3 - 11x - 6 = 0$.

18. Given that -1 and 4 are two roots of the equation $x^3 + 5x^2 + ax + b = 0$, find the third root and the values of a and b.

19. Given that 2 is a repeated root of the equation $2x^3 + px^2 + qx - 4 = 0$, find the third root and the values of p and q.

20. Write down the sums and products of the roots of the following equations.
(a) $x^4 - 6x^3 + 9x^2 - 4 = 0$ (b) $2x^4 - x^3 - 12x + 3 = 0$
(c) $x^5 - 3x^3 + 2x = 0$ (d) $2x^5 - 3x^4 - 2x^3 + 4x^2 = 1$.

21. Solve the equation $3x^4 + 4x^3 - 14x^2 + 4x + 3 = 0$ by writing it in the form $a\left(x + \dfrac{1}{x}\right)^2 + b\left(x + \dfrac{1}{x}\right) + c = 0$.

22. Solve the equation $10x^4 - 37x^3 + 50x^2 - 37x + 10 = 0$ by the method of the previous question.

Exercise 3.7 (miscellaneous)

1. Sketch the following curves, showing any maximum or minimum points.
(a) $y = x^2 + 2x$ (b) $y = 2x^2 + x - 3$
(c) $y = 3 + 4x - 4x^2$ (d) $y = x^2 - 6x + 10$.

2. Show that $x^2 - 2x + 2$ is positive for all values of x. Hence find the range of values of x for which the expression $x^3 - x^2 + 2$ is positive.

3. By means of a suitable substitution, solve the equation
$(x^2 - 3x)^2 - 9(x^2 - 3x) - 10 = 0$.

4. Find the values of k for which the equation $x^2 + 2kx + 3k = 0$ has (a) equal roots, (b) real roots which differ by 4.

5. Given that the equation $x^2 + px + q = 0$ has roots α and β, find an equation whose roots are $\alpha(\alpha - 1)$ and $\beta(\beta - 1)$.

6. If α and β are the roots of the equation $x^2 - x - 3 = 0$, show that $\alpha^3 + \beta^3 = 10$ and find a quadratic equation whose roots are α^2/β and β^2/α.

7. The function $f(x) = x^2 + px + 1$, where p is a constant, is zero when $x = \alpha$ and $x = \beta$; and the function $g(x) = x^2 - 9x + q$, where q is a constant, is zero when $x = \alpha + 2\beta$ and $x = \beta + 2\alpha$. Find p and q, and show that $f(3) = g(3)$. (JMB)

8. (a) The sum of the squares of the roots of the equation $x^2 + px + q = 0$ is 56 and the sum of the reciprocals of the roots is 2. Find the values of p and q. (b) Show that, for all real values of α, the equation $x^2 + (3\alpha - 2)x + \alpha(\alpha - 1) = 0$ has real roots for x. (C)

9. The roots of the equation $9x^2 + 6x + 1 = 4kx$, where k is a real constant, are denoted by α and β. (a) Show that the equation whose roots are $1/\alpha$ and $1/\beta$ is $x^2 + 6x + 9 = 4kx$. (b) Find the set of values of k for which α and β are real. (c) Find also the set of values of k for which α and β are real and positive. (L)

10. Given that a and b are non-zero constants, prove that if the equations $x^2 + ax + b = 0$ and $4x^2 - ax + 6b = 0$ have a common root then $35a^2 + 4b = 0$.

11. A straight line of gradient -2 passes through the point with x-coordinate -1 on the curve $y = 2x^2 - 3x + 8$. Find the coordinates of the point at which the line meets the curve again.

12. The line $y = 2kx$ cuts the curve $y = x^2 - 4x + 1$ at the points A and B. Without finding the coordinates of the points A and B, find the coordinates of the mid-point of the line AB in terms of k.

13. Solve the simultaneous equations $2x - y - 1 = x - 2y = 3x^2 - 8y^2$.

14. Solve the simultaneous equations $3x^2 + y^2 = 7$, $4x^2 - xy + y^2 = 6$.

15. If $P(x) = 2x^5 + 3x^4 - 8x^3 + px^2 + q$ has factors $(x+1)$ and $(x-2)$, find the values of p and q. Obtain also the remaining factors.

16. If the roots of the equation $x^3 + 5x^2 + hx + k = 0$ are α, 2α and $\alpha + 3$, find the values of α, h and k.

17. The quadratic polynomial $P(x)$ leaves a remainder of -6 on division by $(x+1)$, a remainder of -5 on division by $(x+2)$ and no remainder on division by $(x+3)$. Find $P(x)$ and solve the equation $P(x) = 0$. (O&C)

18. Factorise the expression $n^4 + 2n^3 - n^2 - 2n$. Given that n is an integer greater than 1, prove that the expression is divisible by 24.

19. The real numbers a, b, c are such that $b^2 < 4ac$ and $c > 0$. Show that $ax^2 + bx + c > 0$ for all real numbers x and, conversely, that if $ax^2 + bx + c > 0$ for all real x, then $c > 0$ and $b^2 < 4ac$. (AEB 1978)

20. For what range of values of c does the system of equations $x^2 + xy + y^2 = 1$, $2x + y = c$ have real solutions? (O)

21. (a) Solve the equations $xy = 4$, $x^2 + x + y = 6$. (b) Find the value of $x^6 + y^6$ in terms of a and b where $x + y = a$, $x^2 + y^2 = b^2$. Show also that these equations give real values for x and y only if $a^2 \leqslant 2b^2$. (W)

22. If $f(x) \equiv ax^3 + (a+b)x^2 + (a+2b)x + 1$ is exactly divisible by $(x+1)$, express b in terms of a, and find the quotient when the division is carried out, expressing the coefficients in terms of a only. Prove that in this case the equation $f(x) = 0$ has only one real root if $a^2 - 6a + 1 < 0$. Show that this inequality implies that $3 - 2\sqrt{2} < a < 3 + 2\sqrt{2}$.

4 The calculator and numerical work

4.1 Accuracy and types of error

In problems requiring a numerical answer it is sometimes possible to give an exact result, e.g. $1.4 \times 2.3 = 3.22$. In practice, it is much more likely that an approximate solution will be obtained. Such a solution may be of little value unless its degree of accuracy can be estimated. This is done by examining sources of error in the original data and in the method of calculation. In this context possible inaccuracies due to mistakes or blunders are not taken into account. The use of the word *error* is confined to inaccuracies arising in other ways.

The error in calculations based on experimental data depends mainly on the accuracy of the instruments used. If, in an experiment, estimates are made of the errors in the readings, these can be used to find the total possible error in any calculated result.

Example 1 The length and breadth of a rectangular room are 9.7 m and 6.8 m respectively. If the possible error in each of these measurements is ± 0.05 m, find the two values between which the floor area of the room must lie. Find the area and give your answer to an appropriate degree of accuracy.

Minimum area of room $= (9.65 \times 6.75)\,\text{m}^2 = 65.1375\,\text{m}^2$
Maximum area of room $= (9.75 \times 6.85)\,\text{m}^2 = 66.7875\,\text{m}^2$
$\therefore$ the area lies between $65.1375\,\text{m}^2$ and $66.7875\,\text{m}^2$.
Area of room as measured $= (9.7 \times 6.8)\,\text{m}^2 = 65.96\,\text{m}^2$.
Hence, taking possible errors into account the area of the room may be given as $66 \pm 1\,\text{m}^2$.

In some calculations the method itself leads to inaccuracies. For instance, to convert a temperature given in a weather forecast from degrees Celsius to degrees Fahrenheit one method is to 'double and add 30'. This formula is attractive because it is easier to use than the true formula, i.e. 'multiply by 9/5 and add 32', and because for temperatures from $-5°$C to $25°$C the maximum error is only

$\pm 3°F$. Suppose, however, that this conversion formula were used to express an oven temperature of 200°C given in a recipe as 430°F. The error of 38°F in this result is clearly unacceptable. Thus we see that approximate methods must always be used with care, and that if such a process is used in the wrong situation the error in the result may be large.

Consider now a problem such as the division of 37 by 13. We can calculate the answer by long division, giving as many decimal places as necessary, but we cannot obtain an exact answer. For instance, working as far as the third decimal place, $37 \div 13 = 2·846\ldots$ The answer 2·846 is said to be *truncated* (i.e. cut short) to 3 decimal places. Since the true value must lie between 2·846 and 2·847, the *truncation error* lies between 0 and 0·001. The result could also be given as $37 \div 13 = 2·85$, *rounded* to 2 decimal places. This statement implies that 2·845 $\leqslant 37 \div 13 < 2·855$. Hence the *rounding error* lies between $-0·005$ and $+0·005$.

To show how to estimate errors in a longer calculation, we consider the division of 37 by 13 performed using 4-figure logarithm tables.

Number	Logarithm	Possible error
37	1·5682	$\pm 5 \times 10^{-5}$
13	1·1139	$\pm 5 \times 10^{-5}$
2·846	0·4543	$\pm 1 \times 10^{-4}$

Thus the logarithm of $37 \div 13$ lies between 0·4542 and 0·4544. Allowing for rounding errors in the antilogarithm tables, both in the main table and the difference columns, we estimate that $37 \div 13 = 2·846 \pm 0·002$. Hence we may write $37 \div 13 = 2·8$ correct to 1 decimal place, but we cannot be certain that the answer 2·85 is correct to 2 decimal places.

Truncation and rounding errors also arise when working with an electronic calculator. A typical calculator, which displays 8 digits or 6 digits and a 2-digit exponent, produced the following results.

$$37 \div 13 = 2·8461538$$
$$3·7 \div 13 = 0·2846153$$
$$0·037 \div 13 = 2·84615 \times 10^{-3}.$$

We see that these values are truncated rather than rounded. Thus they are subject to a truncation error of between 0 and 1 in the final digit.

Using the same calculator the result

$$53 \div 13 = 4·076923$$

is obtained. The calculator displays 7 digits rather than 8 only because the 8th digit would be a zero. The read-out should be interpreted as 4·0769230 truncated at the final zero.

To estimate error in longer calculations it is necessary to know how numbers are stored in the working register and memory of the calculator. This can be established by tests similar to the following:

Divide 1000 by 7:   142·85714
Subtract 140:   2·8571428.

These results show that a calculator which displays 8 digits may actually work with 10 digits.

In the case of scientific functions many electronic calculators produce rounded values. For example, using a calculator which rounds to 6 significant figures and suppresses unnecessary zeros,

$$\lg 5\!\cdot\!3 = 0\!\cdot\!724276 \quad \lg 5\!\cdot\!2 = 0\!\cdot\!716003$$
$$\lg 53 = 1\!\cdot\!72428 \quad \lg 52 = 1\!\cdot\!716.$$

The maximum rounding error in these values is $\pm 0\!\cdot\!5$ in the 6th significant figure.

Example 2   Allowing for rounding errors, find two values between which $(\lg 7 \div \lg 3)$ must lie. Calculate $\lg 7 \div \lg 3$ and give your answer to an appropriate number of significant figures.

Taking possible rounding errors into account

$$0\!\cdot\!8450975 < \lg 7 < 0\!\cdot\!8450985$$
$$0\!\cdot\!4771205 < \lg 3 < 0\!\cdot\!4771215$$

$$\therefore \quad \text{maximum value of } \frac{\lg 7}{\lg 3} = \frac{0\!\cdot\!8450985}{0\!\cdot\!4771205} = 1\!\cdot\!7712475$$

$$\text{minimum value of } \frac{\lg 7}{\lg 3} = \frac{0\!\cdot\!8450975}{0\!\cdot\!4771215} = 1\!\cdot\!7712417.$$

By calculator, $\lg 7 \div \lg 3 = 1\!\cdot\!7712446$
Hence $\lg 7 \div \lg 3 = 1\!\cdot\!7712$ to 5 s.f.

[In a longer calculation it would probably be unrealistic to give an answer to more than 4 significant figures.]

Thus although a calculator produces more accurate results than 4-figure tables, these results do contain unavoidable rounding and truncation errors. The approximate methods used within the calculator when determining the values of scientific functions may lead to further error, especially at the extremes of the calculator's range. Clearly it is not necessary to make detailed estimates of possible error after every calculation performed. However, it is important to know what degree of accuracy can be expected from the calculating aid one is using, be it mathematical tables, slide rule or calculator. Efforts can then be made to choose methods of working which minimise error.

Exercise 4.1

In questions 1 to 4 find two values between which z must lie, then calculate z giving your answer to an appropriate degree of accuracy.

1. $z = xy$ where $x = 8\!\cdot\!7 \pm 0\!\cdot\!1$, $y = 5\!\cdot\!3 \pm 0\!\cdot\!1$.

2. $z = x^2 + y^2$ where $x = 7\!\cdot\!3 \pm 0\!\cdot\!05$, $y = 6\!\cdot\!4 \pm 0\!\cdot\!05$.

3. $z = \dfrac{1}{x} + \dfrac{1}{y}$ where $x = 4\!\cdot\!16 \pm 0\!\cdot\!01$, $y = 6\!\cdot\!72 \pm 0\!\cdot\!01$.

4. $z = \sqrt{(x-y)}$ where $x = 62 \pm 0.5$, $y = 45 \pm 0.5$.

5.   Find two values between which the product pq must lie given that $p = 4.26$, $q = 2.51$, (a) rounded to 2 decimal places, (b) truncated to 2 decimal places. In each case give the value of pq to an appropriate number of decimal places.

6.   Repeat question 5 for the quotient p/q.

7.   Use 4-figure logarithm tables to perform the following calculations. Estimate the possible errors and give your answers to an appropriate number of significant figures.
(a) 7.34×2.69 (b) $0.581 \div 8.07$
(c) $(3.3)^2 \div 0.76$ (d) $\sqrt{(4.9 \times 32.8)}$.

8.   Answer the following questions about the calculator you are using. If possible, compare its performance with that of other calculators.
(a) How many digits are displayed?
(b) Are unnecessary zeros suppressed?
(c) How many digits are stored in the working register?
(d) In basic function operations, i.e. $+$, $-$, $\times$, $\div$, are the results truncated or rounded?
(e) Are the values of scientific functions, such as $\sqrt{x}$, $\lg x$, $\sin x$, truncated or rounded, and to how many figures?
(f) What degree of accuracy for the various calculator functions is claimed by the manufacturer?

9.   Use an electronic calculator to perform the following calculations. By estimating the possible errors, give your answers to an appropriate number of significant figures.
(a) $3.6 \times 12.7 \div 6.84$ (b) $1/\sqrt{(5.13)}$
(c) $\sin 40° \times \cos 20°$ (d) $\lg 3.6 \div \lg 1.8$.
[The answers to this question will vary according to the type of calculator used.]

4.2 Some functions and their inverses

Consider a function f which is a one-one mapping of a set X onto a set Y. This means that every element of the range Y is the image of exactly one element of the domain X. Under these conditions it is possible to define an *inverse function* f^{-1} with domain Y and range X such that if $f(a) = b$ then $f^{-1}(b) = a$.

Any scientific calculator provides several examples of inverse functions. For instance, if $\lg a = b$ then $10^b = a$. Thus the function 10^x is the inverse of the function $\lg x$. It follows from this relationship that $10^{(\lg x)} = x$ and $\lg (10^x) = x$. [Discrepancies due to rounding errors may occur when testing such results. For instance, using certain calculators it is found that $10^{(\lg 9)} = 9.00001$ and $\lg (10^{0.29}) = 0.289999$.]

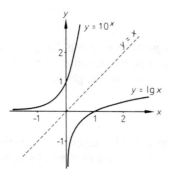

The sketch shows how the graphs $y = \lg x$ and $y = 10^x$ are related. If a point (a, b) lies on the graph $y = \lg x$, then the point (b, a) lies on the graph $y = 10^x$. Hence each graph is the reflection of the other in the line $y = x$.

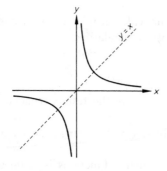

The function $1/x$ is its own inverse, since if $1/a = b$, then $1/b = a$. We find that in this case the graph of $y = 1/x$ is itself symmetrical about the line $y = x$.

The function $f : x \rightarrow x^2$ with domain the set of real numbers has no true inverse, since it is not one-one. For instance, the element 4 in the range is the image of both -2 and $+2$, so it is impossible to assign a single value to $f^{-1}(4)$. However, if the domain of the function $f : x \rightarrow x^2$ is the set of all real numbers greater than or equal to zero, every element of the range is the image of only one element of the domain. Thus we can define the inverse function as $f^{-1} : x \rightarrow \sqrt{x}$.

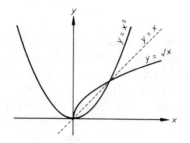

Comparing the graphs of $y = x^2$ and $y = \sqrt{x}$, we again see symmetry about the line $y = x$, but only for $x \geqslant 0$.

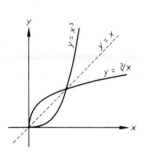

The functions x^y and $x^{1/y}$ are functions of two variables x and y, usually defined only for positive values of x. Using a constant value of y, functions such as x^3 and $\sqrt[3]{x}$ can be studied. From the symmetry of the graphs, we see that x^3 and $\sqrt[3]{x}$ are mutually inverse.

More generally, if $a^k = b$ and $a > 0$, then $b^{1/k} = \sqrt[k]{b} = a$. Hence the function $x^{1/k}$ is the inverse of the function x^k.

The following examples show some of the uses of the calculator functions discussed in this section.

Example 1   Evaluate $\left(2{\cdot}3 + \dfrac{1}{2{\cdot}3}\right)^2$.

By pressing the following keys: 2   ·   3   +   1/x   =   x²

$\left(2{\cdot}3 + \dfrac{1}{2{\cdot}3}\right)^2 = 7{\cdot}4790359$ (truncated to 8 digits).

Example 2   Evaluate $\sqrt[5]{(5{\cdot}6 \times 10^{-4})}$.

By pressing the following keys: .5   ·   6   EXP   4   +/−   x¹ᐟʸ   5   =

$\sqrt[5]{(5{\cdot}6 \times 10^{-4})} = 0{\cdot}223685$ (rounded to 6 s.f.).

Example 3   Find $\log_2 3$.

Using the formula $\log_a x = \log_{10} x / \log_{10} a$ (see §1.6)

   $\log_2 3 = \lg 3 / \lg 2 \approx 1{\cdot}5849616$

Hence $\log_2 3 = 1{\cdot}5850$ (to 5 s.f.).

Alternative method:

$\log_2 3 = x \;\Leftrightarrow\; 2^x = 3$
$\Leftrightarrow\; \lg(2^x) = \lg 3$
$\Leftrightarrow\; x \lg 2 = \lg 3$
$\Leftrightarrow\; x = \lg 3 / \lg 2$

$\therefore$ as before $\log_2 3 = 1{\cdot}5850$ (to 5 s.f.).

Example 4   Solve the equation $2^x \cdot 3^{x-2} = 1$.

$$2^x \cdot 3^{x-2} = 1 \Leftrightarrow 2^x \cdot 3^x \cdot 3^{-2} = 1$$
$$\Leftrightarrow \quad (2 \cdot 3)^x = 3^2$$
$$\Leftrightarrow \quad 6^x = 9$$
$$\Leftrightarrow \quad \lg(6^x) = \lg 9$$
$$\Leftrightarrow \quad x \lg 6 = \lg 9$$
$$\Leftrightarrow x = \lg 9 / \lg 6 \approx 1 \cdot 2262954$$

$$x = 1 \cdot 2263 \quad \text{(to 5 s.f.)}$$

Exercise 4.2

[Give numerical answers to 4 s.f.]

1.   Write down the inverses of the following functions
(a) $f : x \rightarrow 3x$ (b) $f : x \rightarrow \frac{1}{2}x$ (c) $f : x \rightarrow x - 1$
(d) $f : x \rightarrow 1 - x$ (e) $f : x \rightarrow 2x + 3$ (f) $f : x \rightarrow 4(x - 1)$

2.   Write down the equations of the reflections of the following graphs in the line $y = x$.
(a) $y = 2x$ (b) $y = 2/x$ (c) $y = 2 - x$
(d) $y = x - 2$ (e) $y = 2^x$ (f) $y = \lg(x - 2)$.

3.   Using the same pair of axes, draw accurate graphs of the functions x^y and $x^{1/y}$ for $y = 1 \cdot 5$, taking $0 \leqslant x \leqslant 4$.

4.   Evaluate the following expressions

(a) $\sqrt{\left(1 \cdot 9 - \dfrac{1}{1 \cdot 9}\right)}$ (b) $\{(0 \cdot 34)^2 + (0 \cdot 27)^2\}^{-5/2}$

(c) $1/(7 \cdot 1 \times 10^{-3})^2$ (d) $\sqrt[3]{(0 \cdot 027 \times 10^4)}$.

5.   Evaluate   (a) $\log_4 7$,   (b) $\log_3 10$,   (c) $\log_5 2$

6.   Solve the following equations
(a) $2^x = 9$,   (b) $10^{3x} = 15$,   (c) $8 \cdot 7^x = 3 \cdot 2$.

7.   Using the same pair of axes, draw accurate graphs of $y = 3^x$ for $-2 \leqslant x \leqslant 2$ and $y = \log_3 x$ for $0 \cdot 1 \leqslant x \leqslant 10$.

8.   Solve the following equations
(a) $5^{2x} \cdot 5^{1-x} = 40$ (b) $3^x \cdot 4^{x+1} = 100$
(c) $2^{x-1} \cdot 3^{x-2} = 1$ (d) $12^{x-1} = 3^{x+4}$.

9. By taking logarithms of both sides solve the equations
(a) $4^{x+2} = 7^{x-1}$,   (b) $3^{2x} \cdot 4^{1-x} = 13$.

10. Solve the equation $\log_3 x + \log_5 x = 1$.

4.3 Simple flow diagrams

Any mathematical calculation consists of a sequence of steps performed one after the other. A rule or systematic process for carrying out a calculation is sometimes called an *algorithm*. The list of instructions describing how to perform the calculation or algorithm can be referred to as a *program*. In a computer program each step must be within the capabilities of the computer concerned. Programs are usually simpler to follow when they are given in the form of a *flow diagram* or *flow chart*. In a flow diagram the individual steps are written in 'boxes' of various shapes and the order in which the steps are to be performed is indicated by arrows. The shapes used in simple flow diagrams are:

for START and STOP   for other instructions.

The instructions $X: = A + B$ and $LET \ X = A + B$

both mean: 'Calculate X, given that $X = A + B$'.

READ X   'Assign to X the value of x given in the data'

PRINT X   'Write down the value of X'

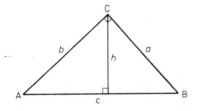

Example 1   In $\triangle ABC$, $\angle C = 90°$ and h is the height of the triangle taking AB as base. Carry out the procedure given in the flow diagram to find h given $a = 6$ and $b = 8$.

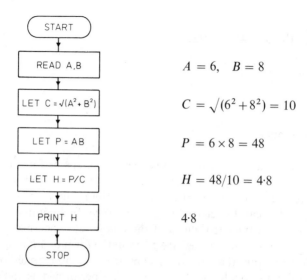

$A = 6, \quad B = 8$

$C = \sqrt{(6^2 + 8^2)} = 10$

$P = 6 \times 8 = 48$

$H = 48/10 = 4{\cdot}8$

$4{\cdot}8$

When writing a computer program, introducing a new variable involves using extra storage space in the computer memory. Thus in a flow diagram as few variables as possible are used. For instance, instead of writing 'LET $X = A + B$', we may write 'LET $A = A + B$' if we do not need to use the original value of A again. To illustrate this approach we give below two alternatives to the flow diagram in Example 1, both designed for use with a small calculator. The letters R and M are used to denote the contents of the working register and the memory.

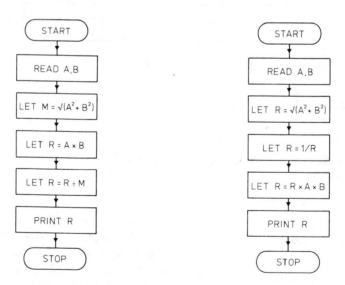

The reader is recommended to test these procedures for himself. In the first case a key sequence similar to the following would be used:

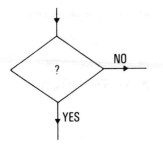

In some calculations the method used depends on the answer to various questions. In a flow diagram these questions are written in *decision boxes*, which have different exits for the answers YES and NO. The use of one of these exits will result in a *jump* to another part of the flow diagram.

Example 2 Construct a flow diagram for a procedure to determine the nature of the roots of the equation $x^2 - 2px + q = 0$ and find their values if real.

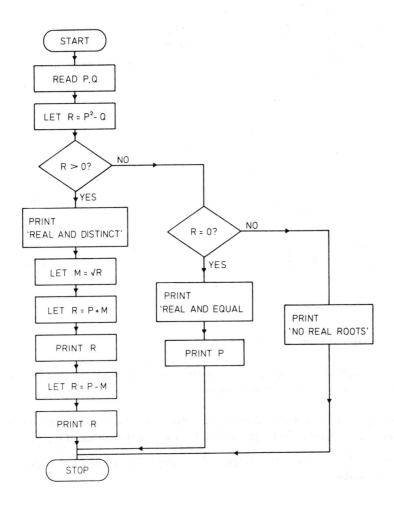

Testing the procedure for $p = 2$ and $q = 3$, we obtain the following results.

Working	Print-out
$P = 2, Q = 3$	
$R = 2^2 - 3 = 1$	
$R > 0$	REAL AND DISTINCT
$M = \sqrt{1} = 1$	
$R = 2 + 1 = 3$	3
$R = 2 - 1 = 1$	1

A flow diagram for a repetitive process may contain *loops* as shown in the next example.

Example 3   Construct a flow diagram for finding the sum to 10 terms, S_{10}, of the series $3 + 6 + 12 + 24 + \dots$ using the relation $S_r = 2S_{r-1} + 3$.

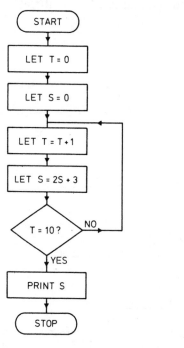

The initial value of S is taken to be 0. Successive values of S and T can be tabulated as follows:

T	S
0	0
1	3
2	9
3	21
4	45
5	93
6	189
7	381
8	765
9	1533
10	3069

Thus the printed value of S is 3069.

In this example T was a *loop counter* used to indicate when the process had been carried out the correct number of times.

Some computer programs require the input of a list of numbers. In this context the instruction 'READ X' means 'assign to X the next number of the list'.

Example 4   Construct a flow diagram for a procedure to find the largest of a set of n real numbers.

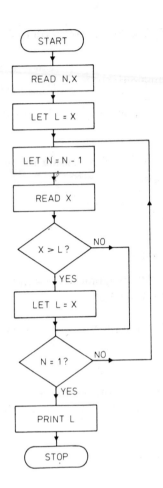

If the given set of numbers is 5, −7, 8, 12, −3, 2, then the values of N, X and L are as follows:

N	X	L
6	5	5
5	−7	5
4	8	8
3	12	12
2	−3	12
1	2	12

Thus the print-out is the number 12, the largest element of the given set.

[Note that this flow diagram contains both a jump and a loop.]

Exercise 4.3

1. Carry out the procedure given in Example 1 for
(a) $a = 2$, $b = 1{\cdot}5$,   (b) $a = 3{\cdot}5$, $b = 12$.

2. Carry out the procedure given in Example 2 for
(a) $p = 1$, $q = -15$ (b) $p = -0{\cdot}5$, $q = 0{\cdot}25$
(c) $p = 2{\cdot}3$, $q = 5{\cdot}3$ (d) $p = 3{\cdot}7$, $q = -4{\cdot}8$.

In questions 3 to 10 construct flow diagrams for carrying out the given processes.

3. Find and print out the total number of seconds in x hours, y minutes and z seconds.

4. Calculate x given that $a + b/x = c$, printing out suitable messages when $a = c$ or $b = 0$.

5. Print out the examination grade corresponding to a mark of $x\%$ given that the

lowest marks in each grade are: grade 1, 75%; grade 2, 60%; grade 3, 50%. Candidates with less than 50% fail.

6. Arrange three real numbers a, b, c in ascending order of magnitude.

7. Determine the number of real roots of the equation $x^3 - ax^2 - bx + ab = 0$ and print out their values.

8. Find the sum to 8 terms of the series $2 + 6 + 18 + 54 + \ldots$ using the relation $S_r = 3S_{r-1} + 2$.

9. Find the nth term of the Fibonacci series
$$1 + 1 + 2 + 3 + 5 + 8 + 13 + \ldots$$

10. Find the sum of the squares of a set of n numbers.

In each of the remaining questions construct a flow diagram for performing the given calculation using a pocket calculator with one memory.

11. Find the coordinates of the minimum point on the curve $y = x^2 + hx + k$.

12. Find the real roots, if any, of the equation $ax^2 + bx + c = 0$.

13. Find the coordinates of the point of intersection of the lines $ax + by = 1$ and $cx + dy = 1$, printing out suitable messages when the lines are parallel or coincident.

14. Find an approximate value for π using the result

$$\frac{\pi^2}{6} \approx \frac{1}{1^2} + \frac{1}{2^2} + \frac{1}{3^2} + \frac{1}{4^2} + \ldots + \frac{1}{n^2}$$

for any given value of n.

4.4 Efficient methods of calculation

When performing a numerical calculation the main aims are to
(1) avoid mistakes
(2) use as few steps as possible
(3) minimise error
[In work with a calculator effective use of the memory may help to achieve all three aims.]

To avoid mistakes numerical answers should be checked whenever possible.

Example 1 The value 1·5 has been obtained for a root of the equation $2x^3 - x^2 + 9 = 0$. Check this result.

Without detailed calculation it is clear that

$$2(1 \cdot 5)^3 - (1 \cdot 5)^2 + 9 > 0$$

∴ $1 \cdot 5$ is not a root of the equation.
[In fact the correct value of the root is $-1 \cdot 5$.]

Example 2   A student using a calculator obtained the result
$38 \cdot 76 \div 0 \cdot 2497 = 6 \cdot 4422 \times 10^{-3}$. Is this answer reasonable?

Rough check: $38 \cdot 76 \div 0 \cdot 2497 \approx 40 \div \frac{1}{4} = 160$
The answer $6 \cdot 4422 \times 10^{-3}$ must be incorrect.

[On some calculators this wrong answer is obtained by inadvertently pressing the
$\boxed{\div}$ key twice.]

The number of steps required in a calculation will depend on the type of calculator, table book or slide rule used. However, when evaluating a polynomial the method called *nested multiplication* is often the most efficient. For instance, rearranging the polynomial $3x^3 - 2x^2 + 4x - 1$ as $\{(3x - 2)x + 4\}x - 1$, we can quickly calculate that its value is 23 when $x = 2$.

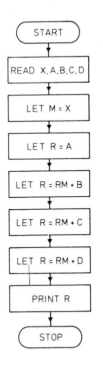

This flow diagram gives the procedure for evaluating the polynomial $ax^3 + bx^2 + cx + d$ using a calculator. In each step the same pattern of operations is repeated. Thus the method of nested multiplication can be summarised as follows:

(i)   Take the coefficient of the highest power of x.
(ii) Multiply by x and add the next coefficient.
(iii) Repeat (ii) until constant term has been added.

When accuracy is an important consideration, it may be possible to reduce error by choosing a suitable method.

Example 3   Evaluate $5\sqrt{2}$ using 4-figure tables.

$\sqrt{2} = 1\cdot414 \pm 0\cdot0005$ $\quad \therefore \quad 5\sqrt{2} = 7\cdot07 \pm 0\cdot0025$
The possible rounding error is reduced by writing
$5\sqrt{2} = \sqrt{50} = 7\cdot071 \pm 0\cdot0005$.

Example 4   Evaluate $2 \div 7 \times 32$ using a calculator.

$2 \div 7 = 0\cdot2857142$ with error between 0 and 1×10^{-7}.
Assuming that the working register carries no additional digits, $2 \div 7 \times 32$
$= 9\cdot1428544$, subject to an error between 0 and 32×10^{-7}.
Hence the maximum number of significant figures that can be given in the
answer is 5,

$$2 \div 7 \times 32 = 9\cdot1429 \quad \text{(to 5 s.f.)}$$

Accuracy may be improved by writing

$$2 \div 7 \times 32 = 2 \times 32 \div 7 = 9\cdot1428571.$$

Since the maximum possible truncation error is now 1×10^{-7}, we know that
$2 \div 7 \times 32 = 9\cdot142857$ (to 7 s.f.)

Example 5   Find $\sqrt{(13\cdot2^2 - 12\cdot6^2)}$ using (a) 4-figure tables, (b) a calculator.

(a) $\sqrt{(13\cdot2^2 - 12\cdot6^2)} = \sqrt{(174\cdot2 - 158\cdot8)} = \sqrt{(15\cdot4)} = 3\cdot924$

Allowing for possible rounding errors of $\pm 0\cdot05$ in the squares of $13\cdot2$ and $12\cdot6$,
the true value of the expression must lie between $\sqrt{(15\cdot3)} \approx 3\cdot912$ and $\sqrt{(15\cdot5)}$
$\approx 3\cdot937$.
Hence $\sqrt{(13\cdot2^2 - 12\cdot6^2)} = 3\cdot9$ (to 2 s.f.)
A more accurate result is obtained by writing

$$\sqrt{(13\cdot2^2 - 12\cdot6^2)} = \sqrt{\{(13\cdot2 + 12\cdot6)(13\cdot2 - 12\cdot6)\}}$$
$$= \sqrt{\{25\cdot8 \times 0\cdot6\}} = \sqrt{(15\cdot48)} = 3\cdot934$$

$\therefore \quad \sqrt{(13\cdot2^2 - 12\cdot6^2)} = 3\cdot93$ (to 3 s.f.)

(b) Since a calculator gives exact values for the squares of $13\cdot2$ and $12\cdot6$,

$\sqrt{(13\cdot2^2 - 12\cdot6^2)} = 3\cdot9344631$ (truncated to 8 digits)

Hence $\sqrt{(13\cdot2^2 - 12\cdot6^2)} = 3\cdot934463$ (to 7 s.f.)

Exercise 4.4

1. Draw flow diagrams for evaluating the polynomials (a) $2x^2 - 7x + 3$,
(b) $x^3 + 3x^2 - 5x - 4$ and (c) $3x^4 - x^3 - 4x + 8$, by the method of nested multiplication. Test your procedures for $x = -1$ and $x = 2$.

2.   Use a calculator to evaluate the following
(a) $3x^2 - 5x + 1$ for $x = 2\cdot8$ and $x = -0\cdot6$
(b) $2x^3 + x^2 - 8x - 3$ for $x = -1\cdot3$ and $x = 2\cdot67$
(c) $x^5 - 3x^3 + 10x + 5$ for $x = 1\cdot54$ and $x = -0\cdot09$
(d) $2x^4 - x^3 - 6x^2 + 3x$ for $x = 3\cdot02$ and $x = -1\cdot4$.

In the remaining questions in the exercise, perform the given calculations by the most direct method, then decide whether there is a way of producing a more accurate result. Consider, if available, the use of 4-figure tables and various types of calculator. Illustrate your answers with flow diagrams where appropriate.

3.   Evaluate   (a) $3\sqrt{10}$,   (b) $\frac{1}{2}\sqrt{6}$,   (c) $4 \div 7\cdot36$.

4.   Evaluate   (a) $\dfrac{8\cdot71}{4\cdot03 \times 5\cdot64}$,   (b) $\dfrac{1}{3\cdot61} \times \dfrac{1}{0\cdot52} \times 7\cdot3$.

5.   Evaluate   (a) $\sqrt{(25\cdot3^2 - 24\cdot7^2)}$,   (b) $\sqrt{(25\cdot3^2 + 24\cdot7^2)}$.

6.   Evaluate   (a) $2\lg 6$,   (b) $(1\cdot76)^4$,   (c) $10^{4\lg 1\cdot76}$.

7.   Find the sum of the first 8 terms of the series $1 + x + x^2 + x^3 + \dots$ given
(a) $x = 2$, (b) $x = -3\cdot4$, (c) $x = 0\cdot8136$.

8.   Find the sum of the series $1 + 2x + 3x^2 + \dots$ as far as the first term which is less than $0\cdot1$, given (a) $x = 0\cdot5$, (b) $x = 0\cdot42$, (c) $x = 0\cdot61$.

[As a rough guide only, answers correct to 4 s.f. are provided for this exercise.]

4.5 Reduction of laws to straight line form

In §2.5 we saw that an equation of the form $y = mx + c$ represents a straight line, where m is the gradient and c is the intercept on the y-axis. If a table of values of variables x and y leads to a straight line graph, we can use the gradient and the intercept on the y-axis to write down the law connecting x and y.

Example 1   Some values of the variables x and y are given in the table below:

x	1	2	3	4	5
y	1·2	2·0	2·8	3·6	4·4

Find a relationship between the variables, giving y in terms of x.

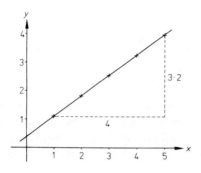

Plotting y against x, a straight line graph is obtained. Hence the relationship between x and y takes the form $y = mx + c$. The intercept on the y-axis is 0.4

$$\therefore \quad c = 0.4.$$

The gradient can be found using the points $(1, 1.2)$ and $(5, 4.4)$

$$\therefore \quad m = \frac{4.4 - 1.2}{5 - 1} = \frac{3.2}{4} = 0.8.$$

Hence the required relationship is $y = 0.8x + 0.4$.

If a graph is drawn from experimental data and the points on it all lie close to a straight line, we often assume that the relationship between the variables is approximately linear. We then draw the line that best fits the data and use points on it to determine the law connecting the variables. When finding the gradient of the line, errors are minimised by choosing two points near the ends of the line whose coordinates are easily read from the graph.

Example 2   The following observations were recorded when weights were suspended at the end of a coiled spring.

Load on spring in grammes, W	20	40	60	80	100
Length of spring in cm, L	8.1	9.0	10.0	10.9	11.8

Find an expression for L in terms of W, assuming that the relationship is linear.

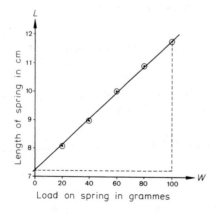

Load on spring in grammes

The points representing the observations all lie close to a straight line, which cuts the vertical axis at $L = 7.2$. Using the points $(0, 7.2)$ and $(100, 11.8)$ on the line, its gradient

$$= \frac{11.8 - 7.2}{100 - 0}$$

$$= \frac{4.6}{100} = 0.046.$$

Hence the relationship between L and W is approximately

$$L = 0.046W + 7.2.$$

[When estimating error in a result of this kind, it is necessary to consider the range of possible positions of the 'best' straight line graph, allowing for experimental errors.]

When variables are connected by a non-linear law, it is sometimes possible to find related variables which obey a linear law.

Example 3   The table below gives values of the variables x and y, which are related by an equation of the form $y^2 = ax^2 + b$. Plot y^2 against x^2 and use your graph to obtain the values of a and b.

x	1	2	3	4	5
y	3·32	4·12	5·20	6·40	7·68

Let $X = x^2$ and $Y = y^2$, then working to 3 s.f. the values of X and Y are:

X	1	4	9	16	25
Y	11·0	17·0	27·0	41·0	59·0

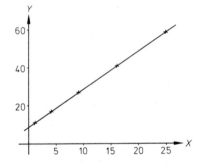

The intercept on the vertical axis is at $Y = 9$. Using the points $(1, 11)$ and $(25, 59)$ the gradient is

$$\frac{59 - 11}{25 - 1} = \frac{48}{24} = 2$$

$$\therefore \quad Y = 2X + 9.$$

Hence the variables x and y are related by the equation $y^2 = ax^2 + b$, where $a = 2$ and $b = 9$.

Logarithms are often used to express experimental laws in linear form.

Example 4   The following values of x and y are believed to obey a law of the form $y = ab^x$, where a and b are constants:

x	1	2	3	4	5
y	3·8	9·2	22·1	53·1	127·4

Show graphically that this is so and estimate the values of a and b correct to 1 decimal place.

Assuming that $y = ab^x$, then $\lg y = \lg (ab^x)$,

i.e. $\lg y = x \lg b + \lg a$.

Hence plotting $\lg y$ against x should produce a straight line graph with gradient $\lg b$ and intercept $\lg a$.

x	1	2	3	4	5
$Y = \lg y$	0·58	0·96	1·34	1·73	2·11

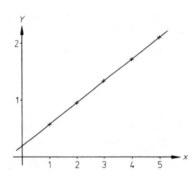

The intercept on the vertical axis is at
$$Y = 0·2$$

$\therefore$ $\lg a \approx 0.2$.

Hence $a = 1·6$ (to 1 d.p.)

Using the points $(1, 0·58)$ and $(5, 2·11)$ on the line, the gradient $= \dfrac{2·11 - 0·58}{5 - 1}$

$$= 0·3825$$

$\therefore$ $\lg b \approx 0·3825$ and hence $b = 2·4$ (to 1 d.p.)

$\therefore$ x and y obey a law of the form $y = ab^x$, where $a = 1·6$ and $b = 2·4$ (to 1 d.p.).

Example 5   The following readings were obtained in an experiment:

x	20	40	60	80	100
y	55	150	280	430	600

Show graphically that, allowing for small errors of observation, there is a relation between y and x of the form $y = ax^k$. Find approximate values for a and k.

Assuming that $y = ax^k$, then $\lg y = \lg (ax^k)$,

i.e. $\lg y = k \lg x + \lg a$.

Hence plotting $\lg y$ against $\lg x$ should produce a straight line graph with gradient k and intercept $\lg a$.

$X = \lg x$	1·301	1·602	1·778	1·903	2·000
$Y = \lg y$	1·740	2·176	2·447	2·633	2·778

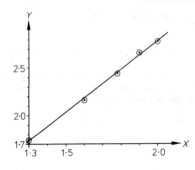

[In order to make full use of the graph paper the X-axis is marked from 1·3 to 2·0 and the Y-axis from 1·7 to 2·8. A suitable scale on both axes would then be 0·1 unit: 2 cm. This means that the true intercept on the Y-axis is not shown. Thus $\lg a$ cannot be read directly from the graph.]

The graph passes through the points (1·3, 1·73) and (2, 2·78)

$$\therefore \quad \text{the gradient} = \frac{2·78 - 1·73}{2 - 1·3} = \frac{1·05}{0·7} = 1·5.$$

Hence X and Y obey a law of the form $Y = 1·5X + \lg a$.
Substituting $X = 2$, $Y = 2·78$, $2·78 = 3 + \lg a$

$$\therefore \quad \lg a = -0·22 \quad \text{and} \quad a \approx 0·60$$

Hence there is a relation between y and x of the form $y = ax^k$, where the approximate values of a and k are 0·60 and 1·5 respectively.

Exercise 4.5

[Give answers to 2 significant figures.]

1. By using the given table of values to draw a graph, find the relationship between the variables x and y in the form $y = mx + c$.

x	0	2	4	7	9
y	0·6	1·2	1·8	2·7	3·3

2. In an electrical experiment the following results were obtained.

V	86	80	70	60	50	40	30
R	282	267	252	231	208	176	154

Determine the law connecting R and V in the form $R = aV + b$.

3. In a test to determine the efficiency of a crane the following values were obtained for the effort E required to raise a load L. Find the law connecting E and L in the form $E = aL + b$.

L	10	20	30	40	50	60	70	80	90	100
E	1·0	1·7	2·1	2·6	3·2	3·8	4·2	4·9	5·5	6·0

4. In an experiment the following observations were recorded:

T	0	12	20	30	42	48	60
θ	7·6	3·9	1·1	−1·9	−5·7	−7·8	−11·8

Find the law connecting T and θ in the form $\theta = T_0 + kT$.

5. The table below gives experimental values of the variables u and v, which are connected by the equation $\dfrac{1}{u} + \dfrac{1}{v} = \dfrac{1}{f}$, where f is a constant. By plotting the reciprocal of v against the reciprocal of u, estimate the value of f.

u	20	25	30	40	50
v	79	45	34	27	23·5

6. The following values for x and y have been found in an experiment:

x	1	2	3	4	5
y	8.29	11·2	8·67	0·84	−12·5

By plotting y/x against x, verify that x and y are connected by a law of the form $y = ax^2 + bx$ and find approximate values for a and b.

7. Two quantities, x and y, are known to be connected by a law of the form $x^n y = k$, where n and k are constants. Using the given table of values, plot $\lg y$ against $\lg x$ and hence find approximate values for n and k.

x	2	3	4	5	6
y	3·54	1·92	1·25	0·90	0·68

8. The given values of x and y are believed to obey a law of the form $y = ab^x$, where a and b are constants. Show graphically that this is so and estimate the values of a and b.

x	1	2	3	4	5
y	0·91	1·27	1·78	2·50	3·50

9. The table below gives experimental values of x and y, which are known to be related by a law of the form $a/x + b/y = 1$. By drawing a suitable graph find approximate values for a and b.

x	1	2	3	4	5
y	−3·25	−19·1	28·8	12·7	9·7

10. The variables x and y tabulated below are believed to satisfy a relationship of the form $y = a(x+2)^n$, where a and n are constants. Show graphically that this is so and obtain approximate values for a and n.

x	0	1	2	3	4	5
y	1·25	0·56	0·31	0·20	0·14	0·10

(AEB 1977)

Exercise 4.6 (*miscellaneous*)

1. The base and height of a triangle are measured as 16 cm and 23 cm respectively. If the possible error in these measurements is ± 0.2 cm, find two values between which the area of the triangle must lie. Find the area and give your answer to an appropriate degree of accuracy.

2. The radius of a circle is 5.95 cm rounded to 2 decimal places. Calculate the circumference by each of the methods given below. By estimating the possible error in each case give the answer to the appropriate number of significant figures.
(a) Use 4-figure tables taking $\lg \pi = 0.4971$.
(b) Take $\pi = 22/7$ and use no calculating aids.
(c) Use a calculator and its value for π.

3. Find the inverses of the following functions.
(a) $f : x \to 1 - 2x$, (b) $g : x \to 5^x$, (c) $h : x \to -1/x$.
4. Find $\log_7 12$.

5. Solve the equation $3^{2x} - 5 \cdot 3^x + 4 = 0$.

6. Solve the equation $\log_3 x = \log_x 5$.

7. Suggest a relationship of the form $z = ap + 10b$, where a and b are integers, given the following experimental data.

p	145	160	170	185	195	200
z	65	170	225	300	360	410

8. Explain how a straight line graph of the form $y = mx + c$ may be drawn to represent the following relationships between variables u and v, where a and n are constants
(a) $u^2 = av^2 + n$　　　　　　　　(b) $v = nu(a + u)$
(c) $u = av^n$　　　　　　　　　　　(d) $v = na^u$

9. The following values of x and y are believed to obey a law of the form $y = kb^x$ where k and b are constants:

x	1	2	3	4	5
y	0.8	2.3	6.7	20.2	60.9

Show graphically that this is so and determine approximate values of k and b.

(AEB 1976)

10. The table shows approximate values of a variable y corresponding to certain values of another variable x. By drawing a suitable linear graph, verify that these values of x and y satisfy approximately a relationship of the form $y = ax^k$. Use your graph to find approximate values of the constants a and k.

x	5	10	15	20	25	30
y	45	63	77	89	100	110

(L)

11. Carry out the procedure given in the flow diagram below, tabulating the successive values of S, T and R as you proceed. State the printed value of S to the number of decimal places that you consider appropriate. (JMB)

Question 11   Question 12

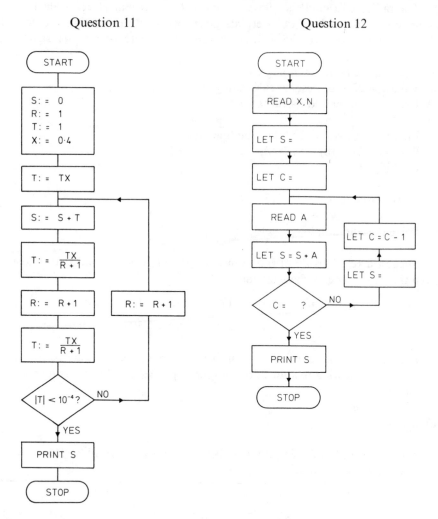

12. Copy and complete the given flow diagram designed to evaluate the polynomial $a_0x^n + a_1x^{n-1} + \ldots + a_{n-1}x + a_n$. Explain the function of the instruction 'READ A'. Demonstrate the use of the procedure to evaluate $2x^3 - 3x^2 + x - 7$ when $x = 2$, tabulating the values assigned to S, C and A throughout.

13. A *part* of a flow diagram is shown below. The purpose of the flow diagram is to find sets of positive integers x, y, z such that $x^2 + y^2 = z^2$, where $x < y$.

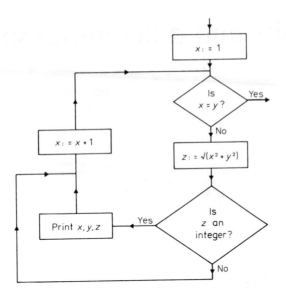

When the current value of y is 12, work through this section of the diagram, showing any values printed.

Complete the flow diagram so that y takes all integer values $1, 2, 3, \ldots, N$ in turn, where N is a positive integer to be read as data at the start.   (C)

14. Draw a flow diagram for a procedure to determine whether a given integer N is prime. [You may assume that N is greater than 1.]

15. Draw a flow diagram for the process of reading 100 positive numbers into a computer, calculating the largest difference between any pair of these numbers and printing out this difference.   (AEB 1975)

16. Construct a flow diagram to calculate the values of $3x^2 - 4x + 2$ using nested multiplication for $x = -1(0.2)2$ (i.e. for $x = -1, -0.8, -0.6, \ldots, 1.6, 1.8, 2$) and to print out the least of these values.

5 Limits and differentiation

5.1 The limit of a function

This chapter deals with the foundations of the study of calculus. The basis of this branch of mathematics is the idea of a limit.

Consider the function $f(x) = x^2 - x - 2$.

x	-1	$-0\cdot5$	$-0\cdot2$	$-0\cdot1$	$-0\cdot01$	$0\cdot01$	$0\cdot1$	$0\cdot2$	$0\cdot5$	1
$f(x)$	0	$-1\cdot25$	$-1\cdot76$	$-1\cdot89$	$-1\cdot9899$	$-2\cdot0099$	$-2\cdot09$	$-2\cdot16$	$-2\cdot25$	-2

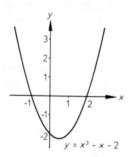

$y = x^2 - x - 2$

The table shows that as the value of x approaches zero, from either above or below, then the value of $f(x)$ approaches -2. Thus -2 is called the limit of $f(x)$ as x approaches zero. Using the symbol $\rightarrow$, which is read 'tends to', we write:

as $x \rightarrow 0$, $f(x) \rightarrow -2$.

In general, if, as we consider values of x closer and closer to $x = a$, the value of a function $f(x)$ approaches a finite value l, then l is called the *limit* of $f(x)$ as x tends to a. This may be written:

$$\lim_{x \to a} f(x) = l \quad \text{or} \quad \text{as } x \to a, f(x) \to l.$$

Example 1 Find the limit as $h \rightarrow 0$ of $2 - 3h + h^2$.

As $h \rightarrow 0$, $2 - 3h + h^2 \rightarrow 2$.

80

Example 2 Find $\lim\limits_{x \to 4} \left(\dfrac{x+3}{x-3} \right)$.

$$\lim_{x \to 4} \left(\frac{x+3}{x-3} \right) = \frac{4+3}{4-3} = 7.$$

As shown in these examples, we find that for many functions arising in elementary mathematics, $f(x) \to f(a)$ as $x \to a$. However, the notion of the limit of a function $f(x)$ as $x \to a$ is most useful when $f(x)$ is undefined at $x = a$.

Example 3 Find $\lim\limits_{x \to 1} \left(\dfrac{x^2+x-2}{x-1} \right)$.

Since there is no definition of division by zero, the function $f(x) = \dfrac{x^2+x-2}{x-1}$

is not defined for $x = 1$.

However, when $x \neq 1$, $f(x) = \dfrac{(x+2)(x-1)}{x-1} = x+2$.

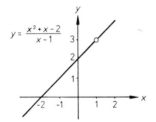

Hence we can obtain values of $f(x)$ as close as we like to 3, by taking values of x closer and closer to 1

$$\therefore \lim_{x \to 1} \left(\frac{x^2+x-2}{x-1} \right) = \lim_{x \to 1} \left(x+2 \right) = 3.$$

If x is any real number, $[x]$ is defined as the greatest integer less than or equal to x, e.g. $[3\frac{3}{4}] = 3$, $[5] = 5$ and $[-1{\cdot}7] = -2$.

Example 4 Discuss the function $f(x) = [x]$ as $x \to 1$.

For values of x close to 1, but less than 1, $[x] = 0$.
For values of x close to 1, but greater then 1, $[x] = 1$
$\therefore$ $f(x) = [x]$ approaches no single value as $x \to 1$.
Hence, although $[x]$ is defined at $x = 1$, where it takes the value 1, the limit of $[x]$ as $x \to 1$ does not exist.

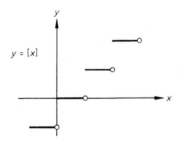

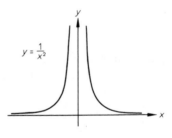

Example 5   Discuss the function $f(x) = 1/x^2$ as $x \to 0$.

As we examine values of x closer and closer to 0, $f(x)$ takes increasingly large values. Thus $f(x)$ has no limit as x tends to 0.

We may describe the behaviour of $f(x) = 1/x^2$ by writing:
$$\text{as } x \to 0, f(x) \to \infty, \text{ where } \to \infty \text{ is read 'tends to infinity'.}$$

However, since a limit is a finite number, we may *not* state that the limit of $f(x)$ is infinity. It is sometimes convenient to say that $f(x)$ is infinite at $x = 0$, but it is *not* usually considered correct to write $f(0) = \infty$.

The concept of a limit may be extended to the value a function $f(x)$ approaches as x takes increasingly large values.

Example 6   Find the limits of $f(x) = 1/x$ as $x \to \infty$ and as $x \to -\infty$.

As x takes increasingly large positive values, $f(x)$ takes smaller and smaller positive values
$\therefore$ as $x \to \infty, f(x) \to 0$ from above.
Similarly, as x takes increasingly large negative values, $f(x)$ takes smaller and smaller negative values
$\therefore$ as $x \to -\infty, f(x) \to 0$ from below.

The following properties of limits are often useful when considering more complicated functions.
If, as $x \to a, f(x) \to l$ and $g(x) \to k$, then
(a) $f(x) + g(x) \to l + k$ (b) $f(x) - g(x) \to l - k$
(c) $f(x) . g(x) \to lk$ (d) $f(x)/g(x) \to l/k \ (k \neq 0)$.

Example 7   Find the limit of $f(x) = \dfrac{3x}{x+3}$ as $x \to \infty$.

Dividing numerator and denominator by x,
$$f(x) = \frac{3x}{x+3} = \frac{3}{1+3/x}.$$
As $x \to \infty, 3/x \to 0$ $\therefore$ as $x \to \infty, f(x) \to 3$.

Exercise 5.1

1.   Find the limit as $x \to 3$ of
(a) $2x + 3$,   (b) $3 - x$,   (c) $x^2 + 1$,   (d) $1 - x^2 + x^4$.

2. Find the limit as $x \to 2$ of

(a) $\dfrac{1}{x+1}$,   (b) $1 + \dfrac{1}{x}$,   (c) $\dfrac{x-4}{x+2}$,   (d) $\dfrac{x^2-4}{x-2}$.

3. Find the limit as $h \to 0$ of

(a) $3-h+h^2$,   (b) $\dfrac{h-2h^2}{h}$,   (c) $\dfrac{3h^3+2h^2}{h^2}$,   (d) $\dfrac{h^3+1}{h+1}$.

4. Find the limit as $x \to 0$ of

(a) $\dfrac{(x+1)^2-(x-1)^2}{2x}$,   (b) $\sqrt{\left(\dfrac{8x-5x^2+3x^3}{1-\{1-2x\}}\right)}$.

5. Discuss the function $f(x) = 1-[x]$ as $x \to 1$.

6. Discuss the function $f(x) = 1/x$ as $x \to 0$.

7. Discuss the values of the following functions as $x \to k$.

(a) $f(x) = (x-k)^2$,   (b) $g(x) = \dfrac{1}{(x-k)^2}$,   (c) $h(x) = \dfrac{1}{x-k}$.

8. Find the limit as $x \to \infty$ of

(a) $\dfrac{1}{x+1}$,   (b) $\dfrac{1}{x^2}$,   (c) $\dfrac{2x}{x-1}$,   (d) $\dfrac{2x^2}{x^2-1}$.

9. Find   (a) $\displaystyle\lim_{x \to -1} \dfrac{3x^2+x-2}{2x^2+5x+3}$,   (b) $\displaystyle\lim_{x \to \infty} \dfrac{3x^2+x-2}{2x^2+5x+3}$,

(c) $\displaystyle\lim_{x \to 1} \dfrac{x^3-1}{2x^3-3x+1}$,   (d) $\displaystyle\lim_{x \to -\infty} \dfrac{x^3-1}{2x^3-3x+1}$.

5.2 The gradient of a curve

In §2.4 gradient was defined as a quantity which indicates the direction of a straight line. It was found that the gradient of the straight line passing through the points (x_1, y_1) and (x_2, y_2) is $(y_2-y_1)/(x_2-x_1)$. However, since the direction of a curve is not constant, there is no similar formula for the gradient of a curve. Instead, the gradient of the curve $y = f(x)$ at the point $x = a$ is defined to be the gradient of the tangent to the curve at that point.

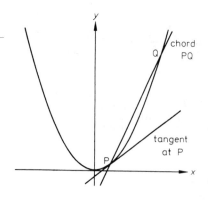

Consider the gradient of the curve $y = x^2$ at the point $P(1,1)$. We cannot find the gradient of the tangent at $(1,1)$ by the methods of Chapter 2, since we have the coordinates of only one point on it. However, we can obtain an approximate value for its gradient, by finding the gradient of a chord PQ, where Q is a point on the curve $y = x^2$ close to P. The table below gives successive approximations to the gradient of the tangent at P, obtained by taking Q closer and closer to P.

Value of x at Q	Value of y at Q	Gradient of PQ
2	4	3
1·5	2·25	2·5
1·1	1·21	2·1
1·01	1·0201	2·01
1·001	1·002001	2·001

This suggests that as Q approaches P, the values of the gradient of PQ approach 2.

To avoid the numerical work above, consider the point Q, where $x = 1+h$, $y = (1+h)^2$, then

$$\text{gradient of } PQ = \frac{(1+h)^2 - 1}{(1+h) - 1} = \frac{2h + h^2}{h} = 2 + h.$$

As $h \to 0$, $Q \to P$ and the gradient of PQ approaches that of the tangent at P.

$\therefore$ gradient of tangent at $P = \lim_{h \to 0} (\text{gradient of } PQ) = \lim_{h \to 0} (2 + h) = 2.$

This limiting process also provides a method for finding the gradient of the curve $y = x^2$ at a general point $P(x, x^2)$.
Let Q be the point $(x+h, \{x+h\}^2)$.

The gradient of PQ is then $\dfrac{(x+h)^2 - x^2}{(x+h) - x} = 2x + h$

$\therefore$ gradient of tangent at $P = \lim_{h \to 0} (2x + h) = 2x.$

Hence the gradient of the curve $y = x^2$ at any point is given by the function $2x$. This function is sometimes referred to as the *gradient function* for the curve.

In general, for the curve $y = f(x)$, if P is the point $(x, f(x))$ and Q is the point $(x+h, f(x+h))$, then the gradient of $PQ = \dfrac{f(x+h) - f(x)}{(x+h) - x} = \dfrac{f(x+h) - f(x)}{h}.$

∴   the gradient of the curve at any point is given by

$$\lim_{h \to 0} \frac{f(x+h)-f(x)}{h}, \quad \text{if this limit exists.}$$

One important application of this principle is to the motion of a particle in a straight line. The velocity of such a particle is the rate at which its displacement from a fixed point is changing.

Example 1   An object moves in a straight line so that its displacement from its starting point after t seconds is s metres, where $s = 20t - 3t^2$. If the object passes through the points P and Q when $t = a$ and $t = a+h$ respectively, find the average velocity of the object as it moves from P to Q. Deduce its velocity at the instant it passes through P.

The displacement PQ (in metres)
$$= \{20(a+h) - 3(a+h)^2\} - \{20a - 3a^2\}$$
$$= 20a + 20h - 3a^2 - 6ah - 3h^2 - 20a + 3a^2$$
$$= 20h - 6ah - 3h^2.$$

The time taken to travel from P to $Q = h$ seconds
∴   the average velocity $= (20 - 6a - 3h)\,\mathrm{m\,s^{-1}}$.
The velocity at P is the limiting value of this average velocity as $h \to 0$.
∴   the velocity of the object as it passes through P is $(20 - 6a)\,\mathrm{m\,s^{-1}}$.

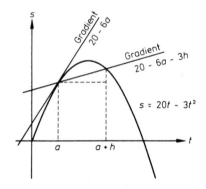

By comparing this example with the earlier work in this section, we see that the velocity of the particle at P is the same as the gradient of the graph $s = 20t - 3t^2$ at the point $t = a$.

Exercise 5.2

1.   Write down a series of approximations to the gradient of the tangent at the point $P(0, 1)$ to the curve $y = 1 - x^2$ by finding the gradient of PQ, where Q is the point on the curve given by $x = 1, 0\cdot5, 0\cdot1, 0\cdot01, 0\cdot001$.

2.   Repeat question 1 for the curve $y = 2x^2 - x + 1$.

3.   If P and Q are the points on the curve $y = x^2 + 3$ with x-coordinates 1 and $1+h$ respectively, find the gradient of PQ. Deduce the gradient of the tangent to the curve at P.

4.   Use the method of question 3 to find the gradients of the following curves at the given points
(a) $y = 3x^2$ at $(1, 3)$,   (b) $y = x - x^2$ at $(2, -2)$,
(c) $y = 2x^2 + x + 1$ at $(0, 1)$,   (d) $y = x^3$ at $(1, 1)$.

5.   Find the gradient of the line joining the points $P(x, 3x^2 - 1)$ and $Q(x + h, 3\{x + h\}^2 - 1)$ on the curve $y = 3x^2 - 1$. By letting h tend to zero, find the gradient function for this curve.

6.   Using the method of question 5 find the gradient functions for the following curves:
(a) $y = x^2 - x$,   (b) $y = 1/x$,   (c) $y = x^3$.

7.   An object moves in a straight line so that its distance from its starting point after t seconds is s metres, where $s = 4t^2$. If the object passes through the points P and Q when $t = a$ and $t = a + h$ respectively, find the average velocity of the object as it moves from P to Q. Deduce its velocity at the instant it passes through P.

8.   Repeat question 7 for (a) $s = 3t$, (b) $s = 12t - 5t^2$.

5.3 Differentiation from first principles

The process of finding the gradient function described in the previous section is called *differentiation* from first principles. For the curve $y = f(x)$ the gradient function is called the derived function or *derivative* of $f(x)$ and denoted by $f'(x)$, so that

$$f'(x) = \lim_{h \to 0} \frac{f(x+h) - f(x)}{h}.$$

For example, it has already been shown that if $f(x) = x^2$, then $f'(x) = 2x$.
    An alternative expression for the derivative of the function $y = f(x)$ is obtained by using the Greek letter δ, 'delta', to mean 'a small increase in'.

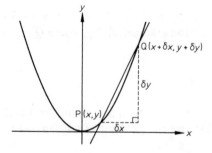

Consider the points $P(x, y)$ and $Q(x + \delta x, y + \delta y)$ on the curve $y = x^2$, where δx is a small increase in x and δy is the corresponding small increase in y.

At P $\qquad y = x^2$

At Q $\qquad y + \delta y = (x + \delta x)^2 = x^2 + 2x\delta x + (\delta x)^2$

$\therefore \qquad\qquad \delta y = 2x\delta x + (\delta x)^2.$

The gradient of $PQ = \dfrac{\delta y}{\delta x} = 2x + \delta x.$

As $\delta x \to 0$, $\dfrac{\delta y}{\delta x} \to 2x$

$\therefore$ the gradient of the tangent at P is $2x$.

In this notation the derivative of y with respect to x is denoted by $\dfrac{dy}{dx}$, so that

$$\frac{dy}{dx} = \lim_{\delta x \to 0} \frac{\delta y}{\delta x}.$$

[Note that in this context d, dy, dx have no separate meaning.]

Both $f'(x)$ and dy/dx are widely used, so differentiation results may be written down using either notation.

Thus,   if $f(x) = x^2$,   then $f'(x) = 2x$

and,   if $y = x^2$,   then $\dfrac{dy}{dx} = 2x.$

We can also think of d/dx as a symbol indicating the operation of differentiation, writing $\dfrac{d}{dx}(x^2) = 2x.$

There is no need to refer to the graph of a function to carry out the process of differentiation.

Example 1   Differentiate from first principles the function $y = x^3$.

Let δx be a small increase in x and let δy be the corresponding small increase in y.

$\qquad\qquad y = x^3$

$\Rightarrow y + \delta y = (x + \delta x)^3 = x^3 + 3x^2\delta x + 3x(\delta x)^2 + (\delta x)^3$

$\Rightarrow \qquad \delta y = 3x^2\delta x + 3x(\delta x)^2 + (\delta x)^3$

$\Rightarrow \qquad \dfrac{\delta y}{\delta x} = 3x^2 + 3x\delta x + (\delta x)^2$

$\therefore \qquad \dfrac{dy}{dx} = \lim_{\delta x \to 0} \dfrac{\delta y}{\delta x} = 3x^2.$

Alternative method

Let
$$f(x) = x^3$$
then
$$f(x+h) = (x+h)^3 = x^3 + 3x^2h + 3xh^2 + h^3$$
$$f(x+h) - f(x) = 3x^2h + 3xh^2 + h^3$$

$$\frac{f(x+h) - f(x)}{h} = 3x^2 + 3xh + h^2$$

$$\therefore \quad f'(x) = \lim_{h \to 0} \frac{f(x+h) - f(x)}{h} = 3x^2.$$

Example 2   Differentiate from first principles $y = 1/x$.

$$y = \frac{1}{x}$$

$$\Rightarrow y + \delta y = \frac{1}{x + \delta x}$$

$$\Rightarrow \quad \delta y = \frac{1}{x + \delta x} - \frac{1}{x} = \frac{x - (x + \delta x)}{x(x + \delta x)} = -\frac{\delta x}{x(x + \delta x)}$$

$$\Rightarrow \quad \frac{\delta y}{\delta x} = -\frac{1}{x(x + \delta x)}$$

$$\therefore \quad \frac{dy}{dx} = \lim_{\delta x \to 0} \frac{\delta y}{\delta x} = -\frac{1}{x^2}.$$

Exercise 5.3

1.   If P is the point (x, y) and Q the point $(x + \delta x, y + \delta y)$ on the curve $y = x^2 + 2$, find the gradient of PQ and hence the gradient of the curve at P.

2.   Repeat question 1 for the curves
(a) $y = 4 - x^2$,   (b) $y = x^2 + 2x$,   (c) $y = \frac{2}{x} + 3$.

In the remaining questions differentiate the given functions from first principles.

3.   (a) $2x^2$ (b) $5x^2$ (c) $2x^2 - 1$.

4.   (a) x (b) $4 - x$ (c) $x + x^2$.

5.   (a) $3x - 1$ (b) $\frac{1}{2}x + 5$ (c) 7.

6.   (a) $-x^3$ (b) $2x^3$ (c) x^4.

7.   (a) $x - \dfrac{1}{x}$ (b) $\dfrac{1}{x^2}$ (c) $\dfrac{1}{x^3}$.

5.4 Differentiation rules

From the results obtained in Exercise 5.3 a table of derivatives of powers of x can be made.

y	$\dfrac{1}{x^3}$	$\dfrac{1}{x^2}$	$\dfrac{1}{x}$	1	x	x^2	x^3	x^4	...
$\dfrac{dy}{dx}$	$-\dfrac{3}{x^4}$	$-\dfrac{2}{x^3}$	$-\dfrac{1}{x^2}$	0	1	$2x$	$3x^2$	$4x^3$	...

A pattern emerges when we rewrite the table:

y	x^{-3}	x^{-2}	x^{-1}	x^0	x^1	x^2	x^3	x^4
$\dfrac{dy}{dx}$	$-3x^{-4}$	$-2x^{-3}$	$-1x^{-2}$	$0x^{-1}$	$1x^0$	$2x^1$	$3x^2$	$4x^3$

It now seems reasonable to assume that for any integer n,

$$\text{if } y = x^n, \quad \text{then } \frac{dy}{dx} = nx^{n-1}.$$

[A formal proof of this statement when n is a positive integer is given in §16.6. Negative values of n are considered in §13.1.]

Some readers may prefer to think of this rule for differentiating powers of x in words:

'multiply by the index and subtract 1 from it'.

Example 1 Find $\dfrac{dy}{dx}$ if (a) $y = x^7$ (b) $y = \dfrac{1}{x^7}$.

(a) If $y = x^7$, then $\dfrac{dy}{dx} = 7x^6$.

[Multiply by the index 7, then subtract 1 to obtain the new index 6.]

(b) If $y = \dfrac{1}{x^7} = x^{-7}$, then $\dfrac{dy}{dx} = -7x^{-8} = -\dfrac{7}{x^8}$.

[In this case the index is -7 and subtracting 1 gives the new index -8.]

It was also found in Exercise 5.3 that, for instance, the derivative of $5x^2$ is $10x$, which is 5 times the derivative of x^2,

i.e. if $y = 5u$ where $u = x^2$, then $\dfrac{dy}{dx} = 10x = 5\dfrac{du}{dx}$.

Other examples show that the derivative of a sum is the same as the sum of the separate derivatives, e.g.   if $y = x^2 + x = u + v$ where $u = x^2$ and $v = x$,

$$\text{then } \frac{dy}{dx} = 2x + 1 = \frac{du}{dx} + \frac{dv}{dx}.$$

These are examples of two general rules which we now prove using the properties of limits.

(1) If $y = ku$, where k is a constant and u is a function of x, then $\dfrac{dy}{dx} = k\dfrac{du}{dx}$.

Proof   Let δx be a small increase in x and let δu and δy be the corresponding increases in u and y.

$$y = ku$$
$$\Rightarrow y + \delta y = k(u + \delta u) = ku + k\delta u$$
$$\Rightarrow \qquad \delta y = k\delta u$$

$$\Rightarrow \qquad \frac{\delta y}{\delta x} = k\frac{\delta u}{\delta x}$$

$$\therefore \quad \frac{dy}{dx} = \lim_{\delta x \to 0}\frac{\delta y}{\delta x} = \lim_{\delta x \to 0} k\frac{\delta u}{\delta x} = k\lim_{\delta x \to 0}\frac{\delta u}{\delta x} = k\frac{du}{dx}.$$

(2) If $y = u + v$, where u and v are functions of x, then

$$\frac{dy}{dx} = \frac{du}{dx} + \frac{dv}{dx}.$$

Proof   Let δx be a small increase in x and let δu, δv and δy be the corresponding increases in u, v and y.

$$y = u + v$$
$$\Rightarrow y + \delta y = (u + \delta u) + (v + \delta v)$$
$$\Rightarrow \qquad \delta y = \delta u + \delta v$$

$$\Rightarrow \qquad \frac{\delta y}{\delta x} = \frac{\delta u}{\delta x} + \frac{\delta v}{\delta x}$$

$$\therefore \quad \frac{dy}{dx} = \lim_{\delta x \to 0}\frac{\delta y}{\delta x} = \lim_{\delta x \to 0}\left(\frac{\delta u}{\delta x} + \frac{\delta v}{\delta x}\right) = \lim_{\delta x \to 0}\frac{\delta u}{\delta x} + \lim_{\delta x \to 0}\frac{\delta v}{\delta x} = \frac{du}{dx} + \frac{dv}{dx}.$$

These results are used in the following examples.

Example 2   Find $\dfrac{dy}{dx}$ if (a) $y = x^3 - 4x^2$, (b) $y = 6 - \dfrac{7}{x^2}$.

(a) If $y = x^3 - 4x^2$, then $\dfrac{dy}{dx} = 3x^2 - 4 \cdot 2x = 3x^2 - 8x$.

(b) If $y = 6 - \dfrac{7}{x^2} = 6 - 7x^{-2}$, then $\dfrac{dy}{dx} = 0 - 7(-2x^{-3}) = \dfrac{14}{x^3}$.

It is necessary to rearrange some functions in order to differentiate them.

Example 3   Find $f'(x)$ when (a) $f(x) = x(x-3)$ (b) $f(x) = \dfrac{x^2-4}{x}$.

(a) $f(x) = x(x-3) = x^2 - 3x$ $\therefore$ $f'(x) = 2x - 3$

(b) $f(x) = \dfrac{x^2+4}{x} = x + \dfrac{4}{x}$ $\therefore$ $f'(x) = 1 - \dfrac{4}{x^2}$.

[Note that differentiating (x^2+4) and x separately then dividing would not give the derivative of the quotient.]

Example 4   Find the gradient of the curve $y = 2x^2 + 3x - 1$ at the point $(1,4)$. Hence find the equation of the tangent to the curve at that point.

If $y = 2x^2 + 3x - 1$, $\dfrac{dy}{dx} = 4x + 3$ $\therefore$ when $x = 1$, $\dfrac{dy}{dx} = 7$.

Hence the gradient of the curve at the point $(1,4)$ is 7. Thus the tangent at this point is the straight line passing through $(1,4)$ with gradient 7

$\therefore$ its equation is $y - 4 = 7(x-1)$,   i.e. $y = 7x - 3$.

In §5.2 we saw that the process of differentiation could be used to find the instantaneous velocity of a moving object.

Example 5   A particle moves along a straight line such that after t seconds its displacement from a fixed point is s metres, where $s = 10t^2 - t^3$. Find (a) an expression for the velocity, $v\,\text{m s}^{-1}$, after t seconds, (b) the velocity after 2 seconds and after 8 seconds.

(a) Since $s = 10t^2 - t^3$, $v = \dfrac{ds}{dt} = 20t - 3t^2$

$\therefore$ the velocity after t seconds is given by $v = 20t - 3t^2$.

(b) When $t = 2$, $v = 20.2 - 3.4 = 28$
$\therefore$ the velocity after 2 seconds is $28\,\text{m s}^{-1}$.

When $t = 8$, $v = 20.8 - 3.64 = 160 - 192 = -32$
$\therefore$ the velocity after 8 seconds is $-32\,\text{m s}^{-1}$.

[The difference in sign shows that the direction of motion after 8 seconds is opposite to the direction after 2 seconds.]

It will be proved in §13.2 that the rule $\dfrac{d}{dx}(x^n) = nx^{n-1}$ can be applied to fractional powers of x.

Example 6   Differentiate (a) $x^{1/4}$, (b) $x^{-5/2}$, (c) $\sqrt{x}$, (d) $\dfrac{1}{\sqrt[3]{x}}$.

(a) If $y = x^{1/4}$, then $\dfrac{dy}{dx} = \tfrac{1}{4}x^{-3/4}$

(b) If $y = x^{-5/2}$, then $\dfrac{dy}{dx} = -\dfrac{5}{2}x^{-7/2}$

(c) If $y = \sqrt{x} = x^{1/2}$, then $\dfrac{dy}{dx} = \tfrac{1}{2}x^{-1/2} = \dfrac{1}{2\sqrt{x}}$

(d) If $y = \dfrac{1}{\sqrt[3]{x}} = x^{-1/3}$, then $\dfrac{dy}{dx} = -\tfrac{1}{3}x^{-4/3} = -\dfrac{1}{3\sqrt[3]{(x^4)}}$

[For revision of surds and indices see §1.5.]

Exercise 5.4

Differentiate the following functions with respect to x:

1. (a) x^5,   (b) x^{12},   (c) $3x^4$,   (d) -5

2. (a) $2x+7$,   (b) $3x^2-4x$,   (c) $2x^3-7x^6$.

3. (a) $3+4x-x^2$,   (b) $\tfrac{1}{2}x^3+\tfrac{1}{3}x^2$,   (c) $5x-4x^4$.

4. (a) $x(2x+3)$,   (b) $(3x-1)^2$,   (c) $x(x+2)^2$.

5. (a) x^{-2},   (b) $1/x^4$,   (c) $1/x^9$,   (d) $2/x^3$.

6. (a) $x^2-\dfrac{1}{x}$,   (b) $x^4+\dfrac{4}{x^2}$,   (c) $\dfrac{1}{2x^2}-\dfrac{1}{3x^3}$.

7. (a) $\dfrac{1+x^2}{x}$,   (b) $\dfrac{2x-5}{x^3}$,   (c) $\dfrac{(x-1)(x+2)}{x}$.

8. Find $f'(x)$, $f'(0)$ and $f'(-2)$ given that
(a) $f(x) = \tfrac{1}{2}(x^3+1)^2$,   (b) $f(x) = 3(x+1)^3$.

9. Find $f'(x)$, $f'(2)$ and $f'(-1)$ given that

(a) $f(x) = \dfrac{1}{x^2}(2x^3+1)$,   (b) $f(x) = \left(\dfrac{x^2+1}{x}\right)^2$.

10. Find the gradients of the following curves for the given values of x

(a) $y = 3x^2-x+2$; $x = 1$,   (b) $y = x^3-2x^2+5x$; $x = -2$,

(c) $y = \dfrac{x^4-1}{2x^3}$; $x = -1$,   (d) $y = \dfrac{(x+2)(1-2x)}{x^2}$; $x = 2$.

11. Find the equations of the tangents to the following curves at the points with given x-coordinates

(a) $y = x^2-3x$; $x = 2$,   (b) $y = 2+x-5x^2$; $x = -1$,

(c) $y = x^3 - 3x + 2$; $x = 0$,

(d) $y = (x^2 - 3)(2x^2 + 5)$; $x = -\frac{1}{2}$,

(e) $y = 1 - \dfrac{1}{x}$; $x = \frac{1}{2}$,

(f) $y = \dfrac{1 - x^3}{x^4}$; $x = 2$.

12. A particle moves along a straight line such that after t seconds its displacement from a fixed point is s metres, where $s = 5t^2 + 8$. Find (a) an expression for the velocity after t seconds, (b) the velocity after 3 seconds and after 5 seconds.

13. Repeat question 12 for $s = 27t - 3t^2$.

14. Given that $A = \pi r^2$, find $\dfrac{dA}{dr}$ when $r = 2$.

15. Given that $pv^3 = 54$, find $\dfrac{dp}{dv}$ when $v = 3$.

16. The gradient of the curve $y = x^2 + ax + b$ at the point $(2, 6)$ is 7. Find the values of a and b.

17. The gradient of the curve $y = ax^2 + b/x^2$ at the point $(\frac{1}{2}, 2\frac{1}{2})$ is -6. Find the values of a and b.

18. Find $\dfrac{dy}{dx}$ if

(a) $y = x^{1/3}$,

(b) $y = x^{3/2}$,

(c) $y = x^{-1/4}$,

(d) $y = \sqrt[5]{x}$,

(e) $y = 1/\sqrt{x}$,

(f) $y = \sqrt[3]{(8x)}$.

19. Differentiate with respect to x

(a) $x^{1/2} + x^{-3/2}$,

(b) $(2x - 1)\sqrt{x}$,

(c) $(1 + \sqrt{x})^2$,

(d) $\dfrac{x - 1}{\sqrt{x}}$,

(e) $\dfrac{x^3 + x^2}{\sqrt{x}}$,

(f) $\left(\sqrt{x} + \dfrac{1}{\sqrt{x}}\right)^3$.

20. Find the equation of the tangent to the curve $y = (\sqrt{x} + 1)\sqrt{x}$ at the point where $x = 4$.

5.5 Elementary curve sketching

In §3.1 it was shown that graphs of the form $y = ax^2 + bx + c$ all have the same general shape. Thus a quick sketch of a curve such as $y = x^2 - 5x + 4$ can be made by finding the points of intersection with the axes and checking the behaviour of the graph for large positive and large negative values of x.

$x = 0 \Rightarrow y = 4$ $\therefore$ the curve cuts the y-axis at $(0, 4)$

$y = 0 \Leftrightarrow x^2 - 5x + 4 = 0 \Leftrightarrow (x - 1)(x - 4) = 0$

$\Leftrightarrow x = 1$ or $x = 4$

$\therefore$ the curve cuts the x-axis at the points $(1, 0)$ and $(4, 0)$.

As $x \to \infty$, $y \to \infty$ and as $x \to -\infty$, $y \to \infty$.

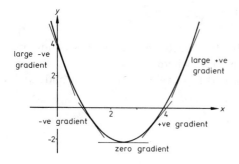

It is clear from the sketch that as x increases, the gradient of the curve moves through negative values to zero and then through positive values.

Differentiating $y = x^2 - 5x + 4$, $\dfrac{dy}{dx} = 2x - 5$.

Hence $2x - 5$ is the gradient function for the curve. As x increases, this does indeed behave as described above.

$$\frac{dy}{dx} = 0 \Leftrightarrow 2x - 5 = 0 \Leftrightarrow x = 2\tfrac{1}{2}$$

$\therefore$ the lowest point in the sketch occurs at $x = 2\tfrac{1}{2}$.

The graph of a more complicated function $y = f(x)$ is shown below.

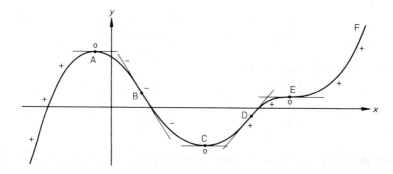

The sign of the gradient as x increases is shown in the diagram. At A, C and E the tangents to the curve are parallel to the x-axis and the gradient is zero. At these points, which are called *stationary points*, y is said to have a *stationary value*. The point A is a *maximum point* and y takes a *maximum value* at A. Similarly, C is called a *minimum point* and y has a *minimum value* there. It is clear from the sketch that these are local maximum and minimum values, since, for instance, the value of y at F is greater than its value at A.

  A and C are both called *turning points*, but although y has a stationary value at E, E is not a turning point. The gradient of the curve is zero at E, but does not change sign as the curve passes through E. Such points are called *points of inflexion*. In general, a point of inflexion is a point at which a curve stops bending in one direction and starts bending the other way. In the diagram B, D and E are all points of inflexion and, as shown, the tangents at these points cross the curve.

> Summarising, a point at which $dy/dx = 0$ is
> (i) a maximum point, if the sign of dy/dx changes from $+$ to $-$
> (ii) a minimum point, if the sign of dy/dx changes from $-$ to $+$
> (iii) a point of inflexion, if the sign of dy/dx does not change.

This provides a new approach to the sketching of graphs.

Example 1 Sketch the graph of $y = 2 + x - x^2$.

$x = 0 \Rightarrow y = 2$
$\therefore$ the curve cuts the y-axis at the point $(0, 2)$.
$y = 0 \Leftrightarrow 2 + x - x^2 = 0 \Leftrightarrow (2 - x)(1 + x) = 0$
$\qquad \Leftrightarrow x = -1 \quad$ or $\quad x = 2$
$\therefore$ the curve cuts the x-axis at the points $(-1, 0)$ and $(2, 0)$.

Differentiating $\quad \dfrac{dy}{dx} = 1 - 2x$

$\therefore \quad \dfrac{dy}{dx} = 0 \Leftrightarrow 1 - 2x = 0 \Leftrightarrow x = \frac{1}{2}$

When $x < \frac{1}{2}$, $\dfrac{dy}{dx} > 0$ and when $x > \frac{1}{2}$, $\dfrac{dy}{dx} < 0$

$\therefore \quad$ when $x = \frac{1}{2}$, y takes the maximum value of $2\frac{1}{4}$.

As $x \to \infty$, $y \to -\infty$ and as $x \to -\infty$, $y \to -\infty$.

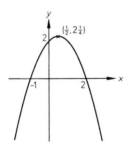

We are now able to sketch the curve and label the maximum point $(\frac{1}{2}, 2\frac{1}{4})$.

Example 2 Sketch the curve $y = x^3 - 6x^2 + 9x$.

$x = 0 \Rightarrow y = 0$
$\therefore$ the curve cuts the y-axis at the origin $(0, 0)$.
$y = 0 \Leftrightarrow x^3 - 6x^2 + 9x = 0 \Leftrightarrow x(x - 3)^2 = 0$
$\qquad \Leftrightarrow x = 0 \quad$ or $\quad x = 3$
$\therefore$ the curve cuts the x-axis at the points $(0, 0)$ and $(3, 0)$.

Differentiating $\dfrac{dy}{dx} = 3x^2 - 12x + 9 = 3(x-1)(x-3)$

$\dfrac{dy}{dx} = 0 \Leftrightarrow (x-1)(x-3) = 0 \Leftrightarrow x = 1 \quad \text{or} \quad x = 3$

[We now determine the nature of the stationary points at $x = 1$ and $x = 3$, by drawing up a table showing the sign of dy/dx when x is less than 1, when x lies between 1 and 3 and when x is greater than 3.]

	$x < 1$	$1 < x < 3$	$x > 3$
$x - 1$	$-$	$+$	$+$
$x - 3$	$-$	$-$	$+$
dy/dx	$+$	$-$	$+$

$\therefore$ when $x = 1$, there is a maximum point $(1, 4)$
when $x = 3$, there is a minimum point $(3, 0)$

As $x \to \infty, y \to \infty$ and as $x \to -\infty, y \to -\infty$.

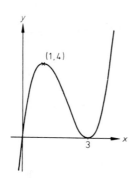

We can now sketch the curve $y = x^3 - 6x^2 + 9x = x(x-3)^2$ and we see that the repeated factor $(x-3)$ leads to a turning point at $x = 3$.

Example 3 Sketch the curve $y = (x+1)(x-3)^3$.

$x = 0 \Rightarrow y = -27$
$\therefore$ the curve cuts the y-axis at the point $(0, -27)$.

$y = 0 \Leftrightarrow (x+1)(x-3)^3 = 0 \Leftrightarrow x = -1 \quad \text{or} \quad x = 3$
$\therefore$ the curve cuts the x-axis at the points $(-1, 0)$ and $(3, 0)$.

By multiplication $\qquad y = x^4 - 8x^3 + 18x^2 - 27$

Differentiating $\qquad \dfrac{dy}{dx} = 4x^3 - 24x^2 + 36x$

$\qquad\qquad\qquad = 4x(x^2 - 6x + 9) = 4x(x-3)^2$.

$\dfrac{dy}{dx} = 0 \Leftrightarrow 4x(x-3)^2 = 0 \Leftrightarrow x = 0 \quad \text{or} \quad x = 3$

	$x < 0$	$0 < x < 3$	$x > 3$
x	$-$	$+$	$+$
$(x-3)^2$	$+$	$+$	$+$
dy/dx	$-$	$+$	$+$

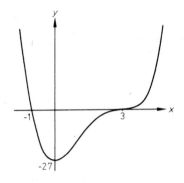

$\therefore$ when $x = 0$, there is a minimum point $(0, -27)$ and when $x = 3$, there is a point of inflexion $(3,0)$. As $x \to \infty$, $y \to \infty$ and as $x \to -\infty$, $y \to \infty$.

Example 4 Sketch the graph $y = \dfrac{2x-5}{x}$.

When $x = 0$ the value of y is undefined
$\therefore$ the curve does not cut the y-axis.

$y = 0 \Leftrightarrow 2x - 5 = 0 \Leftrightarrow x = 2\frac{1}{2}$
$\therefore$ the curve cuts the x-axis at the point $(2\frac{1}{2}, 0)$.

By division $\qquad y = 2 - \dfrac{5}{x}$.

Differentiating $\quad \dfrac{dy}{dx} = \dfrac{5}{x^2}$.

Hence the gradient of the curve is always positive and there are no turning points.
As $x \to 0$ from above, $y \to -\infty$ and as $x \to 0$ from below, $y \to \infty$.
As $x \to \infty$, $y \to 2$ from below and as $x \to -\infty$, $y \to 2$ from above.

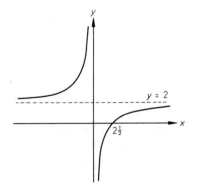

A straight line which a curve approaches more and more closely but never touches is called an *asymptote* to the curve. The curve $y = (2x-5)/x$ has two asymptotes, the y-axis and the line $y = 2$.

Example 5   Sketch the graph $y = x + \dfrac{1}{x}$.

When $x = 0$, the value of y is undefined.

$$y = 0 \Leftrightarrow x + \frac{1}{x} = 0 \Leftrightarrow x^2 + 1 = 0$$

$\therefore$ there is no real value of x for which $y = 0$.

Hence the curve cuts neither of the coordinate axes.

Differentiating $\dfrac{dy}{dx} = 1 - \dfrac{1}{x^2} = \dfrac{x^2 - 1}{x^2}$

$\dfrac{dy}{dx} = 0 \Leftrightarrow x^2 - 1 = 0 \Leftrightarrow x = -1$ or $x = 1$.

When $x = -1$, $y = -2$ and when $x = 1$, $y = 2$.

When x is just less than -1, $\dfrac{dy}{dx} > 0$ and when x is just greater than -1, $\dfrac{dy}{dx} < 0$ $\therefore$ the point $(-1, -2)$ is a maximum point.

When x is just less than 1, $\dfrac{dy}{dx} < 0$ and when x is just greater than 1, $\dfrac{dy}{dx} > 0$ $\therefore$ the point $(1, 2)$ is a minimum point.

As $x \to 0$ from above, $y \to \infty$ and as $x \to 0$ from below, $y \to -\infty$
$\therefore$ the y-axis is an asymptote to the curve.

As $x \to \infty$, $y \to \infty$ and as $x \to -\infty$, $y \to -\infty$.

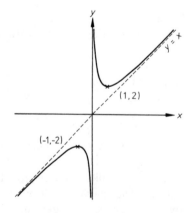

Although we are now ready to sketch the curve, a more accurate sketch will be obtained by noting the fact that as $x \to \pm\infty$, the curve $y = x + \dfrac{1}{x}$ draws closer and closer to the line $y = x$. Hence the line $y = x$ is an asymptote to the curve.

To summarise, here is a list of the steps to take when sketching a curve.
(1) Find the points of intersection with the y-axis by putting $x = 0$.
(2) Find, if possible, the points of intersection with the x-axis by putting $y = 0$.
(3) Find the points at which $dy/dx = 0$.

(4) Determine the nature of these points by considering the sign of dy/dx.

(5) Note any values of x for which y is not defined and examine the behaviour of y as x approaches such values.

(6) Consider the values of y as $x \to \infty$ and as $x \to -\infty$.

(7) If further checks are needed, determine the sign of y for all values of x.

It is important to remember that if differentiating and solving $dy/dx = 0$ is difficult, it is often still possible to sketch a curve fairly quickly using the remaining techniques.

Exercise 5.5

Sketch the following curves showing clearly on each sketch the nature of any stationary points.

1. $y = x^2 - 2x - 3$.

2. $y = 2 - 3x - 2x^2$.

3. $y = (x-1)(x+2)$.

4. $y = 4x^2 - 4x + 3$.

5. $y = x^2(x-3)$.

6. $y = x^3 + 1$.

7. $y = 6x^2 - 12x - x^3$.

8. $y = (x+2)^2(2x-5)$.

9. $y = 2x + x^2 - 4x^3$.

10. $y = 8x^3 + 9x^2 - 15x - 2$.

11. $y = x^4(x+5)$.

12. $y = 4x^3 - 3x^4$.

13. $y = \dfrac{x+1}{x}$.

14. $y = \dfrac{2-3x}{x}$.

15. $y = 9x^3 - 45x^2 + 48x - 11$.

16. $y = x^4 - 2x^3 - 5x^2 + 6$.

17. $y = 4x + \dfrac{9}{x}$.

18. $y = \dfrac{x^3+4}{x^2}$.

19. $y = 2x^5 - 15x^4 + 30x^3 - 64$.

20. $y = x - \sqrt{x} \ (x \geqslant 0)$.

21. The curve $y = x^3 + ax^2 + b$ has a minimum point at $(4, -11)$. Find the coordinates of the maximum point on the curve.

22. Given that the curve $y = x^3 + px^2 + qx + r$ passes through the point $(1,1)$ and has turning points where $x = -1$ and $x = 3$, find the values of p, q and r.

23. Find the conditions that the curve $y = x^3 + ax^2 + bx + c$ should have (i) two distinct stationary points, (ii) one stationary point, (iii) no stationary points.

24. Write down the equations of the two asymptotes to the curve $y = (x^2 + k)/x$, where $k \neq 0$. Sketch the curve for (i) $k > 0$, (ii) $k < 0$.

5.6 The second derivative

For the function $y = f(x)$ the *first derivative* is $f'(x)$ or dy/dx. The *second derivative* is obtained by differentiating again and is written $f''(x)$ or $\dfrac{d^2 y}{dx^2}$, which is shorthand for $\dfrac{d}{dx}\left(\dfrac{dy}{dx}\right)$.

Example 1 If $y = 3x + 2 + \dfrac{1}{x}$, find $\dfrac{d^2 y}{dx^2}$.

Rearranging $y = 3x + 2 + x^{-1}$

Differentiating $\dfrac{dy}{dx} = 3 - x^{-2} = 3 - \dfrac{1}{x^2}$

and $\dfrac{d^2 y}{dx^2} = -(-2)x^{-3} = \dfrac{2}{x^3}$.

Example 2 If $f(x) = x^4 - 6x^2 + 7$, find $f(1)$, $f'(1)$ and $f''(1)$.

$\begin{aligned}
f(x) &= x^4 - 6x^2 + 7 & \therefore \quad f(1) &= 1 - 6 + 7 = 2 \\
f'(x) &= 4x^3 - 12x & \therefore \quad f'(1) &= 4 - 12 = -8 \\
f''(x) &= 12x^2 - 12 & \therefore \quad f''(1) &= 12 - 12 = 0.
\end{aligned}$

If a particle moves in a straight line so that at time t its displacement from a fixed point is s, then the first derivative ds/dt gives the velocity v after t seconds. The second derivative $d^2 s/dt^2$ is also the first derivative of v, dv/dt. Hence it is a measure of the rate at which the velocity is changing, i.e. it represents the acceleration of the particle and we write

$$a = \frac{dv}{dt} = \frac{d^2 s}{dt^2}.$$

Example 3 If a particle moves in a straight line, so that after t seconds its displacement from a fixed point is s metres, where $s = 8t^2 - t^4$, find its velocity and acceleration after 1 second.

Differentiating with respect to t: $v = \dfrac{ds}{dt} = 16t - 4t^3$

$$a = \frac{dv}{dt} = 16 - 12t^2$$

When $t = 1$, $v = 12$ and $a = 4$.

∴   after 1 second the velocity of the particle is $12\,\mathrm{m\,s^{-1}}$ and its acceleration is $4\,\mathrm{m\,s^{-2}}$.

In general, the second derivative, $d^2 y/dx^2$, can be used to examine how the gradient function, dy/dx, changes as x increases.

For the curve $y = 2 + x - x^2$ sketched in Example 1, §5.5,

$$\frac{dy}{dx} = 1 - 2x \quad \text{and} \quad \frac{d^2y}{dx^2} = -2.$$

This means that the graph of this gradient function is a straight line, gradient -2.

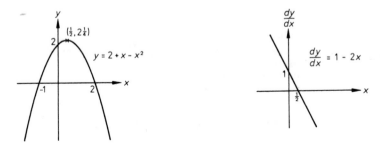

The turning point on the curve corresponds to the point at which the graph of dy/dx cuts the x-axis.

The diagrams below show the graph of a function $y = f(x)$ and the corresponding graph of the gradient function dy/dx or $f'(x)$. The gradient function for this second graph is the second derivative d^2y/dx^2 or $f''(x)$.

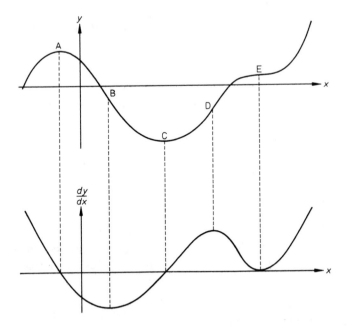

At the maximum point A, the gradient of the curve is decreasing through zero and d^2y/dx^2 is negative. At the minimum point C, the gradient of the curve is increasing through zero and d^2y/dx^2 is positive. Points of inflexion, such as B, D and E, are points where dy/dx reaches either a maximum or a minimum value.

Thus at these points $d^2y/dx^2 = 0$. However, $d^2y/dx^2 = 0$ does not necessarily imply the presence of a point of inflexion.

Summarising, a point where $\dfrac{dy}{dx} = 0$ is a maximum point if $\dfrac{d^2y}{dx^2} < 0$, a minimum point if $\dfrac{d^2y}{dx^2} > 0$. When $\dfrac{d^2y}{dx^2} = 0$, further investigation is needed.

Example 4   Sketch the curve $y = 5x^4 - x^5$.

$x = 0 \Rightarrow y = 0$

$\therefore$ the curve cuts the y-axis at the origin $(0,0)$.

$y = 0 \Leftrightarrow 5x^4 - x^5 = 0 \Leftrightarrow x^4(5-x) = 0$

$\qquad\qquad\qquad \Leftrightarrow x = 0$ or $x = 5$

$\therefore$ the curve cuts the x-axis at the points $(0,0)$ and $(5,0)$.

$\dfrac{dy}{dx} = 20x^3 - 5x^4 = 5x^3(4-x)$, $\dfrac{d^2y}{dx^2} = 60x^2 - 20x^3 = 20x^2(3-x)$

$\therefore \quad \dfrac{dy}{dx} = 0 \Leftrightarrow 5x^3(4-x) = 0 \Leftrightarrow x = 0$ or $x = 4$.

When $x = 4$, $y = 256$ and $\dfrac{d^2y}{dx^2} < 0$ $\therefore$ $(4, 256)$ is a maximum point.

When $x = 0$, $y = 0$ and $\dfrac{d^2y}{dx^2} = 0$ $\therefore$ to determine the nature of this point

we must consider the sign of $\dfrac{dy}{dx}$.

When x is just less than 0, $\dfrac{dy}{dx} < 0$ and when x is just greater than 0,

$\dfrac{dy}{dx} > 0$ $\therefore$ $(0,0)$ is a minimum point.

As $x \to \infty$, $y \to -\infty$ and as $x \to -\infty$, $y \to \infty$.

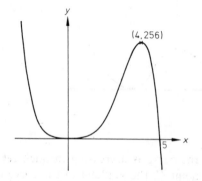

The sketch shows that there must be a point of inflexion between $x = 0$ and $x = 4$.

$\dfrac{d^2y}{dx^2} = 0 \Rightarrow x = 0$ or $x = 3$

and when $x = 3$, $y = 162$

$\therefore$ the coordinates of the point of inflexion are $(3, 162)$.

Exercise 5.6

In questions 1 to 6 find $\dfrac{d^2y}{dx^2}$.

1. $y = 3x^2 + 4x - 1.$

2. $y = 5x^3 - x^2 - x.$

3. $y = x^4 - 2x^3 + \dfrac{1}{x}.$

4. $y = \dfrac{(x^2 - 4)(x + 1)}{x^2}.$

5. $y = \sqrt{x} + \dfrac{1}{\sqrt{x}}.$

6. $y = \dfrac{(x - 2)(2x + 3)}{\sqrt{x}}.$

Sketch the following curves using the second derivative, where possible, to determine the nature of any stationary points.

7. $y = x^3 - 3x.$

8. $y = x^4 - 2x^2 + 1.$

9. $y = x^4 - 2x^3.$

10. $y = 4x - 3 + \dfrac{1}{x}.$

11. $y = 2x^3 - 9x^2 + 12x - 3.$

12. $y = 3x^4 - 4x^3 - 12x^2 + 5.$

13. In each of the following cases find $f(-1), f'(-1)$ and $f''(-1)$
(a) $f(x) = x^5 + 6x^4 - x^2,$ (b) $f(x) = (x^3 - 2)/x.$

14. If $f(x) = x^4 - 4x^3 + 16x - 16,$ show that $f(2) = f'(2) = f''(2) = 0.$ Sketch the graph of the function showing any turning points and points of inflexion.

15. Show that the curve $y = x^3 - 3x^2 + 6x - 4$ has one point of inflexion and find the gradient of the curve at this point. Sketch the curve.

16. A particle moves along a straight line so that after t seconds its displacement from a fixed point A on the line is s metres. If $s = 3t^2(3 - t)$, find (a) expressions for the velocity, v ms^{-1}, and the acceleration, a ms^{-2}, after t seconds, (b) the initial velocity and acceleration of the particle, (c) the velocity and acceleration after 3 seconds.

17. Repeat question 16 for $s = 8t^3 - 33t^2 + 27t.$

18. The curve $y = x^4 + ax^2 + bx + c$ passes through the point $(-1, 16)$ and at that point $\dfrac{d^2y}{dx^2} = -\dfrac{dy}{dx} = 16.$ Find the values of a, b, c and sketch the curve.

19. Sketch the following curves, labelling any turning points and points of inflexion.

(a) $y = x^2 + \dfrac{1}{x},$ (b) $y = \dfrac{1}{x} - \dfrac{1}{x^2},$ (c) $y = \dfrac{1}{x} - \dfrac{1}{x^3}.$

20. Sketch the curve $y = x^4 + 4x$ and show that d^2y/dx^2 is zero at a point which is neither a turning point nor a point of inflexion.

Exercise 5.7   (*miscellaneous*)

1.   Find the limit as $h \to 0$ of

(a) $2 - 3h + 5h^2$,   (b) $\dfrac{h^2 - 4h}{h}$,   (c) $\dfrac{1 - h^2}{1 - 2h}$.

2.   Find the limit as $x \to \infty$ of

(a) $\dfrac{4}{2x - 3}$,   (b) $\dfrac{2x^2 + 1}{x^2 - 7}$,   (c) $\dfrac{x - 1}{x^2 - 3x + 5}$.

3.   (a) Prove that, as x tends to zero, the limit of the expression

$\dfrac{(2 + 3x)^2 - 4(1 + x)^2}{6x}$ is $\dfrac{2}{3}$.

(b) Find the limit, as n tends to infinity, of

$\dfrac{[\sqrt{(4n + 3)} + 2\sqrt{(1 + n)}]}{\sqrt{n}}$.   (AEB 1978)

4.   Differentiate from first principles the functions
(a) $(x + 1)^3$,   (b) $(x + 1)^{-1}$.

5.   Prove that $\sqrt{(a + b)} - \sqrt{a} = \dfrac{b}{\sqrt{(a + b)} + \sqrt{a}}$. Use this result to differentiate the function $y = \sqrt{x}$ from first principles.

6.   Differentiate with respect to x

(a) $(3x + 1)^2$,   (b) $\dfrac{1}{x}(x + 2)(x + 3)$,   (c) $(x + 2)\sqrt{x}$.

7.   Find the equations of the tangents to the curve $y = x^3 - 2x^2 - 3x$ at the points where the curve crosses the x-axis.

8.   Show that the tangent to the curve $y = (1 + x)^2(5 - x)$ at the point where $x = 1$ does not meet the curve again. Show also that the tangent at the point (0.5) cuts the curve at a turning point. Draw a rough sketch of the curve, showing these tangents.

9.   Sketch the following curves indicating the approximate positions of any turning points

(a) $y = x^2 - 4$,   (b) $x^3 - 4x$,   (c) $y = x^4 - 4x^2$,

(d) $y = x - \dfrac{4}{x}$,　　　(e) $y = 1 - \dfrac{4}{x^2}$,　　　(f) $y = \dfrac{1}{x} - \dfrac{4}{x^3}$.

10. Find any maxima, minima or horizontal points of inflexion of the curve
$y = \dfrac{x^3 + 3x - 1}{x^2}$, stating, with reasons, the nature of each point. Sketch the curve,
indicating clearly what happens as $x \to \pm\infty$.　　　　　　　　　　　(SU)

11. A particle moves along a straight line such that after t seconds its displacement from a fixed point on the line is s metres, where $s = 7t + 5t^2 - 2t^3$. Find (a) expressions for the velocity and acceleration of the particle after t seconds, (b) the velocity and acceleration of the particle initially and after 2 seconds.

12. A function $f(x)$ is defined as follows:
　　　　$f(x) = 2x^3 - 9x^2 + 12x$　　　when $0 \leqslant x \leqslant 2$
and　$f(x) = 4(x - 1)(3 - x)$　　　when $2 < x \leqslant 3$.
Sketch the graph of the function.

13. If $f(x) = (x^2 - 1)(x - 3)$ show that the minimum value of $f'(x)$ is -4. Sketch the graphs of the functions $f(x)$ and $f'(x)$.

14. Show that the curve $y = ax^3 + bx^2 + cx + d$ $(a \neq 0)$ always has exactly one point of inflexion and write down its x-coordinate.

15. Show that there are no points of inflexion on the curve
$y = x^4 - 4x^3 + 6x^2 - 4x$, then sketch the curve.

6 Introduction to integration

6.1 The reverse of differentiation

Any graph with gradient function given by $dy/dx = 2$ has gradient 2 at every point on it. Hence the graph must be a straight line with equation $y = 2x + c$,

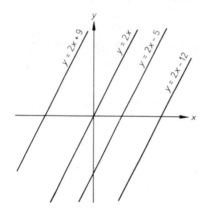

where c is some constant. Conversely, differentiating $y = 2x + c$, we obtain $dy/dx = 2$ for all values of c. Thus the *differential equation*, $dy/dx = 2$, is said to represent the family of all straight lines with gradient 2.

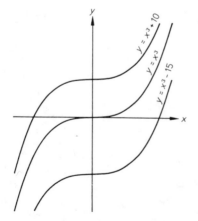

Consider now graphs with gradient function, $dy/dx = 3x^2$. Since $\dfrac{d}{dx}(x^3) = 3x^2$, the result $\dfrac{dy}{dx} = 3x^2$ may be derived from any equation of the form, $y = x^3 + c$, where c is a constant. Thus the equation $dy/dx = 3x^2$ represents the family of curves $y = x^3 + c$, some of which are shown in the diagram.

The operation of obtaining an expression for y in terms of x from the gradient function, dy/dx, is called *integration*. The function $x^3 + c$ is the *integral* of $3x^2$ with respect to x. Since the constant c may take any value, it is called an *arbitrary constant*.

The rule for integrating powers of x is found by reversing the rule for differentiation. We find that to integrate a power of x, we must add 1 to the index, then divide by the new index, i.e.

$$\text{if } \frac{dy}{dx} = x^n, \quad \text{then} \quad y = \frac{x^{n+1}}{n+1} + c \quad \text{provided that } \quad n+1 \neq 0.$$

This rule cannot be applied to the integration of x^{-1}, because there is no power of x whose derivative is x^{-1}.

Example 1   Find y if $\dfrac{dy}{dx}$ is (a) x^4, (b) $\dfrac{1}{x^3}$, (c) $6x$, (d) $x^{1/2}$.

(a) If $\dfrac{dy}{dx} = x^4$, then $y = \dfrac{1}{5}x^5 + c$.

(b) If $\dfrac{dy}{dx} = \dfrac{1}{x^3} = x^{-3}$, then $y = \dfrac{x^{-2}}{-2} + c = -\dfrac{1}{2x^2} + c$.

(c) If $\dfrac{dy}{dx} = 6x = 6.x^1$, then $y = 6.\dfrac{x^2}{2} + c = 3x^2 + c$.

(d) If $\dfrac{dy}{dx} = x^{1/2}$, then $y = \dfrac{x^{3/2}}{\frac{3}{2}} + c = \dfrac{2}{3}x^{3/2} + c$.

Example 2   Find $f(u)$, given that $f'(u) = (u-2)(2u+1)$.

If $\quad f'(u) = (u-2)(2u+1) = 2u^2 - 3u - 2$

then $\quad f(u) = \dfrac{2}{3}u^3 - \dfrac{3}{2}u^2 - 2u + c$.

It is sometimes possible to determine the value of the arbitrary constant in an integration problem, if additional information is given.

Example 3   Find V in terms of h, if $\dfrac{dV}{dh} = 4h^2 + h$ and $V = 2$ when $h = 1$.

$$\frac{dV}{dh} = 4h^2 + h \qquad \Rightarrow \qquad V = \frac{4}{3}h^3 + \frac{1}{2}h^2 + c$$

Since $V = 2$ when $h = 1$, $\qquad 2 = \dfrac{4}{3} + \dfrac{1}{2} + c$

$$\therefore \quad c = 2 - \frac{4}{3} - \frac{1}{2} = \frac{1}{6}$$

Hence $V = \dfrac{4}{3}h^3 + \dfrac{1}{2}h^2 + \dfrac{1}{6} = \dfrac{1}{6}(8h^3 + 3h^2 + 1).$

Example 4   The gradient of a curve is given by $\dfrac{dy}{dx} = 1 - \dfrac{1}{x^2}.$ If the curve passes through the point $(1, 1)$, find its equation.

Integrating $y = x + \dfrac{1}{x} + c$

Since $y = 1$, when $x = 1$, $1 = 1 + 1 + c$ $\therefore c = -1.$

Hence the equation of the curve is $y = x + \dfrac{1}{x} - 1.$

Exercise 6.1

1. Sketch the families of curves with the following gradient functions: (a) 1, (b) $2x$, (c) $4x^3$.

2. Sketch the family of curves represented by each of the following differential equations

(a) $\dfrac{dy}{dx} = -\tfrac{1}{2},$ (b) $\dfrac{dy}{dx} = 1 - 2x,$ (c) $\dfrac{dy}{dx} = 3x^2 - 6x.$

3. Find y if $\dfrac{dy}{dx}$ is   (a) x^2,   (b) $\dfrac{1}{x^2}$,   (c) $8x^3$,   (d) $\dfrac{6}{x^3}$.

In questions 4 to 12 integrate the given functions of x.

4. $3x^2 + 4x.$ 5. $x^5 - 1.$ 6. $(2x + 1)^2.$

7. $6x - \dfrac{1}{6x^3}.$ 8. $\left(x + \dfrac{1}{x}\right)^2.$ 9. $\dfrac{x^2 - 1}{x^4}.$

10. $x^{1/3} + x^{-1/3}.$ 11. $\dfrac{x - 1}{\sqrt{x}}.$ 12. $\sqrt{x^3} - \dfrac{1}{\sqrt{x^3}}.$

13. Find $f(t)$ given that $f'(t)$ is equal to

(a) $(t + 6)(3t - 4),$ (b) $\dfrac{(2t + 1)^2}{t^4},$ (c) $\dfrac{(t - 2)(2t - 1)}{\sqrt{t}}.$

14. Find an expression for y in terms of x, if

$\dfrac{dy}{dx} = (x - 1)(3x - 5)$ and $y = 0$ when $x = 1.$

15. Given that $\dfrac{dy}{dx} = 2\left(x + \dfrac{1}{x^3}\right)$ and that $y = 3$ when $x = 1$, find the value of y when $x = \frac{1}{2}$.

16. The gradient of a curve is given by $\dfrac{dy}{dx} = 3x^2 - 4$. If the curve passes through the point $(2, -1)$, find its equation.

17. The gradient of a certain curve at a typical point (x, y) is $4(1 - x)$. Given that the maximum value of y on the curve is 8, find its equation.

18. Find an expression for A in terms of h, given that

$\dfrac{dA}{dh} = 3\sqrt{h} - \dfrac{1}{2\sqrt{h}}$ and that $A = 10$ when $h = 4$.

19. Find an expression for y in terms of x, given that

$\dfrac{d^2y}{dx^2} = 6x - 10$ and that when $x = 1$, $y = 1$ and $\dfrac{dy}{dx} = 1$.

20. If y is a function of x and $\dfrac{d^2y}{dx^2} = 3 - \dfrac{5}{\sqrt[3]{x}}$, find the value of y when $x = 8$, given that $dy/dx = \frac{1}{2}$ and $y = 5$ when $x = 1$.

6.2 The area under a curve

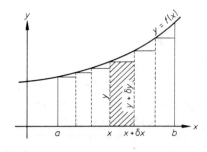

The diagram shows part of a curve $y = f(x)$. The area bounded by the curve, the x-axis and the lines $x = a$ and $x = b$ can be determined by dividing it into strips. These strips are assumed to be approximately rectangular, so that a typical strip of width δx and height y has area $y\,\delta x$. The total area is then estimated by finding the sum of all such areas between $x = a$ and $x = b$. The smaller the width of each strip, the more accurate this estimate will be. Thus, as δx approaches zero, the sum of the areas of the strips approaches the required area under the curve. Since the area is derived from the limit of a sum of terms of the form $y\,\delta x$, as $\delta x \to 0$, the standard notation for this area is

$$\int_a^b y\,dx \quad \text{or} \quad \int_a^b f(x)\,dx, \quad \text{where} \quad \int \text{ is an elongated } S \text{ for sum.}$$

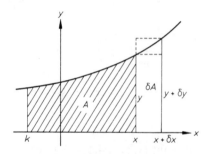

To find a way of evaluating this limit, consider the shaded area, A, between the curve $y = f(x)$ and the x-axis shown in the sketch. Since the value of A depends on the value of x, it may be possible to express A in terms of x.

Let δx be a small increase in x and let δy and δA be the corresponding increases in y and A. The area δA must lie between the areas of the two rectangles shown in the diagram, which have width δx and heights y and $y + \delta y$

$$\text{i.e.} \quad y\,\delta x < \delta A < (y + \delta y)\,\delta x$$

$$\therefore \quad y < \frac{\delta A}{\delta x} < y + \delta y$$

As $\delta x \to 0$, $\delta y \to 0$ and $\dfrac{\delta A}{\delta x} \to \dfrac{dA}{dx}$

$$\therefore \quad \frac{dA}{dx} = y.$$

Hence an expression for A may be obtained by integrating $f(x)$.

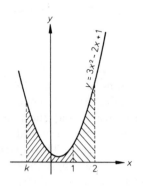

For example, the area function A for the curve $y = 3x^2 - 2x + 1$ is obtained by writing

$$\frac{dA}{dx} = 3x^2 - 2x + 1$$

Thus

$$A = x^3 - x + x + c,$$

where the value of the arbitrary constant c depends on the position of the boundary line $x = k$. However, the size of the area under the curve between $x = 1$ and $x = 2$ does not depend on the values of c and k.

At $x = 2$, $\quad A = 8 - 4 + 2 + c = 6 + c$

At $x = 1$, $\quad A = 1 - 1 + 1 + c = 1 + c$

$\therefore$ the area bounded by the curve $y = 3x^2 - 2x + 1$, the x-axis and the lines $x = 1$ and $x = 2$ is $(6 + c) - (1 + c)$, i.e. 5 square units.

More generally, if the area function for a curve $y = f(x)$ is $A = F(x) + c$, then the area under the curve between $x = a$ and $x = b$ is $F(b) - F(a)$.

Hence
$$\int_a^b y\,dx = \int_a^b f(x)\,dx = F(b) - F(a).$$

The expression $F(b) - F(a)$ is often written $\left[F(x)\right]_a^b$.

Example 1   Find the area bounded by the curve $y = x^2 - x + 3$, the x-axis and the lines $x = 2$, $x = 4$.

Required area $= \displaystyle\int_2^4 y\,dx$

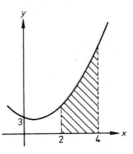

$$= \int_2^4 (x^2 - x + 3)\,dx$$

$$= \left[\tfrac{1}{3}x^3 - \tfrac{1}{2}x^2 + 3x\right]_2^4$$

$$= (\tfrac{1}{3}.4^3 - \tfrac{1}{2}.4^2 + 3.4) - (\tfrac{1}{3}.2^3 - \tfrac{1}{2}.2^2 + 3.2)$$

$$= (64/3 - 8 + 12) - (8/3 - 2 + 6) = 18\tfrac{2}{3}.$$

[We will assume that all areas obtained by integration are measured in square units.]

Example 2   Find the area of the region bounded by the curve $y = 2 + x - x^2$ and the straight line $y = x + 1$.

At the points of intersection of the line and the curve

$$y = 2 + x - x^2 \quad \text{and} \quad y = x + 1$$
$$2 + x - x^2 = x + 1$$
$$\Leftrightarrow \qquad\qquad x^2 = 1$$
$$\Leftrightarrow \quad x = -1 \quad \text{or} \quad x = 1$$

$\therefore$ the points of intersection are $(-1, 0)$ and $(1, 2)$.

The area under the curve from $x = -1$ to $x = 1$

$$= \int_{-1}^{1} (2 + x - x^2)\,dx$$

$$= \left[2x + \tfrac{1}{2}x^2 - \tfrac{1}{3}x^3\right]_{-1}^{1}$$

$$= (2 + \tfrac{1}{2} - \tfrac{1}{3}) - (-2 + \tfrac{1}{2} + \tfrac{1}{3}) = 3\tfrac{1}{3}$$

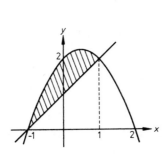

The area of the triangle under the line $= \tfrac{1}{2}.2.2 = 2$.

Hence the area bounded by the given curve and straight line is $3\tfrac{1}{3} - 2$, i.e. $1\tfrac{1}{3}$.

In §6.1 integration was described as the reverse of differentiation. However, we now see that it is also the process used to find areas under curves. Because of this relationship between integration and summation the integral of a function $f(x)$ is denoted by $\int f(x)\,dx$ and referred to as an *indefinite integral*. The expression $\int_a^b f(x)\,dx$ is called a *definite integral*. The constants a and b are the *limits of integration*, a being the lower limit and b the upper.

Example 3   (a) Find $\int (2x-5)\,dx$,   (b) Evaluate $\int_{-1}^2 (2x-5)\,dx$.

(a) $\int (2x-5)\,dx = x^2 - 5x + c$

(b) $\int_{-1}^2 (2x-5)\,dx = \left[x^2 - 5x \right]_{-1}^2 = (2^2 - 5.2) - (\{-1\}^2 - 5.\{-1\}) = -6 - 6$

$$= -12.$$

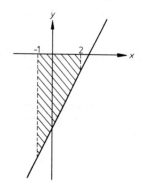

The diagram shows the significance of the negative value of the above definite integral. The line $y = 2x - 5$ is below the x-axis, i.e. y is negative, throughout the interval $x = -1$ to $x = 2$. This means that although the definite integral is negative, it is still numerically equal to the area bounded by the line $y = 2x - 5$, the x-axis and the lines $x = -1, x = 2$.

As shown in the examples, for any given function $f(x)$, the definite integral $\int_a^b f(x)\,dx$ is not itself a function of x, but has a numerical value which depends solely on the limits a and b. If a different variable is used the value is unchanged.

$$\int_a^b f(x)\,dx = \int_a^b f(t)\,dt = \int_a^b f(\theta)\,d\theta.$$

Thus x (or t, or θ) may sometimes be referred to as a *dummy variable*.

Example 4   Evaluate $\int_1^4 \left(\sqrt{u} + \dfrac{1}{\sqrt{u}} \right) du$.

$$\int_1^4 \left(\sqrt{u} + \frac{1}{\sqrt{u}} \right) du = \int_1^4 (u^{1/2} + u^{-1/2})\,du$$

$$= \left[\tfrac{2}{3} u^{3/2} + 2u^{1/2} \right]_1^4$$

$$= (\tfrac{2}{3}.4^{3/2} + 2.4^{1/2}) - (\tfrac{2}{3}.1^{3/2} + 2.1^{1/2})$$

$$= \frac{16}{3} + 4 - \frac{2}{3} - 2 = 6\tfrac{2}{3}.$$

Exercise 6.2

Find the following indefinite integrals

1. (a) $\int (6x+1)\,dx.$ (b) $\int (3x-1)^2\,dx.$ (c) $\int x(4x+1)\,dx.$

2. (a) $\int \left(3 - \dfrac{1}{x^2}\right)dx.$ (b) $\int \left(x - \dfrac{1}{x}\right)^2 dx.$ (c) $\int \dfrac{3x+1}{x^3}\,dx.$

3. (a) $\int 4\sqrt{x}\,dx.$ (b) $\int \dfrac{x-1}{\sqrt{x}}\,dx.$ (c) $\int 4(x^{1/3}+x^{-1/3})\,dx.$

4. (a) $\int \left(3t^2 - \dfrac{1}{t^2}\right)dt.$ (b) $\int u(1-\sqrt{u})\,du.$ (c) $\int \dfrac{s^3+1}{s^3}\,ds.$

Evaluate the following definite integrals

5. (a) $\displaystyle\int_{-1}^{1} (4x+3)\,dx$ (b) $\displaystyle\int_{1}^{3} (3x^2-x)\,dx.$ (c) $\displaystyle\int_{1}^{25} x^{1/2}\,dx.$

6. (a) $\displaystyle\int_{1}^{2} \left(x + \dfrac{1}{x}\right)^2 dx.$ (b) $\displaystyle\int_{-2}^{-1} \dfrac{2x^4+1}{x^4}\,dx.$ (c) $\displaystyle\int_{1}^{4} \dfrac{5x^2-4}{\sqrt{x}}\,dx.$

7. (a) $\displaystyle\int_{-3}^{3} u(u-1)^2\,du.$ (b) $\displaystyle\int_{1}^{27} y^{1/3}\,dy.$ (c) $\displaystyle\int_{1}^{2} \dfrac{3t^4-1}{t^2}\,dt.$

8. Find the areas bounded by the x-axis and the following curves and straight lines.
(a) $y = 4x^2,\ x = 0,\ x = 1,$
(b) $y = x^3 - x,\ x = 1,\ x = 2,$
(c) $y = 18/x^3,\ x = 1,\ x = 3,$
(d) $y = 3\sqrt{x},\ x = 4,\ x = 9.$

9. Find the areas enclosed by the x-axis and the following curves
(a) $y = (1-x)(x+2),$
(b) $y = 3x^2 - x^3,$
(c) $y = 4x^2 - 4x - 3$
(d) $y = x^4 - 8x.$

10. Find the areas of the regions bounded by the following curves and straight lines.
(a) $y = (5-x)(2+x),\ y = 10,$
(b) $y = (x-1)^2,\ y = 9,$
(c) $y = x(3-x),\ y = x,$
(d) $y = (x^2-1)(x-2),\ y = 2-2x,$
(e) $y = 2\sqrt{x},\ 2x-3y+4 = 0,$
(f) $y = 9\left(x + \dfrac{1}{x^2}\right),\ y = 7(x+1).$

11. Find the area of the smaller closed region bounded by the curve $y = x^2$, the straight line $2x+y = 8$ and the x-axis.

12. Sketch the curves $y = x^2$ and $y^2 = x$. Find the coordinates of the points of intersection of the curves. Calculate the area enclosed between the two curves.

6.3 Properties of definite integrals

As shown in the previous section, the definite integral produces positive values for areas above the x-axis but negative values for areas below the x-axis. Thus difficulties may arise when a curve crosses the x-axis inside the range of integration.

Example 1   (a) Evaluate $\displaystyle\int_0^2 x(x-1)\,dx$.   (b) Find the total area enclosed by the curve $y = x(x-1)$, the x-axis and the line $x = 2$.

(a) $\displaystyle\int_0^2 x(x-1)\,dx = \int_0^2 (x^2 - x)\,dx = \left[\tfrac{1}{3}x^3 - \tfrac{1}{2}x^2 \right]_0^2 = (\tfrac{1}{3}.2^3 - \tfrac{1}{2}.2^2) - 0 = \tfrac{2}{3}.$

(b) $y = 0 \Leftrightarrow x(x-1) = 0$
$\Leftrightarrow x = 0$ or $x = 1$

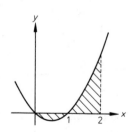

$\therefore$ the curve cuts the x-axis at the points $(0,0)$ and $(1,0)$.

$$\int_0^1 (x^2 - x)\,dx = \left[\tfrac{1}{3}x^3 - \tfrac{1}{2}x^2 \right]_0^1$$

$$= \tfrac{1}{3} - \tfrac{1}{2} = -\frac{1}{6}$$

$\therefore$ the shaded area below the x-axis is $1/6$.

$$\int_1^2 (x^2 - x)\,dx = \left[\tfrac{1}{3}x^3 - \tfrac{1}{2}x^2 \right]_1^2$$

$$= \frac{2}{3} - \left(-\frac{1}{6}\right) = \frac{5}{6}$$

$\therefore$ the shaded area above the x-axis is $5/6$.

Hence the total area enclosed by the curve $y = x(x-1)$, the x-axis and the line $x = 2$ is 1.

[Note that the definite integral gave the difference between the area above the x-axis and the area below it, i.e. $\dfrac{5}{6} - \dfrac{1}{6} = \dfrac{2}{3}$, instead of the sum $\dfrac{5}{6} + \dfrac{1}{6} = 1$, which we required.]

Another type of difficulty arises when we consider a curve such as $y = 1/x^2$.

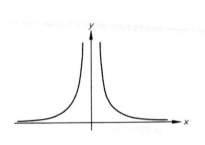

The integral $\displaystyle\int_a^b \frac{1}{x^2}\,dx$ can be used to find areas under the curve as long as a and b are either both positive or both negative. However, there is a break or *discontinuity* in the curve at $x = 0$, as shown in the sketch. It is, therefore, quite meaningless to talk about the area under the curve between such values as $x = -1$ and $x = 2$. Although the integral $\displaystyle\int_{-1}^2 \frac{1}{x^2}\,dx$ can be evaluated in the usual way, the result is not valid and does not represent an area under the curve $y = 1/x^2$.

The area bounded by a curve, the y-axis and lines $y = \alpha$ and $y = \beta$ is given by a definite integral of the form $\displaystyle\int_\alpha^\beta x\,dy$, in which the roles of x and y are interchanged.

Example 2   Find the area bounded by the curve $y = x^2$ and the line $y = 4$.

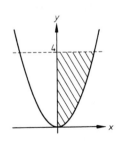

The required area is twice the shaded area shown in the sketch.

Shaded area

$$= \int_0^4 x\,dy = \int_0^4 y^{1/2}\,dy = \left[\tfrac{2}{3}y^{3/2}\right]_0^4$$

$$= 16/3 = 5\tfrac{1}{3}.$$

Hence the area bounded by the curve $y = x^2$ and the line $y = 4$ is $10\tfrac{2}{3}$.

Two general properties of the definite integral may be proved as follows:

Let $\displaystyle\int f(x)\,dx = F(x) + c$,   then

$$\int_b^a f(x)\,dx = \left[F(x)\right]_b^a = F(a) - F(b) = -\{F(b) - F(a)\} = -\left[F(x)\right]_a^b$$

$$= -\int_a^b f(x)\,dx.$$

$$\int_a^b f(x)\,dx + \int_b^c f(x)\,dx = \{F(b) - F(a)\} + \{F(c) - F(b)\} = F(c) - F(a)$$

$$= \int_a^c f(x)\,dx.$$

Summarising:

$$\int_b^a f(x)\,dx = -\int_a^b f(x)\,dx, \quad \int_a^b f(x)\,dx + \int_b^c f(x)\,dx = \int_a^c f(x)\,dx.$$

It may also be shown using the properties of limits that

$$\int_a^b kf(x)\,dx = k\int_a^b f(x)\,dx \quad \text{where } k \text{ is a constant}$$

$$\int_a^b f(x)\,dx + \int_a^b g(x)\,dx = \int_a^b \{f(x)+g(x)\}\,dx.$$

These properties can be used when finding the area between two curves. For instance, in the cases illustrated below, the shaded area is given by

$$\int_a^b \{f(x)-g(x)\}\,dx.$$

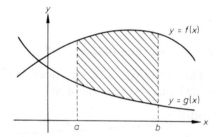
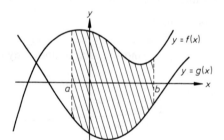

Example 3   Find the area enclosed by the curves $y = 2+x-x^2$ and $y = 2-3x+x^2$.

At the points of intersection of the two curves

$$2+x-x^2 = 2-3x+x^2$$
$$\therefore \quad 4x-2x^2 = 0$$
$$\therefore \quad x(2-x) = 0$$

$\therefore$ either $x = 0, y = 2$ or $x = 2, y = 0$.

Hence the required area

$$= \int_0^2 (2+x-x^2)\,dx - \int_0^2 (2-3x+x^2)\,dx$$

$$= \int_0^2 (4x-2x^2)\,dx$$

$$= \left[2x^2 - \tfrac{2}{3}x^3 \right]_0^2$$

$$= 8-16/3 = 2\tfrac{2}{3}.$$

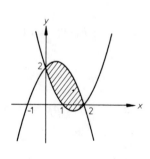

Exercise 6.3

1. Find the total areas of the regions enclosed by the following curves and straight lines
(a) $y = (x+1)(x-3)$, $y = 0$ and $x = -3$,
(b) $y = x^3 - 3x^2$, $y = 0$ and $x = 4$,
(c) $y = x^3 + x^2 - 2x$ and the x-axis,
(d) $y = x^2 - 5 + \dfrac{4}{x^2}$ and the x-axis.

2. Find the area bounded by the curve $y = x^2$ and the line $y = 9$.

3. Find the area bounded by the curve $y^2 = 4 - x$ and the y-axis.

4. Find the area bounded by the curve $y = (x-1)^3$, the y-axis and the line $y = 8$.

5. Find the area bounded by the curve $y = 1/(x+1)^2$, the y-axis and the line $y = 4$.

6. Find the areas of the finite regions bounded by the following curves and straight lines
(a) $y = x(4 - 3x)$ and $y = x$,
(b) $y = x^2 + 2x + 3$ and $y = 2(x+2)$,
(c) $y = 3 + 2x - x^2$ and $y = 3 - 2x$.

7. Sketch the following pairs of curves and find the area of the regions enclosed by them.
(a) $y = 2x^2$, $y = 3 - x^2$,   (b) $y = x^2 - 3x$, $y = 2 - x^2$,
(c) $y = 3x^2 + 9x + 5$, $y = x^2 + x - 1$.

6.4 Volumes of revolution

Consider the volume, V, generated when the shaded area between the curve $y = f(x)$ and the x-axis is rotated completely about the x-axis.

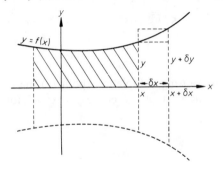
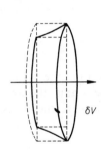

Let δx be a small increase in x and let δy and δV be the corresponding increases in y and V. The volume δV must lie between the volumes of two cylinders or discs of thickness δx and radii y and $y + \delta y$,

i.e. $$\pi y^2 \, \delta x < \delta V < \pi (y + \delta y)^2 \, \delta x$$

$$\therefore \quad \pi y^2 < \frac{\delta V}{\delta x} < \pi (y + \delta y)^2$$

As $\quad \delta x \to 0, \quad \delta y \to 0 \quad$ and $\quad \dfrac{\delta V}{\delta x} \to \dfrac{dV}{dx}$

$$\therefore \quad \frac{dV}{dx} = \pi y^2.$$

Hence the volume generated between $x = a$ and $x = b$ is given by

$$\int_a^b \pi y^2 \, dx.$$

Example 1   Find the volume generated by rotating completely about the x-axis the area bounded by the curve $y = x^3 - 2x^2$ and the x-axis.

$$y = 0 \Leftrightarrow x^3 - 2x^2 = 0 \Leftrightarrow x = 0 \quad \text{or} \quad x = 2$$

$\therefore \quad$ the curve cuts the x-axis at $(0,0)$ and $(2,0)$.

Hence the required volume $= \displaystyle\int_0^2 \pi y^2 \, dx$

$$= \pi \int_0^2 (x^3 - 2x^2)^2 \, dx$$

$$= \pi \int_0^2 (x^6 - 4x^5 + 4x^4) \, dx$$

$$= \pi \left[\frac{1}{7} x^7 - \frac{4}{6} x^6 + \frac{4}{5} x^5 \right]_0^2$$

$$= \pi \left(\frac{1}{7} . 2^7 - \frac{2}{3} . 2^6 + \frac{4}{5} . 2^5 \right) = \pi . 2^7 \left(\frac{1}{7} - \frac{1}{3} + \frac{1}{5} \right) = \frac{128\pi}{105}.$$

Example 2   Find the volumes generated by rotating completely about (a) the y-axis, (b) the x-axis, the area bounded by that part of the curve $y = x^2 + 1$ for which x is positive, the y-axis and the line $y = 2$.

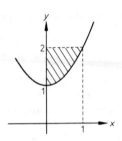

(a) The volume generated by rotation

about the y-axis $= \int_1^2 \pi x^2 \, dy$

$$= \int_1^2 \pi(y-1) \, dy$$

$$= \pi \left[\tfrac{1}{2}y^2 - y \right]_1^2$$

$$= \pi\{(\tfrac{1}{2}.2^2 - 2) - (\tfrac{1}{2} - 1)\} = \tfrac{1}{2}\pi.$$

(b)   [Note that in this case the volume generated by rotation about the x-axis cannot be calculated directly.]

Let V_1 be the volume of the cylinder generated by rotating about the x-axis the rectangle bounded by the axes and the lines $x = 1$, $y = 2$.

Thus $V_1 = \pi . 2^2 . 1 = 4\pi.$

Let V_2 be the volume generated by rotating about the x-axis the area bounded by the curve, the axes and the line $x = 1$.

Thus $V_2 = \int_0^1 \pi y^2 \, dx = \pi \int_0^1 (x^2 + 1)^2 \, dx$

$$= \pi \int_0^1 (x^4 + 2x^2 + 1) \, dx$$

$$= \pi \left[\frac{1}{5}x^5 + \frac{2}{3}x^3 + x \right]_0^1$$

$$= \pi \left(\frac{1}{5} + \frac{2}{3} + 1 \right) = \frac{28\pi}{15}$$

Hence the required volume $= V_1 - V_2 = \left(4 - \frac{28}{15} \right)\pi = \frac{32\pi}{15}.$

Exercise 6.4

[Answers may be left in terms of π.]

1.   Find the volumes of the solids formed by rotating completely about the x-axis the areas bounded by the x-axis and the given curves and lines.

(a) $y = 5x^2$, $x = -1$, $x = 3$,
(b) $y = 3x - x^2$, $x = 1$, $x = 2$,
(c) $y = 2x - 5$, $x = 1$, $x = 4$,
(d) $y = x^3 + 1$, $x = 2$,
(e) $y = 1 + \sqrt{x}$, $x = 4$, $x = 9$,
(f) $y = x + \dfrac{1}{x}$, $x = \tfrac{1}{2}$, $x = 2$.

2.   Find the volumes of the solids formed by rotating completely about the x-axis the areas enclosed by each of the following curves and the x-axis (a) $y = x^2 - 4$, (b) $3y = x^2(3 - x)$.

3. Find the volumes of the solids formed by rotating through 2 right angles about the x-axis the regions bounded by
(a) the curve $y^2 = x^3$ and the lines $x = 1$, $x = 2$,
(b) the curve $y^2 = 2(5-x)$ and the line $x = -1$.

4. Find the volumes of the solids formed by rotating completely about the y-axis the areas enclosed by the y-axis and the following curves and straight lines
(a) $y = \sqrt{x}$, $y = 2$,
(b) $y = x-1$, $y = 0$, $y = 1$,
(c) $y = 1/x$, $y = 1$, $y = 2$,
(d) $y = x^3$, $y = 8$.

5. Find the volumes of the solids formed when the region bounded by the curve $y = x^2 + 3$ and the straight lines $y = 0$, $x = 0$ and $x = 2$ is rotated completely about (a) the x-axis, (b) the y-axis.

6. Find the volumes of the solids formed when the smaller region enclosed by the curve $y^2 = 4-x$, the y-axis and the line $y = 1$ is rotated completely about (a) the x-axis, (b) the y-axis.

7. By considering a solid of revolution generated by the triangle with vertices $(0,0)$, (r,h) and $(0,h)$, find the volume of a cone height h, base radius r.

8. Given that the equation of the circle with radius a and centre the origin is $x^2 + y^2 = a^2$, find by integration the volume of a sphere of radius a.

9. Find the volume generated by rotating completely about the x-axis the finite region bounded by the curve $y^2 = 4x$ and the straight line $y = x$.

10. Find the volume generated by rotating completely about the x-axis the finite region bounded by the curve $y = x^3$, the x-axis and the line $x + 3y = 4$.

6.5 Integration as the limit of a sum

In §6.2 the definite integral $\displaystyle\int_a^b f(x)\,dx$ was introduced as the area bounded by the curve $y = f(x)$, the x-axis and the lines $x = a$ and $x = b$. We arrive at a formal definition of the definite integral by considering the area to be the sum of strips called *elements* of area.

The diagram shows a typical element of area δA. The value of $\displaystyle\int_a^b f(x)\,dx$ is the sum of the areas of all such elements. Using the Greek capital letter 'sigma' Σ to mean 'the sum of', this may be written

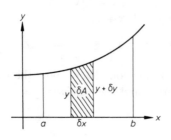

$$\int_a^b f(x)\,dx = \sum_{x=a}^{x=b} \delta A.$$

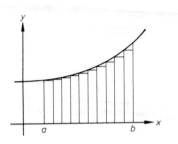

For small values of δx the area δA is approximately equal to the area of a rectangle with height y and width δx. Similarly the total area $\Sigma \delta A$ is approximately equal to the sum of the areas of such rectangles $\Sigma y\,\delta x$. If the number of strips is increased so that $\delta x \to 0$, then the difference between the sums $\Sigma \delta A$ and $\Sigma y\,\delta x$ also approaches zero. Hence the area under the curve $y = f(x)$ is given by the limit as δx tends to zero of $\Sigma y\,\delta x$. Thus the intuitive approach to the definite integral through areas is consistent with the following formal definition:

$$\int_a^b f(x)\,dx = \lim_{\delta x \to 0} \sum_{x=a}^{x=b} f(x)\,\delta x.$$

The fact that the definite integral, considered as the limit of a sum, can be found using a process which is the reverse of differentiation is called the *fundamental theorem of calculus*.

These results indicate that it is often possible to use integration to find a quantity expressed as a sum of elements of the form $f(x)\,\delta x$. For example, the formula for a volume of revolution derived in §6.5 may also be obtained by considering a typical element of volume $\delta V \approx \pi y^2\,\delta x$. The volume generated by rotating about the x-axis the area bounded by the curve $y = f(x)$, the x-axis and the lines $x = a$ and $x = b$, is then given by

$$\lim_{\delta x \to 0} \sum_{x=a}^{x=b} \pi y^2\,\delta x = \int_a^b \pi y^2\,dx.$$

Example 1 Find the volume generated by rotating completely about the line $y = 1$, the area bounded by the curve $y = 1 + 2x - x^2$ and the line $y = 1$.

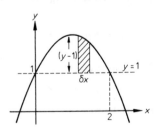

At the points of intersection of the curve $y = 1 + 2x - x^2$ and the line $y = 1$,

$$1 + 2x - x^2 = 1$$
$$2x - x^2 = 0$$
$$x(2 - x) = 0$$

$\therefore$ either $x = 0$ or $x = 2$.

A typical element of the required volume is a disc, radius $(y-1)$ and thickness δx

$\therefore$ element of volume $\approx \pi(y-1)^2\,\delta x$

Hence the required volume $= \lim_{\delta x \to 0} \Sigma \pi (y-1)^2 \, \delta x$

$$= \int_0^2 \pi (y-1)^2 \, dx$$

$$= \pi \int_0^2 (2x - x^2)^2 \, dx$$

$$= \pi \int_0^2 (4x^2 - 4x^3 + x^4) \, dx$$

$$= \pi \left[\frac{4}{3} x^3 - x^4 + \frac{1}{5} x^5 \right]_0^2$$

$$= \pi \left(\frac{4}{3} \cdot 2^3 - 2^4 + \frac{1}{5} \cdot 2^5 \right)$$

$$= \pi \left(\frac{32}{3} - 16 + \frac{32}{5} \right) = \frac{16\pi}{15}.$$

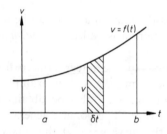

Consider now a particle moving along a straight line, so that its velocity at time t is given by $v = f(t)$. In a short interval of time δt, the displacement of the particle is approximately $v \, \delta t$. Hence the total displacement between $t = a$ and $t = b$ is given by $\lim_{\delta t \to 0} \sum_{t=a}^{t=b} v \, \delta t$.

Thus the total displacement of the particle in any given interval of time is represented by an area under the velocity-time graph and is found by evaluating a definite integral of the form $\int_a^b v \, dt$.

Example 2 A particle moves in a straight line so that after t seconds its velocity in $m\,s^{-1}$ is given by $v = 2t^3 + 3t + 5$. Find the distance moved in the third second of the motion.

Since the velocity of the particle is positive throughout the motion, the distance moved in the third second

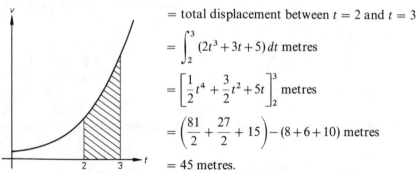

$= $ total displacement between $t = 2$ and $t = 3$

$$= \int_2^3 (2t^3 + 3t + 5) \, dt \text{ metres}$$

$$= \left[\frac{1}{2} t^4 + \frac{3}{2} t^2 + 5t \right]_2^3 \text{ metres}$$

$$= \left(\frac{81}{2} + \frac{27}{2} + 15 \right) - (8 + 6 + 10) \text{ metres}$$

$$= 45 \text{ metres.}$$

[Note that if a particle moving along a straight line changes direction in the interval from $t = a$ to $t = b$ then the total distance travelled in that interval is *not* equal to the total displacement of the particle between $t = a$ and $t = b$.]

Exercise 6.5

1. Find the volume of the solid formed by rotating completely about the line $y = 1$ the area bounded by the curve $y = x^2 + 2$, the y-axis and the lines $x = 1$ and $y = 1$.

2. Find the volume generated by rotating completely about the line $y = 3$, the area enclosed by the curve $y = 4x - x^2$ and the line $y = 3$.

3. A particle moves in a straight line so that after t seconds its velocity is $v \text{ ms}^{-1}$. In each of the following cases sketch the velocity-time graph for the motion and find the total distance moved in the given interval of time.
(a) $v = 10 + 2t$, $t = 0$ to $t = 2$,   (b) $v = 16t - 3t^2$, the 3rd second,
(c) $v = \frac{1}{2}t^2 + 5$, $t = 1$ to $t = 3$,   (d) $v = 18t^2 - 4t^3$, the 4th second,
(e) $v = 5 + 4t - t^2$, $t = 1$ to $t = 6$,   (f) $v = 4t^3 - 16t$, $t = 0$ to $t = 3$.

4. Find the volume of the solid formed by rotating completely about the line $y = 2$ the region enclosed by the curve $y = x(2 - x)$ and the x-axis.

5. Find the volume of the solid formed by rotating through 2 right angles about the line $x = 2$ the area bounded by the curve $y = x^2 - 4x + 5$ and the line $y = 5$.

6. Use integration to find a formula for the volume of a pyramid of height h cm with a square base of side a cm.

7. The horizontal cross-section of a bowl at a height h cm above its base is a regular hexagon of side x cm. Given that $x^2 = 16(h + 1)$, find the volume of liquid in the bowl when it is filled to a depth of 8 cm.

8. By considering the area under the curve $y = \sqrt{x}$ between $x = 0$ and $x = 81$ as a sum of strips of width 1 unit, show that

(a) $\displaystyle\int_0^{81} \sqrt{x}\,dx < \sqrt{1} + \sqrt{2} + \sqrt{3} + \ldots + \sqrt{81},$

(b) $\sqrt{1} + \sqrt{2} + \sqrt{3} + \ldots + \sqrt{80} < \displaystyle\int_0^{81} \sqrt{x}\,dx.$

Deduce that $477 < \sqrt{1} + \sqrt{2} + \sqrt{3} + \ldots + \sqrt{80} < 486$.

9. By considering the area under the curve $y = 1/x^2$ between $x = 1$ and $x = 100$, show that

$$\frac{1}{1^2} + \frac{1}{2^2} + \frac{1}{3^2} + \ldots + \frac{1}{100^2} < 2.$$

6.6 Mean value

The mean value of a function $f(x)$ between $x = a$ and $x = b$ is a type of average of the values of the function in that interval. For example, the mean value of a function $f(t)$ which represents the velocity of a moving object after time t is the same as the average velocity.

Example 1   An object is fired vertically upwards so that after time t its velocity is given by $v = 40 - 10t$. Find the average velocity between $t = 1$ and $t = 3$.

The total displacement between $t = 1$ and $t = 3$

$$= \int_1^3 (40 - 10t) \, dt = \left[40t - 5t^2 \right]_1^3 = 75 - 35 = 40$$

$\therefore$ the average velocity between $t = 1$ and $t = 3$

$$= \frac{40}{3 - 1} = \frac{40}{2} = 20.$$

Similarly, the average or mean velocity of an object moving in a straight line, between $t = a$ and $t = b$, is found by evaluating $\dfrac{1}{(b-a)} \displaystyle\int_a^b v \, dt$. The general definition of mean value is framed along these lines.

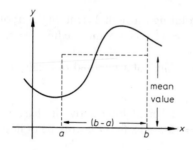

The *mean value* of a function $f(x)$ between $x = a$ and $x = b$ is

$$\frac{1}{(b-a)} \int_a^b f(x) \, dx.$$

Geometrically, this is the height of the rectangle, with base $(b-a)$, whose area is equal to the area under the curve from $x = a$ to $x = b$.

Example 2   The volume, V cm^3, of water in a hemispherical bowl is given by $V = \frac{1}{3}\pi(30h^2 - h^3)$ where h cm is the depth of the water. Find the mean volume of water in the bowl as the depth increases from 0 to 4 cm.

Mean volume (in cm^3) $= \dfrac{1}{4-0} \displaystyle\int_0^4 V \, dh = \dfrac{1}{4} \int_0^4 \dfrac{1}{3}\pi(30h^2 - h^3) \, dh$

$$= \frac{1}{12}\pi \left[10h^3 - \frac{1}{4}h^4 \right]_0^4$$

$$= \frac{1}{12}\pi \left(10 \cdot 4^3 - \frac{1}{4} \cdot 4^4 \right)$$

$$= \frac{1}{12}\pi \cdot 9 \cdot 4^3 = 48\pi.$$

Exercise 6.6

1. A particle is moving in a straight line such that after t seconds its velocity is v m s^{-1}, where $v = 6t + 12t^2$. Find
(a) the mean velocity during the first 2 seconds of the motion,
(b) the mean acceleration between $t = 1$ and $t = 5$.

2. Repeat question 1 for $v = 2t + \dfrac{5}{t^2}$.

3. Find the mean values of the following functions in the given intervals.
(a) $2x^3 - x + 1$, $-1 \leqslant x \leqslant 1$,　　　　(b) $(x+3)(2x-5)$, $1 \leqslant x \leqslant 5$,
(c) $1 - \dfrac{1}{x^2}$, $0 \cdot 1 \leqslant x \leqslant 1$,　　　　(d) $(5x-3)\sqrt{x}$, $1 \leqslant x \leqslant 4$.

4. The volume, V cm^3, of water in a conical vessel is given by $V = \pi h^3/12$, where h cm is the depth of the water. Find the mean volume of water in the bowl as the depth increases from 2 cm to 8 cm.

5. A mass of gas of volume v at pressure p expands according to the law $pv^{3/4} = 30$. Find the mean pressure as the gas expands from $v = 1$ to $v = 16$.

6. Given that $x = 4t - 1$, find the mean values in the interval from $t = 1$ to $t = 4$ of (a) x^2, (b) $1/(x+1)^2$, (c) $(x+1)^{3/2}$.

Exercise 6.7　(*miscellaneous*)

[Answers may be left in terms of π, where appropriate.]

1. Sketch the families of curves with the following gradient functions

(a) $\dfrac{dy}{dx} = 2(1-x)$,　　　(b) $\dfrac{dy}{dx} = \dfrac{1}{2\sqrt{x}}$,　　　(c) $\dfrac{dy}{dx} = 1 + \dfrac{1}{x^3}$.

2. Integrate with respect to x

(a) $4x^5 - x^3$,　　　(b) $4 - \dfrac{3}{x^2}$,　　　(c) $\dfrac{x^4 + 1}{x^3}$.

3. Find the following indefinite integrals

(a) $\displaystyle\int (2-3t)^2 \, dt$,　　　(b) $\displaystyle\int \left(5x^4 - \dfrac{1}{5x^4}\right) dx$,　　　(c) $\displaystyle\int \dfrac{1}{\sqrt[3]{u}} \, du$.

4. Given that $\dfrac{dv}{dt} = (t+1)(3t-7)$ and that $v = 36$ when $t = 5$, show that $v = (t+1)^2 (t-4)$.

5. The gradient of a curve is given by $\dfrac{dy}{dx} = ax + \dfrac{b}{x^2}$. If the curve passes through

the point $(-1, -4)$ and has a turning point at the point $(1, 0)$, find its equation and sketch the curve.

6.   Evaluate   (a) $\int_1^2 \frac{(x+1)(x-2)}{x^4} dx,$ (b) $\int_1^4 \left(\sqrt{x} - \frac{1}{x}\right)^2 dx.$

7.   Find the area enclosed by the curve $y = x - \frac{1}{x^2}$, the x-axis and the line $x = 2$.

8.   Find the area enclosed by the curve $y = 2x^2 + 1$ and the line $y = 3$.

9.   Find the total area of the regions enclosed by the curve $y = x^4 - x^3$, the x-axis and the line $x = 2$.

10. Find the area enclosed by the curve $y^2 - 2y = x$ and the y-axis.

11. Find the equation of the tangent at the point $(1, 0)$ to the curve $y = (x-1)(x^2+1)$. Find the area of the finite region enclosed by the curve and this tangent.

12. Sketch the region in the x-y plane within which the inequalities $y > (x-2)^2$ and $y < x$ are satisfied. Determine the area of this region.

13. Sketch the curves $y = x^2(4x-5)$ and $y = x(3x-4)$ on the same diagram, then find the area of the finite region enclosed between them.

14. When the area bounded by the curve $y = 15x(k-x)$ and the x-axis is rotated completely about the x-axis the volume of the solid generated is 240π. Find the value of k.

15. The region R is bounded by the curve $y = 4/x$, the x- and y-axes and the lines $x = 2$, $y = 4$. Find the volume of the solid formed when R is rotated completely about (a) the x-axis, (b) the y-axis.

16. The region R is bounded by the curve $4y = x^2$ and the lines $x = 4$, $y = 1$. Find the volume of the solid formed when R is rotated completely about (a) the x-axis, (b) the y-axis, (c) the line $y = 1$, (d) the line $x = 4$.

17. The region R is bounded by the curve $y^2 = 2x$ and the line $y = x$. Find the volume of the solid formed when R is rotated completely about (a) the x-axis, (b) the y-axis.

18. The curves $cy^2 = x^3$ and $y^2 = ax$ (where $a > 0$ and $c > 0$) intersect at the origin O and at a point P in the first quadrant. The areas of the regions enclosed by the arcs OP, the x-axis and the ordinate through P are A_1 and A_2 for the two curves; the volumes of the two solids formed by rotating these regions through four right angles about the x-axis are V_1 and V_2 respectively. Prove that $A_1/A_2 = 3/5$ and $V_1/V_2 = 1/2$.

19. Find dy/dx given that $y^2 = x(3-x)^2$. State the values of x for which dy/dx is (a) zero, (b) infinite. Sketch the curve $y^2 = x(3-x)^2$ and find the volume of the solid formed when the area enclosed by the loop of the curve is rotated through $180°$ about the x-axis.

20. Evaluate $\int_{-2}^{4} [x]\,dx$, where $[x]$ is defined as the greatest integer less than or equal to x. (See §5.1.)

21. Evaluate $\int_{-4}^{2} |x|\,dx$, where $|x|$ is defined as the magnitude of x. (See §1.4.)

22. A particle moves in a straight line such that after t seconds its velocity is $v\,\mathrm{ms}^{-1}$, where $v = 4-4t-15t^2$. Find the *total* distance moved by the particle in (a) the first two seconds, (b) the third second of its motion.

23. Show that over the range $0 \leqslant x \leqslant 6$, the mean value of the function $9x(6-x)$ is two-thirds of the maximum value of the function.   (AEB 1975)

24. Find the mean area of vertical cross-sections of a sphere of radius a.

7 Elementary trigonometry

7.1 Trigonometric ratios for any angle

Elementary trigonometry is based on the fact that in any set of similar right-angled triangles the ratios of corresponding pairs of sides remain constant. Using the right-angled triangle in the diagram, the sine and cosine of angle θ are defined as follows:

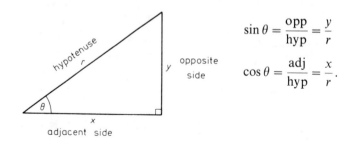

$$\sin \theta = \frac{\text{opp}}{\text{hyp}} = \frac{y}{r}$$

$$\cos \theta = \frac{\text{adj}}{\text{hyp}} = \frac{x}{r}.$$

In more advanced work it is useful to extend these definitions so that they can be applied to angles of any magnitude.

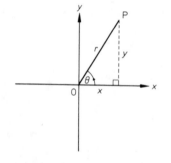

Consider a point P with coordinates (x, y). Let $OP = r$ and let θ be the angle OP makes with the positive direction of the x-axis. Positive angles are measured anti-clockwise from the x-axis and negative angles clockwise.

Sine and cosine may now be defined for any real value of θ by writing:

$$\sin \theta = \frac{y}{r} \qquad \cos \theta = \frac{x}{r}$$

The graphs of $\sin \theta$ and $\cos \theta$ may be plotted directly as shown below. Various points P on a circle of radius 1 unit are used, so that $\sin \theta = y$ and $\cos \theta = x$.

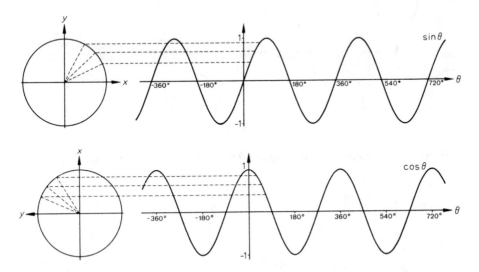

After OP has been rotated through $360°$, $\sin \theta$ and $\cos \theta$ pass through the sets of values they have already taken. For this reason $\sin \theta$ and $\cos \theta$ are described as *periodic functions* with *period* $360°$.

Example 1   Using a sketch graph if necessary, write down all the angles between $-360°$ and $720°$ with the same cosine as $70°$.

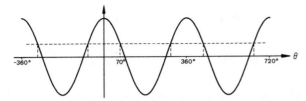

From the sketch the angles with the same cosine as $70°$ are: $-290°, -70°, 70°, 290°, 430°$ and $650°$.

The sine and cosine of an angle can be used to define the remaining trigonometric ratios, tangent, cotangent, secant, cosecant.

$$\tan \theta = \frac{\sin \theta}{\cos \theta} \qquad \cot \theta = \frac{\cos \theta}{\sin \theta}$$

$$\sec \theta = \frac{1}{\cos \theta} \qquad \operatorname{cosec} \theta = \frac{1}{\sin \theta}.$$

Because of their relationship with $\cos\theta$ and $\sin\theta$, $\sec\theta$ and $\mathrm{cosec}\,\theta$ must also be periodic functions with period 360°. Since the range of both $\cos\theta$ and $\sin\theta$ is the set of real numbers from -1 to $+1$, neither $\sec\theta$ nor $\mathrm{cosec}\,\theta$ can take values between -1 and $+1$.

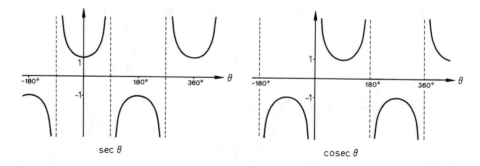

sec θ   cosec θ

However, the functions $\tan\theta$ and $\cot\theta$ take all real values and have a period of 180°.

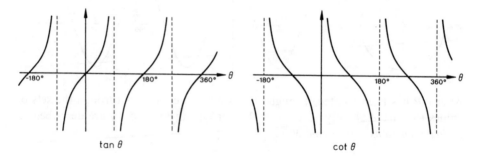

tan θ   cot θ

In terms of the coordinates (x, y) of the point P considered earlier, the six trigonometric ratios may now be defined as follows:

$$\cos\theta = \frac{x}{r} \qquad \sin\theta = \frac{y}{r} \qquad \tan\theta = \frac{y}{x}$$

$$\sec\theta = \frac{r}{x} \qquad \mathrm{cosec}\,\theta = \frac{r}{y} \qquad \cot\theta = \frac{x}{y}$$

The x- and y-axes divide the plane into four regions called *quadrants*.

Second quadrant	First quadrant	$x - \text{ve}$ $y + \text{ve}$	$x + \text{ve}$ $y + \text{ve}$
Third quadrant	Fourth quadrant	$x - \text{ve}$ $y - \text{ve}$	$x + \text{ve}$ $y - \text{ve}$

When P lies in the first quadrant, θ is acute and all the trigonometric ratios are positive. By examining the signs of x and y in the other quadrants, the signs of $\sin\theta$, $\cos\theta$ and $\tan\theta$ for any value of θ can be determined.

$\sin\theta$ +ve	$\sin\theta$ +ve	SIN	ALL
$\cos\theta$ − ve	$\cos\theta$ +ve	+ve	+ve
$\tan\theta$ − ve	$\tan\theta$ +ve		
$\sin\theta$ − ve	$\sin\theta$ − ve	TAN	COS
$\cos\theta$ − ve	$\cos\theta$ +ve	+ve	+ve
$\tan\theta$ +ve	$\tan\theta$ − ve		

The above diagrams show which ratios are positive in each quadrant. There are many mnemonics for the initial letters of All, Sin, Tan, Cos, e.g. All Stations To Crewe, All Steam Tugs Chug.

The trigonometric ratios for any angle θ are numerically equal to the corresponding ratios for the acute angle between OP and the x-axis. The quadrant in which OP lies determines the sign of the ratio, positive or negative.

Example 2 Express the following in terms of ratios of acute angles:
(a) $\sin 215°$, (b) $\cos(-70°)$, (c) $\tan 95°$.

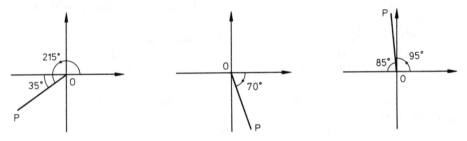

(a) [We note mentally that the acute angle between OP and the x-axis is 35° and that OP lies in the 3rd quadrant.] From the sketch, $\sin 215° = -\sin 35°$.
(b) $\cos(-70°) = \cos 70°$,   (c) $\tan 95° = -\tan 85°$.
[These results can also be obtained using sketch graphs of $\sin\theta$, $\cos\theta$ and $\tan\theta$.]

In general calculators, slide rules or mathematical tables are used to write down values of trigonometric ratios. However, in some cases such aids are not needed.

$$\sin 0° = 0 \qquad \cos 0° = 1 \qquad \tan 0° = 0$$
$$\sin 90° = 1 \qquad \cos 90° = 0 \qquad \tan 90° \text{ is undefined.}$$

[Considering values of θ close to 90° we find that as $\theta \to 90°$ from below, $\tan\theta \to \infty$ and as $\theta \to 90°$ from above, $\tan\theta \to -\infty$.]

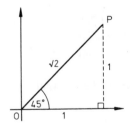

Let P be the point $(1, 1)$, so that $OP = \sqrt{2}$ and $\theta = 45°$.

Hence $\sin 45° = \cos 45° = \dfrac{1}{\sqrt{2}}$

and $\tan 45° = 1$.

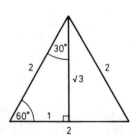

Using an equilateral triangle with sides each 2 units in length and height $\sqrt{3}$ units, we have

$$\sin 30° = \frac{1}{2} \qquad\qquad \sin 60° = \frac{\sqrt{3}}{2}$$

$$\cos 30° = \frac{\sqrt{3}}{2} \qquad\qquad \cos 60° = \frac{1}{2}$$

$$\tan 30° = \frac{1}{\sqrt{3}} \qquad\qquad \tan 60° = \sqrt{3}.$$

Exercise 7.1

1. Express the following in terms of ratios of acute angles
(a) $\cos 165°$, (b) $\sin(-35°)$, (c) $\tan 310°$, (d) $\sin 112°$,
(e) $\tan(-95°)$, (f) $\cos 318°$, (g) $\sin 195°$, (h) $\cos(-237°)$.

2. Express the following in terms of ratios of acute angles
(a) $\tan 124°$, (b) $\sec(-23°)$, (c) $\cot 248°$, (d) $\sin 702°$,
(e) $\operatorname{cosec} 280°$, (f) $\cot(-15°)$, (g) $\cos 186°$ (h) $\sec 99°$.

3. Using sketch graphs if necessary, write down all the angles between $-360°$ and $720°$ with
(a) the same sine as $40°$,   (b) the same cosine as $10°$,
(c) the same tangent as $75°$,   (d) the same secant as $180°$.

4. Find all the values of θ between $0°$ and $360°$ such that
(a) $\cos\theta = \cos 25°$,   (b) $\sin\theta = \sin 50°$,
(c) $\tan\theta = \tan 42\!\cdot\!3°$,   (d) $\sin\theta = -\sin 63\!\cdot\!4°$.

5. Find, without using tables or a calculator, the values of the following
(a) $\sin 120°$, (b) $\cot 30°$, (c) $\cos(-210°)$, (d) $\tan(-45°)$,
(e) $\cos 135°$, (f) $\sin(-300°)$, (g) $\operatorname{cosec} 210°$, (h) $\sec 750°$.

6. Find, without using tables or a calculator, the values of the following
(a) $\cos 270°$, (b) $\tan 120°$, (c) $\sin(-180°)$, (d) $\sec 240°$,
(e) $\cot 450°$, (f) $\cos 540°$, (g) $\operatorname{cosec} 225°$, (h) $\sin(-90°)$.

7. Find the possible values of θ and ϕ between $0°$ and $360°$, given that
(a) $\cos^2\theta + \sin^2\phi = 0$,   (b) $\tan^2\theta + \cot^2\phi = 0$,
(c) $2\sin\theta + \cos\phi = 3$,   (d) $3\sin\theta - 2\cos\phi = 5$.

8. Find the range of possible values of an acute angle θ given that
(a) $\tan 2\theta > 0$,   (b) $\cos 3\theta < 0$,
(c) $\sin 5\theta < 0$,   (d) $\cot 6\theta > 0$.

9.   In each of the following cases sketch the given pair of curves on the same diagram

(a) $y = 2\sin\theta$, $y = \sin 2\theta$,

(b) $y = \frac{1}{2}\cos\theta$, $y = \cos\frac{1}{2}\theta$,

(c) $y = -\sin\theta$, $y = \sin(-\theta)$,

(d) $y = -\cos\theta$, $y = \cos(-\theta)$.

7.2 The inverse functions and simple equations

In the previous section the trigonometric ratios were defined for any given angle. We now consider the reverse process of determining an angle given the value of one of the trigonometric ratios. One approach to this problem is through sketch graphs.

Example 1   Find all the values of θ between $-180°$ and $+180°$ such that $\tan\theta = 1$.

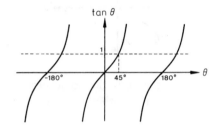

$\tan\theta = 1 = \tan 45°$

$\therefore$ from the sketch

$\theta = -135°$ or $45°$.

[Note that if there is no restriction on the value of θ, the equation $\tan\theta = 1$ has an infinite solution set

$\{\ldots, -315°, -135°, 45°, 225°, 405°, \ldots\}$.]

An alternative approach is to use the notion of the rotating line segment OP discussed in §7.1.

Example 2   Find all the values of θ between $0°$ and $360°$ such that $\sin\theta = -\frac{1}{2}$.

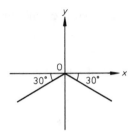

We note mentally that if $\sin\theta$ is $-$ve then OP is in the 3rd or 4th quadrant. If $\sin 30° = \frac{1}{2}$ then the acute angle between OP and the x-axis is $30°$. The sketch shows the two possible positions of OP.

$\sin\theta = -\frac{1}{2} = -\sin 30°$

$\therefore$ $\theta = 210°$ or $330°$.

[More generally, by considering further rotations of OP, the solution set of the equation $\sin\theta = -\frac{1}{2}$ is found to be

$\{\ldots, -150°, -30°, 210°, 330°, 570°, \ldots\}$.]

As illustrated in the above example, if $\sin a = b$ then for any given value of b there are many possible values of a. Thus, in the language of §4.2, the function $f : \theta \rightarrow \sin \theta$ has no true inverse. However, if the domain of the sine function is restricted so that $-90° \leqslant \theta \leqslant 90°$, then an inverse function can be defined. The *inverse sine* of x, written $\sin^{-1} x$ (or arcsin x), is the angle between $-90°$ and $90°$ whose sine is x. Hence if $y = \sin^{-1} x$, then $\sin y = x$ and $-90° \leqslant y \leqslant 90°$. These values, $-90°$ to $+90°$, are sometimes called the *principal values* of the inverse sine. [To avoid confusion between $\sin^{-1} x$ and $(\sin x)^{-1}$, i.e. $1/\sin x$, the symbol $\sin^{-1}$ may be read as 'inverse sine', 'sine inverse' or 'sine minus one', but *not* 'sine to the minus one'.] Similarly, the *inverse cosine* of x, written $\cos^{-1} x$ (or arccos x), is the angle between $0°$ and $180°$ whose cosine is x. Hence if $y = \cos^{-1} x$, then $\cos y = x$ and $0° \leqslant y \leqslant 180°$.

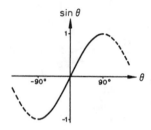
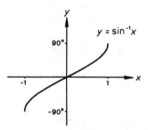

The *inverse tangent* of x, written $\tan^{-1} x$ (or arctan x), is the angle between $-90°$ and $90°$ whose tangent is x. Hence if $y = \tan^{-1} x$, then $\tan y = x$ and $-90° < y < 90°$.

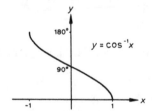
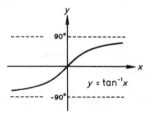

In the following example we demonstrate the use of both calculator inverse functions and mathematical tables.

Example 3   Solve the equation $\cos \theta = -0.4$ for values of θ between $0°$ and $360°$, giving your answers to the nearest tenth of a degree.

Method A   (using calculator inverse cosine function)

We must first find the angle between $0°$ and $180°$ whose cosine is -0.4.
    As a first answer we write down the angle to three decimal points.

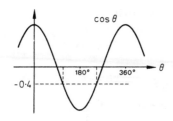

Thus $\cos \theta = -0\cdot4 = \cos 113\cdot578°$
$= 113\cdot6°$ (to nrst $0\cdot1°$)
$\therefore \quad \theta = 113\cdot6°$ or $360° - 113\cdot6°$
i.e. $\quad \theta = 113\cdot6°$ or $246\cdot4°$

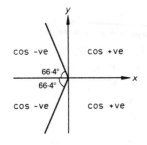

Method B   (using table of cosines)
$\cos \theta = -0\cdot4 = -\cos 66\cdot422°$
$= -\cos 66\cdot4°$ (nrst $0\cdot1°$)
$\therefore \quad \theta = 180° - 66\cdot4°$
$180° + 66\cdot4°$
i.e. $\quad \theta = 113\cdot6°$ or $246\cdot4°$

Example 4   Find the values of θ between $0°$ and $360°$ such that $\tan 2\theta = -2$, giving your answer to the nearest tenth of a degree. [Note that since we require values of θ between $0°$ and $360°$, we must look for values of 2θ between $0°$ and $720°$.]

$\tan 2\theta = -2 = -\tan 63\cdot435°$

$\therefore \quad$ between $0°$ and $720°$, $2\theta = 116\cdot565°, 296\cdot565°, 476\cdot565°$ or $656\cdot565°$.
Hence, between $0°$ and $360°$, $\theta = 58\cdot3°, 148\cdot3°, 238\cdot3°$ or $328\cdot3°$.

The last example concerns a quadratic equation in $\cos \theta$. Since brackets would be rather cumbersome $(\cos \theta)^2$ is usually written $\cos^2 \theta$.

Example 5   Solve the equation $2\cos^2 \theta - 5\cos \theta + 2 = 0$ for values of θ from $0°$ to $360°$.

$$2 \cos^2 \theta - 5 \cos \theta + 2 = 0 \Leftrightarrow (2 \cos \theta - 1)(\cos \theta - 2) = 0$$
$$\Leftrightarrow \cos \theta = \tfrac{1}{2} \text{ or } \cos \theta = 2$$

Since there is no value of θ for which $\cos \theta = 2$, we must have $\cos \theta = \tfrac{1}{2} = \cos 60°$. Hence $\theta = 60°$ or $300°$.

Exercise 7.2

Give answers to the nearest tenth of a degree.

1.   Find all the values of θ between $0°$ and $360°$ such that
(a) $\sin \theta = 0\cdot3$,   (b) $\tan \theta = 1\cdot5$,
(c) $\cos \theta = -0\cdot73$,   (d) $\sin \theta = -0\cdot62$.

2.   Find all the values of θ between $-180°$ and $+180°$ such that
(a) $\cos \theta = 0\cdot6$,   (b) $\sin \theta = 0\cdot8$,
(c) $\tan \theta = -0\cdot62$,   (d) $\cos \theta = -0\cdot342$.

3. Find all the values of θ between $0°$ and $360°$ such that
(a) $\sin \theta = 1/\sqrt{2}$,   (b) $\tan \theta = -1/\sqrt{3}$,   (c) $\cot \theta = 1$,
(d) $\cos \theta = -\frac{1}{2}$,   (e) $\sec \theta = -1$,   (f) $\operatorname{cosec} \theta = 2$.

4. Solve the following equations for $0° \leqslant x \leqslant 360°$
(a) $4 \sin x = 3 \cos x$,   (b) $\sin x + \cos x = 0$,
(c) $4 \tan x = 3 \sec x$,   (d) $2 \tan x \operatorname{cosec} x = 3$.

5. Solve the following equations for $0° \leqslant \theta \leqslant 360°$
(a) $\cos (\theta + 10°) = -0.44$,   (b) $\sin (\theta - 30°) = 0.207$,
(c) $\sin (\theta + 50°) = 0.541$,   (d) $\tan (70° - \theta) = -1.16$,
(e) $\tan 2\theta = 0.7$,   (f) $\cos \frac{1}{3}\theta = 0.34$,
(g) $\sin \frac{1}{2}\theta = 0.83$,   (h) $\sec 2\theta = -1$.

6. Solve the following equations for $-90° \leqslant x \leqslant 90°$
(a) $\sin 3x = 1/\sqrt{2}$,   (b) $\tan 4x = 0$,
(c) $\cot 5x + 1 = 0$,   (d) $2 \cos 3x + 1 = 0$.

7. Solve the following equations for $0° \leqslant x \leqslant 360°$
(a) $4 \sin^2 x = 3$,   (b) $2 \cos^2 x = \cos x$,
(c) $5 \sin x \cos x = \sin x$,   (d) $\sin^2 x - \sin x - 2 = 0$,
(e) $2 \tan^2 x = \sec x \tan x$,   (f) $3 \cos^2 x = 2 \sin x \cos x$.

8. By rearranging as quadratic equations, solve the following for $-180° \leqslant \theta \leqslant 180°$
(a) $2 \tan \theta - \cot \theta = 1$,   (b) $2 \operatorname{cosec} \theta + 7 = 4 \sin \theta$,
(c) $2 \cos \theta + \sec \theta = 3$,   (d) $2 \sin \theta \tan \theta = 3 \sin \theta + 5 \cos \theta$.

9. Solve the given pairs of simultaneous equations, where x and y can take any values from $-180°$ to $+180°$ inclusive.
(a) $x + y = 180°$ (b) $x - y = 120°$
    $\tan (x - y) = \sqrt{3}$ $\sin (x + y) = 0.5$.

10. Draw flow charts for procedures to solve the equations (i) $a \sin x = b$, (ii) $a \cos x = b$, (iii) $a \sin x = b \cos x$, using a pocket calculator. Print out all possible values of x from $0°$ to $360°$ inclusive or an appropriate message if the equation has no solution. Check that your procedures deal with $a = 0$, $b = 0$ and $a = b$.

7.3 Properties of trigonometric ratios

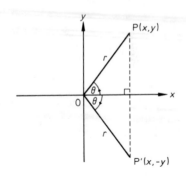

It follows from the definitions of $\sin \theta$, $\cos \theta$ and $\tan \theta$ that
$$\sin (-\theta) = -\sin \theta, \; \cos (-\theta) = \cos \theta,$$
$$\tan (-\theta) = -\tan \theta.$$

These results can also be obtained by considering the graphs of $\sin (-\theta)$, $\cos (-\theta)$ and $\tan (-\theta)$, which are the reflections in the y-axis of the graphs of $\sin \theta$, $\cos \theta$ and $\tan \theta$.

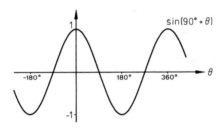

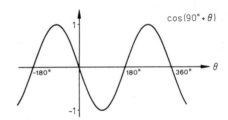

The diagrams above show that
$$\sin (90° + \theta) = \cos \theta \qquad\qquad \cos (90° + \theta) = -\sin \theta$$
By a similar process it is found that
$$\sin (180° + \theta) = -\sin \theta \qquad\qquad \cos (180° + \theta) = -\cos \theta$$
Substituting $-\theta$ for θ in these results
$$\sin (90° - \theta) = \cos (-\theta) \qquad\qquad \cos (90° - \theta) = -\sin (-\theta)$$
$$\therefore \quad \sin (90° - \theta) = \cos \theta \qquad\qquad \therefore \quad \cos (90° - \theta) = \sin \theta$$
Similarly $\quad \sin (180° - \theta) = \sin \theta, \cos (180° - \theta) = -\cos \theta.$

[The graphs of $\sin (90° - \theta)$, $\cos (90° - \theta)$, $\sin (180° - \theta)$ and $\cos (180° - \theta)$ may be obtained by reflection then translation of the graphs of $\sin \theta$ and $\cos \theta$.]

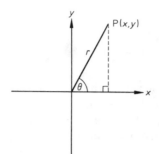

Three basic properties of the trigonometric ratios are derived from Pythagoras'† theorem.

For all values of θ, $x^2 + y^2 = r^2$,

so that $\dfrac{x^2}{r^2} + \dfrac{y^2}{r^2} = 1$.

Thus

$$\boxed{\cos^2\theta + \sin^2\theta = 1.}$$

Dividing by $\cos^2\theta$: $\dfrac{\cos^2\theta}{\cos^2\theta} + \dfrac{\sin^2\theta}{\cos^2\theta} = \dfrac{1}{\cos^2\theta}$

$$\therefore \quad \boxed{1 + \tan^2\theta = \sec^2\theta.}$$

Dividing by $\sin^2\theta$: $\dfrac{\cos^2\theta}{\sin^2\theta} + \dfrac{\sin^2\theta}{\sin^2\theta} = \dfrac{1}{\sin^2\theta}$

$$\therefore \quad \boxed{\cot^2\theta + 1 = \operatorname{cosec}^2\theta.}$$

Such relationships between the trigonometric ratios are called *identities*, because they hold for all values of θ. The word 'equation' is used for relationships which hold for only certain values of θ.

Example 1   If $\sin\theta = 5/13$, where θ is an obtuse angle, find $\cos\theta$ and $\tan\theta$.

$$\cos^2\theta = 1 - \sin^2\theta = 1 - \left(\frac{5}{13}\right)^2 = 1 - \frac{25}{169} = \frac{144}{169}$$

Since θ is obtuse, $\cos\theta = -12/13$

and $\tan\theta = \dfrac{\sin\theta}{\cos\theta} = \dfrac{5}{13}\bigg/\left(-\dfrac{12}{13}\right) = -5/12.$

Alternative method

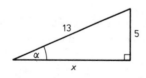

The diagram shows a right-angled triangle constructed so that $\sin\alpha = 5/13$.
By Pythagoras' theorem, $x = 12$.
Hence $\cos\alpha = 12/13$, $\tan\alpha = 5/12$.

†*Pythagoras* (6th cent. B.C.) Greek philosopher. He founded a religious brotherhood which devoted much of its attention to geometry and the theory of numbers. It is believed that Pythagoras or his early followers discovered the theorem that bears his name.

Since θ is an obtuse angle such that $\theta = 5/13$, $\cos \theta = -12/13$ and $\tan \theta = -5/12$.

Example 2 Simplify $\dfrac{\sin^2 \theta}{1 - \cos \theta}$.

$$\frac{\sin^2 \theta}{1 - \cos \theta} = \frac{1 - \cos^2 \theta}{1 - \cos \theta} = \frac{(1 - \cos \theta)(1 + \cos \theta)}{1 - \cos \theta} = 1 + \cos \theta.$$

Example 3 Eliminate θ from the equations

$$x = 1 + 2\cos \theta, \ y = 3 \sin \theta.$$

Rearranging $\qquad \dfrac{x - 1}{2} = \cos \theta, \ \dfrac{y}{3} = \sin \theta$

Since $\cos^2 \theta + \sin^2 \theta = 1$, $\qquad \dfrac{(x-1)^2}{4} + \dfrac{y^2}{9} = 1$.

Example 4 Solve the equation $\sec^2 \theta + \tan \theta - 1 = 0$ for $0° \leqslant \theta \leqslant 360°$.

Substituting $\sec^2 \theta = 1 + \tan^2 \theta$,
$$\sec^2 \theta + \tan \theta - 1 = 0 \Leftrightarrow (1 + \tan^2 \theta) + \tan \theta - 1 = 0$$
$$\Leftrightarrow \qquad \tan^2 \theta + \tan \theta = 0$$
$$\Leftrightarrow \qquad \tan \theta (\tan \theta + 1) = 0$$
$\therefore$ either $\tan \theta = 0 \qquad\qquad$ or $\tan \theta = -1$
$$\theta = 0°, 180°, 360° \qquad\qquad \theta = 135°, 315°.$$
Hence the required solutions are $0°$, $135°$, $180°$, $315°$ and $360°$.

Exercise 7.3

1. Sketch the graphs of the following functions and use them to verify the results given in the text
(a) $\sin(-\theta)$, $\cos(-\theta)$,
(b) $\sin(180° + \theta)$, $\cos(180° + \theta)$,
(c) $\sin(90° - \theta)$, $\cos(90° - \theta)$,
(d) $\sin(180° - \theta)$, $\cos(180° - \theta)$.

2. By drawing sketch graphs, or otherwise, express the following in terms of $\sin \theta$ or $\cos \theta$
(a) $\sin(360° + \theta)$,
(b) $\cos(360° - \theta)$,
(c) $\cos(270° + \theta)$,
(d) $\sin(\theta - 90°)$,
(e) $\cos(\theta - 180°)$,
(f) $\sin(270° - \theta)$.

3. By drawing sketch graphs, or otherwise, express the following in terms of $\tan \theta$ or $\cot \theta$
(a) $\tan(90° + \theta)$,
(b) $\tan(180° + \theta)$,
(c) $\tan(90° - \theta)$,
(d) $\tan(180° - \theta)$,
(e) $\tan(\theta - 360°)$,
(f) $\tan(\theta - 90°)$.

In questions 4 to 7 do not use tables or calculators.

4. If $\cos \theta = 4/5$ and θ is an acute angle, find $\sin \theta$, $\tan \theta$ and $\sec \theta$.

5. If $\tan \theta = -8/15$ and θ is obtuse, find $\sin \theta$, $\sec \theta$ and $\cot \theta$.

6. If $\cot x = 7/24$ and $90° < x < 360°$, find $\sin x$, $\cos x$ and $\operatorname{cosec} x$.

7. If $\sin x = 12/13$ and $0° < x < 90°$, find $\cos x$, $\cos(x+90°)$ and $\tan(x+180°)$.

8. Simplify the following expressions, where $0° < \theta < 90°$:

(a) $\dfrac{\sqrt{(1-\cos^2 \theta)}}{\tan \theta}$,   (b) $\dfrac{\sin \theta}{\sqrt{(1-\sin^2 \theta)}}$,   (c) $\dfrac{\tan \theta}{\sqrt{(1+\tan^2 \theta)}}$.

9. Simplify the following expressions
(a) $\sec^2 \theta - \tan^2 \theta$,

(b) $(\sin^2 \theta - 2)^2 - 4\cos^2 \theta$,

(c) $\dfrac{\cos^4 \theta - \sin^4 \theta}{\cos \theta - \sin \theta}$,

(d) $\dfrac{\sin \theta}{\operatorname{cosec} \theta - \cot \theta}$.

10. Prove the following identities
(a) $\sec \theta - \cos \theta = \sin \theta \tan \theta$,

(b) $\sec^2 \theta + \operatorname{cosec}^2 \theta = \sec^2 \theta \operatorname{cosec}^2 \theta$,

(c) $\dfrac{1-\sin \theta}{1+\sin \theta} = (\sec \theta - \tan \theta)^2$,

(d) $\dfrac{\cot^2 \theta - 1}{\cot^2 \theta + 1} = 1 - 2\sin^2 \theta$.

11. Eliminate θ from the following pairs of equations
(a) $x = \cos \theta$, $y = \sin \theta$,
(b) $x = 1 - 2\sin \theta$, $y = 1 + 3\cos \theta$,
(c) $x = \sec \theta$, $y = \tan \theta$,
(d) $x = \sin \theta + \cos \theta$, $y = \sin \theta - \cos \theta$.

12. Solve for t and θ, where $0° \leqslant \theta \leqslant 360°$, the simultaneous equations $\sin \theta = 2t$, $2\cos \theta + 3t = 0$.

In questions 13 to 22 solve the given equations for $0° \leqslant \theta \leqslant 360°$.

13. $3\cos^2 \theta + 5\sin \theta - 1 = 0$.
14. $8\sin^2 \theta + 2\cos \theta - 5 = 0$.
15. $2\sin \theta = 3\cot \theta$.
16. $\sec \theta \tan \theta = 2$.

17. $\tan^2 \theta = \sec \theta + 5$.
18. $7\sin^2 \theta + \cos^2 \theta = 5\sin \theta$.

19. $2\sec^2 \theta + 3\tan \theta = 4$.
20. $4\cot^2 \theta - 3\operatorname{cosec}^2 \theta = 2\cot \theta$.

21. $\sec \theta = 3\cos \theta + \sin \theta$.
22. $4\sec^2 \theta = 5\tan \theta + 3\tan^2 \theta$

23. Express $2\sin x - \cos^2 x$ in the form $(\sin x + a)^2 + b$ and hence find the maximum and minimum values of the expression.

24. Find the range of the function $f : x \rightarrow \sec^2 x - 2\tan x + 1$.

7.4 Addition formulae

This section deals with expressions for $\sin(A \pm B)$, $\cos(A \pm B)$ in terms of sines and cosines of A and B. [These may also be derived by matrix methods.]

In the diagrams below, OP and OQ make angles A and B respectively with the positive direction of the x-axis, so that $\angle POQ = A - B$. Letting $OP = OQ = 1$, the coordinates of P and Q are $P(\cos A, \sin A)$ and $Q(\cos B, \sin B)$.

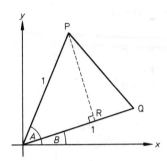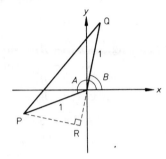

The distance between the points P and Q is given by
$$PQ^2 = (\cos A - \cos B)^2 + (\sin A - \sin B)^2$$
$$= \cos^2 A - 2\cos A \cos B + \cos^2 B + \sin^2 A - 2\sin A \sin B + \sin^2 B$$
$$= (\cos^2 A + \sin^2 A) + (\cos^2 B + \sin^2 B) - 2(\cos A \cos B + \sin A \sin B)$$
$$= 2 - 2(\cos A \cos B + \sin A \sin B).$$
Applying Pythagoras' theorem to $\triangle PQR$, $PQ^2 = PR^2 + RQ^2$
For all values of A and B this leads to
$$PQ^2 = \{\sin(A - B)\}^2 + \{1 - \cos(A - B)\}^2$$
$$= \sin^2(A - B) + \cos^2(A - B) + 1 - 2\cos(A - B)$$
$$= 2 - 2\cos(A - B).$$
Comparing these two expressions for PQ^2, we have

$$\cos(A - B) = \cos A \cos B + \sin A \sin B. \tag{1}$$

Substituting $-B$ for B in this expression

$$\cos(A + B) = \cos A \cos(-B) + \sin A \sin(-B)$$
$$\therefore \quad \cos(A + B) = \cos A \cos B - \sin A \sin B. \tag{2}$$

Substituting $90° - A$ for A in (1)

$$\cos(90° - A - B) = \cos(90° - A)\cos B + \sin(90° - A)\sin B$$
$$\therefore \quad \sin(A + B) = \sin A \cos B + \cos A \sin B. \tag{3}$$

Substituting $-B$ for B in (3)

$$\sin(A - B) = \sin A \cos(-B) + \cos A \sin(-B)$$
$$\therefore \quad \sin(A - B) = \sin A \cos B - \cos A \sin B. \tag{4}$$

Expressions for $\tan(A + B)$ and $\tan(A - B)$ can be deduced from identities (1) to (4).

$$\tan(A + B) = \frac{\sin(A + B)}{\cos(A + B)} = \frac{\sin A \cos B + \cos A \sin B}{\cos A \cos B - \sin A \sin B}.$$

Dividing numerator and denominator by $\cos A \cos B$

$$\tan (A+B) = \left(\frac{\sin A}{\cos A} + \frac{\sin B}{\cos B}\right) \Big/ \left(1 - \frac{\sin A}{\cos A} \cdot \frac{\sin B}{\cos B}\right)$$

$$\therefore \qquad \tan (A+B) = \frac{\tan A + \tan B}{1 - \tan A \tan B}.$$

Similarly,

$$\tan (A-B) = \frac{\tan A - \tan B}{1 + \tan A \tan B}.$$

Summarising: $\sin (A+B) = \sin A \cos B + \cos A \sin B$

$\sin (A-B) = \sin A \cos B - \cos A \sin B$

$\cos (A+B) = \cos A \cos B - \sin A \sin B$

$\cos (A-B) = \cos A \cos B + \sin A \sin B$

$$\tan (A+B) = \frac{\tan A + \tan B}{1 - \tan A \tan B}, \quad \tan (A-B) = \frac{\tan A - \tan B}{1 + \tan A \tan B}.$$

Example 1   Find, without using tables, expressions in surd form for (a) $\sin 15°$, (b) $\tan 105°$.

(a) $\sin 15° = \sin (60° - 45°) = \sin 60° \cos 45° - \cos 60° \sin 45°$

$$= \frac{\sqrt{3}}{2} \cdot \frac{1}{\sqrt{2}} - \frac{1}{2} \cdot \frac{1}{\sqrt{2}} = \frac{\sqrt{3}-1}{2\sqrt{2}} \quad \text{or} \quad \frac{1}{4}(\sqrt{6} - \sqrt{2})$$

(b) $\tan 105° = \tan (60° + 45°) = \dfrac{\tan 60° + \tan 45°}{1 - \tan 60° \tan 45°}$

$$= \frac{\sqrt{3}+1}{1-\sqrt{3}} = \frac{(\sqrt{3}+1)^2}{(1-\sqrt{3})(1+\sqrt{3})} = \frac{3+2\sqrt{3}+1}{1-3} = -2-\sqrt{3}.$$

Example 2   Solve the equation $\cos (\theta - 30°) = 2 \sin \theta$ for values of θ between $0°$ and $360°$.

$$\cos (\theta - 30°) = 2 \sin \theta$$

$$\Rightarrow \qquad \cos \theta \cos 30° + \sin \theta \sin 30° = 2 \sin \theta$$

$$\Rightarrow \qquad \frac{\sqrt{3}}{2} \cos \theta + \frac{1}{2} \sin \theta = 2 \sin \theta$$

$$\Rightarrow \qquad \sqrt{3} \cos \theta = 3 \sin \theta$$

Hence $\tan \theta = \dfrac{1}{\sqrt{3}} = \tan 30°$, so that $\theta = 30°$ or $210°$.

Example 3   Find, without using tables, the values of

(a) $\dfrac{1}{\sqrt{2}} (\cos 75° - \sin 75°)$ (b) $\dfrac{1 + \tan 15°}{1 - \tan 15°}$

(a) $\dfrac{1}{\sqrt{2}}(\cos 75° - \sin 75°) = \cos 45° \cos 75° - \sin 45° \sin 75° = \cos (45° + 75°)$

$= \cos 120° = -\tfrac{1}{2}.$

(b) $\dfrac{1 + \tan 15°}{1 - \tan 15°} = \dfrac{\tan 45° + \tan 15°}{1 - \tan 45° \tan 15°} = \tan (45° + 15°) = \tan 60° = \sqrt{3}.$

Exercise 7.4

Answer questions 1 to 10 without using tables or calculators.

1. Find expressions in surd form for
(a) $\sin 75°$, (b) $\cos 105°$, (c) $\tan (-15°)$, (d) $\cot 75°$.

2. Find the values of
(a) $\sin 80° \cos 70° + \cos 80° \sin 70°$, (b) $\dfrac{1}{\sqrt{2}} \cos 15° - \dfrac{1}{\sqrt{2}} \sin 15°$,

(c) $\cos 105° \cos 15° + \sin 105° \sin 15°$, (d) $\dfrac{\sqrt{3}}{2} \sin 60° + \dfrac{1}{2} \cos 60°$.

3. Find the values of

(a) $\dfrac{\tan 40° + \tan 20°}{1 - \tan 40° \tan 20°}$, (b) $\dfrac{\tan 75° - 1}{\tan 75° + 1}$,

(c) $\cos 75° + \sin 75°$, (d) $\sqrt{3} \cos 15° - \sin 15°$.

4. If $\sin A = 12/13$ and $\sin B = 4/5$, where A and B are acute angles, find $\sin (A + B)$ and $\cos (A + B)$.

5. If $\cos A = 5/7$ and $\sin B = 1/5$, where A is acute and B is obtuse, find $\sin (A - B)$ and $\cos (A - B)$.

6. If $\tan A = 1/2$ and $\tan B = -1/3$, where $180° < A < 360°$ and $-90° < B < 90°$, find $A - B$.

7. If $\tan (A - B) = 2$ and $\tan B = \tfrac{1}{4}$, find $\tan A$.

8. Simplify the following expressions
(a) $\sin (A + B) + \sin (A - B)$, (b) $\cos (A + B) + \cos (A - B)$,
(c) $\cos A \cos (A - B) + \sin A \sin (A - B)$.

9. Solve the equation $\cos 3x \cos 2x - \sin 3x \sin 2x = 0.5$ for $0° \leqslant x \leqslant 180°$.

10. Solve the equation $\sin 5x \cos 3x - \cos 5x \sin 3x = 1$ for $-180° \leqslant x \leqslant 180°$.

11. Solve the following equations for $0° \leqslant \theta \leqslant 360°$
(a) $2 \tan \theta = 3 \tan (45° - \theta)$, (b) $\sin \theta = 2 \sin (60° - \theta)$,
(c) $\cos (\theta - 30°) = \cos (\theta + 30°)$, (d) $\sin (\theta + 45°) = 5 \cos (\theta - 45°)$.

12. Express $\cot(A+B)$ in terms of $\cot A$ and $\cot B$.

13. By writing $A = \tan^{-1}x$ and $B = \tan^{-1}y$, show that under certain conditions
$$\tan^{-1}x + \tan^{-1}y = \tan^{-1}\left(\frac{x+y}{1-xy}\right).$$

14. Use the result of question 13 to show that

(a) $\tan^{-1}\dfrac{1}{3} + \tan^{-1}\dfrac{1}{2} = \tan^{-1}1$,

(b) $\tan^{-1}\dfrac{1}{5} + \tan^{-1}\dfrac{1}{8} = \tan^{-1}\dfrac{1}{3}$.

7.5 Double and half angle formulae

Substituting $B = A$ in the expressions obtained in the previous section for $\sin(A+B)$, $\cos(A+B)$ and $\tan(A+B)$:

$$\sin 2A = \sin(A+A) = \sin A \cos A + \cos A \sin A = 2\sin A \cos A$$
$$\cos 2A = \cos(A+A) = \cos A \cos A - \sin A \sin A = \cos^2 A - \sin^2 A$$
$$\tan 2A = \tan(A+A) = \frac{\tan A + \tan A}{1 - \tan A \tan A} = \frac{2\tan A}{1 - \tan^2 A}.$$

Using the identity $\cos^2 A + \sin^2 A = 1$,

$$\cos 2A = \cos^2 A - (1 - \cos^2 A) = 2\cos^2 A - 1$$
$$\cos 2A = (1 - \sin^2 A) - \sin^2 A = 1 - 2\sin^2 A.$$

Thus we have several useful results:

$$
\begin{array}{ll}
\cos 2A = \cos^2 A - \sin^2 A & \sin 2A = 2\sin A \cos A \\
\qquad\;\; = 2\cos^2 A - 1 & \\
\qquad\;\; = 1 - 2\sin^2 A & \tan 2A = \dfrac{2\tan A}{1 - \tan^2 A}.
\end{array}
$$

Example 1 Solve the equation $1 - 2\sin\theta - 4\cos 2\theta = 0$ for values of θ between $-180°$ and $+180°$.

Substituting $\cos 2\theta = 1 - 2\sin^2\theta$,
$$1 - 2\sin\theta - 4\cos 2\theta = 0 \Leftrightarrow 1 - 2\sin\theta - 4(1 - 2\sin^2\theta) = 0$$
$$\Leftrightarrow \quad 8\sin^2\theta - 2\sin\theta - 3 = 0$$
$$\Leftrightarrow \quad (4\sin\theta - 3)(2\sin\theta + 1) = 0$$
$$\therefore \quad \sin\theta = \tfrac{3}{4} = \sin 48\!\cdot\!6° \quad \text{or} \quad \sin\theta = -\tfrac{1}{2} = -\sin 30°$$
$$\theta = 48\!\cdot\!6°,\ 131\!\cdot\!4° \qquad\qquad \theta = -150°,\ -30°.$$
Hence the required solutions are $-150°$, $-30°$, $48\!\cdot\!6°$, $131\!\cdot\!4°$.

Example 2 Express $\sin 3A$ in terms of $\sin A$.

$$\sin 3A = \sin(2A+A) = \sin 2A \cos A + \cos 2A \sin A$$
$$= 2\sin A \cos A \cos A + (\cos^2 A - \sin^2 A)\sin A$$

$$= 3\sin A\cos^2 A - \sin^3 A$$
$$= 3\sin A(1-\sin^2 A) - \sin^3 A$$
$$\therefore \quad \sin 3A = 3\sin A - 4\sin^3 A.$$

Example 3 Simplify $\dfrac{\sin x}{1+\cos x}$.

$$\sin x = 2\sin\tfrac{1}{2}x\cos\tfrac{1}{2}x, \quad \cos x = 2\cos^2\tfrac{1}{2}x - 1$$

$$\therefore \quad \frac{\sin x}{1+\cos x} = \frac{2\sin\tfrac{1}{2}x\cos\tfrac{1}{2}x}{2\cos^2\tfrac{1}{2}x} = \frac{\sin\tfrac{1}{2}x}{\cos\tfrac{1}{2}x} = \tan\tfrac{1}{2}x.$$

Two further important identities are obtained by rearranging expressions for $\cos 2A$:

$$\boxed{\cos^2 A = \tfrac{1}{2}(1+\cos 2A), \quad \sin^2 A = \tfrac{1}{2}(1-\cos 2A).}$$

Example 4 Express $\cos^4 A$ in terms of cosines of multiples of A.

$$\cos^4 A = (\cos^2 A)^2 = \tfrac{1}{4}(1+\cos 2A)^2 = \tfrac{1}{4}(1+2\cos 2A + \cos^2 2A)$$
$$= \tfrac{1}{4}(1+2\cos 2A + \tfrac{1}{2}\{1+\cos 4A\}) = \tfrac{1}{8}(2+4\cos 2A + 1 + \cos 4A)$$
$$\therefore \quad \cos^4 A = \tfrac{1}{8}(3+4\cos 2A + \cos 4A).$$

If other methods fail, equations can often be solved and identities proved by expressing $\sin\theta$, $\cos\theta$ and $\tan\theta$ in terms of t, where $t = \tan\tfrac{1}{2}\theta$.

Using the formula for $\tan 2A$, $\tan\theta = \dfrac{2t}{1-t^2}$.

$$\cos\theta = \cos^2\tfrac{1}{2}\theta - \sin^2\tfrac{1}{2}\theta = \cos^2\tfrac{1}{2}\theta(1-\tan^2\tfrac{1}{2}\theta) = \frac{1-\tan^2\tfrac{1}{2}\theta}{\sec^2\tfrac{1}{2}\theta} = \frac{1-t^2}{1+t^2}.$$

We may now write $\sin\theta = \cos\theta\tan\theta$ or obtain the result independently,

$$\sin\theta = 2\sin\tfrac{1}{2}\theta\cos\tfrac{1}{2}\theta = 2\tan\tfrac{1}{2}\theta\cos^2\tfrac{1}{2}\theta = \frac{2\tan\tfrac{1}{2}\theta}{\sec^2\tfrac{1}{2}\theta} = \frac{2t}{1+t^2}.$$

Thus
$$\boxed{\sin\theta = \frac{2t}{1+t^2}, \quad \cos\theta = \frac{1-t^2}{1+t^2}, \quad \tan\theta = \frac{2t}{1-t^2}.}$$

The main disadvantage of using these '*t* formulae' to solve trigonometric equations is that polynomial equations involving large powers of t may be produced. It is not always possible to solve these by elementary methods.

Example 5 Solve the equation $5\tan\theta + \sec\theta + 5 = 0$ for values of θ between $0°$ and $360°$.

If $\tan \frac{1}{2}\theta = t$, then provided that $t^2 \neq 1$, the equation becomes

$$5\left(\frac{2t}{1-t^2}\right) + \left(\frac{1+t^2}{1-t^2}\right) + 5 = 0$$
$$10t + (1+t^2) + 5(1-t^2) = 0$$
$$-4t^2 + 10t + 6 = 0$$
$$2t^2 - 5t - 3 = 0$$
$$(t-3)(2t+1) = 0$$

$\therefore$ either $t = 3$ or $t = -\frac{1}{2}$

$\tan \frac{1}{2}\theta = \tan 71{\cdot}565°$ $\tan \frac{1}{2}\theta = -\tan 26{\cdot}565°$

Hence, between $0°$ and $180°$, $\frac{1}{2}\theta = 71{\cdot}565°$ or $153{\cdot}435°$.
Thus the required values of θ are $143{\cdot}1°$ or $306{\cdot}9°$ (nrst $0{\cdot}1°$).

Exercise 7.5

Answer questions 1 to 7 without using tables or calculators.

1. Given that x is an acute angle such that $\cos x = 3/5$, find $\sin 2x$, $\cos 2x$ and $\sin 3x$.

2. Given that θ is an obtuse angle such that $\sin \theta = 2/3$, find $\cos 2\theta$, $\tan 2\theta$ and $\cos 4\theta$.

3. Given that A is an acute angle such that $\tan A = \frac{1}{2}$, find $\sin 2A$, $\tan 2A$ and $\tan 3A$.

4. Given that x is an obtuse angle such that $\cos 2x = 1/8$, find $\cos x$, $\tan x$ and $\sec \frac{1}{2}x$.

5. Given that $\sin A = -0{\cdot}8$ where $0° < A < 270°$, find $\cos \frac{1}{2}A$, $\sin \frac{1}{2}A$ and $\sin \frac{3}{2}A$.

6. Find the values of the following expressions
(a) $(\sin 67\frac{1}{2}° + \cos 67\frac{1}{2}°)^2$,   (b) $\cos^2 15° - \sin^2 15°$,
(c) $\tan 75° \cos^2 75°$,   (d) $1 - 2\sin^2 22\frac{1}{2}°$.

7. Solve the following equations for $0° \leqslant x \leqslant 180°$
(a) $\sin 2x = \sin x$,   (b) $\cos 2x = \cos x$,
(c) $\tan 2x + \tan x = 0$,   (d) $\sin 2x - \tan x = 0$.

In questions 8 to 15 prove the given identities.

8. $\cos^4 A - \sin^4 A = \cos 2A$.   9. $\cot \theta - \tan \theta = 2\cot 2\theta$.

10. $\dfrac{\sin 3A}{\sin A} + \dfrac{\cos 3A}{\cos A} = 4\cos 2A$.   11. $\dfrac{\cos \theta - \sin \theta}{\cos \theta + \sin \theta} = \dfrac{\cos 2\theta}{1 + \sin 2\theta}$.

12. $\dfrac{\sec 2x - 1}{\sec 2x + 1} = \sec^2 x - 1$.   13. $\dfrac{\sin 2\theta - \cos 2\theta + 1}{\sin 2\theta + \cos 2\theta + 1} = \tan \theta$.

14. $\csc x - \cot x = \tan \frac{1}{2} x$.

15. $\tan \left(\frac{1}{2}x + 45°\right) + \cot \left(\frac{1}{2}x + 45°\right) = 2 \sec x$.

16. Solve the following equations for $0° \leqslant x \leqslant 360°$
(a) $\cos 2x = 5 \cos x + 2$,   (b) $3 \cos 2x + 1 = 2 \sin x$,
(c) $4 \sin^2 x - \tan^2 x = 1$,   (d) $4 \sin x = 7 \tan 2x$.

17. By expressing $\cos 2\theta$ and $\sin 2\theta$ in terms of $\tan \theta$, solve the following equations for $0° \leqslant \theta \leqslant 360°$
(a) $\cos 2\theta - 2 \sin 2\theta = 2$,   (b) $5 \cos 2\theta - 2 \sin 2\theta = 2$.

18. Using the substitution $\tan \frac{1}{2}x = t$, solve the following equations for $-180° \leqslant x \leqslant 180°$
(a) $9 \cos x - 8 \sin x = 1$,   (b) $\sin x + 1 = 3 \cos x$,
(c) $2 \cos x - \sin x + 1 = 0$,   (d) $7 \sin x + 9 \cos x = 3$.

19. By squaring $\sin^2 \theta + \cos^2 \theta$, prove that $\sin^4 \theta + \cos^4 \theta = \frac{1}{4}(\cos 4\theta + 3)$. Hence solve the equation $\sin^4 \theta + \cos^4 \theta = \frac{1}{2}$ for $0° < \theta < 360°$.

20. Express $\cos 3\theta$ in terms of $\cos \theta$ and hence solve the equation $\cos 3\theta + 2 \cos 2\theta + 4 \cos \theta + 2 = 0$ for $0° < \theta < 360°$.

7.6 Sum and product formulae

In §7.4 it was shown that

$$\sin (A+B) = \sin A \cos B + \cos A \sin B$$
$$\sin (A-B) = \sin A \cos B - \cos A \sin B$$

Adding $\sin (A+B) + \sin (A-B) = 2 \sin A \cos B$ (1)
Subtracting $\sin (A+B) - \sin (A-B) = 2 \cos A \sin B$ (2)

Similarly $\cos (A+B) = \cos A \cos B - \sin A \sin B$
               $\cos (A-B) = \cos A \cos B + \sin A \sin B$

Adding $\cos (A+B) + \cos (A-B) = 2 \cos A \cos B$ (3)
Subtracting $\cos (A+B) - \cos (A-B) = -2 \sin A \sin B$.   (4)

These results provide a way of expressing a product of sines and cosines as a sum or a difference. Results (1), (3) and (4) cover all possible products.

$$2 \sin A \cos B = \sin (A+B) + \sin (A-B)$$
$$2 \cos A \cos B = \cos (A+B) + \cos (A-B)$$
$$2 \sin A \sin B = -\{\cos (A+B) - \cos (A-B)\}.$$

Example 1 Express as a sum or a difference (a) $2\cos 4\theta \cos \theta$, (b) $\sin 2x \cos 3x$.

(a) $2\cos 4\theta \cos \theta = \cos (4\theta + \theta) + \cos (4\theta - \theta) = \cos 5\theta + \cos 3\theta$.
(b) $\sin 2x \cos 3x = \frac{1}{2}\{\sin (2x + 3x) + \sin (2x - 3x)\}$
$= \frac{1}{2}\{\sin 5x + \sin (-x)\} = \frac{1}{2}\{\sin 5x - \sin x\}$.

Example 2 Find, without using tables, the value of $2\sin 15° \sin 45°$.

$2\sin 15° \sin 45° = -\{\cos (15° + 45°) - \cos (15° - 45°)\}$
$= -\cos 60° + \cos (-30°)$
$= -\frac{1}{2} + \frac{\sqrt{3}}{2} = \frac{1}{2}(\sqrt{3} - 1)$.

Results (1), (2), (3) and (4) can also be used to express sums and differences of sines and cosines as products.

Letting $A + B = X$, $A - B = Y$ so that $A = \frac{1}{2}(X + Y)$, $B = \frac{1}{2}(X - Y)$,

$$\begin{aligned}
\sin X + \sin Y &= 2\sin \tfrac{1}{2}(X + Y)\cos \tfrac{1}{2}(X - Y) \\
\sin X - \sin Y &= 2\cos \tfrac{1}{2}(X + Y)\sin \tfrac{1}{2}(X - Y) \\
\cos X + \cos Y &= 2\cos \tfrac{1}{2}(X + Y)\cos \tfrac{1}{2}(X - Y) \\
\cos X - \cos Y &= -2\sin \tfrac{1}{2}(X + Y)\sin \tfrac{1}{2}(X - Y).
\end{aligned}$$

As these identities. can be used to 'factorise' sums and differences, they are sometimes called the *factor formulae*.

Example 3 Simplify $\dfrac{\sin (A + B) - \sin A}{\cos (A + B) + \cos A}$.

$\sin (A + B) - \sin A = 2\cos \frac{1}{2}\{(A + B) + A\} \sin \frac{1}{2}\{(A + B) - A\}$
$= 2\cos (A + \frac{1}{2}B)\sin \frac{1}{2}B$.

Similarly $\cos (A + B) + \cos A = 2\cos (A + \frac{1}{2}B)\cos \frac{1}{2}B$

$\therefore \quad \dfrac{\sin (A + B) - \sin A}{\cos (A + B) + \cos A} = \dfrac{2\cos (A + \frac{1}{2}B)\sin \frac{1}{2}B}{2\cos (A + \frac{1}{2}B)\cos \frac{1}{2}B} = \tan \tfrac{1}{2}B$.

Example 4 Solve the equation $\sin 5\theta + \sin 3\theta = 0$ for values of θ between $0°$ and $180°$.

$\sin 5\theta + \sin 3\theta = 0 \Rightarrow 2\sin \frac{1}{2}(5\theta + 3\theta)\cos \frac{1}{2}(5\theta - 3\theta) = 0$
$\Rightarrow \qquad\qquad\qquad \sin 4\theta \cos \theta = 0$
$\therefore \qquad \sin 4\theta = 0 \qquad\qquad\qquad\qquad \text{or} \quad \cos \theta = 0$
$\qquad\qquad 4\theta = 0°, 180°, 360°, 540°, 720° \qquad\qquad \theta = 90°$
$\qquad\qquad\quad \theta = 0°, 45°, 90°, 135°, 180°$
Hence the solutions are $\theta = 0°, 45°, 90°, 135°, 180°$.

Example 5   Solve the equation $\cos 5\theta + \cos 3\theta + \cos \theta = 0$ for values of θ between $0°$ and $180°$.

$$\cos 5\theta + \cos 3\theta + \cos \theta = 0$$
$$\Rightarrow \quad (\cos 5\theta + \cos \theta) + \cos 3\theta = 0$$
$$\Rightarrow \quad 2\cos 3\theta \cos 2\theta + \cos 3\theta = 0$$
$$\Rightarrow \quad \cos 3\theta (2\cos 2\theta + 1) = 0$$

$\therefore \quad \cos 3\theta = 0 \qquad \qquad \text{or} \quad \cos 2\theta = -\frac{1}{2}$

$\qquad 3\theta = 90°, 270°, 450° \qquad \qquad 2\theta = 120°, 240°$

$\qquad \theta = 30°, 90°, 150° \qquad \qquad \theta = 60°, 120°.$

Hence the required values of θ are $30°, 60°, 90°, 120°, 150°$.

Exercise 7.6

1. Express as a sum or a difference
(a) $2\sin 2x \cos x$,
(b) $2\cos 3\theta \cos 5\theta$,
(c) $6\sin 3A \sin 2A$,
(d) $\sin 2t \cos 4t$,
(e) $2\cos (A+B)\cos (A-B)$,
(f) $\sin (x+75°)\sin (x+15°)$.

2. Find, without using tables, the values of
(a) $4\sin 75° \cos 45°$,
(b) $20\sin 22\frac{1}{2}° \sin 67\frac{1}{2}°$,
(c) $8\cos 52\frac{1}{2}° \cos 37\frac{1}{2}° \cos 15°$,
(d) $\tan 82\frac{1}{2}° \cot 52\frac{1}{2}°$.

3. Solve the following equations for $0° \leqslant \theta \leqslant 360°$
(a) $2\sin 2\theta \cos \theta = \sin 3\theta$,
(b) $2\cos (\theta + 120°)\cos (\theta + 60°) = 1$,
(c) $2\sin (\theta + 30°) = \sec \theta$,
(d) $\sin (\theta + 30°)\cos (\theta - 40°) = 0\cdot2$.

4. Express as a product
(a) $\sin x + \sin 7x$,
(b) $\cos 3A + \cos 5A$,
(c) $\sin (\theta + \alpha) - \sin (\theta - \alpha)$,
(d) $\cos 3x + \sin (x - 90°)$.

5. Simplify the following expressions
(a) $\dfrac{\cos 3\theta + \cos 7\theta}{\sin 3\theta + \sin 7\theta}$,
(b) $\dfrac{\cos 5x - \cos x}{\sin 5x - \sin x}$,
(c) $\dfrac{\cos A - \cos 2A + \cos 3A}{\sin A - \sin 2A + \sin 3A}$,
(d) $\dfrac{\cos \theta - 2\cos 3\theta + \cos 5\theta}{\cos \theta + 2\cos 3\theta + \cos 5\theta}$.

6. Find the values of
(a) $\dfrac{\cos 50° - \cos 70°}{\sin 70° - \sin 50°}$,
(b) $\dfrac{\sin \theta + \sin (\theta + 120°)}{\cos (60° - \theta) + \cos \theta}$.

7. Solve the following equations for $0° \leqslant x \leqslant 180°$
(a) $\cos 3x = \cos x$,
(b) $\sin 5x = \sin x$,
(c) $\sin (3x + 60°) = \sin x$,
(d) $\sin 3x = \cos x$.

8. Solve the following equations for $0° \leqslant \theta \leqslant 180°$
(a) $\sin \theta + \sin 2\theta + \sin 3\theta = 0$,
(b) $\sin 5\theta - \sin \theta = \cos 3\theta$,
(c) $\cos 3\theta - \sin 3\theta = \cos 2\theta - \sin 2\theta$.

7.7 The expression $a \cos \theta + b \sin \theta$

A sketch of the graph of a function of the form $a \cos \theta + b \sin \theta$ can be made by combining the graphs of $a \cos \theta$ and $b \cos \theta$. The result for the function $\cos \theta + \sin \theta$ is shown below.

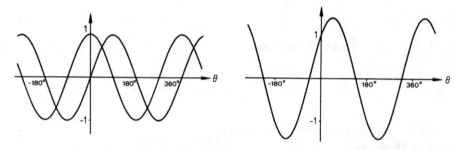

However, the exact form of the graph becomes clear when we apply one of the factor formulae derived in the previous section.

$$\cos \theta + \sin \theta = \cos \theta + \cos (90° - \theta) = 2 \cos 45° \cos (\theta - 45°)$$

$$= 2 \cdot \frac{1}{\sqrt{2}} \cos (\theta - 45°) = \sqrt{2} \cos (\theta - 45°).$$

It seems reasonable to investigate whether the function $a \cos \theta + b \sin \theta$ can be expressed in the form $r \cos (\theta - \alpha)$, where r is positive.

Since $r \cos (\theta - \alpha) = r(\cos \theta \cos \alpha + \sin \theta \sin \alpha)$
$\qquad = (r \cos \alpha) \cos \theta + (r \sin \alpha) \sin \theta$

the expressions $a \cos \theta + b \sin \theta$ and $r \cos (\theta - \alpha)$ will be equal for all values of θ if
$$r \cos \alpha = a, \qquad r \sin \alpha = b$$

$\therefore \quad a^2 + b^2 = r^2 (\cos^2 \alpha + \sin^2 \alpha) = r^2 \quad$ i.e. $r = \sqrt{(a^2 + b^2)}$

and $\quad \dfrac{b}{a} = \dfrac{r \sin \alpha}{r \cos \alpha} = \tan \alpha \qquad$ i.e. $\tan \alpha = b/a$.

Hence the function $a \cos \theta + b \sin \theta$ can be expressed in the form $r \cos (\theta - \alpha)$, where $r = \sqrt{(a^2 + b^2)}$, $\tan \alpha = b/a$.

[Note that since r is assumed positive, the value of α must be chosen so that $\cos \alpha$ has the same sign as a and $\sin \alpha$ the same sign as b.]

Example 1   Express $3 \cos \theta + 4 \sin \theta$ in the form $r \cos (\theta - \alpha)$. Hence find the maximum and minimum values of $3 \cos \theta + 4 \sin \theta$.

If $3\cos\theta + 4\sin\theta = r\cos(\theta - \alpha) = r\cos\alpha\cos\theta + r\sin\alpha\sin\theta$
then $r\cos\alpha = 3$ and $r\sin\alpha = 4$.

Since both $r\cos\alpha$ and $r\sin\alpha$ are positive, we may take $r > 0$ and $0° < \alpha < 90°$.

$\therefore$ $r = \sqrt{(3^2 + 4^2)} = 5$ and $\tan\alpha = 4/3, \alpha = 53\cdot13°$

Hence $3\cos\theta + 4\sin\theta = 5\cos(\theta - 53\cdot1°)$.

Since the maximum and minimum values of $\cos(\theta - 53\cdot1°)$ are $+1$ and -1 respectively, the maximum and minimum values of $3\cos\theta + 4\sin\theta$ must be $+5$ and -5.

Example 2   Solve the equation $3\cos\theta - \sin\theta = 1$ for values of θ between $0°$ and $360°$.

Let $3\cos\theta - \sin\theta = r\cos(\theta + \alpha) = r\cos\alpha\cos\theta - r\sin\alpha\sin\theta$
then $r\cos\alpha = 3$ and $r\sin\alpha = 1$

$\therefore$ $r = \sqrt{(3^2 + 1^2)} = \sqrt{10}$ and $\tan\alpha = 1/3, \alpha = 18\cdot435°$.

Hence the equation $3\cos\theta - \sin\theta = 1$
may be written $\sqrt{10}\cos(\theta + 18\cdot435°) = 1$
$\therefore$ $\cos(\theta + 18\cdot435°) = 1/\sqrt{10} = \cos 71\cdot565°$
$\therefore$ $\theta + 18\cdot435° = 71\cdot565°$ or $288\cdot435°$.

Hence the required solutions are $53\cdot1°$ and $270°$.

In the above example it was more convenient to use $r\cos(\theta + \alpha)$ instead of $r\cos(\theta - \alpha)$, because this led to a value of α between $0°$ and $90°$. The expressions $r\sin(\theta + \alpha)$ and $r\sin(\theta - \alpha)$ may also be used when appropriate.

Exercise 7.7

1.   Express each of the following functions in the form $r\cos(\theta \pm \alpha)$, where $r > 0$ and $0° < \alpha < 90°$. Hence find the maximum and minimum values of the function, giving the values of θ between $-360°$ and $+360°$ for which these occur. Sketch the graph of the function in this interval.
(a) $\cos\theta - \sin\theta$,   (b) $\sqrt{3}\cos\theta - \sin\theta$,
(c) $5\cos\theta + 12\sin\theta$,   (d) $2\cos\theta + \sin\theta$.

2.   Using an appropriate expression of the form $r\cos(\theta \pm \alpha)$, solve the following equations for $0° \leqslant \theta \leqslant 360°$.
(a) $4\cos\theta - 3\sin\theta = 1$,   (b) $2\cos\theta + 5\sin\theta = 4$,
(c) $3\cos\theta + 7\sin\theta + 3 = 0$,   (d) $15\cos\theta - 8\sin\theta + 10 = 0$.

3.   Express $5\sin\theta - 8\cos\theta$ in the form $r\sin(\theta - \alpha)$, where $r > 0$ and $0° < \alpha < 90°$. Hence solve the following equations for $0° \leqslant \theta \leqslant 360°$.
(a) $5\sin\theta - 8\cos\theta = 6$,   (b) $5\sin\theta - 8\cos\theta = 5$.

4.   Solve the following equations for $0° \leqslant x \leqslant 360°$,

(i) by expressing $\cos x$ and $\sin x$ in terms of $t = \tan\frac{1}{2}x$, (ii) using an expression of the form $r\cos(x-\alpha)$.

(a) $3\cos x - 5\sin x = 2$, (b) $4\cos x - 6\sin x = 5$.

5. Construct a flow chart for a procedure to solve the equation $a\cos\theta + b\sin\theta = c$, where $a > 0$, using a pocket calculator. The output of your procedure should be the values of θ, for $0° \leqslant \theta \leqslant 360°$, or a suitable message if there are no solutions. Consider whether accuracy or efficiency would be improved by incorporating in your chart branches or jumps for special cases such as $a = b$ or $b = c$.

Exercise 7.8 (miscellaneous)

1. Find all values of θ from $0°$ to $360°$ inclusive, such that
(a) $\sin\theta = \cos 20°$, (b) $\sin 2\theta = -\sin 50°$,
(c) $\cos\theta = -\cos(-35°)$, (d) $\tan 3\theta = \cot 15°$.

2. Given that $\cos x = -0.8$, where $0° < x < 180°$, find the values of
(a) $\cos(x-180°)$, (b) $\cos(x+90°)$, (c) $\sin(90°-x)$,
(d) $\sin(x+180°)$, (e) $\tan(x+360°)$, (f) $\cot(x-90°)$.

3. Solve the following equations for $0° \leqslant x \leqslant 360°$.
(a) $2\sin x = 3\cos x$, (b) $3\cos x - \sec x = \frac{1}{2}$,
(c) $3\tan x = 5\sin x$, (d) $5\cos 2x + 1 = 11\sin x$.

4. Given that θ is an obtuse angle and that $\cos\theta = k$, find the following in terms of k.
(a) $\cos 2\theta$, (b) $\sin 3\theta$, (c) $\cos\frac{1}{2}\theta$, (d) $\tan\frac{1}{2}\theta$.

5. Eliminate θ from the following pairs of equations.
(a) $x = 1 + \sin\theta$ (b) $x = \tan\theta + \sec\theta$
   $y = \frac{1}{2}\cos\theta$, $y = \sec\theta - \tan\theta$,
(c) $x = \cos\theta$ (d) $x = 1 + \cos\theta$
   $y = \cos 2\theta + 1$, $y = \sin 2\theta$.

6. Solve the following equations for $0° \leqslant x \leqslant 180°$.
(a) $\dfrac{\tan x}{1-\tan^2 x} = \dfrac{1}{5}$, (b) $\dfrac{\tan x}{1+\tan^2 x} = \dfrac{1}{5}$,
(c) $\sin 3x = \sin x$, (d) $\sin 4x + \sin 2x = 0$.

7. Sketch the curves $y = 2\sin x$ and $y = 3\cos x$ for $0° \leqslant x \leqslant 360°$. Find the range of values of x in this interval for which $2\sin x \geqslant 3\cos x$.

8. Given that $\tan A = \frac{2}{3}$ and $\tan B = \frac{1}{2}$, find the value of

$$\frac{\cos 2A + \cos 2B}{\sin 2A - \sin 2B}.$$

9. Sketch the graphs of the following functions for $-360° \leqslant \theta \leqslant 360°$, showing clearly the maximum and minimum values of each function.
(a) $2\cos\theta\sin\theta$,   (b) $2\sin^2\theta$,
(c) $\cos\theta + \sqrt{3}\sin\theta$,   (d) $(\cos\theta + \sqrt{3}\sin\theta)^2$.

10. Prove the following identities

(a) $\dfrac{1-\cos 2A}{1+\cos 2A} = \tan^2 A = \dfrac{\tan^2 A + 1}{\cot^2 A + 1}$,

(b) $\dfrac{\sin\theta}{1+\cos\theta} + \dfrac{1-\cos\theta}{\sin\theta} = 2\tan\tfrac{1}{2}\theta$,

(c) $\dfrac{\cos^4 A + \sin^4 A}{\cos^4 A - \sin^4 A} = \tfrac{1}{2}(\cos 2A + \sec 2A)$,

(d) $\dfrac{\sin(\theta+15°)+\sin(\theta-15°)}{\cos(\theta-15°)+\cos(\theta+15°)} = \tan\theta$.

11. (a) Prove that $15\cos 2\theta + 20\sin 2\theta + 7 \equiv 2(11\cos\theta - 2\sin\theta)(\cos\theta + 2\sin\theta)$. Hence, or otherwise, find all angles θ between $0°$ and $180°$ for which $15\cos 2\theta + 20\sin 2\theta + 7 = 0$, giving your answers to the nearest tenth of a degree.
(b) If $\sin(\theta+\alpha) = \lambda\sin(\theta-\alpha)$, where λ is a numerical constant $(\lambda \neq 1)$, find an expression for $\tan\theta$ in terms of $\tan\alpha$ and λ. In the case when $\lambda = \tfrac{1}{2}$ and $\alpha = 30°$, find the values of θ which lie between $0°$ and $360°$.   (C)

12. (a) Solve the equation $\cos x + \cos 2x + \cos 3x = 0$, giving all values between $0°$ and $360°$ (inclusive).
(b) Express $3\sin x + 4\cos x$ in the form $R\cos(x-\alpha)$, where R is a positive number. Hence, or otherwise, solve the equation $3\sin x + 4\cos x = 2$, giving all values between $0°$ and $360°$.
(c) If x and y are angles between $0°$ and $360°$ (inclusive), and $\sin^2 x + \cos^2 y = 2$, state what the possible values are of x and y.   (SU)

13. Express $\tan 3\theta$ in terms of $\tan\theta$. Hence solve the equation $2\tan 3\theta = 9\tan\theta$ for $0° \leqslant \theta \leqslant 180°$.

14. Write down expressions for $\tan\theta$ and $\sec\theta$ in terms of t, where $t = \tan\dfrac{\theta}{2}$, and show that $\sec\theta + \tan\theta = \tan\left(45° + \dfrac{\theta}{2}\right)$. Find a solution in the interval $0° < \theta < 90°$ of the equation $\sec\theta + \tan\theta = \cot 2\theta$.   (JMB)

15. (a) Express $7\sin x - 24\cos x$ in the form $R\sin(x-\alpha)$, where R is positive and α is an acute angle. Hence or otherwise solve the equation $7\sin x - 24\cos x = 15$, for $0° < x < 360°$.
(b) Solve the simultaneous equations $\cos x + \cos y = 1$, $\sec x + \sec y = 4$, for $0° < x < 180°, 0° < y < 180°$.   (AEB)

16. By expressing $\sin\theta$ and $\cos\theta$ in terms of $\tan\frac{1}{2}\theta$, solve the equation $\sin\theta+2\cos\theta = 1$, for $0° \leqslant \theta \leqslant 360°$. Explain why the method breaks down for the equation $2\sin\theta-\cos\theta = 1$ and find the values of θ, between $0°$ and $360°$, which satisfy this equation.

17. Prove that (a) $\cos A\cos 2A - \cos 2A\cos 5A = \sin 3A\sin 4A$
    (b) $\sin A\cos 2A + \sin 2A\cos 5A = \sin 3A\cos 4A$.

18. (a) Prove the identities $\cos\theta+\cos 3\theta+\cos 5\theta+\cos 7\theta \equiv 4\cos\theta\cos 2\theta\cos 4\theta$ $\equiv \sin 8\theta/2\sin\theta$.
(b) Solve for $0° < \theta < 360°$ the equation $\sec\theta-3\tan\theta = 2$.   (W)

19. Find, to the nearest $0.1°$, the acute angle α for which $4\cos\theta - 3\sin\theta$ $\equiv 5\cos(\theta+\alpha)$. Calculate the values of θ in the interval $-180° \leqslant \theta \leqslant 180°$ for which the function $f(\theta) = 4\cos\theta-3\sin\theta-4$ attains its greatest value, its least value and the value zero.   (JMB)

20. Solve the following equations for $0° \leqslant x \leqslant 180°$.
(a) $2\tan 3x\cot x+1 = 0$,   (b) $3(\sec 2x-\tan 2x) = 2\tan x$,
(c) $\cot 5x+\tan 2x = 0$,   (d) $4\sin x\cos 2x\sin 3x = 1$.

21. Prove that, if the equation $a\sin x+b\cos x = c$ is to have a solution, then $a^2+b^2 \geqslant c^2$. Find all the solutions of the equation $7\sin x+6\cos x = 9$ between $0°$ and $360°$, giving your answers to the nearest tenth of a degree.

22. Write down the values of the following in degrees.
(a) $\sin^{-1}\frac{1}{2}$,   (b) $\cos^{-1}(-1)$,   (c) $\tan^{-1}(-1)$,   (d) $\cot^{-1}\sqrt{3}$,

(e) $\sin^{-1}x+\cos^{-1}x$,   (f) $\tan^{-1}x+\tan^{-1}\frac{1}{x}$,

(g) $\tan^{-1}\frac{2}{3}+\tan^{-1}\frac{1}{5}$,   (h) $\tan^{-1}\frac{5}{2}-\tan^{-1}\frac{3}{7}$.

23. Prove that $\cos 3\theta = 4\cos^3\theta-3\cos\theta$. Hence solve the equation
$$x^3-3x-\sqrt{2} = 0$$
by using a suitable substitution of the form $x = k\cos\theta$, then finding possible values of θ between $0°$ and $180°$. Give your answers in surd form.

24. (a) Find all solutions of the equation $\sin 3\theta = \sin^2\theta$ for which $0° \leqslant \theta < 360°$.
(b) Express $\lambda\sin\theta+(1-\lambda)\cos\theta$ in the form $R\sin(\theta+\phi)$, where $R(R > 0)$ and $\tan\phi$ are to be given in terms of λ. Write down an expression, in terms of λ, for the minimum value of $\lambda\sin\theta+(1-\lambda)\cos\theta$ as θ varies, and show that, for all λ, this minimum is less than or equal to $-\frac{1}{2}\sqrt{2}$.   (C)

8 Trigonometry of triangles

8.1 Review of elementary geometry

[This introductory section contains a review of some of the language and results of elementary geometry which may be required when solving trigonometrical problems.]

In elementary geometrical work the notions of points and distances, straight lines and angles are introduced. Development of these basic ideas leads to the study of plane figures such as triangles and parallelograms. One set of important results involving circles is illustrated here. In diagram (1) angles subtended by the

(1)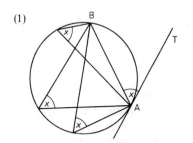

(2)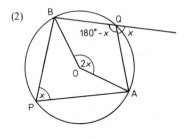

chord AB, lying in the same segment of the circle, are equal. The angle between the tangent AT and the chord AB is equal to any angle in the alternate segment. In diagram (2) the angle subtended by AB at the centre O is twice the angle subtended by AB at P on the circumference of the circle. Opposite angles of the cyclic quadrilateral $APBQ$ are supplementary and any exterior angle is equal to the opposite interior angle.

The ratio properties of similar figures have many applications. One result which may be proved using similar triangles is the angle bisector theorem. In the triangles ABC shown overleaf, BP bisects $\angle A$ internally in diagram (3) and externally in diagram (4).

155

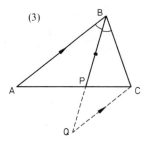

(3)

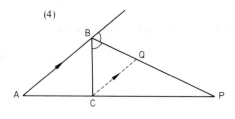

(4)

In both cases we have

$$\frac{AP}{PC} = \frac{AB}{BC}.$$

[The proof uses similar triangles ABP and CQP.]

We now list some further properties of the triangle.

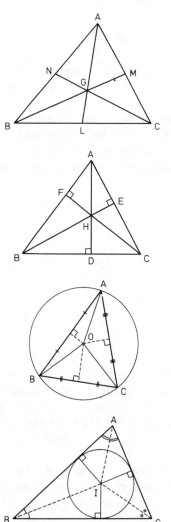

A *median* of a triangle is a line joining a vertex to the mid-point of the opposite side. The medians AL, BM and CN of triangle ABC are concurrent at a point G, called the *centroid* of the triangle. G divides each median in the ratio $2:1$.

An *altitude* of a triangle is a perpendicular from a vertex to the opposite side. The altitudes AD, BE and CF of triangle ABC are concurrent at a point H called the *orthocentre* of the triangle.

The *circumcircle* or circumscribed circle of a triangle is the circle which passes through the vertices of the triangle. The centre of this circle is called the *circumcentre*. It is the point of intersection of the perpendicular bisectors of the sides of the triangle.

The *inscribed circle* or incircle of a triangle is the circle which touches the three sides of the triangle internally. The centre of this circle is called the *incentre*. It is the point of intersection of the bisectors of the angles of the triangle.

Exercise 8.1

1.   The points A, B and C lie on a circle centre O. How are $\angle A$, $\angle B$ and $\angle C$ of the quadrilateral $OABC$ related?

2.   The points P, Q and R lie on a circle and $\angle PRQ$ is obtuse. The tangents to the circle at P and Q intersect at T. If $\angle TPQ = x$, $\angle PRQ = y$ and $\angle PTQ = z$, state the relationships between (a) x and y, (b) y and z.

3.   In triangle PQR the internal bisector of $\angle Q$ meets PR at X and the external bisector of $\angle Q$ meets PR produced at Y. Given that $PQ:QR = 3:2$, find the ratio $PR:XY$. Prove that $\angle XQY$ is a right-angle. Hence show that Q lies on the circle with diameter XY.

4.   A point P moves in a plane containing two fixed points A and B, such that the ratio AP/PB is a constant not equal to 1. Show that the locus of P is a circle. [This locus is called *Apollonius'† circle*.]

5.   By showing that the medians of a triangle divide the triangle into six smaller triangles of equal area, prove that the centroid of the triangle divides each median in the ratio $2:1$.

6.   Given that AD, BE and CF are the altitudes of an acute angled triangle ABC, prove that the orthocentre of $\triangle ABC$ is the incentre of $\triangle DEF$. Find the position of the incentre of $\triangle DEF$ when $\angle A$ is obtuse. [$\triangle DEF$ is called the *pedal triangle* of $\triangle ABC$ because its vertices are the 'feet' of the altitudes.]

8.2 Sine and cosine rules

To establish the sine rule, consider a triangle ABC as shown below. Let h be the height of the perpendicular (or altitude) from C to AB. In both diagrams,

$$\sin A = \frac{h}{b}, \sin B = \frac{h}{a}$$

$$\therefore \quad \frac{a}{\sin A} = \frac{b}{\sin B} \left(= \frac{ab}{h} \right).$$

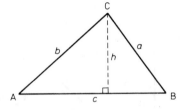
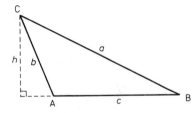

† *Apollonius of Perga* (3rd cent. B.C.) Greek mathematician known as 'the great geometer'. His treatise on conic sections forms the basis of much of the later work on the subject.

Similarly, it may be shown that $\dfrac{b}{\sin B} = \dfrac{c}{\sin C}$, leading to the *sine rule*:

$$\frac{a}{\sin A} = \frac{b}{\sin B} = \frac{c}{\sin C}.$$

Example 1   Solve the triangle ABC in which $A = 40°$, $B = 65°$ and $c = 10$.

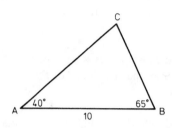

$$A + B + C = 180°$$

$$\therefore \quad C = 180° - 40° - 65° = 75°.$$

By the sine rule,

$$\frac{a}{\sin A} = \frac{b}{\sin B} = \frac{c}{\sin C}$$

$$\therefore \quad \frac{a}{\sin 40°} = \frac{b}{\sin 65°} = \frac{10}{\sin 75°}$$

Hence $\quad a = \dfrac{10 \sin 40°}{\sin 75°} \approx 6\!\cdot\!65, \quad b = \dfrac{10 \sin 65°}{\sin 75°} \approx 9\!\cdot\!38$

$\therefore$ the remaining sides and angles of the triangle are $C = 75°$, $a = 6\!\cdot\!65$ and $b = 9\!\cdot\!38$.

The sine rule can also be applied when two sides and one angle of a triangle are given. However, in this case it may sometimes be possible to construct two triangles satisfying the given conditions. For instance, let us assume that in $\triangle ABC$, the angle A and the sides a and b are given. These diagrams show that the ambiguous case arises when a is less than b, but greater than $b \sin A$.

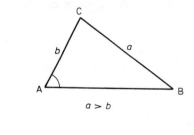

$a > b$

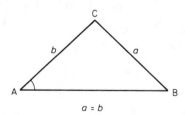

$a = b$

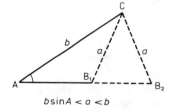

$b \sin A < a < b$

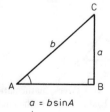

$a = b \sin A$

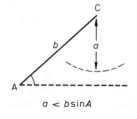

$a < b \sin A$

Example 2 If in triangle ABC, $a = 10$, $b = 16$ and $A = 30°$, find the two possible values of c.

Using the sine rule $\dfrac{a}{\sin A} = \dfrac{b}{\sin B} = \dfrac{c}{\sin C}$

$$\therefore \quad \sin B = \frac{b \sin A}{a} = \frac{16 \sin 30°}{10} = \frac{16 \cdot \frac{1}{2}}{10} = 0 \cdot 8.$$

Since $b > a$, $B > A$ and B may be either acute or obtuse

$$\therefore \quad \text{either} \quad \begin{array}{ll} B = 53 \cdot 13° & \text{or} \quad B = 180° - 53 \cdot 13° = 126 \cdot 87° \\ C = 96 \cdot 87° & \quad\quad C = 23 \cdot 13°. \end{array}$$

If B is acute, $c = \dfrac{a \sin C}{\sin A} = \dfrac{10 \sin 96 \cdot 87°}{\frac{1}{2}} = 20 \sin 96 \cdot 87° \approx 19 \cdot 9$

If B is obtuse, $c = \dfrac{a \sin C}{\sin A} = \dfrac{10 \sin 23 \cdot 13°}{\frac{1}{2}} = 20 \sin 23 \cdot 13° \approx 7 \cdot 86.$

Hence the two possible values of c are $7 \cdot 86$ and $19 \cdot 9$.

Another proof of the sine rule involves the construction of the circumcircle of $\triangle ABC$. In both diagrams, CD is a diameter of the circle, so that $\angle CBD = 90°$. Using the geometrical properties of a circle, in (1) $\angle D = \angle A$ and in (2) $\angle D = 180° - \angle A$.

(1)

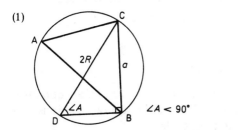

$\angle A < 90°$

(2)

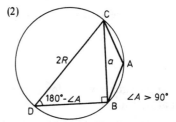

$\angle A > 90°$

Let R be the radius of the circumcircle of $\triangle ABC$, then using $\triangle BCD$, in (1) $\sin A = a/2R$ and in (2) $\sin A = \sin (180° - A) = a/2R$

$$\therefore \quad \text{in both cases,} \qquad \frac{a}{\sin A} = 2R.$$

By similar arguments it may be shown that

$$\boxed{\frac{a}{\sin A} = \frac{b}{\sin B} = \frac{c}{\sin C} = 2R,}$$

which is an extension of the basic sine rule.

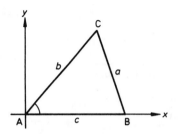
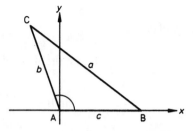

To prove the cosine rule we use Cartesian coordinates and the distance between two points. In both diagrams, the coordinates of B and C are $(c, 0)$ and $(b \cos A, b \sin A)$ respectively,

$$\therefore \quad a^2 = BC^2 = (b \cos A - c)^2 + (b \sin A - 0)^2$$
$$= b^2 \cos^2 A - 2bc \cos A + c^2 + b^2 \sin^2 A$$
$$= b^2 (\cos^2 A + \sin^2 A) + c^2 - 2bc \cos A.$$

Since $\cos^2 A + \sin^2 A = 1$, we have established the *cosine rule*

$$\boxed{a^2 = b^2 + c^2 - 2bc \cos A.}$$

[Note that this proof is valid for an angle A of any magnitude.]

Example 3   Solve the triangle ABC in which $a = 5$, $b = 8$ and $C = 100°$.

Using the cosine rule in the form $c^2 = a^2 + b^2 - 2ab \cos C$,

$$c^2 = 5^2 + 8^2 - 2.5.8 \cos 100° = 89 - 80 \cos 100°$$

$$\therefore \quad c \approx 10 \cdot 1.$$

[The approximate value for c given here is the value $c = 10 \cdot 143561$, obtained by calculator, rounded to 3 significant figures. This more accurate value is stored in the calculator memory for use when finding angle A.]

Using the sine rule, $\dfrac{\sin A}{a} = \dfrac{\sin C}{a}$

$$\therefore \quad \sin A = \frac{a \sin C}{c} = \frac{5 \sin 100°}{c} \quad \text{and} \quad A \approx 29 \cdot 04°.$$

Since $A + B + C = 180°$, $B = 50 \cdot 96°$.

Hence the remaining side and angles of the triangle are $c = 10 \cdot 1$, $A = 29 \cdot 0°$ and $B = 51 \cdot 0°$.

The cosine rule can be rearranged when finding unknown angles:

$$\boxed{\cos A = \frac{b^2 + c^2 - a^2}{2bc}.}$$

Example 4   Solve the triangle ABC in which $a = 8$, $b = 9$ and $c = 10$.

[Since any obtuse angle in the triangle will be opposite the longest side c it may be easier to find angles A and B first.]

Using the cosine rule, $\qquad \cos A = \dfrac{b^2 + c^2 - a^2}{2bc}$

$$\therefore \quad \cos A = \frac{9^2 + 10^2 - 8^2}{2.9.10} = \frac{117}{180} = \frac{13}{20} = 0.65.$$

$$\therefore \quad A \approx 49.458°$$

Using the cosine rule, $\qquad \cos B = \dfrac{a^2 + c^2 - b^2}{2ac}$

$$\therefore \quad \cos B = \frac{8^2 + 10^2 - 9^2}{2.8.10} = \frac{83}{160} = 0.51875$$

$$\therefore \quad B \approx 58.752°.$$

Since $\quad A + B + C = 180°, \quad C \approx 71.790°.$

$\therefore \quad$ the angles of the triangle are $A = 49.5°$, $B = 58.8°$ and $C = 71.8°$.

[Some readers may prefer to find B using the sine rule. However, this does require the use of the approximate or even wrongly calculated value of A. As a check on accuracy, C can also be calculated independently.]

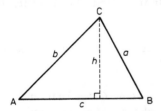

Let $\triangle$ represent the area of $\triangle ABC$ then

$$\triangle = \tfrac{1}{2}ch = \tfrac{1}{2}c.b\sin A$$

$\therefore$ | the area of $\triangle ABC$ is $\tfrac{1}{2}bc\sin A$.

To obtain an expression for $\triangle$ in terms of a, b and c we write

$$\sin^2 A = 1 - \cos^2 A = (1 + \cos A)(1 - \cos A).$$

Using the cosine rule it can then be shown that

$$4b^2c^2 \sin^2 A = (\{b+c\}^2 - a^2)(a^2 - \{b-c\}^2).$$

Hence $\qquad 16\triangle^2 = (a+b+c)(b+c-a)(a+c-b)(a+b-c).$

This result is expressed in a more convenient form using the *semi-perimeter* of the triangle, $s = \tfrac{1}{2}(a+b+c)$.

$$\triangle = \sqrt{\{s(s-a)(s-b)(s-c)\}}$$

This is *Hero's*† *formula* for the area of a triangle.

---

† *Hero (Heron) of Alexandria* (1st cent. A.D.) Greek mathematician. He invented many machines and demonstrated the law of reflection. The derivation of his formula for the area of a triangle appears in *Metrica*, his most important geometrical work.

Exercise 8.2

In questions 1 to 12 solve the triangle ABC.

1. $A = 60°, B = 50°, c = 5.$ 2. $A = 47°, b = 8, C = 72°.$

3. $a = 5, b = 9, C = 51·2°.$ 4. $a = 2·4, B = 45°, c = 5·1.$

5. $a = 6, b = 8, c = 9.$ 6. $a = 3·6, b = 6, c = 4·8.$

7. $A = 67° \ 20', a = 10, b = 10.$ 8. $A = 121°, b = 6·9, c = 4·7.$

9. $a = 6·4, B = 23°, C = 54°.$ 10. $B = 55°, b = 12, c = 9.$

11. $a = 15, b = 8, c = 10.$ 12. $b = 6, c = 3, C = 30°.$

13. In triangle ABC, $a = 4$, $b = 5$ and $A = 48°$. Find the two possible values of c.

14. In triangle ABC, $b = 12$, $c = 9$ and $C = 34·3°$. Find the two possible values of a.

15. In triangle XYZ, $\angle X = 108°$, $\angle Y = 25°$ and $XY = 10\,\text{cm}$. Calculate the area of the triangle.

16. A triangle has sides $13\,\text{cm}$, $7\,\text{cm}$ and $5\sqrt{3}\,\text{cm}$. Without using tables or calculators, find the smallest angle of the triangle and the diameter of the circumcircle.

17. The area of a triangle is $10\,\text{cm}^2$. Two of the sides are of lengths $5\,\text{cm}$ and $7\,\text{cm}$. Calculate the possible lengths of the third side.

18. In quadrilateral $ABCD$, $AB = 7\,\text{cm}$, $BC = 8\,\text{cm}$, $CD = 5\,\text{cm}$ and $\angle ABC = 52°$. Given that $ABCD$ is a cyclic quadrilateral, find the radius of its circumscribing circle and the length of AD.

19. In triangle XYZ, $XY = p$, $XZ = q$ and M is the mid-point of YZ. If $\angle MXY = \theta$ and $\angle MXZ = \phi$, prove that $p \sin \theta = q \sin \phi$.

20. With the usual notation for a triangle ABC, prove that $a \cos B + b \cos A = c$.

21. Use the sine rule to prove, with the usual notation, that in triangle ABC,

$$\frac{b-c}{b+c} = \frac{\sin B - \sin C}{\sin B + \sin C}.$$

Deduce that $\tan \frac{1}{2}(B-C) = \dfrac{b-c}{b+c} \tan \frac{1}{2}(B+C) = \dfrac{b-c}{b+c} \cot \frac{1}{2}A$. Use this formula (called the *tangent rule*) to find B and C given (a) $A = 52°$, $b = 7$, $c = 5$, (b) $A = 40°$, $b = 4·4$, $c = 2$.

22. The triangle *ABC* has incircle of radius *r* with centre at *I*. By considering triangles *IBC, ICA* and *IAB*, find an expression for the area of triangle *ABC* in terms of *r* and the semi-perimeter *s*. Calculate *r* given that $a = 5, b = 7$ and $c = 8$.

8.3 Problems in two dimensions

The next exercise provides further practice in the use of the sine and cosine rules. Some of the terms used in trigonometry problems are defined below.

(1) *Bearings*

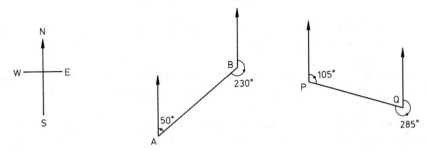

The bearing of *B* from *A* is 050° or N 50° E and that of *A* from *B* is 230° or S 50° W.

The bearing of *Q* from *P* is 105° or S 75° E and that of *P* from *Q* is 285° or N 75° W.

(2) *Angle of depression*

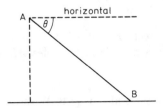

To an observer at *A* the angle of depression of an object at *B* is θ.

(3) *Angle of elevation*

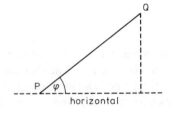

To an observer at *P* the angle of elevation of an object at *Q* is φ.

The angle of elevation of the sun is sometimes called the *sun's altitude*.

Exercise 8.3

1.   A ship A is 6·7 km from a lightship on a bearing of 308°. A second ship B is 8·3 km from the lightship on a bearing of 074°. Calculate the distance AB and the bearing of B from A to the nearest degree.

2. A, B and C are three towns. B is 10 km from A in the direction N 51° E. C is 15 km from B in the direction N 64° W. Calculate the distance and the bearing of A from C.

3.   From a barge moving with constant speed along a straight canal the angle of elevation of a bridge is 10°. After 10 minutes the angle is 15°. How much longer will it be before the barge reaches the bridge, to the nearest second?

4.   In a quadrilateral $PQRS$, $PQ = 10$ cm, $QR = 7$ cm, $RS = 6$ cm, $\angle PQR = 65°$ and $\angle PSR = 98°$. Find the length of PS.

5.   A convex quadrilateral $ABCD$ has $AB = 15$, $BC = CD = 8$, $AD = 7$ and $AC = 13$. Show that the quadrilateral is cyclic and find BD.

6.   A tower stands on a slope which is inclined at an angle of 17·2° to the horizontal. From a point further up the slope and 150 m from the base of the tower the angle of depression of the top of the tower is found to be 9·6°. Find the height of the tower.

7. AB is a level straight road 1500 m long. C and D are the bases of two church spires on opposite sides of the road. The angles BAC, ABC, BAD, ABD are 43°, 57°, 29° and 37° respectively. Find the distance between the spires.

8.   The line ABC is the tangent to a circle at the point B. X, Y and Z are points on the circle, such that $\angle ABX = 24°$, $\angle ABY = 63°$ and $\angle CBZ = 51°$. Given that the radius of the circle is 8 cm, find the area of $\triangle XYZ$.

9.   In triangle ABC the median through A meets BC at M. Given that $AB = 8$ cm, $AM = 5$ cm and $AC = 6$ cm, use the cosine rule in triangles ABM and AMC to find BC. Find also $\angle ABC$.

10. In $\triangle ABC$, $AB = 9$ cm, $AC = 12$ cm, $\angle B = 2\theta$ and $\angle C = \theta$. Without using tables or calculators, find $\cos\theta$ and the length of BC.

8.4 Angles, lines and planes

A line is said to lie in a plane if every point of the line is in the plane. A line is said to be parallel to a plane if the line never meets the plane. If a line neither lies in a plane nor is parallel to it, then the line intersects the plane in a single point. A line which is perpendicular to a plane is perpendicular to every line in the plane

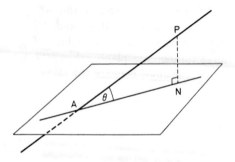

through this point of intersection. In general, a line and a plane intersect as shown in the diagram. If N is the foot of the perpendicular to the plane from a point P on the line, then AN is called the *projection* of AP on the plane. The angle between the line and the plane is defined as the angle θ between the line and its projection on the plane.

Example 1 The prism shown below has three identical rectangular faces. $AD = BE = CF = 12$ cm and the sides of triangles ABC, DEF are each 10 cm long. Find the angle x between the line CD and the plane $ABED$.

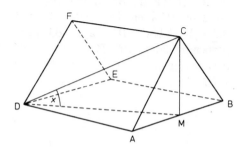

The perpendicular from C to the plane $ABED$ meets the plane at M, the midpoint of AB

$\therefore$ the angle between CD and the plane $ABED$ is $\angle CDM = x$.

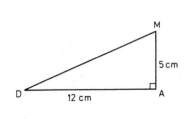
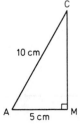
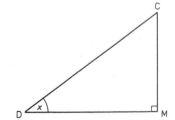

Using Pythagoras' theorem in $\triangle ADM$,

$$DM = \sqrt{(AM^2 + AD^2)} = \sqrt{(5^2 + 12^2)}\,\text{cm} = \sqrt{169}\,\text{cm} = 13\,\text{cm}$$

Using Pythagoras' theorem in $\triangle ACM$,

$$CM = \sqrt{(AC^2 - AM^2)} = \sqrt{(10^2 - 5^2)}\,\text{cm} = \sqrt{75}\,\text{cm} = 5\sqrt{3}\,\text{cm}$$

Hence, in $\triangle CMD$, $\tan x = CM/DM = 5\sqrt{3}/13$ $\quad \therefore \quad x \approx 33\cdot7°$.
The angle between CD and the plane $ABCD$ is $33\cdot7°$.

Two planes are said to be parallel if they do not intersect. Two planes which are not parallel have a line of intersection (or common line) as shown below.

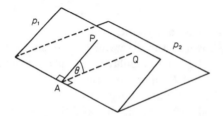

AP lies in the plane p_1 and AQ lies in a second plane p_2. If AP and AQ are *both* perpendicular to the line of intersection of the planes, then $\angle PAQ = \theta$ is the angle between the planes.

[θ is called the *dihedral angle* of the planes.]

Example 2   In a tetrahedron $ABCD$, $AD = BD = CD = 6$ cm and $AB = BC = CA = 8$ cm. Find the angle between the plane ACD and the plane ABC.

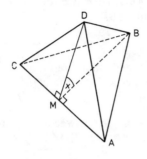

AC is the line of intersection of the planes ACD and ABC. If M is the midpoint of AC, then since $AD = DC$ and $AB = BC$, both DM and BM are perpendicular to AC. Hence the angle between planes ACD and ABC is $\angle BMD = x$.

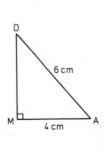
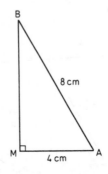
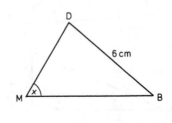

Using Pythagoras' theorem in $\triangle ADM$,

$$DM = \sqrt{(AD^2 - AM^2)} = \sqrt{(36 - 16)}\,\text{cm} = \sqrt{20}\,\text{cm} = 2\sqrt{5}\,\text{cm}.$$

Using Pythagoras' theorem in $\triangle ABM$,

$$BM = \sqrt{(AB^2 - AM^2)} = \sqrt{(64 - 16)}\,\text{cm} = \sqrt{48}\,\text{cm} = 4\sqrt{3}\,\text{cm}.$$

Using the cosine rule in $\triangle BDM$,

$$\cos x = \frac{BM^2 + DM^2 - BD^2}{2BM \cdot DM} = \frac{48 + 20 - 36}{2 \cdot 4\sqrt{3} \cdot 2\sqrt{5}} = \frac{32}{16\sqrt{15}} = \frac{2}{\sqrt{15}}$$

$$\therefore \quad x \approx 58 \cdot 9°.$$

Hence the angle between the planes ACD and ABC is $58 \cdot 9°$.

Two straight lines are said to be parallel if they are coplanar (i.e. lie in the same plane) but do not meet. Two lines which are not parallel and which do not meet are called *skew lines*. Skew lines l_1 and l_2 are shown below. The line l_3 is constructed so that l_3 intersects l_2 and is parallel to l_1. The angle between l_1 and l_2 is defined to be the angle θ between l_2 and l_3. BN is drawn perpendicular to the

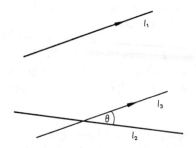
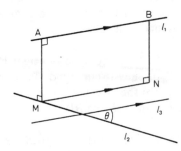

plane containing l_2 and l_3. If M is the point on l_2 such that MN is parallel to l_1, then $ABNM$ is a rectangle. This shows that it is always possible to construct a common perpendicular AM to the two skew lines l_1 and l_2. It can also be shown that AM is the shortest distance between l_1 and l_2.

Example 3   In the prism given in Example 1, find the angle between CD and BE.

Since AD is parallel to BE, the angle between CD and BE is equal to the angle between CD and AD, y.

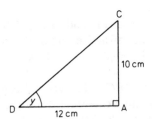

In triangle ACD, $\tan y = \dfrac{10}{12} = \dfrac{5}{6}$,

$\therefore\quad y \approx 39\cdot8°$

Hence the angle between the skew lines CD and BE is $39\cdot8°$.

Exercise 8.4

1.   Triangles ABC and XYZ are two faces of a right triangular prism. $AB = AC = XY = XZ = 7$ cm and $BC = YZ = 10$ cm. The plane of each triangle is perpendicular to the edges AX, BY and CZ, which are of length 15 cm. Find the angles between the plane $BCZY$ and (a) the line AY, (b) the plane ABY and (c) the plane AYZ.

2.   A triangle PQR lies in a horizontal plane and S is a point vertically above P. Given that $PQ = 5$ cm, $PR = 10$ cm, $RS = 14$ cm and $\angle RPQ = 60°$, find the angles between (a) RS and the plane PQR, (b) the planes QRS and PQR, (c) the lines PS and QR.

3. A triangle *ABC* lies in a horizontal plane. The points *X*, *Y* and *Z* are 8 cm vertically above *A*, *B* and *C* respectively. *M* is the mid-point of *YZ*. $AB = 5$ cm, $BC = 6$ cm and $\angle ABC = 90°$. Find the angles between (a) the planes *AYZ* and *ABC*, (b) the line *AM* and the plane *ABC*, (c) the lines *AB* and *XC*.

4. A square *ABCD* of side 10 cm lies in a horizontal plane. A second square *ABXY* lies in a plane inclined at an angle of 40° to the horizontal. Find the angles made with the horizontal by (a) *AX* and (b) the plane *AXD*.

5. *VABCD* is a pyramid of height 15 cm standing symmetrically on a square base *ABCD* of side 16 cm. Find the angles between (a) the edge *VA* and the base *ABCD*, (b) the face *VAB* and the base *ABCD*, (c) the edges *VA* and *CD*, (d) the edge *VA* and the line *BD*.

6. In the tetrahedron *PQRS*, $PQ = QR = RS = SP = 13$ cm and $PR = QS = 10$ cm. Find the angles between (a) the planes *PQR* and *PSR*, (b) the line *QS* and the plane *PSR*.

7. A pyramid *VABCD* stands on a horizontal square base *ABCD*. The faces *VAB*, *VBC*, *VCD*, *VDA* are equilateral triangles of side 2*a*. Find (a) the height of the vertex *V* above the base, (b) the angle between the faces *VAB* and *VCD*, (c) the angle between the faces *VAB* and *VBC*.

8. Each face of a tetrahedron *ABCD* is an equilateral triangle of side 6 units. Without using tables or calculators, find (a) the cosine of the angle between any two faces, (b) the perpendicular distance of any vertex from the opposite face.

9. In a tetrahedron *ABCD*, $AB = AC = AD = 5$ cm and $BC = CD = DB = 8$ cm. Find the angles between (a) the edge *AB* and the face *BCD*, (b) the faces *ABC* and *BCD*, (c) the faces *ABC* and *ACD*, (d) the edge *AD* and the face *ABC*.

10. *ABCDPQRS* is a cuboid on a horizontal base *ABCD*. The vertical edges *PA*, *QB*, *RC* and *SD* are of length 6 cm. $AB = 8$ cm and $BC = 9$ cm. Find the angles between the following pairs of lines (a) *AR* and *BS*, (b) *BD* and *PR*, (c) *AS* and *DR*, (d) *AC* and *BS*.

8.5 Problems in three dimensions

In some problems it is necessary to obtain a general expression connecting the lengths and angles involved.

Example 1   A man at *A* observes a tower *CD* due north of him, with height *h* and angle of elevation α. At a point *B*, due east of the tower and a distance *d* from *A*, the angle of elevation is β. If *A*, *B* and *C* lie in the same horizontal plane, find an expression connecting *h*, *d*, α and β.

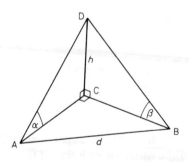

In $\triangle ACD$, $\cot \alpha = \dfrac{AC}{CD} = \dfrac{AC}{h}$

$$\therefore \quad AC = h \cot \alpha.$$

In $\triangle BCD$, $\cot \beta = \dfrac{BC}{CD} = \dfrac{BC}{h}$

$$\therefore \quad BC = h \cot \beta.$$

Using Pythagoras' theorem in $\triangle ABC$,

$$AB^2 = AC^2 + BC^2$$

$$\therefore \quad d^2 = h^2 \cot^2 \alpha + h^2 \cot^2 \beta.$$

Hence the required expression is $d^2 = h^2(\cot^2 \alpha + \cot^2 \beta)$.

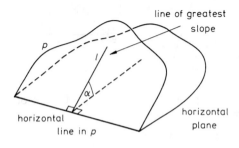

Consider a plane p inclined at an angle α to the horizontal. The diagram shows the plane p and its intersection with a horizontal plane. The inclination to the horizontal of any line l in p which is perpendicular to the line of intersection is α. Since no line in p can be inclined to the horizontal at an angle greater than α, l is called a *line of greatest slope*.

Example 2 On a plane hillside which slopes at an angle $25°$ to the horizontal, there are two straight roads. One lies along a line of greatest slope and the other makes an angle of $15°$ with the horizontal. Find the angle θ at which the roads intersect.

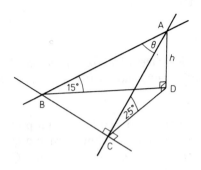

In the diagram AB and AC are segments of the two roads. Their point of intersection A is at a height h above the point D, which lies in the horizontal plane through BC.

In $\triangle ACD$, $\sin 25° = \dfrac{AD}{AC} = \dfrac{h}{AC}$ $\quad \therefore \quad AC = \dfrac{h}{\sin 25°}$

In $\triangle ABD$, $\sin 15° = \dfrac{AD}{AB} = \dfrac{h}{AB}$ $\quad \therefore \quad AB = \dfrac{h}{\sin 15°}$

In $\triangle ABC$, $\cos \theta = \dfrac{AC}{AB} = \dfrac{h}{\sin 25°} \cdot \dfrac{\sin 15°}{h} = \dfrac{\sin 15°}{\sin 25°}$

∴   the angle θ at which the roads intersect is 52·2°.

Problems involving spheres are often solved by drawing a number of circular sections through the sphere.

Example 3   A right pyramid with a square base is inscribed in a sphere of radius 9 cm. If the height of the pyramid is 16 cm, find the length of each side of the base.

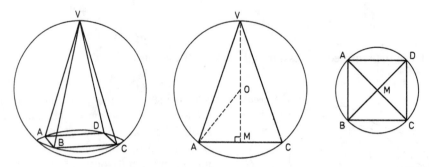

In the diagrams above, O is the centre of the sphere and M is the mid-point of AC, so that VM is the height of the pyramid.

∴ $AO = VO = 9$ cm, $VM = 16$ cm, $OM = VM - VO = 7$ cm.

Using Pythagoras' theorem in $\triangle AOM$,

$$AM^2 = AO^2 - OM^2 = 9^2 - 7^2 = 81 - 49 = 32.$$

Using Pythagoras' theorem in $\triangle ABM$,

$$AB^2 = AM^2 + MB^2 = 2AM^2 = 64.$$

Hence each side of the base of the pyramid is 8 cm long.

Exercise 8.5

1.   Rectangles $ABED$ and $BCFE$ form two vertical faces of a box. Given that $AB = 20$ cm, $BC = 36$ cm, $AD = 15$ cm and $\angle ABC = 90°$, find $\angle DBF$ and the area of $\triangle DBF$.

2.   Three points P, Q and R lie in the same horizontal plane. S, the top of a church spire, is h metres vertically above R. P is due west of the spire and Q is south-east of the spire. The angle of elevation of the spire from both P and Q is 30°. If the distance $PQ = 300$ m, calculate h and find also the greatest angle of elevation of S from any point on PQ.

3.   A clock rests on a stand so that its face makes an angle of 68° with the horizontal. Find the angle made with the horizontal by the minute hand at (a) 08.10, (b) 07.25.

4. A straight path up a hillside, which may be considered as a plane at 25° to the horizontal, makes an angle of 53° with a line of greatest slope. Find the angle the path makes with the horizontal. A second path making an angle of 19° with the horizontal cuts the first path. Find the angle between the two paths.

5. An observer on the ground notes that an aircraft on a bearing of 340° is at a height of 200 m and at an angle of elevation of 45°. The aircraft then travels 400 m due east rising to a height of 300 m. Find the new bearing and angle of elevation of the aircraft.

6. A small boat is due south of a lighthouse and to an observer at the top of the lighthouse, 40 m above sea level, its angle of depression is 30°. After travelling 200 m on a steady course and passing to the east of the lighthouse, the new angle of depression of the boat is 15°. Find the new bearing of the boat from the lighthouse. Find also the bearing of the boat from the lighthouse at its closest point of approach.

7. Two pylons 50 m high stand at points X and Y in the same horizontal plane as an observer at Z. The point Y is 100 m due east of X. The bearings of X and Y from Z are S 70° W and S 30° W respectively. Find the angles of elevation of the tops of the pylons from Z.

8. The triangle ABC lies in a horizontal plane and M is the mid-point of BC. $AM = 5$ m, $AC = 4$ m and $\angle ACB = 90°$. A vertical pole CD of height h metres stands at C. The angle of elevation of D from A is α and the angle of elevation of D from B is β. Given that $\alpha + \beta = 45°$, find h, α and β.

9. A, B and C are three points in a horizontal plane. There are vertical posts AP and BQ at A and B, both of height h. The angles of elevation of P and Q from C are α and β respectively. Angles PCQ and ACB are θ and ϕ respectively. Show that
(a) if $\phi = 90°$, then $\cos \theta = \sin \alpha \sin \beta$,
(b) if $\theta = 90°$, then $\cos \phi = -\tan \alpha \tan \beta$.

10. A, B and C are three points on a plane which slopes at an angle of 30° to the horizontal, B and C being above the horizontal plane containing A. The line of greatest slope through A passes through the mid-point M of BC. $AM = 10$ m, $BC = 16$ m and $\angle AMB = 60°$. Find the lengths of AB and AC. By adding to your diagram a horizontal line through A and lines of greatest slope through B and C, or otherwise, find the angles made by AB and AC with the horizontal plane through A.

11. A right pyramid with a square base is inscribed in a sphere. If the height of the pyramid is 10 cm and the length of each side of the base is 8 cm, find (without using tables or calculators) the radius of the sphere.

Exercise 8.6   (*miscellaneous*)

1. The points A, B, C and D lie on the circumference of a circle. $AB = 7$ cm, $BC = 9$ cm, $CA = 5$ cm and AD is a diameter of the circle. Find $\angle ABC$ and the length of CD.

2. Use the sine rule to prove that if P is a point on the side XZ of $\triangle XYZ$ such that $\angle XYP = \angle PYZ$, then $XP : PZ = XY : YZ$. By letting $\angle XYZ = 30°$, $\angle YZX = 60°$ and $YZ = 2$ units, show that $\cot 15° = 2 + \sqrt{3}$.

3. The points X, Y and Z lie on a circle with centre O. If $XY = 8$ cm, $YZ = 11$ cm and $ZX = 15$ cm, find angles XOY, YOZ and ZOX. Find also the radius of the circle.

4. In the triangle ABC, X, Y and Z are the mid-points of BC, CA and AB, respectively. Using the cosine rule or otherwise, prove that $b^2 + c^2 = 2AX^2 + 2BX^2$. Write down two other similar results, and hence show that $AX^2 + BY^2 + CZ^2 = \frac{3}{4}(a^2 + b^2 + c^2)$.   (JMB)

5. In each of the following cases find (giving reasons for your answers) the number of triangles ABC satisfying the given conditions and, where possible, find by calculation the value(s) of c, leaving your answers in surd form, if you wish.
(i) $a = 5, b = 3, B = 30°$;   (ii) $a = 5, b = 2, B = 30°$;
(iii) $a = 2, b = 3, B = 30°$;   (iv) $a = 2, b = 3, B = 150°$.   (C)

6. Construct a flow diagram for finding $\angle ABC$ in triangle ABC given a, b and $\angle BAC = x°$. Your diagram should produce an appropriate output for all cases.

7. The point P divides the side BC of $\triangle ABC$ internally in the ratio $m : n$. If $\angle B = \theta$, $\angle C = \phi$ and $\angle APC = \alpha$, show that $(m + n)\cot \alpha = n \cot \theta - m \cot \phi$.

8. Using the usual notation in $\triangle ABC$, prove that

$$\frac{a \cos A + b \cos B}{a \cos B + b \cos A} = \cos (A - B).$$

9. With the usual notation in $\triangle ABC$, write down an expression for $\cos A$ in terms of a, b and c. Using this result show that $\sin^2 \frac{1}{2}A = \dfrac{(s - b)(s - c)}{bc}$ and $\cos^2 \frac{1}{2}A = \dfrac{s(s - a)}{bc}$ where $s = \frac{1}{2}(a + b + c)$. Hence find expressions for $\tan^2 \frac{1}{2}A$ and the area of the triangle in terms of a, b, c and s.

10. An isosceles triangle ABC, in which $AB = AC$ and $\angle A = 2\theta$, is inscribed in a circle of radius 5 cm. Prove that the two equal altitudes of the triangle have length $10 \cos \theta \sin 2\theta$ cm. If the sum of the lengths of the three altitudes is 10 cm, find the three angles of the triangle to the nearest degree.   (O & C)

11. From a point A on level ground due S of a church the angle of elevation of the top T of its steeple is $\beta°$. From a point B on the ground whose distance from A is $2a$ on a bearing of N $\theta°$ E $(0 < \theta < 90)$ the angle of elevation of T is also $\beta°$. Show that the height of T above the ground is $a \sec \theta° \tan \beta°$ and find the bearing of the church from B. If D is the point on AB produced distant a from B show that the distance of D from the steeple is $a\sqrt{(3 + \sec^2 \theta°)}$ and that the bearing of the church from D is $S(\theta + \phi)°$ W, where $\sin \phi = \tan \theta / \sqrt{(3 + \sec^2 \theta)}$.   (SU)

12. A, B, C are three landmarks. B is $15\,\text{km}$ from A on a bearing of $140°$ and C is $15\,\text{km}$ from B on a bearing of $216°$. An observer on a ship finds that B is directly behind C and that A bears $015°$. After sailing for 30 minutes directly towards A, the observer finds that C bears due east. Find the speed of the ship.

13. A and B are two points on level ground, and B is a metres due east of A; a tower, h metres high, is also on the same level ground. From A, the tower is in a direction N θ E and from B it is N ϕ W. From the top of the tower, the angle of depression of A is α and of B is β. Prove the following (in any order):
(i) $h \sin (\theta + \phi) = a \cos \phi \tan \alpha$,
(ii) $\cos \phi \tan \alpha = \cos \theta \tan \beta$,
(iii) $h^2(\cot^2 \alpha - \cot^2 \beta) - 2ha \cot \alpha \sin \theta + a^2 = 0$.   (SU)

14.

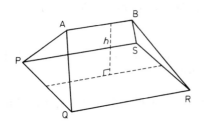

The sketch represents a roof with a horizontal ridge AB at a height h above the (horizontal) ceiling level $PQRS$. $PQRS$ is a rectangle, and AB is parallel to PS and QR. The side of the roof $ABRQ$ makes an angle α with the horizontal, the end APQ makes an angle β with the horizontal, and θ is the angle between AQ and the horizontal. Prove that $\cot^2 \alpha + \cot^2 \beta = \cot^2 \theta$. Prove also that $\cos^2 \gamma + \cos^2 \delta + \sin^2 \theta = 1$, where γ and δ are the angles that AQ makes with PQ and QR respectively.   (SU)

15. A rectangular lamina $ABCD$ has sides $AB = CD = 2a$ and $BC = AD = a$. The mid-point of CD is M and the mid-point of BM is N. The lamina is folded along BM so that the planes BMC and $BMDA$ are at right angles. For this folded figure, find, leaving your answers in surd form if desired, (i) the distance AN, (ii) the distance AC, (iii) the angle ACB, (iv) the tangent of the angle between the planes ADC and ADB.   (O)

16. $ABCD$ is a regular tetrahedron (that is, a triangular pyramid with all its edges equal), and E, F are the midpoints of AB, CD respectively. Find the angles between (i) the planes $ABEF$ and $CDEF$, (ii) the plane $ABEF$ and the line AC, (iii) the lines AC and EF.   (C)

17. A triangle ABC is in a horizontal plane and has $AB = 5$ cm, $BC = 7$ cm and $AC = 8$ cm. The points X, Y and Z are vertically above A, B and C respectively, such that $AX = 10$ cm, $BY = 5$ cm and $CZ = 2$ cm. Find $\angle BAC$ and $\angle YXZ$. If the lines AB and XY, AC and XZ intersect at the points P and Q respectively, show that triangles PAQ and PXQ are isosceles. Find the angle between the planes ABC and XYZ.

18. PTQ is the tangent to a circle at the point T. The points A and B on the circumference of the circle are such that TA and TB make acute angles α and β with TP and TQ respectively. If AB meets the diameter through T at N, prove that $TN = a \sin \alpha \sin \beta / \cos (\alpha - \beta)$, where a is the length of the diameter. If the points C and D on the circumference of the circle are such that TC and TD make acute angles γ and δ with TP and TQ respectively and CD meets the diameter through T at the same point N, prove that $\tan \alpha \tan \beta = \tan \gamma \tan \delta$.   (O)

19. Prove that, for any triangle ABC, with the usual notation and for all values of the angle θ

$$c \sin \theta = a \sin (\theta - B) + b \sin (\theta + A).$$

A quadrilateral $ABCD$ is inscribed in a circle. By applying the above result to the triangle ABC and taking for θ the angle CAD, prove that
$AB . CD + AD . BC = AC . BD.$ (JMB)

20. A road sign is surmounted by a plane equilateral triangle ABC of side 40 cm; its base BC is horizontal and the triangle lies in a vertical plane with BC pointing in the direction N 80° E. The sun is SW at an elevation of 30° and it casts a shadow $A'B'C'$ on a horizontal plane. If M' is the mid-point of $B'C'$, find the length of $A'M'$; hence, or otherwise, find the area of $A'B'C'$.   (O & C)

21. A regular tetrahedron with edges of length $2a$ is inscribed in a sphere. Find the diameter of the sphere.

22. A pyramid has a horizontal square base $ABCD$ of side $2a$. Its vertex V is vertically above the centre O of the base, and the length of VO is $2a$. The line through V perpendicular to the face VAD meets the plane of the base at X. Find (i) the length of VX; (ii) the cosine of the angle between the faces VAB and VAD; (iii) the sine of the angle between the edge VC and the face VAB.   (O)

23. In the region of three fixed buoys A, B and C at sea there is a plane stratum of oil-bearing rock. The depths of the rock below A, B and C are 900 m, 800 m and 1000 m respectively. B is 600 m due east of A and the bearings of C from A and B are 190° and 235° respectively. Calculate (i) the distance BC, (ii) the direction of the horizontal projection of the line of greatest slope of the plane, (iii) the angle this plane makes with the horizontal.   (AEB 1978)

[It may be helpful to consider a horizontal plane at a depth of 900 m.]

9 Vectors

9.1 Introduction to vectors

Some physical quantities, such as temperature, may be completely specified by a number referred to some unit of measurement, e.g. 25°C, 350°F. Such quantities, which have magnitude but which are not related to any definite direction in space, are called *scalar quantities*. Similarly, a number representing the magnitude of some physical quantity is called a *scalar*.

When measuring other quantities, such as wind velocity, it is necessary to give a direction as well as a number and a unit of measurement, e.g. 40 km/h from the north-east. These quantities, which have both magnitude and a definite direction in space, are called *vector quantities*. In general, a *vector* may be described as a number associated with a particular direction in space.

Any vector may be represented by a *directed line segment*, whose direction is that of the vector and whose length represents the magnitude of the vector.

The vector represented here is denoted by **PQ** in bold type (or P̰Q in manuscript). The *magnitude* or modulus of the vector is written |**PQ**| or PQ. The directed line segment joining P to Q is denoted by $\overrightarrow{PQ}$. However, since it is not usually necessary to distinguish between the line segment and the vector it represents, the notation $\overrightarrow{PQ}$ is in common use for both vector and line segment.

The magnitude of a vector is usually positive but may be zero, in which case the vector is called the *zero* or *null vector* and written **0** in bold type. The zero vector is the only vector with indeterminate direction.

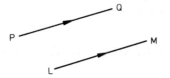

Two vectors are *equal* if they have the same magnitude and direction. Since PQ and LM are parallel and equal in length, **PQ** = **LM**.

175

A *displacement* is one of the simplest examples of a vector quantity. For instance, a car journey from a town A to a town B may be represented by a displacement vector **AB**. Its magnitude is the distance between A and B. Its direction is that of the straight line joining A to B.

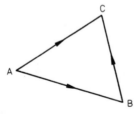

Let us suppose that a man drives from A to B and then from B to C. On another occasion he drives directly from A to C. Since the result of these two journeys is the same, we may write **AB** + **BC** = **AC**.

The displacement **AC** is the *sum* or *resultant* of the displacements **AB** and **BC**.

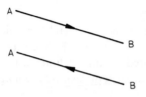

Consider now a journey from A to B, followed by the return journey from B to A. Since the result of these two journeys is a zero displacement, it is reasonable to write, **AB** + **BA** = **0** and **BA** = − **AB**.

All vectors represented by directed line segments may be manipulated in the same way as displacement vectors. For instance, given any three vectors **AB**, **BC** and **AC** represented by the sides of triangle ABC, we may write **AB** + **BC** = **AC**. A vector which has the same magnitude as **AB**, but the opposite direction, is denoted by **BA** or − **AB**.

Example 1   Simplify **EF** + **FG** − **HG**.

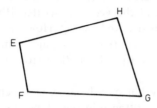

$$\begin{aligned} \mathbf{EF} + \mathbf{FG} - \mathbf{HG} &= \mathbf{EG} - \mathbf{HG} \\ &= \mathbf{EG} + \mathbf{GH} \\ &= \mathbf{EH}. \end{aligned}$$

It is often convenient to denote a vector by a single letter such as **a** in bold type (or $\underset{\sim}{a}$ in manuscript). Its magnitude is then written $|\mathbf{a}|$ or a.

Example 2   In the given diagram **PQ** = **a**, **RS** = **b** and **SQ** = **c**. Find expressions for **RQ**, **PS** and **PR** in terms of **a**, **b** and **c**.

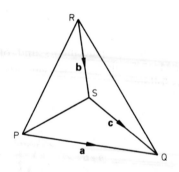

In $\triangle QRS$, $\quad RQ = RS + SQ = \mathbf{b} + \mathbf{c}$

In $\triangle PQS$, $\quad PS = PQ + QS = PQ - SQ$

$\qquad\qquad\qquad\qquad = \mathbf{a} - \mathbf{c}$

In $\triangle PQR$, $\quad PR = PQ + QR = PQ - RQ$

$\qquad\qquad\qquad\qquad = \mathbf{a} - (\mathbf{b} + \mathbf{c})$

$\qquad\qquad\qquad\qquad = \mathbf{a} - \mathbf{b} - \mathbf{c}.$

In general vectors do not have any definite position in space. However, certain vector quantities may be associated with a particular location. For instance, in mechanics, when considering the motion of an object, it may be necessary to take into account the line of action of a force as well as its magnitude and direction. Equal forces with different points of application may produce different effects. A vector which may be represented by any line segment of the appropriate magnitude and direction is sometimes called a *free vector*.

Exercise 9.1

1. Simplify   (a) $\mathbf{PQ + QR + RS}$,   (b) $\mathbf{AB + DE + CD + BC}$,

               (c) $\mathbf{AC - BC}$,   (d) $\mathbf{XY + YX}$,

               (e) $\mathbf{MN - RP - PN}$,   (f) $\mathbf{QT - QR + PR - ST}$.

2. Given that $ABCDEF$ is a regular hexagon, decide whether the following statements are true or false.

(a) $\mathbf{BC = EF}$,   (b) $\mathbf{BF = CE}$,   (c) $|\mathbf{BE}| = |\mathbf{CF}|$,   (d) $|\mathbf{AB}| = |\mathbf{DE}|$,

(e) $\mathbf{AC}$ and $\mathbf{DF}$ are in the same direction,   (f) $\mathbf{AD}$ is parallel to $\mathbf{EF}$.

3. Given that $ABCDPQRS$ is a cuboid lettered so that AP, BQ, CR and DS are edges, decide whether the following statements are true or false.

(a) $\mathbf{PQ = DC}$,   (b) $\mathbf{BR = PD}$,   (c) $|\mathbf{RA}| = |\mathbf{BS}|$,   (d) $|\mathbf{PA}| = |\mathbf{CR}|$,

(e) $\mathbf{DQ}$ and $\mathbf{AR}$ are in the same direction,   (f) $\mathbf{AQ}$ is parallel to $\mathbf{CS}$.

4. In the given diagram $\mathbf{AB = x}$, $\mathbf{BC}$ $= \mathbf{y}$ and $\mathbf{CD = z}$. Find expressions for $\mathbf{AC}$, $\mathbf{BD}$ and $\mathbf{AD}$ in terms of $\mathbf{x}$, $\mathbf{y}$ and $\mathbf{z}$.

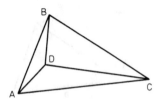

5. In the pentagon $PQRST$ $\mathbf{PQ = a}$, $\mathbf{QR = b}$, $\mathbf{ST = c}$ and $\mathbf{QT = d}$. Find expressions for $\mathbf{PT}$, $\mathbf{RT}$, $\mathbf{RS}$ and $\mathbf{PS}$ in terms of $\mathbf{a}$, $\mathbf{b}$, $\mathbf{c}$ and $\mathbf{d}$.

9.2 Elementary operations on vectors

In the previous section triangles were used to add displacements and other vectors represented by directed line segments. More formally, *vector addition* is a binary operation defined by the *triangle law*, as follows:

If two vectors **a** and **b** are represented by the sides $\overrightarrow{PQ}$ and $\overrightarrow{QR}$ of a triangle, then **a** + **b** is represented by the third side $\overrightarrow{PR}$.

In physics and mechanics it is sometimes more convenient to use the *parallelogram law* for vector addition, which states:

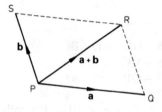

If two vectors **a** and **b** are represented by the sides $\overrightarrow{PQ}$ and $\overrightarrow{PS}$ of a parallelogram, then **a** + **b** is represented by the diagonal $\overrightarrow{PR}$. [Since the line segments $\overrightarrow{PS}$ and $\overrightarrow{QR}$ are equal in magnitude and direction, the two addition laws are equivalent.]

In $\triangle PQR$ shown above, $|\mathbf{a}|$, $|\mathbf{b}|$ and $|\mathbf{a}+\mathbf{b}|$ are represented by the lengths of the sides PQ, QR and PR respectively. For any points P, Q and R we have
$$PR \leqslant PQ + QR.$$

Hence
$$|\mathbf{a}+\mathbf{b}| \leqslant |\mathbf{a}| + |\mathbf{b}|.$$

$[|\mathbf{a}+\mathbf{b}| = |\mathbf{a}| + |\mathbf{b}|$ when P, Q and R are collinear.]

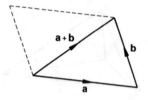
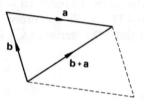

The two diagrams above show that **a** + **b** = **b** + **a**. Thus the *commutative law* holds for vector addition.

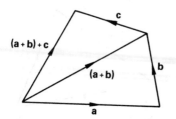

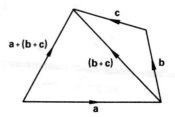

These diagrams show that $(\mathbf{a}+\mathbf{b})+\mathbf{c} = \mathbf{a}+(\mathbf{b}+\mathbf{c})$. Thus the *associative law* holds for vector addition. Hence the sum of a set of vectors is not affected by changes in the order or the grouping of the terms.

The *subtraction* of $\mathbf{b}$ from $\mathbf{a}$ may be defined as the address of $-\mathbf{b}$ to $\mathbf{a}$, i.e. $\mathbf{a}-\mathbf{b} = \mathbf{a}+(-\mathbf{b})$.

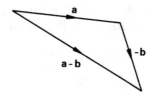

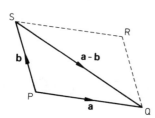

If the vectors $\mathbf{a}$ and $\mathbf{b}$ are represented by the sides $\overrightarrow{PQ}$ and $\overrightarrow{PS}$ of a parallelogram, then $\mathbf{a}-\mathbf{b}$ is represented by the diagonal $\overrightarrow{SQ}$.

Scalar multiples of a vector $\mathbf{a}$, such as $3\mathbf{a}$ and $-2\mathbf{a}$, can be defined using vector addition. For instance, $3\mathbf{a} = \mathbf{a}+\mathbf{a}+\mathbf{a}$ and $-2\mathbf{a} = -(\mathbf{a}+\mathbf{a})$. This leads to the following more general definition of *multiplication* of a vector by a scalar.

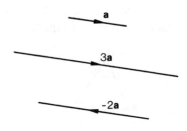

If λ is a scalar and $\mathbf{a}$ is a vector, then the magnitude of $\lambda\mathbf{a}$ is $|\lambda|$ times the magnitude of $\mathbf{a}$, i.e. $|\lambda\mathbf{a}| = |\lambda|a$. The direction of $\lambda\mathbf{a}$ is the same as that of $\mathbf{a}$ if λ is positive, but opposite to that of $\mathbf{a}$ if λ is negative.

[*Division* of a vector by a scalar λ (not equal to zero) is defined as multiplication by $1/\lambda$.]

It follows from the associative and distributive properties of real numbers that

$$\lambda(\mu\mathbf{a}) = (\lambda\mu)\mathbf{a} \quad \text{and} \quad (\lambda+\mu)\mathbf{a} = \lambda\mathbf{a} + \mu\mathbf{a}.$$

We obtain a further distributive law by constructing a pair of similar triangles.

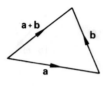
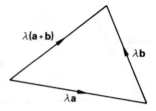

By applying the triangle law of addition to the second triangle, we find that

$$\lambda(\mathbf{a}+\mathbf{b}) = \lambda\mathbf{a}+\lambda\mathbf{b}.$$

We conclude this section with two examples showing vector methods applied to geometrical problems.

Example 1 $ABCD$ is a parallelogram. The points E and F lie on the diagonal BD and $BE = FD$. Prove that $AECF$ is a parallelogram.

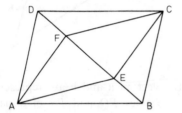

Since $ABCD$ is a parallelogram,

$$\mathbf{AB} = \mathbf{DC}.$$

Since B, E, F and D are collinear and $BE = FD$, $\mathbf{BE} = \mathbf{FD}$.

$$\therefore \quad \mathbf{AE} = \mathbf{AB}+\mathbf{BE} = \mathbf{DC}+\mathbf{FD} = \mathbf{FD}+\mathbf{DC} = \mathbf{FC}.$$

Hence, since AE and FC are parallel and equal in length, $AECF$ must be a parallelogram.

Example 2   Show that the line segment joining the mid-points of any two sides of a triangle is parallel to the third side and equal to half its length.

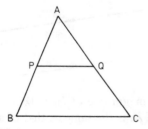

Consider the triangle ABC, where P and Q are the mid-points of the sides AB and AC respectively.

$$\begin{aligned}\mathbf{PQ} &= \mathbf{PA}+\mathbf{AQ} = \tfrac{1}{2}\mathbf{BA}+\tfrac{1}{2}\mathbf{AC}\\ &= \tfrac{1}{2}(\mathbf{BA}+\mathbf{AC})\\ &= \tfrac{1}{2}\mathbf{BC}.\end{aligned}$$

Hence PQ is parallel to BC and half its length.

Exercise 9.2

If $\mathbf{a}$, $\mathbf{b}$ and $\mathbf{c}$ are displacements of 1 km due east, 2 km due north and 1 km in the direction S 30° W respectively, find the magnitudes and directions of the displacements given in questions 1 to 3. (You may leave your answers in surd form.)

1.   (a) 3**a**,   (b) 5**b**,   (c) −**c**,   (d) $-\frac{1}{2}$**b**.

2.   (a) 2**a** + **b**,   (b) **b** + 2**a**,   (c) 2**a** − **b**,   (d) **b** − 2**a**.

3.   (a) **a** + **c**,   (b) **a** + 2**c**,   (c) 2**a** + **c**,   (d) **a** − **c**.

4.   The diagonals of a parallelogram *ABCD* intersect at *M*. If **AB** = **p** and **AD** = **q**, express in terms of **p** and **q** the vectors **AC**, **BD**, **AM** and **MD**.

5.   In a regular hexagon *ABCDEF* **AB** = **a** and **BC** = **b**. Find expressions in terms of **a** and **b** for **DE**, **DC**, **AD** and **BD**.

6.   In the pentagon *PQRST* **PQ** = **a** and **RS** = **b**. If **PT** = 2**RS** and **TS** = 2**PQ**, find expressions for **PR** and **QR** in terms of **a** and **b**.

7.   *PQRS* is a parallelogram with **PQ** = 2**a** and **PS** = **b**. The point *T* is such that **PT** = 2**b**. If the lines *PR* and *QS* intersect at *X* and the lines *RS* and *QT* intersect at *Y*, find expressions for the vectors **TR**, **PY**, **QY** and **XY** in terms of **a** and **b**.

8.   Simplify the following expressions, stating which vector property is being used at each stage.
(a) **x** + (**y** + **x**),   (b) (**u** − 2**v**) + 2(**v** − **w**),   (c) 3(**p** + **q**) + 4**q**.

9.   Given that **HL** + **KN** = **KL** + **HM**, show that the points *M* and *N* are coincident.

10. Given that **OA** + **OC** = **OB** + **OD**, show that the quadrilateral *ABCD* is a parallelogram.

11. If *ABCDEF* is a hexagon in which **AB** = **ED** and **BC** = **FE**, prove that *CDFA* is a parallelogram.

12. If *PQRST* is a pentagon in which **QR** + **RS** = **PT**, prove that **PQ** = **TS**.

13. In a parallelogram *ABCD*, if *M* and *N* are the mid-points of *AB* and *CD* respectively, show that *AMCN* is a parallelogram.

14. Prove that in any quadrilateral *PQRS*,
(a) **PQ** + **RS** = **PS** + **RQ**,   (b) **PR** + **QS** = **PS** + **QR**.

15. In quadrilateral *ABCD* the points *H*, *K*, *L* and *M* are the mid-points of the sides *AB*, *BC*, *CD* and *DA* respectively. Prove that *HKLM* is a parallelogram.

16. Prove that for any vectors **a** and **b** (i) $|\mathbf{a} - \mathbf{b}| \geqslant |\mathbf{a}| - |\mathbf{b}|$, (ii) $|\mathbf{a} - \mathbf{b}| \geqslant |\mathbf{b}| - |\mathbf{a}|$. Deduce that $|\mathbf{a} - \mathbf{b}| \geqslant \left||\mathbf{a}| - |\mathbf{b}|\right|$. Under what conditions does $|\mathbf{a} - \mathbf{b}| = \left||\mathbf{a}| - |\mathbf{b}|\right|$?

9.3 Unit vectors and Cartesian components

Resolving a vector into *component vectors* means expressing it as a sum of two or more non-parallel vectors.

Given two non-parallel vectors **a** and **b**, then for any vector **r** in the same plane it is possible to construct a parallelogram in which one diagonal represents **r** and the sides represent scalar multiples of **a** and **b**. For any particular vectors **a**, **b** and **r** there is only one way of constructing this parallelogram. This suggests that there is a unique expression for **r** of the form $\lambda\mathbf{a} + \mu\mathbf{b}$. Thus a vector **r** may be resolved into component vectors, $\lambda\mathbf{a}$ and $\mu\mathbf{b}$, in the directions of any two non-parallel vectors **a** and **b** in a plane containing **r**.

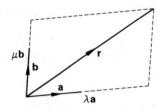 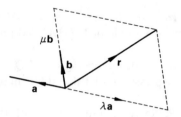

In three-dimensional problems it may be necessary to resolve a vector **r** into component vectors in three given directions. It may be demonstrated that this is possible, provided that the directions are not coplanar, by constructing a parallelepiped with **r** represented by a diagonal.

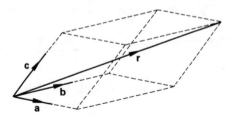

For any vectors **a**, **b** and **c** which are not coplanar, there is a unique expression for **r** of the form $\lambda\mathbf{a} + \mu\mathbf{b} + \nu\mathbf{c}$.

It is often convenient to resolve vectors into component vectors in the directions of the Cartesian coordinate axes. In three-dimensional work the third axis is the *z*-axis, which is constructed perpendicular to the *x*- and *y*-axes to form a *right-handed system*. This means that if we point the thumb of the right hand in the direction of the *z*-axis, then bend the fingers slightly, they will indicate the direction of a rotation from the *x*-axis to the *y*-axis. Two standard ways of drawing these axes are shown below.

Any vector with magnitude 1 is called a *unit vector*. The unit vectors in the directions of the *x*-, *y*- and *z*-axes are denoted by **i**, **j** and **k** respectively.

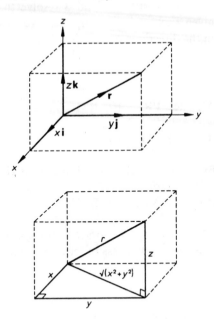

By constructing a cuboid with one diagonal representing **r** and sides parallel to the coordinate axes, we can demonstrate that there is a unique expression for any vector **r** of the form $x\mathbf{i}+y\mathbf{j}+z\mathbf{k}$. The scalars *x*, *y* and *z* are called the *components* of **r** in the directions of **i**, **j** and **k**, or simply the *Cartesian components* of **r**.

Since **i**, **j** and **k** are unit vectors, *x*, *y* and *z* are the lengths of the sides of the cuboid constructed here. Applying Pythagoras' theorem in the two right-angled triangles illustrated, we find that the magnitude of **r** is given by

$$r = \sqrt{(x^2 + y^2 + z^2)}.$$

Example 1 If $\mathbf{a} = 3\mathbf{i}+2\mathbf{j}+7\mathbf{k}$ and $\mathbf{b} = 2\mathbf{i}+4\mathbf{j}-5\mathbf{k}$, find $|\mathbf{a}|$ and $|\mathbf{a}-\mathbf{b}|$.

$|\mathbf{a}| = \sqrt{(3^2+2^2+7^2)} = \sqrt{(9+4+49)} = \sqrt{62}$
$\mathbf{a}-\mathbf{b} = (3\mathbf{i}+2\mathbf{j}+7\mathbf{k})-(2\mathbf{i}+4\mathbf{j}-5\mathbf{k}) = \mathbf{i}-2\mathbf{j}+12\mathbf{k}$
$\therefore \quad |\mathbf{a}-\mathbf{b}| = \sqrt{(1^2+\{-2\}^2+12^2)} = \sqrt{(1+4+144)} = \sqrt{149}.$

Since *a* represents the magnitude of a vector **a**, the vector **a**/*a* has magnitude 1 and the same direction as **a**. Thus the unit vector in the direction of a given vector **a** (sometimes denoted by **â**) may be expressed in the form **a**/*a*.

Example 2 Find the unit vector in the direction of $\mathbf{a} = 2\mathbf{i}+2\mathbf{j}-\mathbf{k}$.

$a = \sqrt{(2^2+2^2+\{-1\}^2)} = \sqrt{(4+4+1)} = \sqrt{9} = 3$
$\therefore$ the unit vector in the direction of **a**

$$= \tfrac{1}{3}(2\mathbf{i}+2\mathbf{j}-\mathbf{k}) = \tfrac{2}{3}\mathbf{i}+\tfrac{2}{3}\mathbf{j}-\tfrac{1}{3}\mathbf{k}.$$

Exercise 9.3

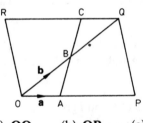

1. In the given diagram $\mathbf{OA} = \mathbf{a}$ and $\mathbf{OB} = \mathbf{b}$. The points *P*, *Q* and *R* are constructed so that $\mathbf{AP} = 2\mathbf{OA}$, $\mathbf{BQ} = \mathbf{OB}$ and *OPQR* is a parallelogram. Express the following vectors as sums of component vectors in the directions of **a** and **b**

(a) **OQ**,   (b) **QR**,   (c) **PR**,   (d) **AB**,   (e) **CQ**,   (f) **OC**.

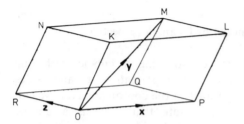

In the parallelepiped shown in the diagram $\mathbf{OP} = \mathbf{x}$, $\mathbf{OM} = \mathbf{y}$ and $\mathbf{OR} = \mathbf{z}$. In questions 2 and 3 resolve the given vectors into component vectors in the directions of of x, y and z.

2. (a) **PQ**. (b) **RQ**, (c) **MN**, (d) **LM**, (e) **PR**, (f) **RM**.

3. (a) **ON**, (b) **KM**, (c) **OK**, (d) **NP**, (e) **KQ**, (f) **LR**.

4. In quadrilateral $ABCD$, $\mathbf{AB} = 2\mathbf{i} - \mathbf{j}$, $\mathbf{BC} = 3\mathbf{i} + 4\mathbf{j}$, $\mathbf{AD} = \mathbf{i} + 5\mathbf{j}$ and M is the mid-point of CD. Express in terms of $\mathbf{i}$ and $\mathbf{j}$ the vectors
(a) **AC**, (b) **BD**, (c) **CD**, (d) **DM**, (e) **AM**, (f) **BM**.

5. In tetrahedron $PQRS$, $\mathbf{PQ} = 3\mathbf{i} - \mathbf{j} - \mathbf{k}$, $\mathbf{PR} = -\mathbf{i} + 4\mathbf{j} + \mathbf{k}$ and $\mathbf{SQ} = 3\mathbf{i} - 4\mathbf{j} - 3\mathbf{k}$. The point T lies on RS produced such that $RS = ST$. Express in terms of $\mathbf{i}$, $\mathbf{j}$ and $\mathbf{k}$ the vectors
(a) **QR**, (b) **PS**, (c) **RS**, (d) **PT**.

6. Find the magnitude of each of the following vectors
(a) $5\mathbf{i} - 12\mathbf{j}$,   (b) $-\mathbf{i} + 2\mathbf{k}$,   (c) $8\mathbf{j} + 6\mathbf{k}$,
(d) $2\mathbf{i} + \mathbf{j} - 2\mathbf{k}$,   (e) $2\mathbf{i} - 6\mathbf{j} + 3\mathbf{k}$,   (f) $-\mathbf{i} + 7\mathbf{j} - 5\mathbf{k}$.

7. Find unit vectors in the direction of
(a) $8\mathbf{i} - 4\mathbf{j} - \mathbf{k}$,   (b) $16\mathbf{i} - 8\mathbf{j} - 2\mathbf{k}$,   (c) $-8\mathbf{i} + 4\mathbf{j} + \mathbf{k}$.

In questions 8 to 10 take $\mathbf{a} = \mathbf{i} - 2\mathbf{j} + \mathbf{k}$, $\mathbf{b} = 3\mathbf{i} - \mathbf{j} - \mathbf{k}$, $\mathbf{c} = \mathbf{i} + 3\mathbf{j} - 2\mathbf{k}$.

8. Find (a) $|\mathbf{a} + \mathbf{b}|$,   (b) $|\mathbf{a} - \mathbf{b}|$,   (c) $|2\mathbf{b} - 2\mathbf{a}|$.

9. Find (a) $|\mathbf{b} + \mathbf{c}|$,   (b) $|\mathbf{a} + \mathbf{b} + \mathbf{c}|$,   (c) $|5\mathbf{a} - \mathbf{b} + 3\mathbf{c}|$.

10. Find unit vectors in the directions of
(a) $2\mathbf{b} - \mathbf{c}$,   (b) $5\mathbf{a} - 5\mathbf{b} + 3\mathbf{c}$,   (c) $5\mathbf{a} - \mathbf{b} + 3\mathbf{c}$.

11. If $\mathbf{a} = \lambda\mathbf{i} - 3\mathbf{j} + 3\lambda\mathbf{k}$ and $|\mathbf{a}| = 7$, find the possible values of λ.

12. Given that $\mathbf{a} = 24\mathbf{i} - 7\mathbf{j}$ and that $|\mathbf{b}| = 15$, find the range of possible values of $|\mathbf{a} + \mathbf{b}|$.

13. Given that $\mathbf{a}$ and $\mathbf{b}$ are two non-zero vectors which are not parallel, prove that if $p\mathbf{a} = q\mathbf{b}$, then $p = q = 0$. Deduce that if $\lambda\mathbf{a} + \mu\mathbf{b} = s\mathbf{a} + t\mathbf{b}$ then $\lambda = s$ and $\mu = t$.

14. Use the result of question 13 to find the values of k and l given that
(a) $k\mathbf{a} + (1 - k)\mathbf{b} = l(\mathbf{a} + \mathbf{b})$,   (b) $k(\mathbf{a} + \mathbf{b}) + 2\mathbf{b} = l(2\mathbf{a} + \mathbf{b})$.

9.4 Position vectors

Taking a fixed point O as origin, the position of any point P can be specified by giving the vector **OP**, which is then called the *position vector* of P.

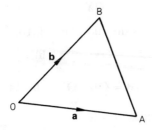

If A and B are two points with position vectors **a** and **b** respectively, then

$$\mathbf{AB} = \mathbf{AO} + \mathbf{OB} = -\mathbf{OA} + \mathbf{OB} = \mathbf{OB} - \mathbf{OA}$$
$$\therefore \quad \mathbf{AB} = \mathbf{b} - \mathbf{a}.$$

Hence the distance between the points A and B is given by $|\mathbf{b} - \mathbf{a}|$.

Consider now the mid-point M of AB.

$$\mathbf{AM} = \mathbf{MB}$$
$$\Rightarrow \mathbf{OM} - \mathbf{OA} = \mathbf{OB} - \mathbf{OM}$$
$$\Rightarrow \quad 2\mathbf{OM} = \mathbf{OA} + \mathbf{OB}$$
$$\Rightarrow \quad \mathbf{OM} = \tfrac{1}{2}(\mathbf{OA} + \mathbf{OB})$$

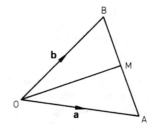

$\therefore$ | the position vector of the mid-point of AB is $\frac{1}{2}(\mathbf{a} + \mathbf{b})$.

Example 1 Prove that the diagonals of a parallelogram bisect each other.

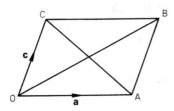

In parallelogram $OABC$, let **a** and **c** be the position vectors of A and C, then the position vector of the mid-point of AC is $\frac{1}{2}(\mathbf{a} + \mathbf{c})$.

By the parallelogram law, $\mathbf{OB} = \mathbf{a} + \mathbf{c}$.
$\therefore$ the position vector of the mid-point of OB is $\frac{1}{2}(\mathbf{a} + \mathbf{c})$.
Since the diagonals have the same mid-point, they must bisect each other.

If P is a point on the straight line through given points A and B, then there must exist some scalar t, such that $\mathbf{AP} = t\mathbf{AB}$. It is often helpful to assume a relationship of this form when dealing with collinear points.

Example 2   In the given diagram X and Y are the mid-points of OA and AB respectively. If **a** and **b** are the position vectors of the points A and B, find the position vector of the point Z.

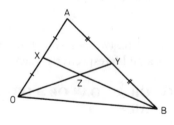

Since X is the mid-point of OA,

$$OX = \tfrac{1}{2}\mathbf{a}.$$

Since Y is the mid-point of AB,

$$\mathbf{OY} = \tfrac{1}{2}(\mathbf{a} + \mathbf{b}).$$

$$\mathbf{XB} = \mathbf{OB} - \mathbf{OX} = \mathbf{b} - \tfrac{1}{2}\mathbf{a}.$$

Since Z lies on XB, for some scalar t, $\mathbf{XZ} = t\mathbf{XB} = t(\mathbf{b} - \tfrac{1}{2}\mathbf{a})$.

$\therefore$ $\mathbf{OZ} = \mathbf{OX} + \mathbf{XZ} = \tfrac{1}{2}\mathbf{a} + t(\mathbf{b} - \tfrac{1}{2}\mathbf{a}) = \tfrac{1}{2}(1 - t)\mathbf{a} + t\mathbf{b}$.

However, since Z also lies on OY, its position vector $\mathbf{OZ}$ must also be a scalar multiple of $(\mathbf{a} + \mathbf{b})$

$$\therefore \quad \tfrac{1}{2}(1 - t) = t \quad \text{and} \quad t = \tfrac{1}{3}.$$

Hence the position vector of Z is $\tfrac{1}{3}(\mathbf{a} + \mathbf{b})$.

When working in a Cartesian coordinate system, the point P with coordinates (x, y, z) has position vector $x\mathbf{i} + y\mathbf{j} + z\mathbf{k}$, where **i**, **j** and **k** are unit vectors in the directions of the coordinate axes.

Example 3   The position vectors of points A and B with respect to an origin O are $\mathbf{a} = \mathbf{i} - 2\mathbf{j} + 7\mathbf{k}$ and $\mathbf{b} = 5\mathbf{i} + \mathbf{j} - 5\mathbf{k}$. Find (i) the distance between the origin and the point A, (ii) the distance between the points A and B, (iii) the position vector of the mid-point of AB.

(i)   the distance between the origin and the point A
$$= |\mathbf{a}| = \sqrt{(1^2 + \{-2\}^2 + 7^2)} = \sqrt{(1 + 4 + 49)} = \sqrt{54}.$$

(ii) $\mathbf{AB} = \mathbf{b} - \mathbf{a} = (5\mathbf{i} + \mathbf{j} - 5\mathbf{k}) - (\mathbf{i} - 2\mathbf{j} + 7\mathbf{k}) = 4\mathbf{i} + 3\mathbf{j} - 12\mathbf{k}$
$\therefore$ the distance between the points A and B
$$= |\mathbf{b} - \mathbf{a}| = |4\mathbf{i} + 3\mathbf{j} - 12\mathbf{k}| = \sqrt{(16 + 9 + 144)} = 13.$$

(iii)   the position vector of the mid-point of AB
$$= \tfrac{1}{2}(\mathbf{a} + \mathbf{b}) = \tfrac{1}{2}(1 + 5)\mathbf{i} + \tfrac{1}{2}(-2 + 1)\mathbf{j} + \tfrac{1}{2}(7 - 5)\mathbf{k} = 3\mathbf{i} - \tfrac{1}{2}\mathbf{j} + \mathbf{k}.$$

Exercise 9.4

1.   In a regular hexagon $OABCDE$ the position vectors of A and B relative to O are **a** and **b** respectively. Find expressions in terms of **a** and **b** for the vectors **AB** and **BC**. Find also the position vectors of the points C, D and E.

2.   In a quadrilateral $ABCD$ the point M is the mid-point of the diagonal BD. Given that the position vectors of A, B and C are **a**, **b** and **c** respectively and that

ABCM is a parallelogram, find in terms of **a**, **b** and **c** the vectors **BA**, **BC** and **BM**. Hence find the position vector of the point *D*.

3. The points *A*, *B*, *C*, *D* have position vectors **a**, **b**, **c** and **d** respectively. Prove that the lines joining the mid-points of opposite edges of the tetrahedron *ABCD* bisect each other and give the position vector of the point of intersection.

4. Write down, in terms of unit vectors **i** and **j** in the directions of the *x*- and *y*-axes, the position vectors of the points $A(2,3)$, $B(5,-1)$ and $C(4,4)$. If *D* is the fourth vertex of the parallelogram *ABCD*, find the vector **AD**. Hence find the position vector and the coordinates of the point *D*.

5. The points *A* and *B* have position vectors **a** and **b** respectively. In each of the following cases, find the distance between *A* and *B*, the unit vector in the direction of **AB** and the position vector of the mid-point *M* of *AB*.
(a) $\mathbf{a} = 3\mathbf{i} - 5\mathbf{j} - \mathbf{k}$, $\mathbf{b} = \mathbf{i} - \mathbf{j} + 3\mathbf{k}$,
(b) $\mathbf{a} = -7\mathbf{i} + 13\mathbf{j}$, $\mathbf{b} = 5\mathbf{i} - 5\mathbf{j} - 4\mathbf{k}$,
(c) $\mathbf{a} = \mathbf{i} - \mathbf{j} + 3\mathbf{k}$, $\mathbf{b} = \mathbf{i} + 2\mathbf{j} + 2\mathbf{k}$,
(d) $\mathbf{a} = 5\mathbf{i} + 4\mathbf{j} + \mathbf{k}$, $\mathbf{b} = -\mathbf{i} + \mathbf{j} - 2\mathbf{k}$.

6. Write down in terms of **i**, **j** and **k** the position vectors of the points $A(2, -1, 5)$, $B(0, 2, -1)$ and $C(-2, 4, 3)$. Hence find the length of *AB* and the coordinates of the mid-point of *BC*. Find also the coordinates of the point *D*, given that it is the fourth vertex of the parallelogram *ABCD*.

7. The points *X*, *Y* and *Z* have position vectors $-2\mathbf{i} + 2\mathbf{j} - 3\mathbf{k}$, $2\mathbf{i} + 4\mathbf{j} - 5\mathbf{k}$ and $-4\mathbf{i} + \mathbf{j} - 2\mathbf{k}$ respectively. By finding **XY** and **XZ**, show that *X*, *Y* and *Z* are collinear.

8. Use vector methods to show that the points $X(3, -4, 0)$, $Y(-1, 8, -8)$ and $Z(6, -13, 6)$ are collinear.

9. The points *A*, *B* and *C* have position vectors $\mathbf{a} = -\mathbf{i} + 2\mathbf{j} + 3\mathbf{k}$, $\mathbf{b} = 8\mathbf{i} + 7\mathbf{j} - 9\mathbf{k}$ and $\mathbf{c} = 2\mathbf{i} - 3\mathbf{j} - \mathbf{k}$. Prove that the triangle *ABC* is right-angled and find its area.

10. The points *A*, *B*, *C* and *D* have position vectors **a**, **b**, 4**b** and $k(\mathbf{a} - \mathbf{b})$ respectively. Find the value of *k* given that
(a) **AD** is parallel to **b**,   (b) **BD** is parallel to **a**,
(c) **CD** is parallel to $\mathbf{a} + \mathbf{b}$,   (d) *A*, *C* and *D* are collinear.

11. The points *P* and *Q* have position vectors **p** and **q** with respect to an origin *O*. *X* is the mid-point of *PQ* and *Y* is the point on *OX* such that $OY = 2YX$. If *PY* meets *OQ* at *Z*, find expressions in terms of **p** and **q** for **OX**, **OY** and **PY**. If $\mathbf{PZ} = k\mathbf{PY}$ find **OZ** in terms of **p**, **q** and *k*. Deduce that the position vector of *Z* is $\frac{1}{2}\mathbf{q}$.

12. The points *A* and *B* have position vectors **a** and **b** with respect to an origin *O*.

The points P, Q and R are defined such that $3\mathbf{OP} = \mathbf{OA}$, $3\mathbf{OQ} = 2\mathbf{OB}$, $2\mathbf{PR} = \mathbf{RQ}$ and S is the point of intersection of AB and OR produced. Find the position vector of R. If $\mathbf{AS} = k\mathbf{AB}$ and $\mathbf{OS} = l\mathbf{OR}$, find the values of k and l. Hence write down the position vector of S.

13. The points A, B, P and Q are defined as in question 12. X is the point of intersection of AQ and BP. Y is the point of intersection of AB and OX produced. Find the position vectors of X and Y.

Exercise 9.5   (miscellaneous)

1. Given that $\mathbf{SR} + \mathbf{QT} = \mathbf{SQ} + \mathbf{PT}$, show that the points P, Q and R are collinear.

2. If P is any point and D, E, F are the mid-points of the sides BC, CA and AB respectively of the triangle ABC, show that $\mathbf{PA} + \mathbf{PB} + \mathbf{PC} = \mathbf{PD} + \mathbf{PE} + \mathbf{PF}$.

3. In a tetrahedron $PQRS$, if H and K are the mid-points of the edges PQ and RS respectively, show that

$$\mathbf{PR} + \mathbf{QR} + \mathbf{PS} + \mathbf{QS} = 4\mathbf{HK}.$$

4. (i) In the quadrilateral $ABCD$, X and Y are the mid-points of the diagonals AC and BD respectively. Show that (a) $\vec{BA} + \vec{BC} = 2\vec{BX}$,

$$\text{(b) } \vec{BA} + \vec{BC} + \vec{DA} + \vec{DC} = 4\vec{YX}.$$

(ii) The point P lies on the circle through the vertices of a rectangle $QRST$. The point X on the diagonal QS is such that $\vec{QX} = 2\vec{XS}$. Express $\vec{PX}$, $\vec{QX}$ and $(\vec{RX} + \vec{TX})$ in terms of $\vec{PQ}$ and $\vec{PS}$.   (L)

5. If $\mathbf{a}$, $\mathbf{b}$ and $\mathbf{c}$ are vectors representing the edges of a parallelepiped, find vectors representing the four diagonals.

6. In a tetrahedron $OABC$ $\mathbf{OA} = \mathbf{a}$, $\mathbf{OB} = \mathbf{b}$ and $\mathbf{OC} = \mathbf{c}$. The points P and Q are constructed such that $\mathbf{OA} = \mathbf{AP}$ and $2\mathbf{OB} = \mathbf{BQ}$. The point M is the mid-point of PQ. Find, in terms of $\mathbf{a}$, $\mathbf{b}$ and $\mathbf{c}$, expressions for (a) $\mathbf{AB}$, (b) $\mathbf{PQ}$, (c) $\mathbf{CQ}$, (d) $\mathbf{QM}$, (e) $\mathbf{MB}$, (f) $\mathbf{OM}$.

7. If $\mathbf{a} = 2\mathbf{i} - 4\mathbf{j} + 2\mathbf{k}$ and $\mathbf{b} = 3\mathbf{i} + 4\mathbf{j} - 5\mathbf{k}$, find (a) the magnitudes of $\mathbf{a}$, $\mathbf{b}$ and $5\mathbf{a} + 2\mathbf{b}$, (b) the unit vector in the direction of $\mathbf{a} + 2\mathbf{b}$.

8. If P and Q are the points with position vectors

$$\mathbf{p} = -\mathbf{i} + 3\mathbf{j} + 2\mathbf{k} \text{ and } \mathbf{q} = \mathbf{i} - 7\mathbf{j} + 4\mathbf{k}$$

respectively, find (a) the distance PQ, (b) the unit vector in the direction of PQ, (c) the position vector of the mid-point of PQ.

9. Given that $\mathbf{a} = 6\mathbf{i} + (p - 10)\mathbf{j} + (3p - 5)\mathbf{k}$ and that $|\mathbf{a}| = 11$, find the possible values of p.

10. Given that the points A, B and C have position vectors $\mathbf{a} = \mathbf{i} - 2\mathbf{j} + 2\mathbf{k}$, $\mathbf{b} = 3\mathbf{i} - \mathbf{k}$ and $\mathbf{c} = -\mathbf{i} + \mathbf{j} + 4\mathbf{k}$, prove that the triangle ABC is isosceles.

11. In the parallelogram $OPQR$, the position vectors of P and R with respect to O are $6\mathbf{i}+\mathbf{j}$ and $3\mathbf{i}+4\mathbf{j}$ respectively. M and N are the mid-points of the sides PQ and QR respectively. Find (a) the unit vector in the direction of $\overrightarrow{MN}$, (b) the magnitude of the vector $\overrightarrow{OM}$.   (L)

12. In triangle OAB, $\mathbf{OA} = \mathbf{a}$ and $\mathbf{OB} = \mathbf{b}$. Points X, Y and Z are constructed so that $2\mathbf{AX} = 3\mathbf{AB}$, $\mathbf{OB} = \mathbf{BY}$ and Z is the point of intersection of OX and AY. If $\mathbf{AZ} = k\mathbf{AY}$, express $\mathbf{OZ}$ in terms of $\mathbf{a}$, $\mathbf{b}$ and k. Show that $\mathbf{OZ}$ is a scalar multiple of $3\mathbf{b}-\mathbf{a}$. Hence find the value of k and the position vector of Z.

13. In a parallelogram $OABC$, M is the mid-point of AB and P is the point of intersection of OM and AC. If the position vectors of A and B with respect to O as origin are $\mathbf{a}$ and $\mathbf{b}$, find the position vector of P.

14. The points $A(3, -2,0)$, $B(-2,5, -4)$ and $C(-1,0,4)$ are three vertices of a rhombus. Use vector methods to find the fourth vertex and the area of the rhombus.

15. Prove that three distinct points A, B, C with position vectors $\mathbf{a}$, $\mathbf{b}$, $\mathbf{c}$ are collinear if and only if there exist non-zero scalars l, m, n such that $l\mathbf{a}+m\mathbf{b}+n\mathbf{c} = 0$ where $l+m+n = 0$. Find the condition that the point with position vector $s\mathbf{a}+t\mathbf{b}$ should lie on the straight line which passes through A and B.

16. In triangle ABC $\mathbf{AB} = \mathbf{p}$ and $\mathbf{AC} = \mathbf{q}$. If G is the point of intersection of the medians of the triangle, find an expression for $\mathbf{AG}$ in terms of $\mathbf{p}$ and $\mathbf{q}$. Use your result to find the position vector of G, given that $\mathbf{a}$, $\mathbf{b}$ and $\mathbf{c}$ are the position vectors of A, B and C respectively.

10 Forces and equilibrium

10.1 Mathematical models

When applying mathematics to a practical problem, we try to set up what is called a mathematical model. In many cases this will take the form of a set of equations. However, it may also include diagrams, graphs or flow charts. As real situations are often very complicated, a mathematical model is usually a simplification of the original problem. Many scientific laws used in setting up systems of equations represent no more than good approximations to the true relationships between the quantities involved. Some factors governing a situation may be neglected altogether if their effect is likely to be small.

To set up a mathematical model we must first decide what factors are to be taken into account and what mathematical relationships are to be assumed, bearing in mind the degree of accuracy required in the result. We then find ways of giving numerical values to the variables involved, choosing suitable units of measurement.

To illustrate the process we now consider the problem of estimating the amount of petrol which will be consumed during a car journey. For a rough estimate we can use the distance, d miles, to be travelled and an approximate rate of petrol consumption for the car, c miles per gallon, based on past experience. We then assume that the amount of petrol used, P gallons, is proportional to the distance travelled and write $P = d \div c$. Clearly this single equation represents one of the simplest possible models which can be constructed to solve this problem. Many factors have been ignored, such as the load the car will be carrying, the type of road to be used, the traffic and weather conditions.

In elementary mechanics we set up simple models of practical situations in order to study the action of forces on various types of body. In early work we simplify the bodies involved as much as possible and take into account only the most important forces on them. For example, walls, floors and table tops are regarded as flat surfaces. A light rope or cable is assumed to be straight when pulled taut, whereas in practice, unless it is vertical, it hangs in a slight curve. We now list some of the other conventions observed when solving problems in mechanics.

A *particle* is a body of dimensions so small that its position in space can be represented by a single point.

A *lamina* is a flat object whose thickness is negligible but whose area is not.

A *rigid body* is one in which any pair of particles of matter remain a fixed distance apart. This means that rigid bodies are assumed to retain their shape when touching or colliding. For instance, when a sphere rests on a plane surface the area of contact is regarded as a single point.

A *light* string is taken to be weightless. If a body is suspended from one or more light strings, these can be represented by straight lines. Unless otherwise stated strings are assumed *inextensible*, i.e. of constant length.

A *smooth* surface is one which offers negligible frictional resistance to the motion of a body sliding across it. Wheels, pulleys, pivots and joints can also be described as smooth or frictionless.

For most purposes the surface of the earth is assumed to be flat and horizontal. Effects due to air currents and air resistance are usually ignored.

Exercise 10.1

Discuss the setting up of mathematical models in relation to various practical situations. Some problems for consideration are suggested below.

1. To decide which is the quickest way to travel from home to work or college in the rush hour.

2. To estimate the weekly cost of running a car.

3. To budget for the household bills expected in the next three months.

4. To plan orders of crisps, fruit, etc., for a college snack bar.

5. To organise a school time-table.

6. To plan a revision course for candidates taking a mathematics examination.

7. To decide whether buying a domestic appliance such as a freezer or dish-washer will save a family (a) time, (b) money.

8. To evaluate possible ways of saving energy at home or school, such as installing double glazing or switching off unnecessary lights.

10.2 The action of forces

The effect of a *force* is to set a body in motion or to change the speed and direction of its motion. If the forces on a stationary body cancel each other out, then the body is said to be in *equilibrium*. The two main branches of mechanics are *dynamics*, which is the study of the action of forces on bodies in motion, and *statics*, which is the study of forces in equilibrium. Closely related to dynamics is *kinematics*, which is the study of motion without reference to the forces involved.

A force may be completely specified by stating its magnitude, its direction and either its point of application or its line of action. Thus a force is a vector quantity associated with a particular point or line in space.

A force which is important in most problems is the gravitational attraction between the earth and any object close to it. This is the force which causes a body to fall to the ground and is called the *weight* of the body. Weight must not be confused with mass, which is not a force. The *mass* of a body is the amount of matter it contains. The weight of an object of fixed mass takes slightly different values at different points on the earth's surface. The gravitational pull on an object is much smaller on the surface of the moon, although its mass is unchanged.

When a body is attached to a string then the force acting along the string is called the *tension* in the string. The tension in a light string is assumed constant throughout its length. It is also possible to produce tension in rods and springs.

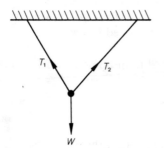

This diagram shows the forces on a particle of weight W, suspended from a beam by two light strings.

Another type of force called a *reaction* must be taken into account when two bodies are in contact. For a body resting on a smooth surface this force is referred to as the *normal reaction*, because its line of action is normal (i.e. perpendicular) to the surface. When two rough surfaces are in contact the total reaction is the sum of a normal reaction and a frictional force. [Friction will be considered further in Chapter 18.]

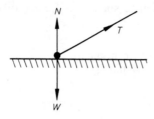

Shown here are the forces on a particle being pulled by a string across a smooth horizontal surface. The normal reaction is denoted by N.

Consider now a particle at rest on a rough inclined plane. As well as the normal reaction there will be a frictional force opposing the tendency of the particle to slide down the plane.

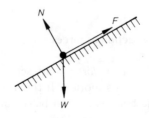

Force and other basic quantities are now usually measured in SI units, i.e. according to the Système International d'Unités. The SI unit of force is the newton, defined as the force needed to give a mass of 1 kilogramme an acceleration of 1 metre per second per second. [Students not familiar with the unit may find it helpful to remember that a cricket ball weighs approximately 1·5 N and a heavyweight boxer 800 N or more.]

We now give a table showing some of the SI units which are being used in this course.

Quantity	Name of unit	Symbol
length	metre	m
mass	kilogramme	kg
time	second	s
area	square metre	m^2
volume	cubic metre	m^3
velocity	metre per second	$m\,s^{-1}$ or m/s
acceleration	metre per second squared	$m\,s^{-2}$ or m/s^2
force	newton	N

[Note that 's' should never be added to a symbol to make a plural, because it is the symbol for second. For instance, 3 metres cannot be written as 3 ms because this means 3 milliseconds. Similarly the space in symbols for derived units is important, since without the space the symbol may have a different meaning. For example, $m\,s^{-1}$ means 'metre per second' but ms^{-1} means 'per millisecond'.]

Exercise 10.2

Draw diagrams to show the forces acting on a particle in each of the following situations.

1. At rest on a smooth horizontal plane.

2. At rest suspended from a fixed point by a string.

3. Sliding down a smooth inclined plane.

4. Sliding down a rough inclined plane.

5. Pulled across a rough horizontal plane by a string at an angle of 30° to the plane.

6. Held at rest on a smooth plane inclined at 25° to the horizontal by a string parallel to the plane.

7. Pulled up a rough plane inclined at 25° to the horizontal by a string parallel to the plane.

8. Swinging at the end of a string, when the string is at an angle of 35° to the vertical.

9. Sliding down the smooth inner surface of a hemispherical bowl.

10.3 Resultant forces

Since forces are vector quantities, they can be represented by directed line segments and added using either the triangle law or the parallelogram law as stated in §9.2. Thus, if F_1 and F_2 are two forces acting on a particle, their sum or resultant **R** may be represented by either the third side of a vector triangle or the diagonal of a parallelogram. A double arrow is used to emphasise that **R** is a resultant force rather than a third force acting on the particle.

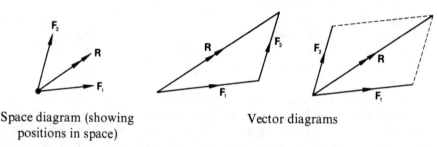

Space diagram (showing   Vector diagrams
positions in space)

The magnitude and direction of the resultant can be found either by accurate scale drawing or by calculation. Although it is usually more convenient to use a vector triangle for this purpose, the parallelogram form of the vector diagram has the advantage that it shows the two forces and their resultant acting at a single point.

Example 1   Find the resultant of two forces of magnitudes 5 N and 4 N acting at right angles.

Let R be the magnitude of the resultant and θ the angle it makes with the 5 N force.

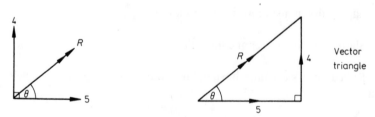

Vector
triangle

Using Pythagoras' theorem in the vector triangle,

$$R^2 = 5^2 + 4^2 = 25 + 16 = 41 \qquad \therefore \quad R \approx 6{\cdot}4$$
$$\tan \theta = 4/5 = 0{\cdot}8 \qquad\qquad\qquad \therefore \quad \theta \approx 38{\cdot}7°$$

The resultant is a force of magnitude 6·4 N at an angle 38·7° to the 5 N force.

Example 2 Two forces which act on a particle have magnitudes 2 N and 3 N. If the angle between their directions is 65°, find the resultant force on the particle.

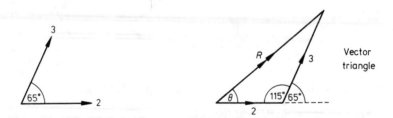

By the cosine rule, $R^2 = 2^2 + 3^2 - 2.2.3 \cos 115°$
$$= 13 - 12 \cos 115°$$
$$\therefore \quad R \approx 4.25$$

By the sine rule, $\dfrac{\sin \theta}{3} = \dfrac{\sin 115°}{R}$

$$\therefore \quad \sin \theta = \dfrac{3 \sin 115°}{R} \quad \text{and} \quad \theta \approx 39.8°$$

Hence the resultant is a force of magnitude 4·25 N at an angle 39·8° to the 2 N force.

[Note that to achieve maximum accuracy the unrounded value of R should be used to find θ. The given results were obtained using a calculator, the value $R = 4.251048811$ being stored in the memory.]

When the resultant of more than two forces is required a vector polygon can be constructed.

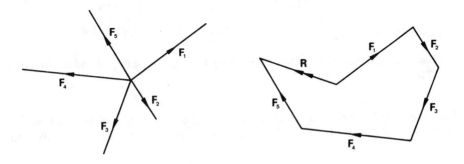

Example 3 Three forces, each of magnitude 10 N, act horizontally in the directions due west, N 30° E and S 50° E. Find the magnitude and direction of their resultant.

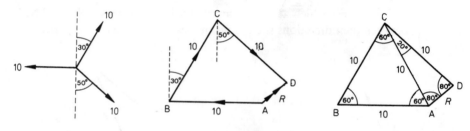

In the vector diagram, $\angle ABC = 60°$ and $AB = BC = 10$

$\therefore$ $\triangle ABC$ is an equilateral triangle.

Hence in $\triangle ACD$, $\angle ACD = 20°$ and $AC = CD = 10$

$\therefore$ $\triangle ACD$ is an isosceles triangle in which $\angle A = \angle D = 80°$ and $R = 2.10 \cos 80° \approx 3\cdot47$.

Thus the resultant force has magnitude 3·47 newtons and direction N 50° E.

The resultant of a set of forces can be found more quickly if the forces are expressed in terms of their Cartesian components. (See §9.3.)

Example 4   Find the magnitude and direction of the resultant of the forces represented by $2\mathbf{i} - 5\mathbf{j}$, $3\mathbf{j}$ and $\mathbf{i} + \mathbf{j}$, where $\mathbf{i}$ and $\mathbf{j}$ are unit vectors in the directions due east and due north respectively.

Let $\mathbf{R}$ be the resultant of the given forces and let θ be the angle between $\mathbf{R}$ and $\mathbf{i}$.

$$\mathbf{R} = (2\mathbf{i} - 5\mathbf{j}) + 3\mathbf{j} + (\mathbf{i} + \mathbf{j}) = (2+1)\mathbf{i} + (-5+3+1)\mathbf{j}$$
$$= 3\mathbf{i} - \mathbf{j}$$

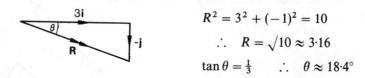

$$R^2 = 3^2 + (-1)^2 = 10$$
$$\therefore \quad R = \sqrt{10} \approx 3\cdot16$$
$$\tan\theta = \tfrac{1}{3} \quad \therefore \quad \theta \approx 18\cdot4°$$

Hence the resultant is represented by $3\mathbf{i} - \mathbf{j}$. Its magnitude is 3·16 and its direction S 71·6° E.

Example 5   If $\mathbf{i}$, $\mathbf{j}$ and $\mathbf{k}$ are unit vectors parallel to the x-, y- and z-axes, find the magnitude of the resultant of the forces represented by $3\mathbf{i} - 2\mathbf{j} + 4\mathbf{k}$, $\mathbf{i} - 3\mathbf{k}$ and $2\mathbf{i} + 5\mathbf{j} + \mathbf{k}$.

The resultant is given by

$$(3\mathbf{i} - 2\mathbf{j} + 4\mathbf{k}) + (\mathbf{i} - 3\mathbf{k}) + (2\mathbf{i} + 5\mathbf{j} + \mathbf{k}), \quad \text{i.e.} \quad 6\mathbf{i} + 3\mathbf{j} + 2\mathbf{k}$$

$\therefore$ the magnitude of the resultant is

$$|6\mathbf{i} + 3\mathbf{j} + 2\mathbf{k}| = \sqrt{(6^2 + 3^2 + 2^2)} = \sqrt{49} = 7.$$

Example 6   Two forces **P** and **Q** act in the directions of the vectors $4\mathbf{i}+3\mathbf{j}$ and $\mathbf{i}-2\mathbf{j}$ respectively and the magnitude of **P** is 25 N. Given that the magnitude of the resultant of **P** and **Q** is also 25 N, find the magnitude of **Q**.

$$|4\mathbf{i}+3\mathbf{j}| = \sqrt{(4^2+3^2)} = \sqrt{25} = 5$$

∴   the force **P** can be represented in magnitude and direction by the vector $5(4\mathbf{i}+3\mathbf{j})$, i.e. $20\mathbf{i}+15\mathbf{j}$.

Since **Q** can be represented by a vector of the form $\lambda(\mathbf{i}-2\mathbf{j})$, we may write

$$\mathbf{P}+\mathbf{Q} = (20\mathbf{i}+15\mathbf{j})+\lambda(\mathbf{i}-2\mathbf{j}) = (20+\lambda)\mathbf{i}+(15-2\lambda)\mathbf{j}$$

∴   the magnitude of the resultant of **P** and **Q**

$$= \sqrt{\{(20+\lambda)^2+(15-2\lambda)^2\}} = 25$$

∴ $400+40\lambda+\lambda^2+225-60\lambda+4\lambda^2 = 625$
$$5\lambda^2-20\lambda = 0$$
∴   assuming λ is non-zero, $\lambda = 4$

Hence $\mathbf{Q} = 4(\mathbf{i}-2\mathbf{j})$ and $|\mathbf{Q}| = 4\sqrt{\{1^2+(-2)^2\}} = 4\sqrt{5}$.

Thus the magnitude of the force **Q** is $4\sqrt{5}$ N.

Exercise 10.3

1.   The resultant of two forces **P** and **Q** is of magnitude R newtons and acts at an angle θ to the direction of **P**. Find R and θ, given that the magnitudes of **P** and **Q** and the angle between their directions are respectively
(a) 12 N, 5 N; 90°,   (b) 4 N, 7 N; 90°,
(c) 8 N, 11 N; 120°,   (d) 6 N, 10 N; 52°,
(e) 14 N, 9 N; 108°,   (f) 2 N, 7 N; 23°.

2.   Find the angle between the lines of action of two forces of magnitudes 7 N and 11 N, given that their resultant is of magnitude 8 N.

3.   The resultant of two forces **F** and **G** is a force of 20 N acting at an angle of 45° to the force **G**. Given that the magnitude of **G** is 12 N, find the magnitude and direction of **F**.

4.   Two forces of magnitudes 7 N and 9 N act at an angle θ where $\sin\theta = 0.4$. Find the magnitudes of the two possible resultants.

5.   Find the resultant of three horizontal forces whose magnitudes and directions are 16 N in the direction N 70° W, 8 N in the direction N 20° E and 11 N in the direction S 70° E.

6.   Four forces, each of magnitude 12 N, act horizontally in the directions N 30° E, S 60° E, S 15° E and S 75° W. Find the magnitude and direction of their resultant.

7.   Horizontal forces of magnitudes 12 N, 9 N and 6 N act on a particle in the directions due north, due east and S 60° E respectively. Find the magnitude and direction of their resultant.

8.   Find the magnitude and direction of the resultant of the forces represented by each of the following sets of vectors, where i and j are unit vectors in the directions due east and due north respectively.
(a) $2i-j, 3i, -i+4j,$ (b) $5i-2j, 2i+3j, -j,$
(c) $7i+4j, 2i-5j, -3i-2j,$ (d) $i+j, -3i-2j, -3i-4j.$

9.   If i, j and k are unit vectors parallel to the x-, y- and z-axes, find the magnitude of the resultant of the forces represented by
(a) $2i-3j, 5i+j-3k$ and $-3i+4j-k,$
(b) $i+6j-3k, 8i-j-5k$ and $3i+3j-k,$
(c) $2j-5k, 5i-k$ and $2i+3j+k.$

10. Find the magnitude of the resultant of two forces of magnitudes 15 N and $4\sqrt{2}$ N acting in the directions of the vectors $3i-4j$ and $i+j$ respectively.

11. Two forces, each of magnitude $5P$ newtons, act in the directions of the vectors $7i+j$ and $i-j$. Show that their resultant acts in the direction of the vector $3i-j$. Given that the magnitude of the resultant is 20 newtons, find the value of P in surd form.

12. Two forces $\mathbf{F}$ and $\mathbf{G}$ act parallel to the vectors $3i+j$ and $i+2j$ respectively. If the resultant is a force of 10 newtons in the direction of the vector $3i-4j$, find the magnitudes of $\mathbf{F}$ and $\mathbf{G}$ in surd form.

10.4 Resolution into components

If p and q are two non-parallel lines in a plane, then any force $\mathbf{F}$ acting in that plane can be resolved into component forces $\mathbf{P}$ and $\mathbf{Q}$ in directions parallel to p and q by constructing a vector triangle or parallelogram as shown below.

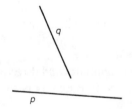 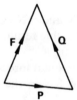 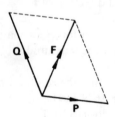

Consider now two forces and their resolution into components parallel to the lines p and q. Let $\mathbf{F}_1 = \mathbf{P}_1+\mathbf{Q}_1$, $\mathbf{F}_2 = \mathbf{P}_2+\mathbf{Q}_2$ and let the resultant of $\mathbf{F}_1$ and $\mathbf{F}_2$ be $\mathbf{R}$.

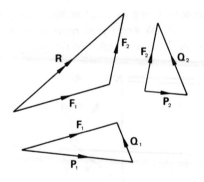

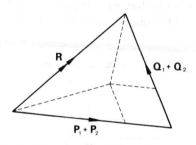

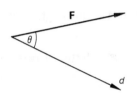

As shown in the diagrams,

$$R = F_1 + F_2 = (P_1 + Q_1) + (P_2 + Q_2) = (P_1 + P_2) + (Q_1 + Q_2).$$

Thus the components of a resultant force are the sums of the corresponding components of the original forces. This property of forces and their components is often used to find resultant forces.

For most practical purposes forces are resolved into two components at right angles.

Example 1   Find the magnitudes of the horizontal and vertical components of a force of 6 N acting at an angle of 63° to the horizontal.

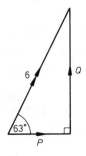

Let P and Q newtons be the magnitudes of the required components,

then $P = 6\cos 63° \approx 2.72$
and $Q = 6\sin 63° \approx 5.35$

$\therefore$ the magnitudes of the horizontal and vertical components are 2·72 N and 5·35 N respectively.

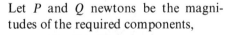

If a force F acts at an angle θ to a given direction d, then the quantity $F\cos\theta$ is called the *resolved part* of F in the direction d, or simply the component of F in the direction d. If θ is acute, $F\cos\theta$ is positive, but if θ is obtuse, $F\cos\theta$ is negative. Thus the sign of the resolved part indicates whether the direction of the component force is the same as d or opposite to d.

Example 2   A particle is sliding down a smooth plane inclined at an angle α to the horizontal. Find the components of the weight **W** and the normal reaction **N** down the plane and in the direction of an upward perpendicular to the plane.

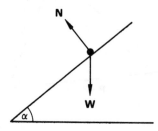
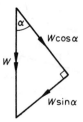

The components of **W** are $W \sin \alpha$ down the plane and $-W \cos \alpha$ along an upward perpendicular to the plane.

Since **N** acts at right angles to the plane, its component down the plane is 0 and its component along an upward perpendicular is N.

The resultant of a set of forces can now be found by resolving in two perpendicular directions.

Example 3   A small package of weight 20 N is attached to three light strings lying in the same vertical plane. If the tensions in the strings are as shown in the given diagram, find the resultant force on the package.

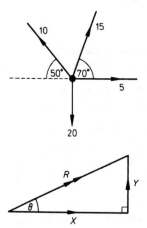

Let the resultant **R** have horizontal and vertical components X and Y. Resolving horizontally and vertically,

$$\rightarrow X = 5 + 15 \cos 70° - 10 \cos 50°.$$
$$\uparrow Y = 15 \sin 70° + 10 \sin 50° - 20$$

By calculator $X = 3 \cdot 70242,$
$$Y = 1 \cdot 755835.$$

$$\therefore \quad R = \sqrt{(X^2 + Y^2)} \approx 4 \cdot 10$$

and $\tan \theta = Y/X, \quad \theta \approx 25 \cdot 4°.$

Hence the resultant is a force of magnitude 4·10 N at an angle 25·4° to the horizontal.

It is sometimes convenient to express a set of forces in terms of perpendicular unit vectors **i** and **j**.

Example 4   A set of horizontal forces of magnitudes 3 N, 5 N, 6 N and 10 N act on a particle in directions due east, due south, N 30° W and N 60° E respectively. Find the magnitude and direction of the resultant.

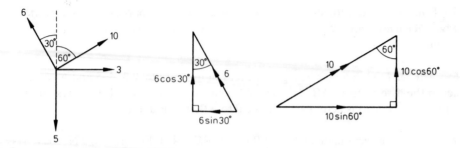

Let **i** and **j** be unit vectors in the directions due east and due north. The given set of forces can then be represented by

$$3\mathbf{i}, \; -5\mathbf{j}, \; -6\sin 30°\mathbf{i}+6\cos 30°\mathbf{j} \quad \text{and} \quad 10\sin 60°\mathbf{i}+10\cos 60°\mathbf{j}.$$

Hence the resultant **R** is given by

$$\mathbf{R} = 3\mathbf{i}+(-5\mathbf{j})+(-3\mathbf{i}+3\sqrt{3}\mathbf{j})+(5\sqrt{3}\mathbf{i}+5\mathbf{j}) = 5\sqrt{3}\mathbf{i}+3\sqrt{3}\mathbf{j}$$

$$R^2 = (5\sqrt{3})^2+(3\sqrt{3})^2 = 75+27 = 102$$

$$\therefore \quad R \approx 10\cdot1$$

$$\tan\theta = \frac{3\sqrt{3}}{5\sqrt{3}} = 0\cdot6 \quad \therefore \quad \theta \approx 30°.$$

The resultant is a force of magnitude 10·1 N and direction N 59° E.

Exercise 10.4

1. Find the magnitudes of the horizontal and vertical components of
(a) a force of 20 N acting at 40° to the horizontal,
(b) a force of 7 N acting at 78° to the horizontal,
(c) a force of 12 N acting at 36° to the vertical.

2. A particle is being pulled up a smooth plane, inclined at an angle α to the horizontal, by a string which makes an acute angle β with the plane. Find the resolved parts of the weight of the particle **W**, the normal reaction **N** and the tension in the string **T** (a) vertically upwards, (b) up the plane, (c) in the direction of a downward perpendicular to the plane.

3. A set of coplanar forces of magnitudes 12, 9, 15 and 16 newtons act on a particle in directions south-west, north, north-west and east respectively. Find the magnitude and direction of their resultant.

4. A particle is being pulled across a smooth horizontal plane by three strings. The directions of the strings are N 60° W, due north and N 30° E. If the tensions in the strings are of magnitude 15 N, 12 N and 10 N respectively, find the resultant force on the particle.

5.   Four horizontal forces of equal magnitude act on a particle in the directions whose bearings are 050°, 100°, 250° and 290°. Calculate the bearing of the resultant.

6.   *ABCDEF* is a regular hexagon. Forces of magnitude 2, 3, 4, 5 and 6 newtons act in the directions of **AB**, **AC**, **AD**, **AE** and **AF** respectively. If their resultant is a force of *R* newtons acting at an angle θ to **AB**, find *R* and θ.

7.   *ABCD* is a rectangle in which *AB* is 3 m, *BC* is 4 m. Forces 3 N, 2 N, 10 N, 5 N act in the directions of **AB**, **AD**, **AC**, **BD** respectively. Find their resultant, giving the angle it makes with **AB**.

8.   *ABCDE* is a regular pentagon. Find the magnitude of the resultant of forces 2 N, 3 N, 1 N, 4 N, 5 N in the directions of **AB**, **BC**, **CD**, **DE**, **EC** respectively.

9.   A vertical flagpole of height 24 m is held in position by three straight wires fixed to its top and to points on the horizontal ground. The wires are of lengths 25 m, 30 m and 26 m and their fixing points are at bearings 060°, 150° and 270° respectively from the base of the pole. If the tensions in the wires are each 50 N, find (a) the vertical components of the tensions, (b) the horizontal components of the tensions. Hence find (c) the resultant force downwards caused by the wires, (d) the resultant horizontal force on the flagpole.

10.5 A particle in equilibrium

Forces on a particle are said to be in *equilibrium* if their resultant is zero. Thus when forces in equilibrium are resolved into components in two fixed directions, the sums of the components in each direction will be zero. This fact provides a basic approach to problem solving.

Example 1   A set of horizontal forces of magnitudes 5, 4 and 7 newtons act on a particle in the directions N 50° W, due north and S 80° E respectively. Find the magnitude and direction of a fourth force which holds the particle in equilibrium.

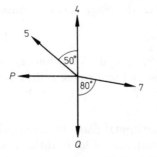

Let the components in the directions due west and due south of the fourth force **F** be *P* and *Q* respectively. Resolving in the directions west and south,

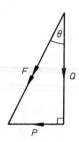

$$\leftarrow P + 5\sin 50° - 7\sin 80° = 0$$
$$\downarrow Q + 7\cos 80° - 5\cos 50° - 4 = 0$$

$$\therefore \quad P = 7\sin 80° - 5\sin 50°$$
$$Q = 4 + 5\cos 50° - 7\cos 80°.$$

By calculator $P = 3\!\cdot\!063436,$
$$Q = 5\!\cdot\!998404$$

$$\therefore \quad F = \sqrt{(P^2 + Q^2)} \approx 6\!\cdot\!74 \quad \text{and} \quad \tan\theta = P/Q, \quad \theta \approx 27\!\cdot\!1°.$$

Hence the fourth force is of magnitude 6·74 newtons in the direction S 27·1° W.

Example 2   A light inextensible string is attached to two points A and B on a horizontal beam. A smooth ring C of weight 10 N, which can move freely on the string, is held in equilibrium by a horizontal force P acting in the vertical plane containing A, B and C. Given that $\angle BAC = 36°$ and $\angle ABC = 48°$, find the magnitude of P and the tension T in the string.

[Note that since the ring is smooth the tension is the same in both parts of the string.]

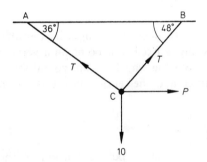

Resolving horizontally and vertically,

$$\rightarrow P + T\cos 48° - T\cos 36° = 0$$
$$\uparrow T\sin 48° + T\sin 36° - 10 = 0$$

$$\therefore \quad T = \frac{10}{\sin 48° + \sin 36°} \approx 7\!\cdot\!51$$

and $P = T(\cos 36° - \cos 48°) \approx 1\!\cdot\!05.$

Hence the magnitude of the horizontal force P is 1·05 N and the tension in the string is 7·51 N.

In §10.3 we saw that forces can be added by constructing a vector polygon. If a set of forces is in equilibrium the length of the side representing their resultant must be zero. Hence a set of forces in equilibrium can be represented by the sides of a closed polygon, often referred to as the *polygon of forces*.

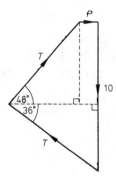

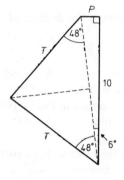

Example 2 could have been solved by drawing a polygon of forces. Using the right-angled triangles shown in the diagram,

$$T \sin 48° + T \sin 36° = 10$$
and $$T \cos 48° + P = T \cos 36°$$

$\therefore$ as before we find that

$$T = \frac{10}{\sin 48° + \sin 36°},$$

$$P = T(\cos 36° - \cos 48°).$$

However, considering instead the angle properties of the figure, we could obtain the equivalent results,

$$P = 10 \tan 6° \text{ and}$$
$$2T \cos 48° \cos 6° = 10.$$

Of special interest are problems involving three forces in equilibrium, because in this case the vector diagram is simply a *triangle of forces*.

Example 3   A particle of weight $4W$ is attached to the end of a light inextensible string of length 2 m. The other end of the string is fixed to a point on a vertical wall. The particle is being held 1·2 m away from the wall by a horizontal force. Express the magnitudes of the horizontal force and the tension in the string in terms of W.

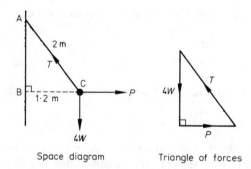

Space diagram   Triangle of forces

Let T be the tension in the string and P the horizontal force.

By Pythagoras' theorem in $\triangle ABC$,

$$AB^2 = AC^2 - BC^2 = 2^2 - (1·2)^2 = 2·56 \qquad \therefore \quad AB = 1·6$$

As shown in the diagrams, $\triangle ABC$ is similar to the triangle of forces

$$\therefore \quad \frac{T}{2} = \frac{P}{1 \cdot 2} = \frac{4W}{1 \cdot 6}.$$

Hence $T = 5W$ and $P = 3W$.

The magnitudes of the horizontal force and the tension in the string are $3W$ and $5W$ respectively.

Consider now three forces of magnitudes P, Q and R acting on a particle. If the angles between the forces are α, β and γ, then the exterior angles of the triangle of forces are also α, β and γ.

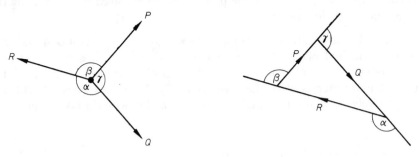

By the sine rule, $\dfrac{P}{\sin(180° - \alpha)} = \dfrac{Q}{\sin(180° - \beta)} = \dfrac{R}{\sin(180° - \gamma)}$

$$\therefore \quad \boxed{\dfrac{P}{\sin \alpha} = \dfrac{Q}{\sin \beta} = \dfrac{R}{\sin \gamma}}$$

This result is called *Lami's† Theorem*. It is often useful in solving three force problems in which angles rather than lengths are known.

Example 4   A small block of weight 20 N is at rest on a smooth plane inclined at an angle of 25° to the horizontal. The block is being held in position by a light rope. Given that the angle between the rope and the plane is 15°, find the magnitudes of the tension T in the rope and of the reaction R with the plane.

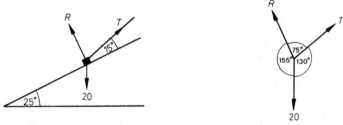

† *Lami, Bernard* (1640–1715) French scholar. He made a substantial contribution to the philosophical and mathematical theories of his time. His work on equilibrium appears in the *Traité de Mécanique* (1679).

By Lami's theorem, $\dfrac{T}{\sin 155°} = \dfrac{R}{\sin 130°} = \dfrac{20}{\sin 75°}$

$\therefore \quad T = \dfrac{20 \sin 155°}{\sin 75°} \approx 8\!\cdot\!75 \quad \text{and} \quad R = 20 \sin 130°/\sin 75° \approx 15\!\cdot\!9$

The tension in the rope is 8·75 N and the normal reaction with the plane is 15·9 N.

[This problem could, of course, also be solved by resolving the forces parallel and perpendicular to the plane or by drawing a triangle of forces.]

Problems involving more than one particle are often solved by considering the forces on each particle separately.

Example 5   A light inextensible string is suspended from two points A and D in the same horizontal plane. Particles of weight W and $2W$ are attached to the string at points B and C respectively. If AB and CD are inclined at 60° to the horizontal, find in terms of W the tensions T_1, T_2, T_3 in the strings AB, BC and CD respectively. Find also the angle θ, which BC makes with the horizontal.

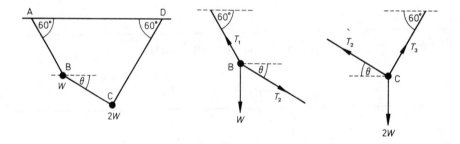

Resolving the forces on the particle at B

horizontally $\qquad\qquad\qquad \rightarrow T_2 \cos\theta - T_1 \cos 60° = 0 \qquad\qquad\qquad (1)$

vertically $\qquad\qquad\qquad \uparrow T_1 \sin 60° - T_2 \sin\theta - W = 0 \qquad\qquad\qquad (2)$

Resolving the forces on the particle at C

horizontally $\qquad\qquad\qquad \rightarrow T_3 \cos 60° - T_2 \cos\theta = 0 \qquad\qquad\qquad (3)$

vertically $\qquad\qquad\qquad \uparrow T_3 \sin 60° + T_2 \sin\theta - 2W = 0 \qquad\qquad\qquad (4)$

Adding (1) to (3): $\qquad T_3 \cos 60° - T_1 \cos 60° = 0 \qquad\qquad \therefore \quad T_1 = T_3$

Adding (2) to (4): $\quad T_1 \sin 60° + T_3 \sin 60° - 3W = 0$

$\therefore \quad T_1 + T_3 = 3W/\sin 60° = 2\sqrt{3}W \quad \text{and} \quad T_1 = T_3 = \sqrt{3}W$

Using (1) $T_2 \cos\theta = \sqrt{3}W \cdot \cos 60° = \dfrac{\sqrt{3}}{2}W$

Using (2) $T_2 \sin\theta = \sqrt{3}W \cdot \sin 60° - W = \dfrac{1}{2}W$

By squaring and adding these results:

$$T_2^2 = T_2^2(\cos^2\theta + \sin^2\theta) = \tfrac{3}{4}W^2 + \tfrac{1}{4}W^2 = W^2$$

Dividing: $\tan\theta = T_2\sin\theta/T_2\cos\theta = \dfrac{1}{2}W \bigg/ \dfrac{\sqrt{3}}{2}W = 1/\sqrt{3}$

$$\therefore \quad T_2 = W \quad \text{and} \quad \theta = 30°.$$

Hence the tensions in the strings AB, BC and CD are $\sqrt{3}W$, W and $\sqrt{3}W$ respectively and BC is inclined at 30° to the horizontal.

Exercise 10.5

1.   A set of horizontal forces of magnitude 20 N, 12 N and 30 N act on a particle in the directions due south, due east and N 40° E respectively. Find the magnitude and direction of a fourth force which holds the particle in equilibrium.

2.   A set of four horizontal forces in equilibrium act in directions whose bearings are 020°, 110°, 230° and 320°. If the magnitudes of the forces are 8, 12, P and Q newtons respectively, find the values of P and Q.

3.   A smooth ring C of weight 15 N can move freely on a light inextensible string attached to two fixed points A and B in the same horizontal line. A horizontal force of magnitude P holds the ring in equilibrium, such that AC is vertical and $\angle ABC = 30°$. Find the value of P using a polygon of forces.

4.   A set of six forces in equilibrium act in the directions of the sides $\vec{AB}$, $\vec{BC}$, $\vec{CD}$, $\vec{DE}$, $\vec{EF}$ and $\vec{FA}$ of a regular hexagon. If the magnitudes of the forces are 7, 4, 10, 5, P and Q newtons respectively, use a polygon of forces to find the values of P and Q.

5.   A body of weight 20 newtons is held at rest on a smooth plane inclined at 60° to the horizontal by a force of magnitude P newtons. Find P given that the force is (a) horizontal, (b), vertical, (c) parallel to the plane.

6.   One end of a light inextensible string of length 75 cm is fixed to a point on a vertical pole. A particle of weight 12 N is attached to the other end of the string. The particle is held 21 cm away from the pole by a horizontal force. Find the magnitude of this force and the tension in the string.

7.   A small object of weight 4 N is suspended from a fixed point by a string. The object is held in equilibrium, with the string at an angle of 25° to the vertical, by a horizontal force. Find the magnitude of this force and the tension in the string.

8.   A particle P of weight 50 N is hanging in equilibrium supported by two strings inclined at 20° and 40° to the vertical. Find the tensions in the strings.

9.   A particle of weight $10W$ is held at rest on a smooth plane inclined at 30° to the horizontal by a light string. Find the tension in the string and the reaction

between the particle and the plane if (a) the string is parallel to the plane, (b) the string is at an angle of 30° to the plane.

10. A particle of weight 2 N hangs on a string. The particle is pulled to one side by a force of magnitude 3 N acting at an angle of 20° below the horizontal. Find the tension in the string and the inclination of the string to the vertical.

11. A particle P of weight $10W$ is supported by two light inextensible strings attached to fixed points X and Y which lie in the same horizontal plane. Given that $PX = 36$ cm, $PY = 48$ cm and $XY = 60$ cm, find the tension in each string.

12. A light inextensible string passes through a smooth ring fixed at a point P. Particles A and B of weights 10 N and 20 N respectively are attached to the ends of the string. The system is held in equilibrium by a horizontal force acting on the particle A. Find the magnitude of this force. Find also the magnitude and direction of the resultant force on the ring due to the tension in the string.

13. A light inextensible string ABC is fixed at the point A. Two particles each of weight W are attached to the string at B and C. The system is held in equilibrium by a horizontal force of magnitude $2W$ acting on the particle at C. Find (a) the tensions in BC and AB, (b) the inclinations of BC and AB to the vertical.

14. Two smooth straight rods XY and XZ are fixed at 45° to the vertical and at right angles to each other. Two small beads A and B, each of weight W, are free to slide on the rods XY and XZ respectively. The beads are connected by a light inextensible string, to the mid-point of which is attached a particle C of weight $2W$. Find the reaction between the bead A and the rod XY, the tension in the string AC and its inclination to the horizontal when the system is in equilibrium.

15. $ABCD$ is a light inextensible string. The ends A and D are fixed so that the line AD is horizontal. Particles of weight $3W$ and $4W$ are attached to the string at the points B and C respectively. Find the tensions in the strings AB, BC and CD given that (a) $\angle DAB = 45°$ and $\angle ABC = 150°$, (b) $\angle DAB = 45°$ and the strings BC and DC are equally inclined to the horizontal.

Exercise 10.6   (miscellaneous)

1. Find the angle between the lines of action of two forces of magnitudes 15 N and 9 N, given that their resultant is a force of magnitude 20 N.

2. Two forces of magnitude $3P$ and $4P$ newtons act on a particle. The magnitude of their resultant is 20 newtons. When one of the forces is reversed the magnitude of the resultant is halved. Calculate the value of P and the original angle between the forces.

3. Two forces of magnitude 5 N and 3 N act on a particle. Their resultant acts at an angle θ to the direction of the 5 N force. Find the range of possible values of θ.

4.   Given that **i** and **j** are unit vectors in the directions due east and due north respectively, find the magnitude and direction of the resultant of the forces represented by the vectors $3\mathbf{i} - \mathbf{j}$, $-2\mathbf{i}$, $5\mathbf{i} + 7\mathbf{j}$ and $\mathbf{i} - 2\mathbf{j}$.

5.   Two concurrent forces **X** and **Y** have resultant **R**. If $X = R$ and the angle between **R** and **Y** is α, prove (a) $Y = 2X \cos \alpha$, (b) the angle between **X** and **Y** is $180° - \alpha$.

6.   A force of magnitude 10 N parallel to the vector $4\mathbf{i} + 3\mathbf{j}$ is the resultant of two forces parallel respectively to the vectors $2\mathbf{i} + \mathbf{j}$, $\mathbf{i} + \mathbf{j}$. Find the magnitudes of these two forces.   (L)

7.   The point O is the centre of a regular hexagon $ABCDEF$. Forces of magnitude 7, 3, 10, 9, 6 and 2 newtons act on a particle at O in the directions of **OA**, **OB**, **OC**, **OD**, **OE** and **OF** respectively. Find the magnitude of the resultant and show that it acts in the direction of **AC**.

8.   Find the bearing of the resultant of five horizontal forces of equal magnitude which act in the directions whose bearings are 070°, 130°, 160°, 220° and 310°.

9.   A set of four forces on a particle are in equilibrium. If **i** and **j** are unit vectors in the directions due east and due north respectively, three of the forces are represented by the vectors $\mathbf{i} + \mathbf{j}$, $3\mathbf{i} - 2\mathbf{j}$ and $2\mathbf{i} + 3\mathbf{j}$. Find the magnitude and direction of the fourth force.

10. A particle is in equilibrium under the action of a set of five horizontal forces. If four of the forces are of magnitude 6 N, 11 N, 8 N and 15 N acting in the directions N 72° W, N 15° W, N 35° E and S 63° E respectively, find the magnitude and direction of the fifth force.

11. A particle P of weight $10W$ is hanging in equilibrium from two strings PA and PB. If $\angle APB = 90°$ and the tensions in the strings are T and $2T$, find T in terms of W.

12. A particle P of weight 50 N is supported by two light inextensible strings attached to fixed points A and B. Given that A and B lie 2 m apart in the same horizontal plane and that $AP = 2$ m, $PB = 1$ m, find the tension in each string.

13. A small block of weight 20 N is held at rest on a smooth plane inclined at 20° to the horizontal by a string which makes an angle θ with the plane. If the tension in the string is 10 N, find the value of θ and the magnitude of the reaction between the block and the plane.

14. Show that the least force required to keep a particle of weight W in equilibrium on a smooth plane inclined at an angle α to the horizontal is of magnitude $W \sin \alpha$ and parallel to the plane.

15. The ends of a light inextensible string of length 80 cm are fixed to points A and B, which are 48 cm apart. A smooth ring C, of weight 5 N, which can move freely along the string, hangs in equilibrium. Find the tension in the string if (a) AB is horizontal, (b) AB is inclined at 60° to the horizontal.

16. One end of a light inextensible string is attached to a fixed point on a smooth straight wire inclined at 30° to the horizontal. The other end of the string is attached to a light ring which slides freely on the wire. A smooth bead of weight W, which is threaded on the string, is also free to slide. Find, in terms of W, the tension in the string when the system is in equilibrium.

17. A particle of weight 2 N is attached to a light string ABC at the point B. The ends A and C are threaded through two fixed smooth rings at points X and Y respectively, where the line XY makes an angle θ with the horizontal. A particle of weight 3 N is then attached to the string at A and a particle of weight 4 N at C. Show that equilibrium is possible only if Y lies above X and $1/4 < \sin\theta < 11/16$.

18. Two rods OA, OB are fixed in a vertical plane with O uppermost, each rod making an acute angle α with the vertical. Two smooth rings of equal weight W, which can slide one on each rod, are connected by a light inextensible string, upon which slides a third smooth ring of weight $2W$. Show that, in the symmetrical position of equilibrium, the angle between the two straight pieces of the string is 2β, where $\tan\beta = 2\cot\alpha$. Show that the tension in the string is $W\sqrt{(1+4\cot^2\alpha)}$ and that the reaction between the rings and the rods is $2W\operatorname{cosec}\alpha$. (W)

19. Four packages, each of weight 20 N, are fastened to a light rope at points B, C, D and E. The rope is attached to two fixed points A and F, which lie in the same horizontal plane. The points B and C are 1·2 m and 2·2 m respectively below the line AF and $AB = BC = CD = DE = EF$. Find the tension in the rope CD and the length of the rope.

20. Construct a flow chart for a procedure to find the resultant of a set of n horizontal forces, printing out its magnitude and bearing. Assume that the data is provided in the form of a list giving the magnitude x newtons and the bearing $y°$ of each force, where $x > 0$ and $0 \leqslant y < 360$.

11 Relative motion

11.1 Velocity and acceleration as vectors

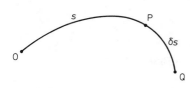

The diagram shows the path of a moving particle. The motion begins at the point O and after time t the particle reaches the point P. The *distance* travelled is the length s of the path OP. The *speed* of the particle is the rate at which it moves along its path. If the particle moves a further distance δs to a point Q in time δt, then its average speed in the motion from P to Q is $\delta s / \delta t$.

Hence the speed of the particle as it passes through $P = \displaystyle\lim_{\delta t \to 0} \frac{\delta s}{\delta t} = \frac{ds}{dt}$.

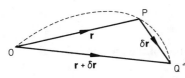

The *displacement* of the particle from O at time t is the vector $\mathbf{r}$ represented by the line segment $\overrightarrow{OP}$. The *velocity* of the particle is the rate of change of its displacement. Thus the average velocity of the particle in its motion from P to Q can be expressed as $\delta \mathbf{r} / \delta t$. Hence the velocity of the particle as it passes through

$P = \displaystyle\lim_{\delta t \to 0} \frac{\delta \mathbf{r}}{\delta t} = \frac{d\mathbf{r}}{dt}$.

By considering the magnitude and direction of $\overrightarrow{PQ}$ as $\delta t \to 0$, we see that the magnitude of the velocity is the speed of the particle and the direction of the velocity is the direction of motion.

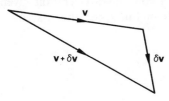

The *acceleration* of a particle is defined as the rate of change of its velocity. If the velocities of a particle at times t and $t + \delta t$ are $\mathbf{v}$ and $\mathbf{v} + \delta \mathbf{v}$ respectively, then the acceleration at

time $t = \displaystyle\lim_{\delta t \to 0} \frac{\delta \mathbf{v}}{\delta t} = \frac{d\mathbf{v}}{dt}$.

[Note that unless a particle is moving in a straight line the magnitude of its acceleration does *not* correspond to the rate of change of its speed and the direction of the acceleration is *not* the same as the direction of motion.]

It is clear from these definitions that distance and speed are scalar quantities, whereas displacement, velocity and acceleration are vector quantities. We now give some examples illustrating the relationships between them. Unless otherwise stated we will assume that the magnitudes of vector quantities are measured in the appropriate SI units.

Example 1   A particle moving with uniform velocity $4\mathbf{i}+3\mathbf{j}$ passes through the point A whose position vector with respect to the origin O is $3\mathbf{i}-\mathbf{j}$. Find (a) the speed of the particle, (b) the distance of the particle from the origin 3 seconds after it passes through A.

(a) The speed of the particle
$$= |4\mathbf{i}+3\mathbf{j}| = \sqrt{(4^2+3^2)} = 5.$$
(b) After 3 seconds the displacement of the particle from A
$$= 3(4\mathbf{i}+3\mathbf{j}) = 12\mathbf{i}+9\mathbf{j}$$
$\therefore$ after 3 seconds its position vector
$$= (12\mathbf{i}+9\mathbf{j})+(3\mathbf{i}-\mathbf{j}) = 15\mathbf{i}+8\mathbf{j}.$$
Hence the distance of the particle from the origin
$$= |15\mathbf{i}+8\mathbf{j}| = \sqrt{(15^2+8^2)} = \sqrt{289} = 17.$$

Example 2   A particle moves for 2 seconds with velocity $6\mathbf{i}+3\mathbf{j}-2\mathbf{k}$, then for 3 seconds with velocity $\mathbf{i}-2\mathbf{j}-2\mathbf{k}$. Find (a) the average velocity and (b) the average speed of the particle during the motion.

(a) The total displacement of the particle
$$= 2(6\mathbf{i}+3\mathbf{j}-2\mathbf{k})+3(\mathbf{i}-2\mathbf{j}-2\mathbf{k}) = 15\mathbf{i}-10\mathbf{k}$$
$\therefore$ the average velocity during the 5 seconds of the motion is $3\mathbf{i}-2\mathbf{k}$.
(b) The distance moved in the first 2 seconds
$$= 2|6\mathbf{i}+3\mathbf{j}-2\mathbf{k}| = 2\sqrt{(6^2+3^2+2^2)} = 14.$$
The distance moved in the next 3 seconds
$$= 3|\mathbf{i}-2\mathbf{j}-2\mathbf{k}| = 3\sqrt{(1^2+2^2+2^2)} = 9$$
Hence the particle moves 23 m in 5 seconds
$\therefore$ the average speed during the motion is $4{\cdot}6\,\mathrm{m\,s}^{-1}$.

Example 3 A particle moves in a plane with constant acceleration given by $-2\mathbf{i}+\mathbf{j}$. If the initial velocity of the particle is $3\mathbf{i}$, find its speed after 1 second and after 3 seconds.

The velocity of the particle after 1 second
$$= 3\mathbf{i}+(-2\mathbf{i}+\mathbf{j}) = \mathbf{i}+\mathbf{j}$$
$\therefore$ the speed of the particle after 1 second
$$= |\mathbf{i}+\mathbf{j}| = \sqrt{(1^2+1^2)} = \sqrt{2}.$$

The velocity of the particle after 3 seconds
$$= 3\mathbf{i} + 3(-2\mathbf{i} + \mathbf{j}) = -3\mathbf{i} + 3\mathbf{j}$$
$\therefore$ the speed of the particle after 3 seconds
$$= |-3\mathbf{i} + 3\mathbf{j}| = 3|-\mathbf{i} + \mathbf{j}| = 3\sqrt{2}.$$

Resultant velocities and accelerations can be found by the methods used to add forces in Chapter 10.

Example 4   An aircraft is heading due south with a velocity of 300 km/h. It is also being blown by a wind with velocity 45 km/h from the direction S 70°W. Find the velocity of the aircraft relative to the ground.

Vector diagram

Let R km/h be the resultant velocity and θ the angle it makes with due south.

By the cosine rule, $R^2 = 300^2 + 45^2 - 2.300.45\cos 70°$
$\therefore$ $R \approx 288$

By the sine rule, $\dfrac{\sin \theta}{45} = \dfrac{\sin 70°}{R}$

$\therefore$ $\sin \theta = \dfrac{45 \sin 70°}{R}$ and $\theta \approx 8.5°$.

Hence the velocity of the aircraft relative to the ground is 288 km/h in the direction S 8·5° E.

Sums of three or more velocities can be found using a vector polygon. However, it is often more convenient to resolve the velocities in suitable directions and then add the components.

Example 5   A ship travelling through the water at 15 km/h is steering a course N 20°E. It is also being carried by a current with velocity 3 km/h to the north west. If a man is walking across the deck of the ship directly from port to starboard at 5 km/h, find his velocity relative to the shore.

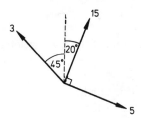

Let U and V be the components of the man's velocity along the perpendicular to the course steered by the ship, then resolving in those directions,
$$\nearrow \ U = 15 + 3\cos 65°$$
$$\rightarrow \ V = 5 - 3\sin 65°$$

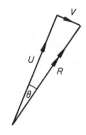

$$R = \sqrt{(U^2 + V^2)} \approx 16\cdot4$$
$$\tan\theta = V/U \quad \therefore \quad \theta \approx 8°$$

Hence the man's velocity relative to the shore is 16·4 km/h in the direction N 28° E.

Exercise 11.1

1. A particle starts from a point A with position vector $\mathbf{i} - 4\mathbf{j}$ and moves with velocity $12\mathbf{i} + 5\mathbf{j}$. Find the speed of the particle and its position vector after 2 seconds.

2. A particle moves from A to B in 5 seconds. If the position vectors of A and B are $7\mathbf{i} - 12\mathbf{j}$ and $-3\mathbf{i} + 8\mathbf{j}$ respectively, find the average velocity of the particle.

3. A particle starts from the origin and moves with velocity $6\mathbf{i} - 3\mathbf{j} + 6\mathbf{k}$ for 2 seconds, then with velocity $12\mathbf{i} + 9\mathbf{j}$ for 1 second. Find the position vector of the particle. Find also the average velocity and the average speed of the particle during its motion.

4. A particle with position vector $40\mathbf{i} + 10\mathbf{j} + 20\mathbf{k}$ moves with constant speed $5\,\mathrm{m\,s^{-1}}$ in the direction of the vector $4\mathbf{i} + 7\mathbf{j} + 4\mathbf{k}$. Find its distance from the origin after 9 seconds.

5. A particle moves with constant acceleration given by the vector $\mathbf{i} + \mathbf{j}$. If the initial velocity of the particle is $2\mathbf{i} + 5\mathbf{j}$, find its velocity after 3 seconds. Given that the speed of the particle after k seconds is $15\,\mathrm{m\,s^{-1}}$, find the value of k.

6. The initial velocity of a particle moving with constant acceleration is $2\mathbf{i} - 4\mathbf{j}$. After 2 seconds the particle is moving in the opposite direction with double the speed. Find the acceleration of the particle.

7. If $\mathbf{i}$ and $\mathbf{j}$ are unit vectors in the directions due east and due north respectively, find the magnitude and direction of the resultant of the velocities represented by each of the following sets of vectors
(a) $5\mathbf{i} + 3\mathbf{j}, 4\mathbf{i}, 3\mathbf{i} + 2\mathbf{j}$,
(b) $\mathbf{i} - 2\mathbf{j}, 3\mathbf{i} - \mathbf{j}, 2\mathbf{i} + 5\mathbf{j}$,
(c) $7\mathbf{i} - 5\mathbf{j}, -\mathbf{i} + 2\mathbf{j}, 2\mathbf{i} - \mathbf{j}$,
(d) $-2\mathbf{i} + \mathbf{j}, \mathbf{i} - 6\mathbf{j}, -3\mathbf{i} - 4\mathbf{j}$.

8. Find the magnitudes and directions of the resultants of the given velocities
(a) $7\,\mathrm{m\,s^{-1}}$ due north and $10\,\mathrm{m\,s^{-1}}$ N 75°E;
(b) $50\,\mathrm{km/h}$ N 40°E and $60\,\mathrm{km/h}$ S 25°E;
(c) $12\,\mathrm{m\,s^{-1}}$ N 20°E, $6\,\mathrm{m\,s^{-1}}$ due west and $5\,\mathrm{m\,s^{-1}}$ due north;
(d) $30\,\mathrm{km/h}$ S 45°W, $50\,\mathrm{km/h}$ S 30°E and $70\,\mathrm{km/h}$ due east.

9. A stretch of river has straight parallel banks 100 m apart. A boat is being rowed at $4\,\mathrm{m\,s^{-1}}$ in a direction perpendicular to the banks. If the river is flowing at $1\cdot2\,\mathrm{m\,s^{-1}}$, find (a) the magnitude of the resultant velocity of the boat, (b) the distance downstream that the boat is carried as it crosses from one bank to the other.

10. A ship travelling through the water at 12 km/h is steering a course S 75° E. It is also being carried by a current with velocity 2 km/h due E. If a passenger is walking across the deck of the ship directly from starboard to port at 4 km/h, find his velocity relative to the shore.

11.2 Relative velocity

If A and B are two moving objects, then the apparent velocity of B when observed from A is called the velocity of B relative to A, i.e.

> velocity of B relative to A = velocity of B − velocity of A.

Denoting the velocities of A and B by $\mathbf{v}_A$ and $\mathbf{v}_B$ respectively, the velocity of B relative to A is $\mathbf{v}_B - \mathbf{v}_A$ and may be found using a vector triangle.

Space diagram   Vector diagram

Example 1   A car is travelling due north on a straight stretch of motorway at 90 km/h. The car is observed by the driver of a lorry travelling north-east on an approach road at 60 km/h. Find the apparent speed of the car.

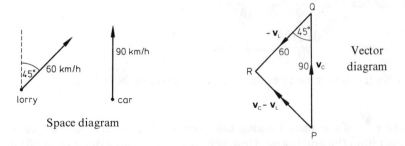

Space diagram

If the velocities of the car and the lorry are $\mathbf{v}_C$ and $\mathbf{v}_L$ respectively, then the velocity of the car relative to the lorry is $\mathbf{v}_C - \mathbf{v}_L$.

Using the cosine rule in the vector triangle PQR,

$$PR^2 = 90^2 + 60^2 - 2 \cdot 90 \cdot 60 \cos 45° \qquad \therefore \quad PR \approx 63 \cdot 7.$$

Hence the apparent speed of the car is $63 \cdot 7 \, \text{km/h}$.

It follows from the definition of relative velocity that for moving objects A and B,

velocity of B = velocity of B relative to A + velocity of A.

The advantage of using this result when solving problems is that it involves addition rather than subtraction of vectors.

Example 2   A man in a boat wishes to reach a small island due west of his present position. He knows that there is a current with a velocity of $3 \, \text{km/h}$ in the direction S 28°E. Given that the boat moves at $10 \, \text{km/h}$ through still water, find the course the man must steer to reach the island.

The true velocity of the boat

= velocity of boat relative to water + velocity of current.

If the boat is to reach the island its true velocity must be in the direction due west. Relative to the water the velocity of the boat is of magnitude $10 \, \text{km/h}$. Let θ be the angle between these velocities.

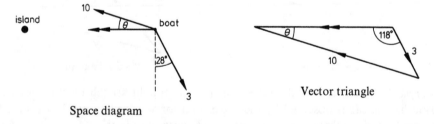

Space diagram   Vector triangle

By the sine rule, $\dfrac{\sin \theta}{3} = \dfrac{\sin 118°}{10}$

$$\therefore \quad \sin \theta = \frac{3 \sin 118°}{10} \quad \text{and} \quad \theta \approx 15 \cdot 4°$$

Hence the man must steer the boat in the direction N 74·6° W.

Example 3   To a man walking due west at $5 \, \text{km/h}$ the wind appears to be blowing from the north-west. However, to a man running due east at $10 \, \text{km/h}$ the wind appears to be blowing from the north-east. Find the true velocity of the wind.

Direction of walker

Direction of runner

Apparent direction
of wind to walker

Apparent direction
of wind to runner

True magnitude
and direction
of wind

Velocity of wind
= velocity of wind relative to walker + velocity of walker
= velocity of wind relative to runner + velocity of runner

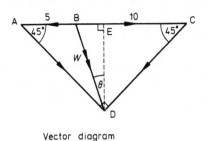

Vector diagram

From the symmetry of the diagram,

$$BE = 2{\cdot}5 \quad \text{and} \quad DE = 7{\cdot}5.$$

Using Pythagoras' theorem in $\triangle BDE$,

$$W^2 = (2{\cdot}5)^2 + (7{\cdot}5)^2$$
$$\therefore \quad W \approx 7{\cdot}9$$
$$\tan \theta = \frac{2{\cdot}5}{7{\cdot}5} = \frac{1}{3} \quad \therefore \quad \theta \approx 18{\cdot}4°$$

Hence the velocity of the wind is 7·9 km/h from the direction N 18·4° W.

Exercise 11.2

1. A man is walking due south at 5 km/h. The wind is blowing from the west at 8 km/h. Find the magnitude and direction of the velocity of the wind relative to the man.

2. A ship is sailing south-east at 20 km/h and a second ship is sailing due west at 25 km/h. Find the magnitude and direction of the velocity of the first ship relative to the second.

3. If particles A and B have velocity vectors $7\mathbf{i} - 5\mathbf{j}$ and $-2\mathbf{i} + 7\mathbf{j}$ respectively, find the magnitude of the velocity of B relative to A and the angle its direction makes with the direction of $\mathbf{i}$.

4. Particles A and B have velocity vectors $3\mathbf{i} - 11\mathbf{j}$ and $5\mathbf{i} + \mathbf{j}$ respectively. The velocity of a particle C relative to A is $-2\mathbf{i} + 7\mathbf{j}$. Find, in vector form, the velocity of C and the velocity of B relative to C.

5. A boat A travels due west at a speed of 30 km/h. The velocity of a boat B relative to A is 14 km/h due south. Find the speed of boat B and the direction in which it is moving.

6.   To an observer on a ship P steaming at $20\,\text{km/h}$ in the direction due west, a ship Q appears to be steaming at $20\,\text{km/h}$ in the direction N $30°$W. Find the true velocity of Q.

7.   To a cyclist riding due east at $15\,\text{km/h}$ the wind appears to be blowing from the north-east at $12\,\text{km/h}$. Find the magnitude and direction of the true velocity of the wind.

8.   To the driver of a car travelling due north at $40\,\text{km/h}$ the wind appears to be blowing at $50\,\text{km/h}$ from the direction N $60°$E. The wind velocity remains constant, but the speed of the car is increasing. Find its speed when the wind appears to be blowing from the direction N $30°$E.

9.   A river with straight parallel banks $400\,\text{m}$ apart flows due south at $3\cdot5\,\text{km/h}$. Find the direction in which a boat, travelling at $12\cdot5\,\text{km/h}$ relative to the water, must be steered in order to cross the river from east to west along a course perpendicular to the banks. Find also the time taken to make the crossing.

10.   A stretch of river has straight parallel banks $120\,\text{m}$ apart. A man rowing at $5\,\text{m\,s}^{-1}$ relative to the water and heading directly across the river reaches the other bank $54\,\text{m}$ downstream. Find the speed of the current and the direction in which the man would have to row in order to cross the river at right angles to the banks.

11.   A helicopter whose speed in still air is $60\,\text{km/h}$ is to fly to a village $80\,\text{km}$ away in the direction N $40°$W. The wind is blowing at $10\,\text{km/h}$ from the west. Find the direction in which the helicopter must fly and the time taken to reach the village to the nearest minute.

12.   A plane flies in a straight line from A to B, where $AB = 580\,\text{km}$ and the bearing of B from A is N $50°$E. The speed of the plane in still air is $300\,\text{km/h}$. Given that the wind is blowing from the direction N $40°$W and that the flight takes 2 hours, calculate the speed of the wind and the course that should be set for the return journey, if the wind velocity remains unchanged.

11.3 Relative displacement

Some problems involve the positions as well as the velocities of objects in motion. In such cases it is usually best to select one of the objects and work with displacements and velocities relative to it.

   If A and B are two moving objects then the displacement of B relative to A is given by the displacement vector **AB**. The velocity of B relative to A can be considered to be the rate of change of **AB**. The two objects can meet only if the relative velocity is parallel to the relative displacement.

Example 1   A motor-boat capable of 30 km/h wishes to intercept a yacht 5 km away on a bearing N 28° E. If the yacht is travelling at 20 km/h in the direction N 53° W, what course should the motor-boat take? Find also the time taken to reach the yacht.

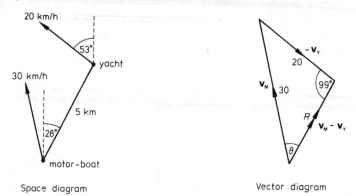

Space diagram   Vector diagram

If the velocities of the yacht and the motor-boat are v_Y and v_M respectively, then the velocity of the motor-boat relative to the yacht is $v_M - v_Y$ and must lie in the direction N 28° E. Let R be the magnitude of this relative velocity and θ the angle it makes with the course of the motor-boat.

By the sine rule, $\dfrac{\sin \theta}{20} = \dfrac{\sin 99°}{30}$

$\therefore \quad \sin \theta = \dfrac{20 \sin 99°}{30}$ and $\theta \approx 41 \cdot 182°$.

Hence the motor-boat must be steered in the direction N 13·2° W.
The remaining angle in the vector triangle is 39·818°.

$\therefore \quad$ by the sine rule, $\dfrac{R}{\sin 39 \cdot 818°} = \dfrac{30}{\sin 99°}$

$\therefore \quad R = \dfrac{30 \sin 39 \cdot 818°}{\sin 99°} \approx 19 \cdot 45.$

Thus on the course N 13·2° W the motor-boat can approach the yacht at a relative speed of 19·45 km/h

$\therefore \quad$ the time taken in minutes $= \dfrac{5}{19 \cdot 45} \times 60 \approx 15 \cdot 4.$

Hence the motor-boat can reach the yacht in 15·4 minutes.

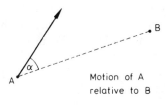

Motion of A
relative to B

We consider now two objects A and B with given speeds and positions. We assume that the direction of B is such that the objects can never meet. If α is the angle between the velocity of A relative to B and the initial displacement vector **AB**, then *A* will approach *B* as closely as possible when α takes its smallest possible value.

Example 2   A ship *A* is sailing due north at 20 km/h and a ship *B* is sailing due east at 15 km/h. At 12.00 hours *A* is 75 km due south of *B*. Find the shortest distance between the ships and the time at which they are closest, assuming that their velocities remain unchanged.

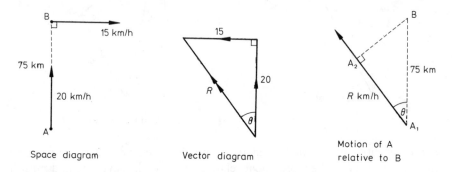

Space diagram   Vector diagram   Motion of A relative to B

Let the velocity of *A* relative to *B* be *R* km/h at an angle θ west of north.

Using the vector diagram, $R = \sqrt{(15^2 + 20^2)} = 25$

$$\therefore \quad \sin\theta = \frac{15}{25} = \frac{3}{5} \quad \text{and} \quad \cos\theta = \frac{20}{25} = \frac{4}{5}.$$

Let A_1 be the initial position of *A* and let A_2 be its position when closest to *B*. By considering the motion of *A* relative to *B*, we see that A_2B must be perpendicular to *A*'s relative path

∴   the shortest distance between the two ships is $75\sin\theta$ km, i.e. 45 km.

Relative to *B* the apparent distance between A_1 and A_2 is $75\cos\theta$ km, i.e. 60 km.

∴   the time *A* takes to move from A_1 to A_2

$$= \frac{\text{relative distance}}{\text{relative speed}} = \frac{60}{25}\text{hours} = 2 \text{ hours } 24 \text{ min.}$$

Hence the ships are closest together at 14.24 hours.

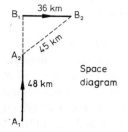

Space diagram

[A further diagram can be drawn to show the true paths of *A* and *B*. The true distance A_1A_2 is 48 km and the distance moved by *B* is 36 km.]

Example 3 An aircraft *A* flying at a speed of 350 km/h is 50 km due west of another aircraft *B*. If *B* is flying at the same height on a course S 20°E at a speed of 400 km/h, find the direction in which *A* must fly to approach *B* as closely as possible. Find also the shortest distance between the two aircraft.

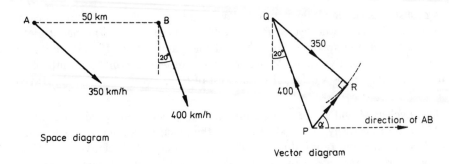

Space diagram

Vector diagram

In the vector diagram $\vec{QR}$ represents the velocity of *A*, $\vec{QP}$ represents the velocity of *B* and $\vec{PR}$ represents the velocity of *A* relative to *B*. Considering all possible directions for $\vec{QR}$, the smallest possible value of α occurs when $\angle PRQ$ is a right angle.

Hence $\angle PQR = \cos^{-1}(350/400) \approx 28\cdot955°$

∴ *A* must fly in the direction S 49° E to approach *B* as closely as possible.

Using the vector diagram, $\alpha = 49°$.

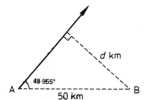

Let *d* km be the shortest distance between the aircraft, then by considering the direction of *A*'s motion relative to *B*,

$d = 50 \sin 48\cdot955° \approx 37\cdot7$

∴ the shortest distance between the aircraft is 37·7 km.

Exercise 11.3

1. A destroyer must intercept a vessel 40 km away on a bearing of 110°. The vessel is sailing on a course of 200° at a speed of 20 km/h. If the speed of the destroyer is 30 km/h, find, to the nearest degree, the course it must steer. Find also the time taken for the destroyer to reach the vessel.

2. A man walking due north wishes to intercept a cyclist travelling due west at 12 km/h. The bearing of the cyclist from the man is N 60°E. Find the speed at which the man must walk if he does not change his direction. If the man walks at 8 km/h, find the direction in which he must go to intercept the cyclist as quickly as possible. What is the minimum speed at which the man can walk in order to intercept the cyclist?

3. A ship P is sailing due west at $16\,\text{km/h}$ and a ship Q is sailing due north at $24\,\text{km/h}$. At a certain instant Q is $10\,\text{km}$ due west of P. Find the velocity of P relative to Q and the shortest distance between the ships if they continue on these courses.

4. A cyclist A is riding along a straight level road on a bearing $030°$ at a steady speed of $5\,\text{m s}^{-1}$. A second cyclist B is travelling due east along another straight level road at a steady speed of $6\,\text{m s}^{-1}$. Find the magnitude and the direction of the velocity of A relative to B. At a certain instant A is $500\,\text{m}$ due south of B. Find the time that elapses before A is due west of B.   (JMB)

5. Two cyclists A and B are approaching a road junction at speeds of $20\,\text{km/h}$ and $15\,\text{km/h}$ respectively. A is $100\,\text{m}$ south of the junction riding due north, when B reaches the junction riding due east. Find the velocity of B relative to A. Find also the shortest distance between A and B, assuming their velocities remain unchanged.

6. At 10.00 hours a plane A passes through a point X travelling due south at $360\,\text{km/h}$. At the same instant a plane B is $50\,\text{km}$ due west of X travelling at $450\,\text{km/h}$ towards X. Find the magnitude and direction of the velocity of B relative to A and the shortest distance between A and B. Find also the time at which the planes are closest and the bearing of B from A at that time.

7. Two planes A and B, flying at the same height, are travelling at $500\,\text{km/h}$ in the directions due north and due east respectively. At a certain instant A is $100\,\text{km}$ from B on a bearing of $150°$. Find the velocity of B relative to A and the shortest distance between the planes if neither changes course.

8. (In this question the units of time and distance are the second and metre respectively.) Two particles A and B start simultaneously from points which have position vectors $3\mathbf{i}+4\mathbf{j}$ and $8\mathbf{i}+4\mathbf{j}$ respectively. Each moves with constant velocity, the velocity of B being $\mathbf{i}+4\mathbf{j}$. (i) If the velocity of A is $5\mathbf{i}+2\mathbf{j}$, show that the least distance between A and B subsequently is $\sqrt{5}$ metres. (ii) If the speed of A is $5\,\text{m/s}$ and the particles collide, find the velocity of A.   (L)

9. A destroyer sights a ship travelling with constant velocity $5\mathbf{j}$, whose position vector at the time of sighting is $2000(3\mathbf{i}+\mathbf{j})$ relative to the destroyer, distances being measured in m and speeds in m s^{-1}. The destroyer immediately begins to move with velocity $k(4\mathbf{i}+3\mathbf{j})$, where k is a constant, in order to intercept the ship. Find k and the time to interception. Find also the distance between the vessels when half the time to interception has elapsed.   (O&C)

10. A ship P sailing at $16\,\text{km/h}$ is $10\,\text{km}$ due south of a second ship Q. If Q is sailing in the direction S $70°$E at a speed of $20\,\text{km/h}$, find the direction in which P must sail to approach Q as closely as possible. Find also the shortest distance between the ships.

Exercise 11.4   (*miscellaneous*)

[1 knot = 1 nautical mile per hour.]

1.   During a race between two yachts, *A* and *B*, there is a wind of 9 knots blowing from due north. The resultant velocity of *A* is 6 knots on a bearing of 060°. Find the direction of the wind relative to *A*. At the same time, the resultant velocity of *B* is 6 knots on a bearing of 300°. Find, correct to the nearest degree, the direction of the wind relative to *B* and, in knots correct to one decimal place, the velocity of *A* relative to *B*.   (L)

2.   A yacht *A* is sailing due east at 9 knots and a second yacht *B* is sailing on a bearing of 030° at 6 knots. At a certain instant a third yacht *C* appears to an observer on *A* to be sailing due south and appears to an observer on *B* to be sailing on a bearing of 150°. Find the speed of the yacht *C* and the bearing on which it is sailing.   (L)

3.   A river flows at 5 m/s from west to east between parallel banks which are at a distance 300 metres apart. A man rows a boat at a speed of 3 m/s in still water. (i) State the direction in which the boat must be steered in order to cross the river from the southern bank to the northern bank in the shortest possible time. Find the time taken and the actual distance covered by the boat for this crossing. (ii) Find the direction in which the boat must be steered in order to cross the river from the southern bank to the northern bank by the shortest possible route. Find the time taken and the actual distance covered by the boat for this crossing.

(AEB 1978)

4.   A river flows at a constant speed of 5 m/s between straight parallel banks which are 300 m apart. A boat, which has a maximum speed of $3\frac{1}{4}$ m/s in still water, leaves a point *A* on one bank and sails in a straight line to the opposite bank. Find, graphically or otherwise, the least time the boat can take to reach a point *B* on the opposite bank where *AB* = 500 m and *B* is downstream from *A*. Find also the least time the boat can take to cross the river. Find the time taken to sail from *A* to *B* by the slowest boat capable of sailing directly from *A* to *B*. (L)

5.   A helicopter flies with constant airspeed 100 knots from position *A* to position *B*, which is 50 nautical miles north east of *A*, and then flies back to *A*. Throughout the whole flight the wind velocity is 30 knots from the west. Find, by drawing or calculation, the course set for each of the two legs of the flight. Find also the total time of flight from *A* to *B* and back.   (L)

6.   A motorist *A* is travelling in a direction N 10°E at a constant speed of 60 km h⁻¹. The wind is blowing from the direction S 80°E at a constant speed of 25 km h⁻¹. Find the magnitude and direction of the velocity of the wind relative to the motorist. A second motorist *B* is travelling at a constant speed of 50 km h⁻¹. The velocity of *B* relative to *A* is in a direction S 30°W. Find graphically, or otherwise, the two possible directions in which *B* can be travelling.

(C)

7. Three fixed buoys are at the vertices A, B and C of an equilateral triangle of side 12 km. The point B is due north of A, and C is to the east of the line AB. A steady current of speed 3 km h^{-1} flows from west to east. A boat which has a top speed of 8 km h^{-1} in still water does the triangular journey $ABCA$ at top speed. Find, graphically or otherwise, the time taken on each leg of the journey, giving your answers to the nearest minute. (C)

8. A helicopter, with a maximum speed in still air of 120 km h^{-1}, is to fly between airports A and B. B is 238 km from A in a direction N $\theta°$ W where $\theta = \cos^{-1} \dfrac{12}{13}$. The helicopter leaves A at 12.00 hours in a wind blowing at 50 km h^{-1} from the west. What is the earliest possible time of arrival at B, and the corresponding direction in which it must head? At 13.00 hours the helicopter receives a radio message not to proceed but to return to A as soon as possible. Neglecting the time to change course, what is the earliest time it can be back at A? (W)

9. A small boat fitted with an outboard engine is 10 nautical miles due east of a trawler. The trawler. is travelling at 15 knots on a course N $\theta°$ E $[\theta = \sin^{-1}(24/25)]$. By considering the trawler's velocity components due east and due north, or otherwise, show that the minimum speed necessary for the boat to intercept the trawler is 4·2 knots, and find the corresponding direction in which it must steer. If the boat travels at 9 knots, show that it can steer in either of two directions to reach the trawler and calculate the shorter time needed to do so. (W)

10. [In this question the units of distance and time are the metre and second respectively.] At time $t = 0$ a particle A is at the origin and a particle B is at the point with position vector 10i. Each particle is moving with constant velocity, the velocity of B being $-4\mathbf{i} + 3\mathbf{j}$. Subsequently the particles collide. Find (a) the least speed at which A can travel, (b) the velocity of A if the collision occurs at time $t = 2$. A third particle C is moving in the same plane. When the velocity of A is $\mathbf{i} + 3\mathbf{j}$ the velocity of C relative to A is in the direction of the vector $3\mathbf{i} + 2\mathbf{j}$ and relative to B is in the direction of the vector $7\mathbf{i} + 3\mathbf{j}$. Find the speed of C. (L)

11. A motorist A and a cyclist B are travelling along straight roads which cross at right angles at a point O. A is travelling at a constant speed of 48 km h^{-1} towards the east, and B at 14 km h^{-1} towards the north. At the moment that B is passing through O, A is 400 m from O and has not yet passed O. Calculate: (i) the velocity of B relative to A in magnitude and direction; (ii) the least distance between car and cycle; (iii) the distances of the car and cycle from O when they are nearest to one another. (SU)

12. A cruiser is moving due east at 30 km/h. Relative to the cruiser a frigate is moving on a course of 210° (S 30° W) at 48 km/h. Using a graphical method, or otherwise, find the magnitude and the direction of the velocity of the frigate relative to a coastguard who is recording the paths of these ships from a lighthouse. At 1300 hours the frigate is 10 km due east of the cruiser. If both ships

maintain their speeds and courses, find the time at which the distance between them is least and their actual distance apart at this instant. Find also the time at which the frigate is due south of the cruiser.   (AEB 1977)

13. A ship A whose full speed is 40 km/h is 20 km due west of a ship B which is travelling uniformly with speed 30 km/h in a direction due north. The ship A travels at full speed on a course chosen so as to intercept B as soon as possible. Find the direction of this course and calculate to the nearest minute the time A would take to reach B. When half of this time has elapsed the ship A has engine failure and thereafter proceeds at half speed. Find the course which A should then set in order to approach as close as possible to B, and calculate the distance of closest approach (in kilometres to 2 decimal places).   (JMB)

14. A ship A is travelling with constant velocity 20 km/h due east and a ship B has constant velocity 15 km/h in a direction 30° east of north. At noon B is 30 km due south of A. Find the magnitude of the velocity of B relative to A, and show that the direction of this relative velocity is approximately 44° west of north. Find, to the nearest minute, the time at which A and B are closest. At the time of closest approach a boat leaves A to intercept B. Find the least speed at which it must travel.   (JMB)

15. Construct a flow chart for a procedure in which the input is the speeds and courses of two ships, A and B, in the form x km/h on a bearing $y°$. Assuming that initially ship A is 20 km due south of ship B, the following output is required:
(a) the time and distance of closest approach or an appropriate message if the ships are not moving closer to each other;
(b) the course A should steer to intercept B as soon as possible and the time of interception;
(c) if interception is not possible, the course A should steer to approach B as closely as possible.
[It may be helpful to consider velocity components due east and due north.]

12 Velocity and acceleration

12.1 Elementary kinematics

We consider now the motion of a particle in a straight line. Since in this type of motion the displacements, velocities and accelerations are parallel, there is no need to use vector notation. In general the displacement of the particle from a fixed point on the line after t seconds is denoted by s m, the velocity by v m s^{-1} and the acceleration by a m s^{-2}. As shown in Chapter 5, $v = \dfrac{ds}{dt}$ and $a = \dfrac{dv}{dt} = \dfrac{d^2s}{dt^2}$. Thus if a displacement-time graph is plotted the velocity is represented by the gradient of the curve. When the gradient is positive the motion of the particle is in the positive sense along the straight line and when the gradient is negative motion is in the opposite direction. At stationary points on the curve the velocity is zero and the particle is instantaneously at rest. Similarly the gradient of the velocity-time graph represents the acceleration of the particle.

Consider, for instance, the motion of a particle in a straight line as represented by the graphs given below. The first graph shows the displacement s from a fixed point O at time t and the second is the corresponding velocity-time graph.

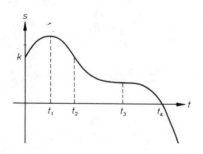

At the start of the motion the particle is at a distance k from O, moving away from O with positive velocity. It comes to rest at time t_1, then starts moving towards O. Its speed increases to a maximum value at $t = t_2$, then decreases so that the particle is again at rest when $t = t_3$. It then continues to move with negative velocity as t increases, passing through O at time t_4.

Considering the gradient of the velocity-time graph, we see that a negative acceleration produces the initial

226

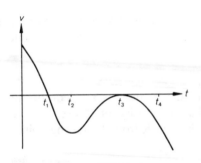

reduction in speed, the change of direction at time t_1, then an increasing speed towards O. The minimum point on the graph at $t = t_2$ corresponds to a maximum speed of the particle in the negative direction. The positive acceleration from $t = t_2$ to $t = t_3$ brings the particle to rest. The motion then continues with velocity and acceleration both negative and hence with increasing speed in the negative direction.

Example 1   A particle is moving in a straight line, so that after t seconds its displacement s metres from a fixed point O is given by $s = 9t + 3t^2 - t^3$. Find (a) the initial displacement, velocity and acceleration of the particle, (b) the time at which the particle is instantaneously at rest, (c) the greatest speed attained by the particle in the first 4 seconds of the motion.

Differentiating,
$$v = \frac{ds}{dt} = 9 + 6t - 3t^2$$

$$a = \frac{dv}{dt} = 6 - 6t.$$

(a) When $t = 0$, $s = 0$, $v = 9$ and $a = 6$
∴   initially the particle is at O moving with velocity $9\,\mathrm{m\,s^{-1}}$ and acceleration $6\,\mathrm{m\,s^{-2}}$ (both in the positive direction along the line of motion).
(b) $v = 0 \Leftrightarrow 9 + 6t - 3t^2 = 0 \Leftrightarrow 3(3 - t)(1 + t) = 0$
∴ $v = 0$ when $t = 3$ (assuming $t \geqslant 0$).
Hence the particle is instantaneously at rest after 3 seconds.
(c) $a = 0 \Leftrightarrow 6 - 6t = 0 \Leftrightarrow t = 1$.
When $t < 1$, $a > 0$ and   when $t > 1$, $a < 0$
∴   when $t = 1$, the velocity reaches a maximum value of $12\,\mathrm{m\,s^{-1}}$.

The velocity then decreases during the remainder of the motion. Since the particle is at rest when $t = 3$, it must move with increasing speed in the negative direction for $t > 3$. When $t = 4$, $v = -15$ ∴   the greatest speed attained by the particle in the first 4 seconds of the motion is $15\,\mathrm{m\,s^{-1}}$.

[These results are illustrated in the sketch graphs below.]

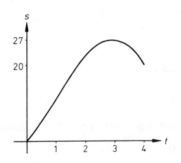

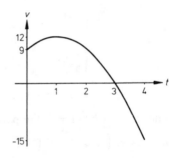

When the acceleration of a particle is represented by a known function of the time t, then expressions for the velocity and displacement can often be found by integration.

Example 2   A vehicle moves in a straight line so that its acceleration after t seconds is given by $a = 2t$. Initially it has a velocity of $20\,\mathrm{m\,s^{-1}}$. Find an expression for the displacement s metres of the vehicle from its starting point.

$$a = \frac{dv}{dt} = 2t$$

Integrating,
$$v = t^2 + c$$

but $v = 20$ when $t = 0$, thus $c = 20$

$$\therefore \qquad v = \frac{ds}{dt} = t^2 + 20.$$

Integrating,
$$s = \tfrac{1}{3}t^3 + 20t + k$$

but $s = 0$ when $t = 0$, thus $k = 0$

$$\therefore \qquad s = \tfrac{1}{3}t^3 + 20t.$$

In §6.5 it was shown that the total displacement of a particle in any given interval of time is represented by the area under the velocity-time graph and that this can be found using a definite integral.

Example 3   A particle moves in a straight line so that after t seconds its velocity in $\mathrm{m\,s^{-1}}$ is given by $v = (2t^2 - 25)(t - 1)$. Find (a) the displacement of the particle from its starting point after 3 seconds, (b) the total distance moved by the particle in the first 3 seconds of the motion.

(a) The displacement of the particle from its starting point after 3 seconds (in metres)

$$= \int_0^3 v\,dt = \int_0^3 (2t^2 - 25)(t - 1)\,dt$$

$$= \int_0^3 (2t^3 - 2t^2 - 25t + 25)\,dt$$

$$= \left[\frac{1}{2}t^4 - \frac{2}{3}t^3 - \frac{25}{2}t^2 + 25t \right]_0^3$$

$$= \left(\frac{1}{2}.81 - \frac{2}{3}.27 - \frac{25}{2}.9 + 25.3 \right) - 0$$

$$= 40\tfrac{1}{2} - 18 - 112\tfrac{1}{2} + 75 = -15.$$

(b) Since $v = 0$ when $t = 1$, the particle is instantaneously at rest after 1 second.

For $0 < t < 1$, $v > 0$ and for $1 < t < 3$, $v < 0$.

Hence the direction of motion changes once in the first 3 seconds, i.e. when $t = 1$.

$$\int_0^1 v \, dt = \left[\frac{1}{2}t^4 - \frac{2}{3}t^3 - \frac{25}{2}t^2 + 25t \right]_0^1$$

$$= \left(\frac{1}{2} - \frac{2}{3} - \frac{25}{2} + 25 \right) - 0 = \frac{1}{2} - \frac{2}{3} - 12\frac{1}{2} + 25 = 12\frac{1}{3}$$

$$\int_1^3 v \, dt = \left[\frac{1}{2}t^4 - \frac{2}{3}t^3 - \frac{25}{2}t^2 + 25t \right]_1^3$$

$$= \left(\frac{1}{2}.81 - \frac{2}{3}.27 - \frac{25}{2}.9 + 25.3 \right) - \left(\frac{1}{2} - \frac{2}{3} - \frac{25}{2} + 25 \right)$$

$$= -15 - 12\frac{1}{3} = -27\frac{1}{3}$$

∴ the total distance travelled in the first 3 seconds of the motion is $39\frac{2}{3}$ metres.

Exercise 12.1

Each of the questions in this exercise concerns a particle P moving along a straight line such that after t seconds its displacement from a fixed point A is s m, its velocity is $v \, \text{m s}^{-1}$ and its acceleration is $a \, \text{m s}^{-2}$.

1. Find expressions in terms of t for v and a, given that
   (a) $s = 6 - 5t - 3t^3$, (b) $s = (2t - 1)(t + 2)$.

2. Find expressions in terms of t for v and s, given that
   (a) $a = 12t$ and when $t = 0$, $v = 7$ and $s = 0$,
   (b) $a = -10$ and when $t = 0$, $v = 20$ and $s = 5$,
   (c) $a = 24t(t - 1)$ and when $t = 1$, $v = s = 0$.

3. If $v = 2t^3 - 9t^2$, find the values of t for which $a = 0$. Given that $s = 20$ when $t = 0$, find the value of s when $t = 2$.

4. If $v = 12 + 3t^2$, find the value of v when $a = 12$. Find also the distance travelled in the first two seconds of the motion.

5. The particle P starts from rest at A and moves such that $a = 6 - 2t$. If the particle later comes to rest at a point B, find the distance AB and the greatest speed attained while moving from A to B.

6. Given that the initial velocity of the particle P is $9 \, \text{m s}^{-1}$ and that $a = 2t - 10$, find the times when the particle is instantaneously at rest and the maximum speed attained by the particle between these two times.

7. Given that $a = 3t$ and that $v = 5$ when $t = 0$, find the value of v when $t = 4$ and the distance travelled in the first four seconds of the motion.

8. If $s = t^3 - 12t$, find the initial velocity of the particle P and the distance between its positions when $t = 0$ and $t = 4$. Show that P changes direction between $t = 0$ and $t = 4$, then find the total distance travelled in this interval of time.

9. Given that $v = 25 - 4t^2$, calculate (a) the value of t when the particle is instantaneously at rest, (b) the distance travelled by the particle in the third second.

10. Given that $s = t^3 - 2t^2 - 15t$, find (a) the initial velocity and acceleration of P, (b) the times at which P passes through A, (c) the total distance travelled in the first five seconds of the motion.

11. The particle P starts from rest at the point A and is again instantaneously at rest after 3 seconds. If $a = k - 6t$, where k is constant, find (a) expressions for v and s in terms of t, (b) the distance P travels before returning to A.

12. Given that $s = 6t^2 - t^3$, sketch the displacement-time and velocity-time graphs for the motion of the particle P from $t = 0$ to $t = 6$. Find, in the interval $0 \leqslant t \leqslant 6$,
(a) the times at which the particle is at A,
(b) the greatest value of the distance AP,
(c) the total distance travelled,
(d) the greatest speed attained.

13. Repeat question 12 for $s = t^2 - 5t + 4$.

14. Repeat question 12 for $s = 4t^3 - 33t^2 + 54t$.

15. Given that when $t = 1$, $s = 0$ and that when $t > 1$, $v = t^2 + \dfrac{9}{t^2}$, find the values of a, v and s when $t = 3$. Find also the minimum speed of the particle for $t > 1$.

16. If the particle P starts from rest and moves such that $v = 4t^2(3 - t)$, find (a) the distance travelled by the particle before it next comes to rest, (b) the maximum speed attained by the particle in the first three seconds of its motion, (c) the maximum acceleration in the first two seconds of its motion.

12.2 Motion with constant acceleration

We now obtain general equations for the motion of a particle in a straight line with constant acceleration. These will then be used to solve a variety of problems.

Consider a particle with initial velocity u and constant acceleration a. Let its displacement from its initial position be s and its velocity v at time t.

$$\frac{dv}{dt} = a \quad \text{where } a \text{ is constant}$$

Integrating

$$v = at + c$$

When $t = 0$, $v = u$ $\qquad \therefore \quad c = u$

Thus $\qquad v = u + at \qquad\qquad (1)$

and $\qquad \dfrac{ds}{dt} = u + at$

Integrating $\qquad s = ut + \tfrac{1}{2}at^2 + k$

When $t = 0$, $s = 0$ $\qquad \therefore \quad k = 0$

Thus $\qquad s = ut + \tfrac{1}{2}at^2 \qquad\qquad (2)$

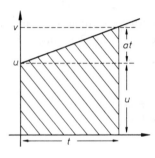

Results (1) and (2) can also be obtained by considering the velocity-time graph for the motion. As shown in the sketch this is a straight line with gradient a. The displacement s is represented by the shaded area under the line.

Rearranging (1)
Substituting in (2)

$$at = v - u$$
$$s = ut + \tfrac{1}{2}(v - u)t$$
$$\therefore \quad s = \tfrac{1}{2}(u + v)t \qquad\qquad (3)$$

Hence

$$2as = (u + v)at = (u + v)(v - u) = v^2 - u^2$$
$$\therefore \quad v^2 = u^2 + 2as. \qquad\qquad (4)$$

Eliminating u from equations (1) and (2),

$$s = vt - \tfrac{1}{2}at^2. \qquad\qquad (5)$$

We now have five equations connecting the quantities s, u, v, a and t. When using them it is advisable to check that the acceleration is constant and that the units of measurement are consistent.

$$\boxed{\begin{aligned} v &= u + at \\ s &= ut + \tfrac{1}{2}at^2 \\ s &= vt - \tfrac{1}{2}at^2 \\ s &= \tfrac{1}{2}(u + v)t \\ v^2 &= u^2 + 2as. \end{aligned}}$$

Example 1   A particle moving in a straight line accelerates uniformly from rest to a speed of $60\,\mathrm{m\,s^{-1}}$ in 10 seconds. Find the distance travelled in this time.

Given: $u = 0, v = 60, t = 10$.

Using the equation $s = \frac{1}{2}(u+v)t$
$$s = \frac{1}{2}(0+60)10 = 300.$$

Hence the particle travelled a distance of 300 metres.

Example 2   A train travelling in a straight line is brought to rest in 1 minute with a uniform retardation of $0 \cdot 5\,\mathrm{m\,s^{-2}}$. Find the speed at which the train was travelling in km/h.

Given: $v = 0, a = -0 \cdot 5, t = 60$.
Using the equation $v = u + at$
$$0 = u - 0 \cdot 5 \times 60 \qquad \therefore \quad u = 30.$$

Hence the speed of the train $= 30\,\mathrm{m\,s^{-1}}$
$$= \frac{30 \times 60 \times 60}{1000}\,\text{km/h} = 108\,\text{km/h}.$$

Example 3   A particle moving in a straight line with constant acceleration travels 10 m in 2 seconds, then a further 30 m in the following 2 seconds. Find the acceleration of the particle.

When $s = 10, t = 2$ and when $s = 40, t = 4$

$\therefore$ using the equation $s = ut + \frac{1}{2}at^2$, $10 = 2u + 2a$
$$40 = 4u + 8a$$

Simplifying $u + a = 5$
$$u + 2a = 10 \qquad\qquad \therefore \quad a = 5.$$

Hence the acceleration of the particle is $5\,\mathrm{m\,s^{-2}}$.

Although the constant acceleration formulae are very useful, some problems are better solved by drawing a sketch of the velocity-time graph.

Example 4   A train takes 10 minutes to travel a distance of 12 km between two stations. It starts from rest at the first station, accelerating uniformly until it reaches a speed of V km/h. It travels at this speed for 5 minutes then decelerates uniformly, coming to rest at the second station. Find the value of V.

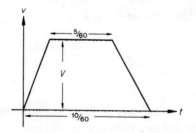

Consider the velocity-time graph for the motion, taking the units of distance and time as kilometres and hours.

Since the area of the trapezium under the graph represents the distance travelled,

$$\frac{1}{2}\left(\frac{5}{60} + \frac{10}{60}\right)V = 12, \quad \text{i.e.} \; \frac{1}{8}V = 12.$$

Hence the value of V is 96.

An object falling freely under gravity has a downward acceleration usually denoted by g. The magnitude of this acceleration varies slightly over the surface of the earth. In $\mathrm{m\,s^{-2}}$ the variation is from about 9·78 at the equator to 9·83 at the poles. [On the moon the value of g is 1·62.] Although the pull of gravity on an object decreases as its distance from the earth's surface increases, this decrease can, in general, be regarded as negligible. Thus in most problems g is taken to be a constant with value 9·8 or 10 depending on the accuracy required.

Example 5   A boy standing at the edge of a cliff throws a stone vertically upwards with a velocity of $10\,\mathrm{m\,s^{-1}}$. If the stone hits the sea below with a velocity of $40\,\mathrm{m\,s^{-1}}$, find the height of the cliff above sea-level and the maximum height above sea-level reached by the stone. [Take $g = 10$.]

Taking the upward direction as positive, $u = 10$, $v = -40$ and $a = -10$.

Using the equation
$$v^2 = u^2 + 2as$$
$$1600 = 100 + 2(-10)s$$
$$1500 = -20s \qquad \qquad \therefore \quad s = -75.$$

Hence the height of the cliff above sea-level is 75 m.

When the stone reaches its maximum height its velocity is zero
$$\therefore \quad u = 10, v = 0 \text{ and } a = -10.$$

Using the equation
$$v^2 = u^2 + 2as$$
$$0 = 100 + 2(-10)s \qquad \qquad \therefore \quad s = 5.$$

Hence the maximum height reached by the stone is 5 m above the cliff, i.e. 80 m above sea-level.

Example 6   A particle is projected vertically upwards with a speed of $30\,\mathrm{m\,s^{-1}}$. After 3 seconds another particle is projected vertically upwards from the same point with a speed of $25\,\mathrm{m\,s^{-1}}$. Find the height at which the two particles collide. [Take $g = 10$.]

Let us assume that the collision occurs at a height of h m and T seconds after the projection of the second particle, then taking the upward direction as positive,

for the first particle: $s = h, u = 30, a = -10, t = T + 3$
for the second particle: $s = h, u = 25, a = -10, t = T$.

Using the equation $s = ut + \frac{1}{2}at^2$ in both cases

$$h = 30(T+3) + \frac{1}{2}(-10)(T+3)^2$$
$$= 30T + 90 - 5T^2 - 30T - 45$$

i.e. $$h = 45 - 5T^2 \qquad (1)$$

and $$h = 25T + \tfrac{1}{2}(-10)T^2$$

i.e. $$h = 25T - 5T^2. \qquad (2)$$

Hence $25T = 45$ so that $T = 9/5$.

Substituting in (1) $\qquad h = 45 - 5\left(\dfrac{9}{5}\right)^2 = 28 \cdot 8$

$\therefore$ the particles collide at a height of 28·8 metres.

Exercise 12.2

[Take g as $10\,\mathrm{m\,s}^{-2}$.]

In questions 1 to 10 a particle is moving in a straight line with constant acceleration a and initial velocity u. At time t its displacement from its initial position is s and its velocity is v.

1.   If $u = 5$, $a = 2$, $t = 3$, find v and s.

2.   If $v = 7$, $u = 5$, $a = 1$, find s and t.

3.   If $u = 10$, $v = -2$, $t = 4$, find s and a.

4.   If $s = 15$, $t = 5$, $u = 13$, find v and a.

5.   If $s = -5$, $v = 2$, $u = -3$, find a and t.

6.   If $t = 3$, $a = -8$, $s = 36$, find v and u.

7.   If $u = 50$, $s = 300$, $a = -4$, find t and v.

8.   If $a = 0{\cdot}4$, $t = 15$, $v = 7$, find u and s.

9.   If $v = -18$, $s = -64$, $t = 8$, find a and u.

10. If $s = 45$, $a = 40$, $v = 60$, find u and t.

11. A car accelerates uniformly from rest to $60\,\mathrm{km/h}$ in 30 seconds. Find the distance it travels in this time.

12. A cyclist travels $1{\cdot}25\,\mathrm{km}$ as he accelerates uniformly at a rate of $k\,\mathrm{m\,s}^{-2}$ from a speed of $15\,\mathrm{km/h}$ to $30\,\mathrm{km/h}$. Find the value of k.

13. A particle is projected vertically upwards from a point O with a speed of $25\,\mathrm{m\,s}^{-1}$. Find the maximum height reached by the particle and the time that elapses before it returns to O.

14. A boy on a bridge throws a pebble vertically upwards at a speed of $6\,\mathrm{m\,s}^{-1}$. After 2 seconds it hits the water below. Find the speed at which the pebble hits the water and its initial height above the water.

15. The points O, A, B and C lie in a straight line such that $AB = 28\,\mathrm{m}$ and $BC = 72\,\mathrm{m}$. A particle moving with constant acceleration starts from rest at O and passes through A, B and C, its velocities at B and C being $9\,\mathrm{m\,s}^{-1}$ and $15\,\mathrm{m\,s}^{-1}$ respectively. Find the velocity of the particle at A and the time it takes to travel from A to C.

16. A particle moving in a straight line with constant acceleration travels $10\,\mathrm{m}$ in 2 seconds, then a further $22\,\mathrm{m}$ in the next 2 seconds. Find the further distance travelled in 2 more seconds and the speed of the particle at the end of this 6-second interval of time.

17. A train, being brought to rest with uniform retardation, travels $30\,\mathrm{m}$ in 2 seconds, then a further $30\,\mathrm{m}$ in 4 seconds. Find the retardation of the train and the additional time it takes to come to rest.

18. A particle is projected vertically upwards from a point A. Given that it rises $15\,\mathrm{m}$ in the third second of its motion, find its initial speed and the maximum height above A that it reaches.

19. A bus sets off from a bus station P. It accelerates uniformly for T_1 seconds, covering a distance of $300\,\mathrm{m}$. It travels at a speed of $V\,\mathrm{km/h}$ for T_2 seconds, covering a further distance of $1250\,\mathrm{m}$. It then decelerates uniformly for T_3 seconds, coming to rest at a bus stop Q. If the total time taken is 3 minutes and $2T_1 = 3T_3$, find the distance from P to Q and the values of T_1, T_2, T_3 and V.

20. A train stops at a station A. It then accelerates at $0.1\,\mathrm{m\,s}^{-2}$ for 5 minutes, reaching a speed of $V\,\mathrm{m\,s}^{-1}$. It continues at this speed for 12 minutes, then the brakes are applied for 3 minutes, bringing the train to rest with uniform retardation at a station B. Find the value of V and the distance AB.

21. A train stops at two stations $24\,\mathrm{km}$ apart. It takes 3 minutes to accelerate uniformly to a speed of $40\,\mathrm{m\,s}^{-1}$ then maintains this speed until it comes to rest with uniform retardation in a distance of $1200\,\mathrm{m}$. Find the time taken for the journey.

22. Two stations A and B are $10\,\mathrm{km}$ apart. A train travelling at a constant speed of $144\,\mathrm{km/h}$ passes station A at 10.00 hours. At a distance $d\,\mathrm{km}$ from B the brakes are applied, producing a constant retardation of $0.4\,\mathrm{m\,s}^{-2}$. If the train comes to rest at station B, find the value of d and the time at which the train reaches B.

23. A particle accelerates from rest at a constant rate of $3\,\mathrm{m\,s}^{-2}$ to a speed of $V\,\mathrm{m\,s}^{-1}$. It continues to move at that speed for a certain time, then decelerates at a constant rate of $1.5\,\mathrm{m\,s}^{-2}$. If the total time taken is one minute and the total distance travelled is $1\,\mathrm{km}$, find the value of V.

24. A car travelling along a straight level road at a constant speed of 54 km/h passes a second car as it starts to accelerate from rest at a uniform rate of $0.5\,\mathrm{m\,s^{-2}}$. Find the time that elapses and the distance covered when the second car draws level with the first.

25. A ball is thrown vertically upwards from a point A with a speed of $20\,\mathrm{m\,s^{-1}}$. At the same instant a second ball is dropped from a point B which is 60 m vertically above A. Find the time which elapses before the two balls meet and their height above A at this instant.

26. A stone is dropped from the top of a tower. After one second another stone is thrown vertically downwards from the same point at a speed of $15\,\mathrm{m\,s^{-1}}$. If the stones reach the ground simultaneously, find the height of the tower.

12.3 Projectiles

Any particle which is given a non-zero initial velocity and then moves freely under gravity may be described as a *projectile*. In elementary work it is assumed that projectiles move in a vertical plane with constant downward acceleration g. The point of projection is taken as the origin O of a Cartesian coordinate system with the x-axis horizontal and the y-axis vertical. If at time t the projectile is at the point (x, y) then its displacement from O has horizontal component x and vertical component y. Thus the horizontal and vertical components of the velocity are dx/dt and dy/dt. Using a dot to indicate differentiation with respect to time these components are usually denoted by $\dot{x}$ and $\dot{y}$. Similarly the horizontal and vertical components of the acceleration are $\ddot{x}$ and $\ddot{y}$ respectively.

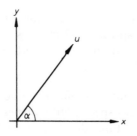

Consider a particle projected with initial velocity u at an angle α to the horizontal, the horizontal and vertical components of this velocity being $u\cos\alpha$ and $u\sin\alpha$ respectively. Since the particle has a constant downward acceleration of magnitude g, $\ddot{x} = 0$ and $\ddot{y} = -g$.

Thus the constant acceleration formulae derived in the previous section can now be applied to both horizontal and vertical motion. Using the equations $v = u + at$ and $s = ut + \frac{1}{2}at^2$,

$$\dot{x} = u\cos\alpha \qquad\qquad \dot{y} = u\sin\alpha - gt$$
$$x = ut\cos\alpha \qquad\qquad y = ut\sin\alpha - \tfrac{1}{2}gt^2.$$

Most projectile problems can be solved using one or more of these four equations.

The equation of the *trajectory* or path of the particle is obtained by eliminating t from the expressions for x and y.

Writing $t = \dfrac{x}{u\cos\alpha}$, $y = u \cdot \dfrac{x}{u\cos\alpha} \cdot \sin\alpha - \dfrac{1}{2}g\left(\dfrac{x}{u\cos\alpha}\right)^2$

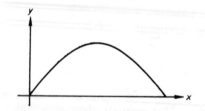

$$\therefore \quad y = x\tan\alpha - \frac{gx^2\sec^2\alpha}{2u^2}.$$

As shown in the sketch this is the equation of a parabola.

Example 1   A particle is projected from a point O on a horizontal plane with a speed of $50\,\mathrm{m\,s^{-1}}$ at an angle of $40°$ to the horizontal. Find (a) the velocity of the particle after 2 seconds, (b) the greatest height reached by the particle. [Take $g = 9\cdot8$.]

The equations for the motion of the particle are

$\dot{x} = 50\cos 40°$ $\dot{y} = 50\sin 40° - gt$
$x = 50t\cos 40°$ $y = 50t\sin 40° - \tfrac{1}{2}gt^2$.

(a)

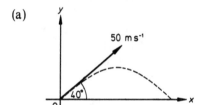

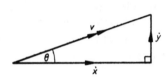

Let the velocity of the particle after 2 seconds be $v\,\mathrm{m\,s^{-1}}$ at an angle θ to the horizontal.

When $t = 2$, $\dot{x} = 50\cos 40°$ $\dot{y} = 50\sin 40° - 9\cdot8 \times 2$
$\therefore \quad v = \sqrt{(\dot{x}^2 + \dot{y}^2)} \approx 40\cdot3$
and $\tan\theta = \dot{y}/\dot{x}$, $\theta \approx 18\cdot1°$.

[These results were obtained using a calculator, storing the value $\dot{y} = 12\cdot5394$ in the memory.]

Thus the velocity after 2 seconds is $40\cdot3\,\mathrm{m\,s^{-1}}$ at an angle $18\cdot1°$ above the horizontal.

(b) When the particle reaches its greatest height, $\dot{y} = 0$

$\therefore \quad 50\sin 40° - gt = 0$ and $t = 50\sin 40°/g$

For this value of t,

$$y = \frac{(50\sin 40°)^2}{g} - \frac{1}{2}g\left(\frac{50\sin 40°}{g}\right)^2 = \frac{(50\sin 40°)^2}{2g} \approx 52\cdot7$$

$\therefore$ the greatest height reached by the particle is $52\cdot7\,\mathrm{m}$.

Example 2   A stone is thrown horizontally from the edge of a cliff 40 m above the sea. Given that the stone travels 60 m horizontally before it hits the water, find the time for which it is in the air and its initial speed. [Take $g = 9\cdot8$.]

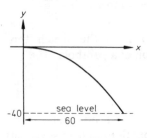

Let the initial speed of the stone be $u\,\mathrm{m\,s^{-1}}$, then after t seconds,

$$x = ut \quad y = -\tfrac{1}{2}gt^2.$$

Let the time for which the stone is in the air be T seconds, then when $t = T$, $x = 60$ and $y = -40$

$\therefore \quad 60 = uT$ and $-40 = -\tfrac{1}{2}\times9\cdot8T^2.$

Hence
$$T = \sqrt{\left(\frac{40}{4\cdot9}\right)} = \sqrt{\left(\frac{400}{49}\right)} = \frac{20}{7} = 2\frac{6}{7}$$

and
$$u = \frac{60}{T} = 60 \times \frac{7}{20} = 21$$

$\therefore$ the stone is in the air for $2\frac{6}{7}$ seconds and its initial speed is $21\,\mathrm{m\,s^{-1}}$.

If a particle is projected from a point O on a horizontal plane and strikes the plane again at the point P, then the distance OP is called the *range* of the projectile. The time which the particle takes to travel from O to P is called the *time of flight*.

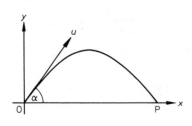

Consider again a particle projected from O with initial velocity u and angle of projection α. Let the range of the particle on the horizontal plane through O be R and the time of flight be T. The horizontal and vertical displacements of the particle at time t are given by

$$x = ut \cos \alpha, \quad y = ut \sin \alpha - \tfrac{1}{2}gt^2$$

When $t = T$, $x = R$ and $y = 0$

$\therefore \quad R = uT \cos \alpha, \quad 0 = uT \sin \alpha - \tfrac{1}{2}gT^2$

Since $T \neq 0$, $u \sin \alpha - \tfrac{1}{2}gT = 0$ $\therefore$ $T = 2u \sin \alpha / g$

Thus
$$R = u \cos \alpha \cdot \frac{2u \sin \alpha}{g} = \frac{u^2}{g}\cdot 2 \sin \alpha \cos \alpha = \frac{u^2 \sin 2\alpha}{g}$$

Hence the range of the particle is $u^2 \sin 2\alpha / g$ and its time of flight is $2u \sin \alpha / g$.

If u is fixed, but α can be varied, the maximum range of the particle is given by $\sin 2\alpha = 1$, i.e. $\alpha = 45°$.

Hence the particle has a maximum range of u^2/g when projected at $45°$ to the horizontal.

Example 3   A particle projected from a point O on horizontal ground, with speed $35 \, \text{m s}^{-1}$ and angle of projection θ, strikes the ground again at a point $75 \, \text{m}$ from O. Find the two possible values of θ. [Take $g = 9 \cdot 8$.]

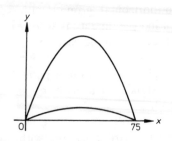

Taking O as origin, after t seconds

$$x = 35t \cos \theta$$
$$y = 35t \sin \theta - \tfrac{1}{2}gt^2$$

$\therefore \quad y = 0$ when $35t \sin \theta - \tfrac{1}{2}gt^2 = 0$

i.e. when $t = 0$ or $t = 70 \sin \theta / g$

$\therefore$ the time of flight of the particle is $70 \sin \theta / g$.

Since the range of the particle is $75 \, \text{m}$,

$$75 = 35 \cos \theta \, . \, \frac{70 \sin \theta}{g} = \frac{(35)^2}{9 \cdot 8} \, . \, 2 \sin \theta \cos \theta$$

Hence

$$\sin 2\theta = 2 \sin \theta \cos \theta = \frac{75 \times 9 \cdot 8}{(35)^2} = 0 \cdot 6$$

Since θ is an acute angle, $2\theta = 36 \cdot 87°$ or $143 \cdot 13°$

$\therefore$ the two possible values of θ are $18 \cdot 4°$ and $71 \cdot 6°$.

Example 4   A particle is projected with speed $14 \, \text{m s}^{-1}$ at an angle θ to the horizontal. Find the values of θ for which the particle just clears an obstacle $5 \, \text{m}$ high and $10 \, \text{m}$ from the point of projection. [Take $g = 9 \cdot 8$.]

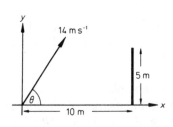

Taking the point of projection as origin, after t seconds

$$x = 14t \cos \theta$$
$$y = 14t \sin \theta - \tfrac{1}{2}gt^2$$

If the particle just clears the obstacle, then for some value of t, $x = 10$ and $y = 5$

$\therefore$ $10 = 14t \cos \theta$ and $5 = 14t \sin \theta - 4 \cdot 9t^2$.

Eliminating t from these equations,

$$5 = 14 \, . \, \frac{10}{14 \cos \theta} \, . \, \sin \theta - 4 \cdot 9 \left(\frac{10}{14 \cos \theta} \right)^2$$

$\Leftrightarrow \qquad 5 = 10 \tan \theta - \dfrac{5}{2} \sec^2 \theta$

$\Leftrightarrow \qquad 2 = 4 \tan \theta - (1 + \tan^2 \theta)$

$\Leftrightarrow \qquad \tan^2 \theta - 4 \tan \theta + 3 = 0$

$\Leftrightarrow \qquad (\tan \theta - 1)(\tan \theta - 3) = 0.$

Thus either $\tan \theta = 1$ or $\tan \theta = 3$.

Hence the particle just clears the obstacle if $\theta = 45°$ or $\theta = 71 \cdot 6°$.

Exercise 12.3

[In numerical questions take g as $9\cdot8\,\mathrm{m\,s}^{-2}$.]

1. A particle is projected from a point on horizontal ground with speed $24\cdot5\,\mathrm{m\,s}^{-1}$ at an angle of 30° to the horizontal. Find the time it takes to reach its greatest height. Find also the time of flight and the range of the particle.

2. A particle is projected horizontally with speed $30\,\mathrm{m\,s}^{-1}$ from a point $19\cdot6\,\mathrm{m}$ above horizontal ground. Find the horizontal distance it travels before striking the ground and its velocity as it strikes the ground.

3. A particle is projected from a point O on a horizontal plane with speed $35\,\mathrm{m\,s}^{-1}$ at an angle of elevation θ, where $\tan\theta = 2$. Find the maximum height above the plane of the particle and the equation of its trajectory.

4. A golf ball is given initial velocity $30\,\mathrm{m\,s}^{-1}$ at an angle α to the horizontal, where $\tan\alpha = 4/3$. The ball strikes the ground after T seconds at a horizontal distance of $90\,\mathrm{m}$ from the point of projection and $k\,\mathrm{m}$ below it. Find the values of T and k. Find also the velocity of the ball when it strikes the ground.

5. A particle is projected from a point O $24\cdot5\,\mathrm{m}$ above a horizontal plane. After 5 seconds it hits the plane at a point whose horizontal distance from O is $100\,\mathrm{m}$. Find the horizontal and vertical components of the initial velocity of the particle and the greatest height it reaches above the plane.

6. A boy throws a ball horizontally from a point $4\cdot9\,\mathrm{m}$ above horizontal ground. What is the minimum speed at which the ball must be thrown to clear a fence $2\cdot4\,\mathrm{m}$ high at a horizontal distance of $8\,\mathrm{m}$ from the point of projection? Find the distance beyond the fence of the point at which the ball strikes the ground if projected at this minimum speed.

7. A particle is projected from a point on a horizontal plane with velocity u and angle of projection α. The range of the particle is R and its time of flight T. Show that the particle reaches its greatest height h at time $\frac{1}{2}T$ and that $h = \frac{1}{4}R\tan\alpha$.

8. A gun fires shells with speed $210\,\mathrm{m\,s}^{-1}$. Find the maximum range of the gun over horizontal ground. Find also the smaller angle of elevation at which a shell should be fired to hit a target at a distance of $3\cdot6\,\mathrm{km}$.

9. A particle is projected from a point O on horizontal ground with an initial velocity whose horizontal and vertical components are $3u$ and $5u\,\mathrm{m\,s}^{-1}$ respectively. Find the equation of the trajectory of the particle. Given that it just clears an obstacle $5\,\mathrm{m}$ high and $9\,\mathrm{m}$ from O, find the value of u and the distance from O of the point at which the particle strikes the ground.

10. A ball thrown with speed $u\,\mathrm{m\,s}^{-1}$ and angle of projection α just clears the top of a fence $3\cdot6\,\mathrm{m}$ high and a horizontal distance of $9\cdot6\,\mathrm{m}$ from the point of

projection. Find u and α given that as the ball passes over the fence it is moving in a horizontal direction.

11. A projectile with initial speed $u\,\mathrm{m\,s^{-1}}$ has a maximum range of 80 m on a horizontal plane. Find the value of u. Given that the angle of projection is α, find the set of values of α for which the range is greater than 40 m.

12. The horizontal and vertical components of the initial velocity of a projectile are u and $3u$ respectively. Find in terms of u and g (a) the range of the projectile on the horizontal plane through the point of projection, (b) the times at which the particle is moving in a direction at $45°$ to the horizontal.

13. A particle is projected with speed u and angle of projection α. If after T seconds the velocity of the particle is perpendicular to its initial velocity, find an expression for T in terms of u, α and g. Comment on the significance of your answer in the cases $\alpha = 0°$ and $\alpha = 90°$.

14. A gun is fired from a vehicle moving along a straight horizontal road at speed U. Relative to the vehicle shells are fired with speed V and elevation θ. Find an expression for the range of the gun along the road ahead of the vehicle.

15. A particle is fired with speed V from a point O of a horizontal plane, the angle of elevation being $\arctan\left(\frac{2}{3}\right)$. The particle strikes the plane at a point A. Find the distance OA. Show that the same point A could have been reached by firing the particle with the same speed from O at an angle of elevation $\arctan\left(\frac{3}{2}\right)$. Find the difference in the times of flight for the two trajectories.   (O&C)

16. A projectile is fired with initial speed V at an angle θ to the horizontal. It passes through two points A and B at the same height h above the point of projection. Find the speed of the projectile at A, and the time it takes to travel from A to B. Given that $AB = 2h$ and $V^2 = 4gh$, find the angle θ.   (L)

17. Prove that two particles projected from the same point at the same time, but with different velocities, can never collide while moving freely under gravity.

18. Two particles are projected with the same speed from a point A, and land at B on the same horizontal level as A. If the greatest heights of the particles are in the ratio $4:1$, find the angles of projection and the ratio of AB to the maximum range with the same speed of projection.

19. A projectile is fired from a point P with speed $V\,\mathrm{m\,s^{-1}}$ and angle of elevation θ, where $\cos\theta = 0{\cdot}8$. At the same instant a target moving vertically upwards with constant speed $13\,\mathrm{m\,s^{-1}}$ passes through a point Q in the same horizontal plane as P and at a distance of 480 m from P. Given that the projectile hits the target, find the value of V. Find also the speed at which the projectile is travelling when it hits the target.

20. A particle is projected from a point A on horizontal ground such that the horizontal and vertical components of its initial velocity are u_1 and u_2 respectively. At the same instant a particle is projected from a point B at a height h vertically above A. The horizontal and vertical components of its initial velocity are v_1 and v_2. Find the time of flight of the first particle and show that the particles will collide before hitting the ground, provided that $u_1 = v_1$ and $2u_2(u_2 - v_2) > gh$.

12.4 Differentiation of position vectors

In §11.1 it was shown that if a moving particle has position vector $\mathbf{r}$, velocity $\mathbf{v}$ and acceleration $\mathbf{a}$, then

$$\mathbf{v} = \frac{d\mathbf{r}}{dt} \text{ or } \dot{\mathbf{r}} \quad \text{and} \quad \mathbf{a} = \frac{d\mathbf{v}}{dt} = \frac{d^2\mathbf{r}}{dt^2} \text{ or } \ddot{\mathbf{r}}.$$

In general, the basic rules for differentiation still hold when vectors are involved.

Thus $$\frac{d}{dt}\left(\mathbf{r}_1 + \mathbf{r}_2\right) = \frac{d\mathbf{r}_1}{dt} + \frac{d\mathbf{r}_2}{dt}, \qquad [\text{See §5.4, rule (2).}]$$

and if $\mathbf{r}$ is of the form $f(t)\mathbf{c}$, where $\mathbf{c}$ is a constant vector,

then $$\frac{d\mathbf{r}}{dt} = \frac{d}{dt}\left\{f(t)\mathbf{c}\right\} = \frac{df}{dt}\mathbf{c}. \qquad [\text{See §5.4, rule (1).}]$$

Hence if a position vector $\mathbf{r}$ is expressed in terms of the standard set of unit vectors $\mathbf{i}, \mathbf{j}$ and $\mathbf{k}$, then

$$\boxed{\begin{aligned} \mathbf{r} &= x\mathbf{i} + y\mathbf{j} + z\mathbf{k} \\ \mathbf{v} &= \dot{\mathbf{r}} = \dot{x}\mathbf{i} + \dot{y}\mathbf{j} + \dot{z}\mathbf{k} \\ \mathbf{a} &= \ddot{\mathbf{r}} = \ddot{x}\mathbf{i} + \ddot{y}\mathbf{j} + \ddot{z}\mathbf{k}. \end{aligned}}$$

Using these results we can extend some of the earlier work in this chapter to motion in two and three dimensions.

Example 1   A particle moves so that after t seconds its position vector $\mathbf{r}$ is given by $\mathbf{r} = (4t-1)\mathbf{i} + t^2\mathbf{j} + (15t - t^3)\mathbf{k}$. Find vector expressions for its velocity and acceleration at time t. Find also its speed after 3 seconds.

$$\mathbf{r} = (4t-1)\mathbf{i} + t^2\mathbf{j} + (15t - t^3)\mathbf{k}$$
$$\therefore \quad \mathbf{v} = \dot{\mathbf{r}} = 4\mathbf{i} + 2t\mathbf{j} + (15 - 3t^2)\mathbf{k}$$
$$\therefore \quad \mathbf{a} = \ddot{\mathbf{r}} = 2\mathbf{j} - 6t\mathbf{k}.$$

Hence the velocity and acceleration at time t are given by

$$\mathbf{v} = 4\mathbf{i} + 2t\mathbf{j} + (15 - 3t^2)\mathbf{k} \quad \text{and} \quad \mathbf{a} = 2\mathbf{j} - 6t\mathbf{k}.$$

When $t = 3$, $\mathbf{v} = 4\mathbf{i} + 6\mathbf{j} - 12\mathbf{k}$

$$\therefore \quad |\mathbf{v}| = \sqrt{\{4^2 + 6^2 + (-12)^2\}} = \sqrt{(16 + 36 + 144)} = \sqrt{196} = 14.$$

Hence after 3 seconds the speed of the particle is $14\,\mathrm{m\,s^{-1}}$.

Example 2   A particle moves so that at time t its velocity is given by $\mathbf{v} = 10t\mathbf{i} + (3 - 6t^2)\mathbf{k}$. Given that when $t = 1$ the particle is at the point with position vector $\mathbf{j} + \mathbf{k}$, find an expression for its position vector $\mathbf{r}$ at time t.

$$\mathbf{v} = \dot{\mathbf{r}} = 10t\mathbf{i} + (3 - 6t^2)\mathbf{k}$$

Integrating
$$\mathbf{r} = 5t^2\mathbf{i} + (3t - 2t^3)\mathbf{k} + \mathbf{c}$$

where $\mathbf{c}$ is a constant vector.

When $t = 1$, $\mathbf{r} = \mathbf{j} + \mathbf{k}$ $\therefore \quad \mathbf{j} + \mathbf{k} = 5\mathbf{i} + \mathbf{k} + \mathbf{c}$

$$\text{i.e.} \quad \mathbf{c} = -5\mathbf{i} + \mathbf{j}$$

Hence
$$\mathbf{r} = 5t^2\mathbf{i} + (3t - 2t^3)\mathbf{k} - 5\mathbf{i} + \mathbf{j}$$

i.e.
$$\mathbf{r} = (5t^2 - 5)\mathbf{i} + \mathbf{j} + (3t - 2t^3)\mathbf{k}.$$

Consider now a particle with initial velocity $\mathbf{u}$ and constant acceleration $\mathbf{a}$.

$$\frac{d\mathbf{v}}{dt} = \mathbf{a}, \text{ where } \mathbf{a} \text{ is constant}$$

Integrating $\mathbf{v} = \mathbf{a}t + \mathbf{c}$, where $\mathbf{c}$ is constant

When $t = 0$, $\mathbf{v} = \mathbf{u}$ $\therefore \quad \mathbf{c} = \mathbf{u}$.

Thus
$$\mathbf{v} = \frac{d\mathbf{r}}{dt} = \mathbf{u} + \mathbf{a}t.$$

Integrating again and taking the initial position of the particle as origin,
$$\mathbf{r} = \mathbf{u}t + \tfrac{1}{2}\mathbf{a}t^2.$$

Hence $\mathbf{v} = \mathbf{u} + \mathbf{a}t$ $\mathbf{r} = \mathbf{u}t + \tfrac{1}{2}\mathbf{a}t^2.$

Eliminating $\mathbf{a}$, we also have $\mathbf{r} = \tfrac{1}{2}(\mathbf{u} + \mathbf{v})t$.

These results are the vector forms of the constant acceleration formulae derived in §12.2.

If the acceleration due to gravity is denoted by the constant vector $\mathbf{g}$, then by substituting $\mathbf{a} = \mathbf{g}$ we obtain the vector equations governing the motion of a projectile, namely

$$\mathbf{v} = \mathbf{u} + \mathbf{g}t \quad \text{and} \quad \mathbf{r} = \mathbf{u}t + \tfrac{1}{2}\mathbf{g}t^2.$$

Using the notation of §12.3 and horizontal and vertical unit vectors $\mathbf{i}$ and $\mathbf{j}$, let $\mathbf{r} = x\mathbf{i} + y\mathbf{j}$, $\mathbf{v} = \dot{x}\mathbf{i} + \dot{y}\mathbf{j}$, $\mathbf{u} = (u\cos\alpha)\mathbf{i} + (u\sin\alpha)\mathbf{j}$ and $\mathbf{g} = -g\mathbf{j}$, then

$$\dot{x}\mathbf{i} + \dot{y}\mathbf{j} = (u\cos\alpha)\mathbf{i} + (u\sin\alpha - gt)\mathbf{j}$$

and
$$x\mathbf{i} + y\mathbf{j} = (ut\cos\alpha)\mathbf{i} + (ut\sin\alpha - \tfrac{1}{2}gt^2)\mathbf{j}.$$

Clearly in this form the vector equations correspond exactly to the results derived in the previous section.

Example 3   A particle is projected from a point O with initial velocity represented by the vector $\mathbf{i}+2\mathbf{j}$. Find vector expressions for its velocity $\mathbf{v}$ and its displacement $\mathbf{r}$ from O at time t. [Take $g = 10$.]

Using the equations $\mathbf{v} = \mathbf{u}+\mathbf{g}t$ and $\mathbf{r} = \mathbf{u}t+\frac{1}{2}\mathbf{g}t^2$,
where $\mathbf{u} = \mathbf{i}+2\mathbf{j}$ and $\mathbf{g} = -10\mathbf{j}$,

$$\mathbf{v} = (\mathbf{i}+2\mathbf{j})-10t\mathbf{j} = \mathbf{i}+(2-10t)\mathbf{j}$$

and $\qquad \mathbf{r} = (\mathbf{i}+2\mathbf{j})t+\frac{1}{2}(-10\mathbf{j})t^2 = t\mathbf{i}+(2t-5t^2)\mathbf{j}.$

[Note that although vector equations can be used in any projectile problem, in elementary work many students will prefer the methods of §12.3.]

Exercise 12.4

[Use SI units and take $g = 10$.]

1.   A particle moves such that its position vector at time t is $\mathbf{r}$. Find vector expressions for its velocity and acceleration at time t given that
(a) $\mathbf{r} = 3t\mathbf{i}-2\mathbf{j}+2t^3\mathbf{k}$,   (b) $\mathbf{r} = (t^2+1)\mathbf{i}+t^4\mathbf{j}-6t\mathbf{k}$.

2.   A particle moves such that at time t its position vector is $\mathbf{r} = t^2\mathbf{i}+(2-3t)\mathbf{j}$ $+3t^2(t-2)\mathbf{k}$. Find the initial speed of the particle and its speed after 2 seconds.

3.   A particle moves such that at time t its velocity is given by $\mathbf{v} = 3\mathbf{i}+4t\mathbf{j}$ $+(1-3t^2)\mathbf{k}$. Given that when $t = 2$ the position vector of the particle is $6(\mathbf{j}-\mathbf{k})$, find an expression for its position vector at time t.

4.   A particle moves such that at time t its acceleration is given by $\mathbf{a} = 6t\mathbf{i}-4\mathbf{k}$. Initially the particle has velocity vector $\mathbf{i}+\mathbf{j}$ and position vector $3\mathbf{j}$. Find expressions for its velocity and position vectors at time t.

In questions 5 to 8 the directions of the unit vectors $\mathbf{i}$ and $\mathbf{j}$ are horizontal and vertically upward respectively.

5.   A particle is projected from a point O with initial velocity represented by the vector $3\mathbf{i}+2\mathbf{j}$. Find vector expressions for its velocity $\mathbf{v}$ and its displacement $\mathbf{r}$ from O at time t.

6.   A particle is projected from a point O on a horizontal plane with initial velocity represented by the vector $40\mathbf{i}+25\mathbf{j}$. Find the greatest height reached by the particle, its range and time of flight.

7.   A particle is projected from a point O with speed $20\,\mathrm{m\,s^{-1}}$ at an angle of $60°$ to the horizontal. Find vector expressions for its velocity $\mathbf{v}$ and its displacement $\mathbf{r}$ from O at time t.

8.   A stone is thrown from the top of a cliff with initial velocity vector $8\mathbf{i}-\mathbf{j}$. If the stone hits the sea below after 3 seconds, find the height of the cliff and the horizontal distance travelled by the stone. Find also the speed at which the stone is travelling when it hits the water.

Exercise 12.5 (miscellaneous)

[In numerical questions take g as $10\,\mathrm{m\,s^{-2}}$.]

1. A particle is moving in a straight line such that its displacement from a fixed point after t seconds is s metres, where $s = 2t^3 - 13t^2 + 20t$. Find the total distance travelled and the maximum speed attained by the particle in the first 4 seconds of its motion. Find also the range of values of t for which the acceleration of the particle is negative.

2. A particle moving in a straight line starts from rest at a point A. Its acceleration at time t is $(45 + 12t - 9t^2)\,\mathrm{m\,s^{-2}}$. If the particle comes to rest instantaneously at a point B, find the distance AB and the time the particle takes to reach B. Find also the maximum acceleration and the maximum speed attained in that time.

3. Two particles A and B move along a straight line starting from a point O such that at time t their displacements from O are s_1 and s_2 respectively. If $s_1 = 5t^2(t+7)$ and $s_2 = t^3(t+3)$, find (a) the time at which the particles again meet, (b) an expression for the velocity of A relative to B, (c) the range of values of t for which A is travelling at a greater speed than B.

4. Due to track repairs a train retards uniformly, with retardation $1\,\mathrm{m\,s^{-2}}$, from a speed of $40\,\mathrm{m\,s^{-1}}$ at A to a speed of $10\,\mathrm{m\,s^{-1}}$ at B. The train travels from B to C, a distance of $3\cdot5\,\mathrm{km}$, at a constant speed of $10\,\mathrm{m\,s^{-1}}$ and then accelerates uniformly, with acceleration $0\cdot2\,\mathrm{m\,s^{-2}}$ so that its speed at D is $40\,\mathrm{m\,s^{-1}}$. Sketch the velocity-time graph for the journey from A to D, and show that the distance from A to D is $8\,\mathrm{km}$. Show that the journey from A to D takes $330\,\mathrm{s}$ more than it would if the train travelled at a constant speed of $40\,\mathrm{m\,s^{-1}}$ from A to D. (C)

5. Two trains, P and Q, travel by the same route from rest at station A to rest at station B. Train P has constant acceleration f for the first third of the time, constant speed for the second third and constant retardation f for the last third of the time. Train Q has constant acceleration f for the first third of the distance, constant speed for the second third and constant retardation f for the last third of the distance. Show that the times taken by the two trains are in the ratio $3\sqrt{3}:5$. (L)

6. A bus starts from rest and moves along a straight road with constant acceleration f until its speed is V; it then continues at constant speed V. When the bus starts, a car is at a distance b behind the bus and is moving in the same direction with constant speed u. Find the distance of the car behind the bus at time t after the bus has started (i) for $0 < t < V/f$, (ii) for $t > V/f$. Show that the car cannot overtake the bus during the period $0 < t < V/f$ unless $u^2 > 2fb$. Find the least distance between the car and the bus in the case when $u^2 < 2fb$ and $u < V$. State briefly what will happen if $u^2 < 2fb$ and $u > V$. (JMB)

7. A man descending in a lift at constant speed $V\,\mathrm{m\,s^{-1}}$ throws a ball vertically upwards with speed $U\,\mathrm{m\,s^{-1}}$ relative to the lift. Find the time that elapses before the ball (a) has zero velocity relative to the lift, (b) returns to the man's hand.

8. A batsman hits a cricket ball with initial speed 25 m/s at an elevation of 50°. Find the greatest height attained, the range on level ground and the time of flight. Show that it will clear the roof of a pavilion 10 m high which is at a distance of 50 m from the batsman. (L)

9. A particle P, projected from a point A on horizontal ground, moves freely under gravity and hits the ground again at B. Referred to A as origin, AB as x-axis and the upward vertical at A as y-axis, the equation of the path of P is $y = x - x^2/40$, where x and y are measured in metres. Calculate (i) the distance AB, (ii) the greatest height above AB attained by P, (iii) the magnitude and the direction of the velocity of P at A, (iv) the time taken by P to reach B from A. Calculate the coordinates of the points on the path of P at the two instants when the speed of P is 15 m/s. (AEB 1978)

10. A particle P is projected from a point A on a horizontal plane with speed V m/s at an angle of elevation α. The particle strikes the horizontal plane again at B, where $AB = 160$ metres. Show that $V^2 \sin 2\alpha = 1600$. A second particle Q is projected from A with the same speed V m/s at an angle of elevation $\alpha - 30°$ and strikes the horizontal plane at C, where C is the mid-point of AB. Find the values of V and α and show that the time of flight of Q is $8 \sin 15°$ seconds. Given that P and Q are projected from A simultaneously, find, at the instant when Q reaches C, the angle between the tangent to the path of P and the horizontal. (AEB 1978)

11. A particle is projected from a point O with a speed of $10\,\mathrm{m\,s^{-1}}$ and angle of elevation θ. Given that the projectile just clears an obstacle of height h at a horizontal distance D from O, express in the form of a quadratic equation in $\tan \theta$ the relationship connecting θ, h and D. Show that provided that $D^2 + 20h < 100$, there are two distinct possible values of θ. Show further that if $h = 0$ the sum of these two values is 90°.

12. A point O lies on a horizontal plane, and the point A is at a height h vertically above O. A particle is projected from A with speed V at an angle α above the horizontal. Taking O as the origin and Oy vertically upwards, show that the equation of the path of the particle can be written in the form $y = h - \dfrac{gx^2}{2V^2} + x \tan \alpha - \dfrac{gx^2}{2V^2} \tan^2 \alpha$. The particle hits the plane at the point $B(r,0)$. In the case when $V^2 = gh$ derive a quadratic equation for $\tan \alpha$ in terms of h and r, and show that $r \leqslant \sqrt{3h}$. For the same value of V show that $r = \sqrt{3h}$ when $\alpha = 30°$. (JMB)

13. A particle is fired from the origin O with speed V at an angle of inclination α above the x-axis, which is horizontal. At the instant when the particle is at the highest point of its path, a second particle is fired from O with speed U at an angle of inclination β above the horizontal so that the particles collide at a point on the x-axis. Show that $\tan \beta = \frac{1}{4}\tan \alpha$. Show also that, if $\tan \alpha = 3$, then $8U^2 = 5V^2$.

14. *A* and *B* are points distant *a* apart on horizontal ground. A ball is thrown from *A* towards *B* with velocity *u* at an angle 2α ($< 90°$) to the horizontal, and simultaneously a second ball is thrown from *B* towards *A* with velocity *v* at an angle α to the horizontal. If the balls collide, show that $v = 2u\cos\alpha$ and hence that they collide after a time $t = a/u(2\cos 2\alpha + 1)$. Show also that (i) the two trajectories would have the same maximum height above *AB*, (ii) the range of the ball from *B* would always exceed the range of the ball from *A*. (W)

15. A particle *P* is set in motion at a point *A* and then moves such that at time *t* its position vector is $\mathbf{r} = 3(t-2)^2\mathbf{i} + 2\mathbf{j} - 4t(t-4)\mathbf{k}$. By considering the vector **AP**, show that the particle moves in a straight line and find a unit vector parallel to the direction of motion. Find also the time that has elapsed and the position vector of *P* when it comes instantaneously to rest.

16. [In this question the directions of the unit vectors **i** and **j** are horizontal and vertically upwards.] A particle projected from the origin *O* with initial velocity represented by the vector $u\mathbf{i} + v\mathbf{j}$ passes through the points *A* and *B* with position vectors $20(\mathbf{i} + \mathbf{j})$ and $25\mathbf{i}$ respectively. Find the values of *u* and *v*. Show that a particle projected from *O* with velocity vector $v\mathbf{i} + u\mathbf{j}$ also passes through *B*.

13 Further differentiation

13.1 Composite functions

If f and g are functions defined on the set of real numbers, such that $f: x \to u$ and $g: u \to y$, then the *composite function* gf is defined by writing $gf: x \to y$. Thus $gf(x)$, the image of x under gf, is obtained by applying f followed by g.

For example, if $f: x \to x-2$ and $g: u \to 3u$, then $f(5) = 3$ and $g(3) = 9$
∴ $gf(5) = 9$.

In this case a general expression for $gf(x)$ can be obtained by writing $gf(x) = g(u)$, where $u = f(x) = x-2$. Hence $gf(x) = 3u = 3(x-2)$, i.e. $gf: x \to 3(x-2)$.

[Note that since $gf(x) = g(f(x))$, the function gf may also be described as a *function of a function*.]

Example 1   If $f(x) = x^2$ and $g(x) = x+1$, find $fg(x)$, $gf(x)$, $f^2(x)$ and $g^2(x)$.

$$\begin{aligned} fg(x) &= f(u) & \text{where} \quad u &= g(x) \\ \text{i.e.} \qquad fg(x) &= u^2 & \text{where} \quad u &= x+1 \\ \therefore \qquad fg(x) &= (x+1)^2 = x^2+2x+1. \end{aligned}$$

Similarly
$$\begin{aligned} gf(x) &= g(x^2) = x^2+1 \\ f^2(x) &= ff(x) = f(x^2) = (x^2)^2 = x^4 \\ g^2(x) &= g(x+1) = (x+1)+1 = x+2. \end{aligned}$$

It is sometimes convenient to reverse the process and consider a function as a composite function. If y is a given function of x, then this may be done by expressing y in terms of u, where u is a function of x. For instance, given $y = (2x+1)^5$, then $y = u^5$ where $u = 2x+1$.

In general, for a composite function of the form $y = fg(x)$, we may write $y = f(u)$ where $u = g(x)$.

To obtain a rule for differentiating functions of this type, let δx be a small increase in x and let δu and δy be the corresponding increases in u and y.

As δx, δu and δy tend to zero,

$$\frac{\delta y}{\delta x} \to \frac{dy}{dx}, \quad \frac{\delta y}{\delta u} \to \frac{dy}{du} \quad \text{and} \quad \frac{\delta u}{\delta x} \to \frac{du}{dx}$$

$\therefore$ since $\dfrac{\delta y}{\delta x} = \dfrac{\delta y}{\delta u} \cdot \dfrac{\delta u}{\delta x}$ we find that $\boxed{\dfrac{dy}{dx} = \dfrac{dy}{du} \cdot \dfrac{du}{dx}}$.

This formula is known as the *chain rule*.

Example 2   Find $\dfrac{dy}{dx}$ if $y = (2x+1)^5$.

Let $\qquad\qquad y = u^5 \qquad$ where $\qquad u = 2x+1$

then $\qquad\qquad \dfrac{dy}{du} = 5u^4 \qquad\qquad\qquad \dfrac{du}{dx} = 2$

$\therefore$ $\qquad\qquad \dfrac{dy}{dx} = \dfrac{dy}{du} \cdot \dfrac{du}{dx} = 5u^4 . 2 = 10(2x+1)^4.$

Example 3   Find $\dfrac{dy}{dx}$ if $y = \sqrt{\left(1 - \dfrac{1}{x^2}\right)^3}$

Let $\qquad y = \sqrt{(u^3)} = u^{3/2} \qquad$ where $\qquad u = 1 - \dfrac{1}{x^2} = 1 - x^{-2}$

then $\qquad \dfrac{dy}{du} = \dfrac{3}{2}u^{1/2} = \dfrac{3}{2}\sqrt{u} \qquad\qquad \dfrac{du}{dx} = 2x^{-3} = \dfrac{2}{x^3}$

$\therefore$ $\qquad \dfrac{dy}{dx} = \dfrac{dy}{du} \cdot \dfrac{du}{dx} = \dfrac{3}{2}\sqrt{u} \cdot \dfrac{2}{x^3} = \dfrac{3}{x^3}\sqrt{\left(1 - \dfrac{1}{x^2}\right)}.$

A student who is confident that he understands this technique may write down solutions in one of the following ways.

Example 4   Differentiate $\dfrac{1}{1+x^3}$.

Either   If $\qquad y = \dfrac{1}{1+x^3} = (1+x^3)^{-1}$

then $\qquad \dfrac{dy}{dx} = -1 . (1+x^3)^{-2} . 3x^2 = -\dfrac{3x^2}{(1+x^3)^2}$

or $\dfrac{d}{dx}\left(\dfrac{1}{1+x^3}\right) = \dfrac{d}{dx}\{(1+x^3)\}^{-1} = -1 . (1+x^3)^{-2} . \dfrac{d}{dx}(1+x^3)$

$$= -(1+x^3)^{-2} . 3x^2 = -\frac{3x^2}{(1+x^3)^2}.$$

This rule can be extended to 'chains' of three or more functions.

Example 5 Find $\dfrac{dy}{dx}$ if $y = \{1 + (x^2 - 1)^3\}^{1/3}$.

Let $\qquad y = u^{1/3}, \qquad\qquad u = 1 + v^3, \qquad\qquad v = x^2 - 1$

$$\frac{dy}{du} = \tfrac{1}{3}u^{-2/3} \qquad\qquad \frac{du}{dv} = 3v^2 \qquad\qquad \frac{dv}{dx} = 2x$$

$\therefore\qquad \dfrac{dy}{dx} = \dfrac{dy}{du} \cdot \dfrac{du}{dv} \cdot \dfrac{dv}{dx} = \tfrac{1}{3}u^{-2/3} . 3v^2 . 2x = 2xv^2/u^{2/3}.$

Hence $\qquad\qquad \dfrac{dy}{dx} = 2x(x^2 - 1)^2/\{1 + (x^2 - 1)^3\}^{2/3}.$

The differentiation rule $\dfrac{d}{dx}(x^n) = nx^{n-1}$ can now be proved for negative values

of n, assuming that the rule holds for positive values of n and that

$$\frac{d}{dx}\left(\frac{1}{x}\right) = -\frac{1}{x^2}.$$

Let $\qquad\qquad y = x^n = x^{-m} \qquad$ where m is positive,

then $\qquad\qquad y = u^m \qquad$ where $\quad u = x^{-1}$

$$\frac{dy}{du} = mu^{m-1} \qquad\qquad \frac{du}{dx} = -x^{-2}$$

$\therefore\qquad \dfrac{dy}{dx} = \dfrac{dy}{du} \cdot \dfrac{du}{dx} = mu^{m-1} . (-x^{-2}) = -mx^{-m+1} . x^{-2}$

Hence $\qquad\qquad \dfrac{dy}{dx} = -mx^{-m-1} = nx^{n-1}.$

Exercise 13.1

1. If $f(x) = x^2 - 1$ and $g(x) = 2x + 1$, find
    (a) $gf(x)$, $\qquad\qquad$ (b) $fg(x)$, $\qquad\qquad$ (c) $f^2(x)$, $\qquad\qquad$ (d) $g^2(x)$.

2. If $f(x) = 2x$, $g(x) = x - 1$ and $h(x) = x^2$, find
    (a) $fgh(x)$, $\qquad\qquad$ (b) $f^2g(x)$, $\qquad\qquad$ (c) $hgf(x)$, $\qquad\qquad$ (d) $gh^2(x)$.

3. If $f(x) = 2x - 3$ and $fg(x) = 2x + 1$, find $g(x)$.

4. If $f(x) = x - 1$ and $gf(x) = 3 + 2x - x^2$, find $g(x)$.

5. Express y in terms of u, given that $u = 2x - 1$ and
    (a) $y = \sqrt{(2x - 1)}$ $\qquad\qquad$ (b) $y = (x - \tfrac{1}{2})^5$
    (c) $y = 1/(1 - 2x)$ $\qquad\qquad$ (d) $y = x(2x - 1)^8$.

Differentiate with respect to x:

6. $(x + 3)^6$. $\qquad\qquad\qquad$ 7. $(2x - 1)^4$. $\qquad\qquad\qquad$ 8. $(5 - 3x)^7$.

9. $(3x-2)^{-2}$.   10. $(x^3-1)^5$.   11. $(4-x^2)^{-4}$.

12. $(1+3x)^{1/2}$.   13. $(6x+1)^{-1/3}$.   14. $(3x^2-1)^{5/2}$.

15. $\left(x-\dfrac{1}{x}\right)^{-3}$.   16. $\dfrac{1}{\sqrt{(1-x^2)}}$.   17. $\dfrac{1}{(1+\sqrt{x})^2}$.

18. Find the gradient of the curve $y=(2x^2-1)^3$ at the point where $x=-1$. Hence find the equation of the tangent to the curve at this point.

19. Find the equation of the tangent to the curve $y=\sqrt{\left(1+\dfrac{6}{x}\right)}$ at the point where $x=2$.

20. A particle moves along a straight line so that after t seconds its displacement, s metres, from a fixed point on the line is given by $s^3=3t-2$. Find the velocity and acceleration of the particle when $t=1$.

21. An object moves along a straight line such that after t seconds its velocity is $v \, \mathrm{m\,s}^{-1}$, where $v=t+4/(t+1)$. Find an expression for its acceleration after t seconds. Find also the minimum speed of the object during its motion.

22. Differentiate the following functions with respect to x by substituting $u=x+1$.

(a) $x(x+1)^5$,   (b) $\dfrac{x-1}{x+1}$,   (c) $\dfrac{x}{\sqrt{(x+1)}}$.

23. Differentiate with respect to x:

(a) $\sqrt{\{(2x-3)^4-1\}}$,   (b) $\dfrac{1}{4+\sqrt{(4x^2+1)}}$.

24. Find the turning points on the curve $y=x(5-x)^4$ and determine their nature. Sketch the curve.

25. Find any turning points on the curve $y=4/(x^2-2x+5)$ and show that the x-axis is an asymptote to the curve. Hence sketch the curve.

13.2 Functions defined implicitly

In previous sections we have considered functions defined explicitly by equations of the form $y=f(x)$. Sometimes a function $f:x\to y$ is defined *implicitly* using an equation such as $x^2+2y-y^2=5$.

[Note that since a function is a rule assigning one and only one value of y to every value of x, restrictions on the values of x and y are often required to complete such a definition.

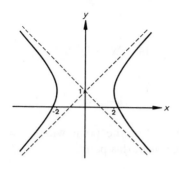

The sketch of the curve $x^2 + 2y - y^2 = 5$ shows that in this case possible restrictions would be $|x| > 2$, $y > 1$. However, in general an expression for dy/dx can be obtained without considering such restrictions.]

Given $x^2 + 2y - y^2 = 5$, dy/dx is found by differentiating both sides of the equation with respect to x.

$$\frac{d}{dx}(x^2) = 2x, \quad \frac{d}{dx}(2y) = 2\frac{dy}{dx}, \quad \frac{d}{dx}(5) = 0.$$

The derivative of y^2 is obtained using the chain rule:

$$\frac{d}{dx}(y^2) = \frac{d}{dy}(y^2)\cdot\frac{dy}{dx} = 2y\frac{dy}{dx}.$$

Thus $x^2 + 2y - y^2 = 5 \Rightarrow 2x + 2\frac{dy}{dx} - 2y\frac{dy}{dx} = 0$

$$\Rightarrow \qquad \frac{dy}{dx} - y\frac{dy}{dx} = -x$$

$\therefore$ provided that $y \neq 1$, $\dfrac{dy}{dx} = \dfrac{x}{y-1}$.

Example 1   Given that $(x+1)^2 + (y-2)^2 = 1$, find $\dfrac{dy}{dx}$.

Differentiating with respect to x:

$$2(x+1) + 2(y-2)\frac{dy}{dx} = 0$$

$\therefore$ $(y-2)\dfrac{dy}{dx} = -(x+1)$.

Hence provided that $y \neq 2$ $\dfrac{dy}{dx} = -\dfrac{x+1}{y-2}$.

Example 2   Find the gradient of the curve $x^3 + y^3 = 4y^2$ at the point $(2, 2)$.

Differentiating with respect to x: $3x^2 + 3y^2\dfrac{dy}{dx} = 8y\dfrac{dy}{dx}$

Substituting $x = 2$, $y = 2$:
$$12 + 12\frac{dy}{dx} = 16\frac{dy}{dx}$$

$$\therefore \quad \frac{dy}{dx} = 3$$

Hence at the point $(2, 2)$ the gradient of the curve is 3.

Using this approach it may be proved that if $\dfrac{d}{dx}(x^n) = nx^{n-1}$ for integral values of n, then the result also holds for fractional values of the form $n = p/q$, where p and q are integers.

Let $\quad y = x^n = x^{p/q} \quad$ then $\quad y^q = x^p$.

Differentiating with respect to x, $\quad qy^{q-1}\dfrac{dy}{dx} = px^{p-1}$

$$\therefore \quad \frac{dy}{dx} = \frac{px^{p-1}}{qy^{q-1}} = \frac{px^p y}{qxy^q} = \frac{px^p x^n}{qx \cdot x^p} = nx^{n-1}.$$

Given an equation connecting x and y, dy/dx is a measure of the rate of change of y as x increases. Similarly dx/dy is the rate of change of x as y increases. To establish the relationship between these derivatives, let δx and δy be corresponding small increases in x and y, then $\dfrac{\delta y}{\delta x} \cdot \dfrac{\delta x}{\delta y} = 1$.

As $\quad \delta x \to 0, \delta y \to 0, \dfrac{\delta y}{\delta x} \to \dfrac{dy}{dx} \quad$ and $\quad \dfrac{\delta x}{\delta y} \to \dfrac{dx}{dy}$

$$\therefore \quad \frac{dy}{dx} \cdot \frac{dx}{dy} = 1 \quad \text{so that} \quad \frac{dx}{dy} = 1 \Big/ \frac{dy}{dx}.$$

Consider, for example, the relation $x^2 - y = 0$ for positive values of x and y.

$$y = x^2 \qquad \text{and} \qquad x = y^{1/2}$$

$$\frac{dy}{dx} = 2x \qquad\qquad \frac{dx}{dy} = \tfrac{1}{2}y^{-1/2} = \frac{1}{2x}$$

$$\frac{d^2 y}{dx^2} = 2 \qquad\qquad \frac{d^2 x}{dy^2} = -\frac{1}{4}y^{-3/2} = -\frac{1}{4x^3}$$

Hence $\quad \dfrac{dx}{dy} = 1 \Big/ \dfrac{dy}{dx} \quad$ but $\quad \dfrac{d^2 x}{dy^2} \neq 1 \Big/ \dfrac{d^2 y}{dx^2}$.

Thus although it is tempting to think of derivatives such as dy/dx as fractions, higher derivatives clearly do *not* behave like fractions.

Exercise 13.2

Find $\dfrac{dy}{dx}$ in terms of x and y when

1. $y^2 = 6x$. 　　　　　 2. $x^2 + y^2 = 5$. 　　　　　 3. $4x^2 + y^2 = 4$.

4. $y^3 = x^2 + 1$.   5. $(y-1)^2 = 2x$.   6. $3y^2 = x(x^2 - 3)$.

7. $x^2 + y^2 - 2x + y + 1 = 0$ 8. $(x+y)^3 = 3x + 5$.

9. Find the gradient of the curve $(x-3)^2 + (y+4)^2 = 5$ at the point $(1, -3)$.

10. Find the gradient of the curve $y(y+2) = 4x - 1$ at each of the points on it where $x = 4$.

11. Find the equation of the tangent to the curve $y^2 = 3x + 1$ at the point $(1, -2)$.

12. Find the equation of the tangent to the curve $x^2 - y^2 = 9$ at the point $(5, 4)$.

13. Find the x-coordinates of the stationary points on the curve $x^3 + 4x^2 + 3y^2 + 5x - 2 = 0$.

14. Find the coordinates of the stationary points on the curve $x^2 + y^2 - 6x - 8y = 0$.

13.3 Products and quotients

Consider a function of x, y, which can be expressed as a product of two other functions of x, u and v. Let δx be a small increase in x and let δu, δv and δy be the corresponding small increases in u, v and y, then

$$y = uv$$
$$\Rightarrow \quad y + \delta y = (u + \delta u)(v + \delta v) = uv + u\delta v + v\delta u + \delta u \delta v$$
$$\Rightarrow \qquad \delta y = u\delta v + v\delta u + \delta u \delta v$$
$$\Rightarrow \qquad \frac{\delta y}{\delta x} = u\frac{\delta v}{\delta x} + v\frac{\delta u}{\delta x} + \delta u \frac{\delta v}{\delta x}.$$

As $\delta x \to 0$, $\delta u \to 0$, $\dfrac{\delta y}{\delta x} \to \dfrac{dy}{dx}$, $\dfrac{\delta u}{\delta x} \to \dfrac{du}{dx}$, $\dfrac{\delta v}{\delta x} \to \dfrac{dv}{dx}$

$$\therefore \qquad \boxed{\frac{dy}{dx} = u\frac{dv}{dx} + v\frac{du}{dx}.}$$

Example 1   Find $\dfrac{dy}{dx}$ if $y = (x^2 + 4)(x^5 + 7)$.

Let $y = uv$ where $u = x^2 + 4$ $v = x^5 + 7$

$$\frac{du}{dx} = 2x \qquad\qquad \frac{dv}{dx} = 5x^4$$

$$\therefore \quad \frac{dy}{dx} = u\frac{dv}{dx} + v\frac{du}{dx} = (x^2+4).5x^4 + (x^5+7).2x = 5x^6 + 20x^4 + 2x^6 + 14x$$

$$\therefore \quad \frac{dy}{dx} = 7x^6 + 20x^4 + 14x.$$

Clearly, this result could have been obtained by writing $y = x^7 + 4x^5 + 7x^2 + 28$, then differentiating. The next two examples show how the product rule and the chain rule can be used together to differentiate more complicated functions.

Example 2 Find $\dfrac{dy}{dx}$ if $y = (x^2-1)^3(3x+1)^4$.

Let $\quad u = (x^2-1)^3 \qquad\qquad\qquad v = (3x+1)^4$

then $\quad \dfrac{du}{dx} = 3(x^2-1)^2.2x \qquad\qquad \dfrac{dv}{dx} = 4(3x+1)^3.3$

$$= 6x(x^2-1)^2 \qquad\qquad\qquad = 12(3x+1)^3.$$

$$\therefore \quad \frac{dy}{dx} = u\frac{dv}{dx} + v\frac{du}{dx} = (x^2-1)^3.12(3x+1)^3 + (3x+1)^4.6x(x^2-1)^2$$

$$= 6(x^2-1)^2(3x+1)^3\{2(x^2-1)+x(3x+1)\}$$
$$= 6(x^2-1)^2(3x+1)^3(5x^2+x-2).$$

Example 3 Find $\dfrac{dy}{dx}$ if $y = (x+4)\sqrt{(x^2-1)}$.

Let $\quad u = x+4 \qquad\qquad\qquad v = \sqrt{(x^2-1)} = (x^2-1)^{1/2}$

$$\frac{du}{dx} = 1 \qquad\qquad \frac{dv}{dx} = \tfrac{1}{2}(x^2-1)^{-1/2}.2x = \frac{x}{\sqrt{(x^2-1)}}$$

$$\therefore \quad \frac{dy}{dx} = u\frac{dv}{dx} + v\frac{du}{dx} = (x+4).\frac{x}{\sqrt{(x^2-1)}} + \sqrt{(x^2-1)}.1$$

$$= \frac{x(x+4)}{\sqrt{(x^2-1)}} + \frac{x^2-1}{\sqrt{(x^2-1)}}$$

$$= (2x^2+4x-1)/\sqrt{(x^2-1)}.$$

Consider now a quotient, $y = u/v$, where u and v are functions of x. Let δx be a small increase in x and let δu, δv and δy be the corresponding small increases in u, v and y, then

$$y = \frac{u}{v}$$

$$\Rightarrow \quad y+\delta y = \frac{u+\delta u}{v+\delta v}$$

$$\Rightarrow \quad \delta y = \frac{u+\delta u}{v+\delta v} - \frac{u}{v} = \frac{v(u+\delta u)-u(v+\delta v)}{v(v+\delta v)} = \frac{v\delta u - u\delta v}{v(v+\delta v)}$$

$$\Rightarrow \quad \frac{\delta y}{\delta x} = \frac{v\dfrac{\delta u}{\delta x} - u\dfrac{\delta v}{\delta x}}{v(v+\delta v)}.$$

As $\delta x \to 0$, $\delta v \to 0$, $\dfrac{\delta y}{\delta x} \to \dfrac{dy}{dx}$, $\dfrac{\delta u}{\delta x} \to \dfrac{du}{dx}$ and $\dfrac{\delta v}{\delta x} \to \dfrac{dv}{dx}$

$$\therefore \qquad \frac{dy}{dx} = \frac{v\dfrac{du}{dx} - u\dfrac{dv}{dx}}{v^2}.$$

Example 4   Find $\dfrac{dy}{dx}$ if $y = \dfrac{x}{x+1}$.

Method 1 using the quotient rule:

Let $y = \dfrac{u}{v}$ where $u = x$ $\qquad\qquad v = x + 1$

$$\frac{du}{dx} = 1 \qquad\qquad \frac{dv}{dx} = 1$$

$$\therefore \quad \frac{dy}{dx} = \frac{v\dfrac{du}{dx} - u\dfrac{dv}{dx}}{v^2} = \frac{(x+1).1 - x.1}{(x+1)^2} = \frac{x+1-x}{(x+1)^2} = \frac{1}{(x+1)^2}.$$

Method 2 using the product rule:

Let $y = uv$ where $u = x \qquad\qquad v = (x+1)^{-1}$

$$\frac{du}{dx} = 1 \qquad\qquad \frac{dv}{dx} = -1.(x+1)^{-2}.1$$
$$= -1/(x+1)^2$$

$$\therefore \quad \frac{dy}{dx} = u\frac{dv}{dx} + v\frac{du}{dx} = x.\frac{-1}{(x+1)^2} + \frac{1}{(x+1)}.1$$
$$= \frac{-x+(x+1)}{(x+1)^2} = \frac{1}{(x+1)^2}$$

Example 5   Find $\dfrac{dy}{dx}$ if $y = \sqrt{\left(\dfrac{x+1}{x^2+1}\right)}$.

$$y = \sqrt{\left(\frac{x+1}{x^2+1}\right)} \qquad y^2 = \frac{x+1}{x^2+1}.$$

Differentiating: $\qquad 2y\dfrac{dy}{dx} = \dfrac{(x^2+1).1 - (x+1).2x}{(x^2+1)^2}$

$$\therefore \quad \frac{dy}{dx} = \frac{1}{2y}.\frac{(x^2+1-2x^2-2x)}{(x^2+1)^2} = \left(\frac{x^2+1}{x+1}\right)^{1/2}.\frac{(1-2x-x^2)}{2(x^2+1)^2}.$$

Hence $\qquad\qquad \dfrac{dy}{dx} = \dfrac{1-2x-x^2}{2(x+1)^{1/2}(x^2+1)^{3/2}}.$

[This function can also be differentiated by writing $y = \sqrt{u}$ where $u = (x+1)/(x^2+1)$ and using the chain rule.]

The next two examples show the use of the product formula when y is an implicit function of x.

Example 6 Find $\dfrac{dy}{dx}$ if $y^3 - xy^2 - x^3 = 1$.

Differentiating with respect to x:

$$\frac{d}{dx}(y^3) - \frac{d}{dx}(xy^2) - \frac{d}{dx}(x^3) = 0$$

$$3y^2 \frac{dy}{dx} - \left(x \cdot 2y\frac{dy}{dx} + y^2 \cdot 1\right) - 3x^2 = 0$$

$$(3y^2 - 2xy)\frac{dy}{dx} - y^2 - 3x^2 = 0$$

$$\therefore \quad \frac{dy}{dx} = \frac{3x^2 + y^2}{3y^2 - 2xy}.$$

Example 7 If $y^3 + x^3 = 3x + 7$ show that

$$y^2 \frac{d^2y}{dx^2} + 2y\left(\frac{dy}{dx}\right)^2 + 2x = 0.$$

Differentiating with respect to x: $3y^2 \dfrac{dy}{dx} + 3x^2 = 3$

$$\therefore \quad y^2 \frac{dy}{dx} + x^2 = 1.$$

Differentiating again with respect to x:

$$y^2 \frac{d}{dx}\left(\frac{dy}{dx}\right) + \frac{dy}{dx}\cdot\frac{d}{dx}(y^2) + 2x = 0$$

$$y^2 \frac{d^2y}{dx^2} + \frac{dy}{dx}\cdot 2y\frac{dy}{dx} + 2x = 0$$

$$\therefore \quad y^2 \frac{d^2y}{dx^2} + 2y\left(\frac{dy}{dx}\right)^2 + 2x = 0.$$

Exercise 13.3

Differentiate the following with respect to x, simplifying your answers when possible.

1. $(x+3)(x^4-1)$.

2. $(x-1)(x^2+x-6)$.

3. $(3x^2-1)(2-5x^3)$.

4. $(x^2-2)(x^2+2+1/x^2)$.

5. $x^2(x+3)^3$.

6. $x(3x-1)^5$.

7. $(x+1)^2(x+2)^5$.

8. $(2x+3)^3(1-x)^4$.

9. $x^2\sqrt{(1+x^2)}$.

10. $(x+3)\sqrt{(1-x^3)}$.

11. $\dfrac{x-2}{x+2}$.

12. $\dfrac{x^2}{2x+1}$.

13. $\dfrac{x+3}{2-3x}$.

14. $\dfrac{x^2+1}{x^3+1}$.

15. $\dfrac{x^2-2}{x^2+1}$.

16. $\dfrac{x^2+x-1}{1-x^2}$.

17. $\dfrac{x^2-2}{(x+2)^2}$.

18. $\dfrac{x^2}{(x-1)^3}$.

19. $\dfrac{(2x+1)^2}{(4x+3)^3}$.

20. $\dfrac{x}{\sqrt{(1-x^2)}}$.

21. $\dfrac{x^3}{\sqrt{(1-2x^2)}}$.

22. $\dfrac{\sqrt{(x^2+4)}}{x+1}$.

23. $\left(\dfrac{x-1}{2-x}\right)^2$.

24. $\sqrt{\left(\dfrac{2x}{x-1}\right)}$.

25. $\sqrt{\left(\dfrac{1+x^2}{x}\right)}$.

26. Find $\dfrac{dy}{dx}$ in terms of x and y, given that (a) $x^2-xy+y^2=1$, (b) $x^2y+y^3=4$, (c) $x^2+y^2-6xy+3x-2y+5=0$.

27. Find the gradient of the curve $x^2+2xy-2y^2+x=2$ at the point $(-4,1)$.

28. Find the equation of the tangent to the curve $(x+1)y^2=x-2$ at the point $(-2,2)$.

29. Find $\dfrac{d^2y}{dx^2}$ given that

(a) $y=\dfrac{1}{x^2+1}$,

(b) $y=\dfrac{x}{x-1}$,

(c) $y=\sqrt{(x^2+1)}$.

30. Differentiate the following with respect to x

(a) $\dfrac{(x^2+1)(x-1)^2}{2x-1}$,

(b) $\dfrac{x\sqrt{(x-1)}}{x+1}$,

(c) $\sqrt{\left\{\dfrac{x^3+1}{(x+1)^3}\right\}}$.

31. If $x^2-y^2=1$, prove that $y\dfrac{d^2y}{dx^2}+\left(\dfrac{dy}{dx}\right)^2=1$.

32. Find the values of $\dfrac{dy}{dx}$ and $\dfrac{d^2y}{dx^2}$ at the point $(1,3)$ on the curve $3x^2+y^2=4y$.

33. If $y = \sqrt{\left(\dfrac{6x}{x+2}\right)}$, find the values of $\dfrac{dy}{dx}$ and $\dfrac{d^2y}{dx^2}$ when $x = 4$.

34. If $y^2 - 2xy = 2x$, prove that $(x-y)\dfrac{d^2y}{dx^2} + 2\dfrac{dy}{dx} = \left(\dfrac{dy}{dx}\right)^2$.

13.4 Maxima and minima

In Chapter 5 it was shown that maximum and minimum points on a curve $y = f(x)$ can be found by examining points at which $dy/dx = 0$. Thus any maximum or minimum values of the function $f(x)$ occur when $f'(x) = 0$. The nature of such stationary values is determined either by studying the behaviour of $f'(x)$ or by finding the sign of the second derivative $f''(x)$.

Example 1 Find the maximum and minimum values of the function $f(x) = \dfrac{4x}{x^2+4}$ and sketch its graph.

Differentiating, $\quad f'(x) = \dfrac{4\{(x^2+4).1 - x.2x\}}{(x^2+4)^2} = \dfrac{4(4-x^2)}{(x^2+4)^2}$

$$f'(x) = 0 \Leftrightarrow 4 - x^2 = 0 \Leftrightarrow x = -2 \quad \text{or} \quad x = 2.$$

When x is just less than -2, $f'(x) < 0$ and when x is just greater than -2, $f'(x) > 0$
∴ when $x = -2$, $f(x)$ takes the minimum value $f(-2) = -1$.
When x is just less than 2, $f'(x) > 0$ and when x is just greater than 2, $f'(x) < 0$
∴ when $x = 2$, $f(x)$ takes the maximum value $f(2) = 1$.

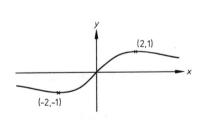

Hence the curve $y = \dfrac{4x}{x^2+4}$ has a minimum point $(-2, -1)$ and a maximum point $(2, 1)$. The curve cuts the axes only at the origin. As $x \to \infty$, $y \to 0$ from above and as $x \to -\infty$, $y \to 0$ from below,
∴ the x-axis is an asymptote to the curve.

These methods can also be applied to finding maximum and minimum values in practical problems.

Example 2 A carton of volume $V \, m^3$ is made from a piece of cardboard as shown below. If the area of cardboard used is $6 \, m^3$, find expressions for h and V in terms of x and the value of x which produces a box of maximum volume.

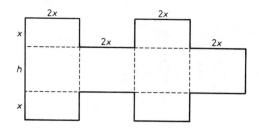 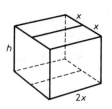

The area of cardboard used $= (4.2x.h + 4.2x.x)\,\text{m}^2$
$$= 8x(h+x)\,\text{m}^2$$

$\therefore \quad 8x(h+x) = 6 \quad \text{and} \quad h+x = 6/8x = 3/4x$

$\therefore \quad$ in terms of x, $\quad h = \dfrac{3}{4x} - x$.

The volume of the box $= V\,\text{m}^3 = 4hx^2\,\text{m}^3$

$\therefore \quad$ in terms of x, $V = 4x^2\left(\dfrac{3}{4x} - x\right) = 3x - 4x^3$.

Differentiating with respect to x,

$$\frac{dV}{dx} = 3 - 12x^2, \quad \frac{d^2V}{dx^2} = -24x$$

$\dfrac{dV}{dx} = 0 \Leftrightarrow 3 - 12x^2 = 0 \Leftrightarrow x^2 = \tfrac{1}{4}$

$\therefore \quad$ the only positive value of x for which $\dfrac{dV}{dx} = 0$ is $x = \tfrac{1}{2}$.

When $x = \tfrac{1}{2}$, $\dfrac{d^2V}{dx^2} < 0$ and $V = 1$

$\therefore \quad x = \tfrac{1}{2}$ gives a maximum value for V of 1.

It is important to note the greatest value of a function in a given interval is not always a 'maximum value'.

Example 3   Find the greatest value of $y = x^3 - 5x^2 + 7x$ in the interval from $x = 0$ to $x = 4$.

Differentiating, $\dfrac{dy}{dx} = 3x^2 - 10x + 7 = (3x - 7)(x - 1)$

$$\frac{d^2y}{dx^2} = 6x - 10$$

$\dfrac{dy}{dx} = 0 \Leftrightarrow (3x - 7)(x - 1) = 0 \Leftrightarrow x = 1 \text{ or } x = 2\tfrac{1}{3}$.

When $x = 1$, $\dfrac{d^2 y}{dx^2} < 0$ $\therefore$ y takes a maximum value.

When $x = 2\frac{1}{3}$, $\dfrac{d^2 y}{dx^2} > 0$ $\therefore$ y takes a minimum value.

However, in the interval from $x = 2\frac{1}{3}$ to $x = 4$, dy/dx is positive, so that the value of y must be increasing.
When $x = 1$, $y = 1 - 5 + 7 = 3$.
When $x = 4$, $y = 64 - 80 + 28 = 12$.

Hence although the function has a 'local' maximum value of 3 when $x = 1$, its greatest value in the interval $x = 0$ to $x = 4$ is 12.

Exercise 13.4

In questions 1 to 6 find any maximum or minimum values of the given function and sketch its graph.

1. $(2x - 5)^4$.

2. $x(x + 4)^3$.

3. $x^2(x - 3)^4$.

4. $\dfrac{2x - 1}{x^2 + 2}$.

5. $\dfrac{3x}{x^2 + x + 1}$.

6. $\dfrac{x^2 - 1}{x^2 + 1}$.

7.   Find the greatest and least values of $y = x^3 + x^2 - 5x - 4$ in the interval from $x = -3$ to $x = 3$.

8.   Find the greatest and least values of $y = x^2 + 16/x$ in the interval from $x = \frac{2}{3}$ to $x = 3$.

9.   Find the range of values taken by y on the curve $y = 12/(x^2 + 3)$. Find also the coordinates of the points of inflexion on the curve.

10.   Repeat question 9 for the curve $y = 12x/(x^2 + 3)$.

11.   The line $x = t$ meets the curves $y = x(x - 3)$ and $y = 5x - x^2$ at the points A and B. Find the maximum length of AB as t varies between 0 and 4.

12.   A square of side x cm is cut from each of the corners of a rectangular piece of cardboard 15 cm by 24 cm. The cardboard is then folded to form an open box of depth x cm. Show that the volume of the box is $(4x^3 - 78x^2 + 360x)$ cm³. Find the value of x for which the volume is a maximum.

13.   A man wishes to fence in a rectangular enclosure of area 200 m². One side of the enclosure is formed by part of a wall already in position. What is the least possible length of fencing required for the other three sides?

14.   A lump of modelling clay of volume 72 cm³ is moulded into the shape of a cuboid with edges of lengths x cm, $2x$ cm and y cm. Find the minimum surface area of this cuboid.

15. A right circular cylinder of height $2h$ is contained in a sphere of radius R, the circular edges of the cylinder touching the sphere. The volumes of the cylinder and the sphere are denoted by V and W respectively. Express V in terms of R and h. By finding the maximum value of V, as h varies, show that $V/W \leqslant 1/\sqrt{3}$.   (W)

16. A closed hollow right-circular cone has internal height a and the internal radius of its base is also a. A solid circular cyclinder of height h just fits inside the cone with the axis of the cylinder lying along the axis of the cone. Show that the volume of the cylinder is $V = \pi h(a-h)^2$. If a is fixed, but h may vary, find h in terms of a when V is maximum.   (JMB)

17. $ABCDE$ is a pentagon of fixed perimeter P cm. Its shape is such that ABE is an equilateral triangle and $BCDE$ is a rectangle. If the length of AB is x cm, find the value of P/x for which the area of the pentagon is a maximum.

18. A right circular cylinder is of radius r cm and height pr cm. The total surface area of the cylinder is S cm^2 and its volume is V cm^3. Find an expression for V in terms of p and S. If the value of S is fixed, find the value of p for which V is a maximum.

13.5 Related rates of change

The chain rule can be used to solve many practical problems involving the rates of change of more than two variables.

Example 1   A piece of paper is burning round the circumference of a circular hole. After t seconds, the radius, r cm, of the hole is increasing at the rate of 0.5 cm s^{-1}. Find the rate at which the area, A cm^2, of the hole is increasing when $r = 5$.

The rate of increase of the radius of the hole in cm s^{-1} is given by

$$\frac{dr}{dt} = 0.5$$

The area of the hole in cm^2 is $A = \pi r^2$

$$\therefore \quad \frac{dA}{dr} = 2\pi r = 10\pi \quad \text{when} \quad r = 5.$$

The rate of increase of the area of the hole in cm^2 s^{-1} is

$$\frac{dA}{dt} = \frac{dA}{dr} \cdot \frac{dr}{dt} = 10\pi \times 0.5 = 5\pi.$$

Hence the area of the hole is increasing at a rate of 5π cm^2 s^{-1} when the radius is 5 cm.

Example 2   Water is being poured into a conical vessel at a rate of $10\,\mathrm{cm^3\,s^{-1}}$. After t seconds, the volume, $V\,\mathrm{cm^3}$, of water in the vessel is given by $V = \dfrac{1}{6}\pi x^3$, where $x\,\mathrm{cm}$ is the depth of the water. Find, in terms of x, the rate at which the water is rising.

Since $V = \dfrac{1}{6}\pi x^3$ where x is a function of t, $\dfrac{dV}{dt} = \dfrac{dV}{dx}\cdot\dfrac{dx}{dt}$.

The rate at which the volume is increasing is the rate at which the water is being poured into the vessel

$$\therefore \quad \frac{dV}{dt} = 10.$$

By differentiation $\dfrac{dV}{dx} = \frac{1}{2}\pi x^2$.

Hence $10 = \frac{1}{2}\pi x^2 \cdot \dfrac{dx}{dt}$, i.e. $\dfrac{dx}{dt} = \dfrac{20}{\pi x^2}$

$\therefore$ the rate at which the water is rising is $20/\pi x^2 \,\mathrm{cm\,s^{-1}}$.

Example 3   A particle moves along a straight line so that its velocity, $v\,\mathrm{m\,s^{-1}}$, when it is $s\,\mathrm{m}$ from a fixed point, is given by $v = s^2 + 3$. Find an expression for its acceleration, $a\,\mathrm{m\,s^{-2}}$, in terms of s.

$$a = \frac{dv}{dt} = \frac{dv}{ds}\cdot\frac{ds}{dt} = v\frac{dv}{ds}.$$

By differentiation, $\dfrac{dv}{ds} = 2s$

$$\therefore \quad a = v\frac{dv}{ds} = 2s(s^2 + 3).$$

Exercise 13.5

[Answers may be left in terms of π where appropriate.]

1.   The side of a square is increasing at the rate of $3\,\mathrm{cm/s}$. Find the rate of increase of the area when the length of the side is $10\,\mathrm{cm}$.

2.   The volume of a cube is increasing at the rate of $12\,\mathrm{cm^3/s}$. Find the rate of increase of the length of an edge when the volume of the cube is $125\,\mathrm{cm^3}$.

3.   The area of a circle is increasing at the rate of $12\,\mathrm{cm^2/s}$. Find the rate of increase of the circumference when the radius is $3\,\mathrm{cm}$.

4.   The area of a rectangle $x\,\mathrm{cm}$ by $y\,\mathrm{cm}$ is constant. If $dx/dt = 2$ when $3x = y$, find the corresponding value of dy/dt.

5.   If $M = (2p+3)^4$, find $\dfrac{dM}{dt}$ when $p = 1$, given that $\dfrac{dp}{dt} = 2$.

6.   If $r = \dfrac{1+\theta}{1+\theta^2}$, find $\dfrac{d\theta}{dt}$ when $\theta = 2$, given that $\dfrac{dr}{dt} = 14$.

7.   A particle moves along a straight line so that its velocity, $v\,\text{m s}^{-1}$, when it is s m from a fixed point, is given by $v = (1+s^2)/s$. Find an expression for the acceleration of the particle in terms of s.

8.   A particle moves along a straight line so that after t seconds its displacement s m from a fixed point satisfies the equation $s^2 + s = t$. Find an expression for its acceleration after t seconds in terms of s.

9.   The volume of a spherical bubble is increasing at the rate of $6\,\text{cm}^3/\text{s}$. Find the rate at which the surface area of the bubble is increasing when the radius is 3 cm.

10.   The radius of the base of a right circular cylinder is r cm and its height is $2r$ cm. Find (a) the rate at which its volume is increasing, when the radius is 2 cm and is increasing at 0.25 cm/s, (b) the rate at which the total surface area is increasing when the radius is 5 cm and the volume is increasing at $5\pi\,\text{cm}^3/\text{s}$.

11.   A horizontal trough is 4 m long and 1 m deep. Its cross-section is an isosceles triangle of base 1.5 m with its vertex downwards. Water runs into the trough at the rate of $0.03\,\text{m}^3\,\text{s}^{-1}$. Find the rate at which the water level is rising after the water has been running for 25 seconds.

12.   Water runs into a conical vessel fixed with its vertex downwards at the rate of $3\pi\,\text{cm}^3/\text{s}$, filling the vessel to a depth of 15 cm in a time of one minute. Find the rate at which the depth of water in the vessel is increasing when the water has been running for 7.5 seconds.

13.6 Small changes and errors

If δx is a small change in x and δy is the corresponding small change in y,

$$\text{then} \quad \frac{\delta y}{\delta x} \approx \frac{dy}{dx}, \quad \text{i.e.} \quad \boxed{\delta y \approx \frac{dy}{dx}\,\delta x.}$$

Example 1   In an experiment, the diameter, x cm, of a sphere is measured and the volume, $V\,\text{cm}^3$, calculated using the formula $V = \dfrac{1}{6}\pi x^3$. If the diameter is found to be 10 cm with a possible error of 0·1 cm, estimate the possible error in the volume calculated from this reading.

$$\frac{dV}{dx} = \tfrac{1}{2}\pi x^2 = 50\pi \quad \text{when} \quad x = 10$$

$\therefore$ if $\delta x = 0\cdot 1$ then $\delta V \approx \dfrac{dV}{dx}\delta x \approx 50\pi \times 0\cdot 1 \approx 5\pi.$

Hence the possible error in the volume is $5\pi\,\text{cm}^3$.

Example 2 The time, T seconds, taken for one complete swing of a pendulum, length l m, is given by $T = 2\pi\sqrt{\dfrac{l}{g}}$, where g is constant. If a 1% error is made in measuring the length of a pendulum, estimate the percentage error in the value of T.

$$T = 2\pi\sqrt{\frac{l}{g}} = \frac{2\pi}{\sqrt{g}}\cdot l^{1/2} \qquad \therefore \quad \frac{dT}{dl} = \frac{2\pi}{\sqrt{g}}\cdot\tfrac{1}{2}l^{-1/2} = \frac{\pi}{\sqrt{(gl)}}$$

$\therefore \quad \delta T \approx \dfrac{dT}{dl}\delta l = \dfrac{\pi}{\sqrt{(gl)}}\cdot\delta l$ and $\dfrac{\delta T}{T} \approx \dfrac{\pi}{\sqrt{(gl)}}\cdot\dfrac{1}{2\pi}\sqrt{\dfrac{g}{l}}\cdot\delta l = \dfrac{1}{2}\cdot\dfrac{\delta l}{l}.$

Since $\dfrac{\delta l}{l} = \dfrac{1}{100}, \quad \dfrac{\delta T}{T} \approx \dfrac{1}{200}$

$\therefore$ the error in the value of T is approximately 0·5%.

In the next example we find an approximate value for $y + \delta y$ rather than just δy.

Example 3 Find an approximate value for $\sqrt[3]{(1003)}$.

Let $y = \sqrt[3]{x} = x^{1/3},$ then $\dfrac{dy}{dx} = \tfrac{1}{3}x^{-2/3}$

$$\therefore \quad \delta y \approx \frac{dy}{dx}\delta x = \tfrac{1}{3}x^{-2/3}\delta x.$$

Substituting $x = 1000$ and $\delta x = 3, \quad \delta y \approx \dfrac{1}{3} \times \dfrac{1}{100} \times 3 = 0\cdot 01.$

Hence $\sqrt[3]{(1003)} \approx 10 + 0\cdot 01 = 10\cdot 01.$

Exercise 13.6

[Answers may be left in terms of π where appropriate.]

1. If $y = x^2(x - 2)$, find the approximate increase in y when x increases from 3 to 3·02.

2. If $y = 2x^3 + 1$, find the approximate increase in x which causes y to increase from 3 to 3·06.

3. If $T = 2\pi \sqrt{\dfrac{l}{10}}$, find the approximate decrease in T if l is reduced from 2·5 to 2·4.

4. The radius of a circle is 6 cm. Find the approximate reduction in the area of the circle, if the radius is reduced by 0·1 cm.

5. A cylindrical hole is said to be 25 cm deep and 6 cm in diameter. Find the error in the calculated volume if there is an error of (a) 0·1 cm in the diameter, (b) 0·3 cm in the depth.

6. If $y = 3x^4$, find the approximate percentage increase in y when x in increased by 0·5%.

7. If a 2% error is made in measuring the diameter of a sphere, find approximately the resulting percentage errors in the volume and surface area of the sphere.

8. The length of a rectangle is twice its breadth. If the area of the rectangle increases by 5%, find the percentage increase in its perimeter.

9. The pressure P and volume V of a certain mass of gas are connected by the formula $PV = k$ where k is constant. If the pressure increases by 1%, what is the approximate change in the volume?

10. If $y = x^{3/2}$, find $\dfrac{dy}{dx}$ when $x = 4$. Hence find approximate values for (a) $(4\cdot01)^{3/2}$, (b) $(3\cdot98)^{3/2}$.

11. Find approximately the values of $7x^2 - 3x + \dfrac{4}{x}$ when (a) $x = 2\cdot01$, (b) $x = 1\cdot98$.

12. Find approximately the values of

(a) $\sqrt{(4\cdot004)}$, (b) $\dfrac{1}{5\cdot05}$, (c) $\sqrt[3]{(124\cdot7)}$, (d) $\dfrac{1}{\sqrt{(0\cdot96)}}$.

Exercise 13.7 (miscellaneous)

1. Differentiate with respect to x

(a) $\dfrac{1}{(5 + 2x - x^2)^3}$,

(b) $\sqrt{(2x^3 + 5)}$.

2. If $y = \left(1 - \dfrac{1}{x}\right)^4$, prove that $(x^2 - x)\dfrac{dy}{dx} = 4y$.

3. Find the equation of the tangent to the curve $y = 1/(x-1)^2$ at the point where $x = 2$.

4. Find the gradient of the curve $x^2 + y^2 = 10y$ at each of the points where $x = 3$.

5. Find the equation of the tangent to the curve $(x+3)^2 - 4(y-2)^2 = 9$ at the point $(2, 4)$.

6. Differentiate with respect to x:

(a) $(2x+3)(x^2+1)^4$,

(b) $(x^2-2)\sqrt{(2x-1)}$,

(c) $\dfrac{4x^2 + 3x + 5}{(x+1)^2}$,

(d) $\sqrt{\left(\dfrac{x+5}{x+3}\right)}$.

7. Find $\dfrac{dy}{dx}$ and $\dfrac{d^2y}{dx^2}$ when $y = \sqrt{(1+4x^2)}$.

8. If $x^4 - x^2y^2 + y^4 = 5$, find $\dfrac{dy}{dx}$ in terms of x and y.

9. Find the values of $\dfrac{dy}{dx}$ and $\dfrac{d^2y}{dx^2}$ at the point $(1, -2)$ on the curve $3x^2 + 2xy + y^2 = 3$.

10. Prove that if $y = \sqrt{(3x^2 + 2)}$, then $y\dfrac{d^2y}{dx^2} + \left(\dfrac{dy}{dx}\right)^2 = 3$.

11. Find any stationary values of the function $x^2(x-5)^3$ and sketch its graph.

12. Find any turning points and points of inflexion on the graph $y = x/(x^2+1)$. Hence sketch the curve.

13. In each of the following cases, express f', the derivative of f (with respect to x) in terms of g'. The number a is constant. (i) $f(x) = g(x+g(a))$. (ii) $f(x) = g(a+g(x))$. (iii) $f(x) = g(x^2)$.   (AEB 1975)

14. Given that $(x+y) = (x-y)^2$, prove by differentiating implicitly, or otherwise, that $1 - \dfrac{dy}{dx} = \dfrac{2}{2x-2y+1}$. Hence, or otherwise, prove that $\dfrac{d^2y}{dx^2} = \left(1 - \dfrac{dy}{dx}\right)^3$.   (JMB)

15. A piece of wire 80 cm in length is cut into three parts, two of which are bent into equal circles and the third into a square. Find the radius of the circles if the sum of the enclosed areas is a minimum.   (AEB 1975)

16. The diagonals of a rhombus are of lengths $2x$ cm and $(10-x)$ cm. As x varies, find (a) the maximum area of the rhombus, (b) the minimum length of its perimeter.

17. The height h and the base radius r of a right circular cone vary in such a way that the volume remains constant. Find the rate of change of h with respect to r at the instant when h and r are equal.

18. A right circular cone is of height 4 cm. The radius of the base is increasing at the rate of 0·5 cm/s. When the radius is 3 cm, find the rate of increase of (a) the volume of the cone, (b) the curved surface area.

19. Find approximate values for

(a) $(8·003)^{2/3}$,   (b) $\dfrac{1}{4·98}$,   (c) $\sqrt[4]{(16·032)}$.

20. Find approximately the values of $x^3 + 2x - \dfrac{16}{x^2}$ when

(a) $x = 4·02$,   (b) $x = 3·96$.

21. Find the percentage increase in the circumference of a circle, which will result in an increase in the area of the circle of approximately 3%.

22. In an experiment the value of a quantity f is calculated using the formula $\dfrac{1}{f} = \dfrac{1}{u} + \dfrac{1}{v}$. When $u = 20$, the value of v is found to be 30 with a possible error of 0·5. Find the corresponding error in the value of f.

23. A right pyramid having a square base is inscribed in a sphere of radius R, all five vertices of the pyramid lying on the sphere. The height of the pyramid is x; show that the four vertices forming the base of the pyramid lie on a circle of radius r, where $r^2 = 2Rx - x^2$. Hence, or otherwise, show that the volume V, of the pyramid is given by the formula $V = \frac{2}{3}x^2(2R - x)$. If R is fixed but x may vary, find the greatest possible value of V.   (C)

24. Water starts running into an empty basin at the rate of 6π cm^3/s. The basin is in the shape of the surface formed when the curve $4y = x^2$ is rotated completely about the y-axis. Show that when the depth of the water is y cm, the volume of water in the basin is $2\pi y^2$ cm^3. Find the rate at which the water level is rising when the water has been running for 3 seconds.

25. A hemispherical bowl of radius a cm is initially full of water. The water runs out of a small hole at the bottom of the bowl at a constant rate which is such that it would empty the bowl in 24 s. Given that, when the depth of the water is x cm, the volume of water is $\frac{1}{3}\pi x^2(3a - x)$ cm^3, prove that the depth is decreasing at a rate of $a^3/\{36x(2a - x)\}$ cm/s. Find after what time the depth of water is $\frac{1}{2}a$ cm, and the rate at which the water level is then decreasing.   (O&C)

14 Trigonometric functions

14.1 Circular measure

In elementary work, angles are usually measured in degrees, but it is often more convenient to use another unit called a radian.

An angle of 1 *radian* at the centre of a circle is subtended by an arc equal in length to the radius of the circle. Since the circumference of the circle is an arc of length $2\pi r$, it subtends an angle of 2π radians at the centre

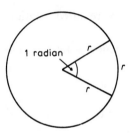

$$\therefore \quad \boxed{\text{one complete revolution} = 360° = 2\pi \text{ radians.}}$$

Hence $\quad 1 \text{ rad} = \dfrac{360°}{2\pi} \approx 57.3° \quad$ and $\quad 1° = \dfrac{2\pi}{360} \text{ rad} \approx 0.0175 \text{ rad}.$

Because of their relation to circular arcs, radians are referred to as *circular measure* and sometimes denoted by ^{c}, e.g. $2\pi^c = 360°$. This symbol and the abbreviation rad are used only when it is necessary to distinguish between radians and degrees. Usually it is assumed that an angle is measured in radians unless otherwise stated.

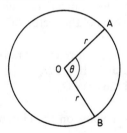

Consider now two points A and B on a circle, centre O and radius r. Let $\angle AOB = \theta$ (in radians). Since an arc of length r subtends an angle of 1 radian at O, the length of arc AB must be $r\theta$.

269

Since the area of the circle is πr^2, the area of sector AOB must be $\dfrac{\theta}{2\pi} \times \pi r^2$, i.e. $\frac{1}{2}r^2\theta$.

$$\therefore \quad \boxed{\text{length of arc } AB = r\theta, \quad \text{area of sector } AOB = \tfrac{1}{2}r^2\theta.}$$

Solutions of trigonometric equations are sometimes given in radians.

Example 1   Solve the equation $\cos 2\theta + \cos \theta + 1 = 0$ where $0 \leqslant \theta \leqslant 2\pi$.

$$\cos 2\theta + \cos \theta + 1 = 0 \Leftrightarrow (2\cos^2 \theta - 1) + \cos \theta + 1 = 0$$
$$\Leftrightarrow \qquad\qquad 2\cos^2 \theta + \cos \theta = 0$$
$$\Leftrightarrow \qquad\qquad \cos \theta(2\cos \theta + 1) = 0$$

$$\therefore \quad \cos \theta = 0 \qquad \text{or} \qquad \cos \theta = -\tfrac{1}{2} = -\cos \tfrac{1}{3}\pi$$
$$\theta = \frac{1}{2}\pi, \frac{3}{2}\pi \qquad\qquad\qquad \theta = \frac{2}{3}\pi, \frac{4}{3}\pi$$

Hence the required solutions are $\dfrac{1}{2}\pi, \dfrac{2}{3}\pi, \dfrac{4}{3}\pi$ and $\dfrac{3}{2}\pi$.

[The equivalent solutions in degrees are 90°, 120°, 240° and 270°.]

Exercise 14.1

1. Express the following angles in degrees:
(a) π rad,   (b) $\pi/4$ rad,   (c) $\pi/6$ rad,   (d) $4\pi/3$ rad.

2. Express the following angles in radians (as multiples of π):
(a) 90°,   (b) 60°,   (c) 150°,   (d) 315°.

3. Find, without using tables
(a) $\sin \dfrac{3\pi}{2}$,   (b) $\cos \dfrac{2\pi}{3}$,   (c) $\tan \dfrac{5\pi}{4}$,   (d) $\sec\left(-\dfrac{\pi}{6}\right)$.

4. Give the values in radians of
(a) $\cos^{-1} \dfrac{1}{\sqrt{2}}$,   (b) $\tan^{-1}(-\sqrt{3})$,   (c) $\sin^{-1}\dfrac{1}{2}$,   (d) $\cos^{-1}(-1)$.

5. An arc PQ subtends an angle of 1·5 radians at the centre O of a circle with radius 10 cm. Find the length of the arc PQ and the area of the sector POQ.

6. P and Q are points on the circumference of a circle centre O and radius 4 cm. The tangent to the circle at P meets OQ produced at T. If $\angle POQ = \frac{1}{4}\pi$ radians, find the area of the region bounded by PT, TQ and the arc PQ.

7. AB is a diameter of a circle radius r and C is a point on the circumference

such that $\angle ABC = \theta$ radians. Find expressions for the area and the perimeter of the region bounded by AB, BC and the arc AC.

8. A and B are points on the circumference of a circle centre O, radius $5\,\text{cm}$. If $\angle AOB = \theta$ radians, find the values of θ for which (a) the area of sector OAB is $15\,\text{cm}^2$, (b) the length of the perimeter of sector OAB is $12\,\text{cm}$, (c) the chord AB is of length $5\sqrt{3}\,\text{cm}$, (d) the area of triangle OAB is $6\cdot25\,\text{cm}^2$.

9. AB is an arc of a circle with centre O and radius $12\,\text{cm}$. A second circle with centre O and radius $(12+x)\,\text{cm}$ cuts OA and OB produced at the points C and D. If $\angle AOB = 0\cdot8$ radians and the difference between the areas of the sectors AOB and COD is $72\,\text{cm}^2$, find the value of x.

10. Solve the following equations in radians for $0 \leqslant \theta \leqslant 2\pi$
(a) $2\sin\theta = 1$,   (b) $4\cos^2\theta = 1$,   (c) $\sec^2\theta = 2\tan\theta$.

11. Solve the following equations in radians for $0 \leqslant \theta \leqslant \pi$
(a) $\sin 2\theta = \cos 4\theta$,   (b) $\sin 6\theta = \cos 3\theta$.

12. Sketch on the same diagram the graphs $y = \sqrt{2}\cos x$ and $y = \sin 2x$, for $-\pi \leqslant x \leqslant 2\pi$. Find the ranges of values of x in that interval for which $\sin 2x > \sqrt{2}\cos x$.

In questions 13 to 15 you may use a calculator, giving your answers to 3 significant figures. Check that the calculator is adjusted for work in radians.

13. A chord AB subtends an angle of $0\cdot75\,\text{rad}$ at the centre of a circle of radius $20\,\text{cm}$. Find the area and the length of the perimeter of the minor segment of the circle cut off by AB.

14. AB is the diameter of a circle radius $12\,\text{cm}$ and P is a point on the circumference such that $\angle PAB = 0\cdot6\,\text{rad}$. The tangent to the circle at P cuts AB produced at T. Find the area of the region bounded by BT, TP and arc PB.

15. A chord AB of a circle of radius $5\,\text{cm}$ subtends an angle θ at the centre O. A second chord CD is parallel to AB and on the same side of O as AB. If $AB = 2\,\text{cm}$ and $\angle COD = \theta + \tfrac{1}{2}\pi$, find θ and the area of the part of the circle lying between AB and CD.

14.2 General solutions of trigonometric equations

The trigonometric equation $\sin\theta = 0$ has an infinite solution set $\{\ldots, -2\pi, -\pi, 0, \pi, \ldots\}$. Since all the solutions are of the form $n\pi$ where n is an integer, $\theta = n\pi$ is said to be the *general solution* of the equation. We now consider three more general cases.

(1)　$\cos\theta = \cos\alpha$

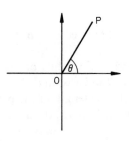

Geometrically, the roots of this equation are represented by the possible positions of a line OP, which rotates about the origin as shown in the diagram. In every complete revolution of OP, θ passes through two values which satisfy $\cos\theta = \cos\alpha$. These values correspond with the positions of OP given by $\theta = \alpha$ and $\theta = -\alpha$. Hence every root takes the form $\alpha + 2n\pi$ or $-\alpha + 2n\pi$ where n is an integer, $\therefore$ the general solution of the equation $\cos\theta = \cos\alpha$ is

$$\theta = 2n\pi \pm \alpha, \quad \text{where } n \text{ is an integer.}$$

(2)　$\sin\theta = \sin\alpha$

In this case, since $\sin(\pi - \alpha) = \sin\alpha$, the two possible positions of OP in every complete revolution are given by $\theta = \alpha$ and $\theta = \pi - \alpha$, $\therefore$ every root of the equation takes either the form $\alpha + 2n\pi$ or $(\pi - \alpha) + 2n\pi$, where n is an integer. Hence the general solution of the equation $\sin\theta = \sin\alpha$ is

$$\theta = 2n\pi + \alpha \quad \text{or} \quad 2n\pi + (\pi - \alpha) \quad \text{where } n \text{ is an integer.}$$

This means that θ is either an even multiple of π plus α or an odd multiple of π minus α. Hence the general solution of the equation $\sin\theta = \sin\alpha$ can also be written in the form

$$\theta = n\pi + (-1)^n\alpha, \quad \text{where } n \text{ is an integer.}$$

(3)　$\tan\theta = \tan\alpha$

The function $\tan\theta$ takes every real value exactly once in every rotation of OP through π radians. Since one solution of the equation is $\theta = \alpha$, the general solution of $\tan\theta = \tan\alpha$ is

$$\theta = n\pi + \alpha \quad \text{where } n \text{ is an integer.}$$

Summarising:

$$\boxed{\begin{aligned} \cos\theta = \cos\alpha &\Rightarrow \theta = 2n\pi \pm \alpha \\ \sin\theta = \sin\alpha &\Rightarrow \theta = n\pi + (-1)^n\alpha \\ \tan\theta = \tan\alpha &\Rightarrow \theta = n\pi + \alpha \end{aligned}}$$

where in every case n is an integer.

There are a few special cases in which these general forms can be simplified.

$$\cos\theta = 0 \quad \Rightarrow \theta = (2n+1)\frac{\pi}{2} \qquad\qquad \sin\theta = 0 \quad \Rightarrow \theta = n\pi$$

$$\cos\theta = 1 \quad \Rightarrow \theta = 2n\pi \qquad\qquad\qquad \sin\theta = 1 \quad \Rightarrow \theta = (4n+1)\frac{\pi}{2}$$

$$\cos\theta = -1 \Rightarrow \theta = (2n+1)\pi \qquad\qquad \sin\theta = -1 \Rightarrow \theta = (4n-1)\frac{\pi}{2}$$

Example 1   Find the general solution of the equation $\sin 2\theta - \sqrt{3}\cos\theta = 0$.

$$\sin 2\theta - \sqrt{3}\cos\theta = 0 \Leftrightarrow 2\sin\theta\cos\theta - \sqrt{3}\cos\theta = 0$$
$$\Leftrightarrow \quad \cos\theta(2\sin\theta - \sqrt{3}) = 0$$

$\therefore$ either $\cos\theta = 0$ or $\sin\theta = \dfrac{\sqrt{3}}{2} = \sin\dfrac{\pi}{3}$

$$\theta = n\pi + \frac{\pi}{2} \qquad\qquad \theta = n\pi + (-1)^n\frac{\pi}{3}$$

Hence the general solution of the equation is

$$\theta = (2n+1)\frac{\pi}{2} \quad\text{or}\quad n\pi + (-1)^n\frac{\pi}{3}, \quad\text{where } n \text{ is an integer.}$$

Example 2   Find the general solution of the equation $\sin 3\theta = \sin\theta$.

$$\sin 3\theta = \sin\theta$$

$\therefore$ either $3\theta = 2n\pi + \theta$ or $3\theta = 2n\pi + \pi - \theta$
$\qquad\qquad\quad 2\theta = 2n\pi \qquad\qquad\quad 4\theta = (2n+1)\pi$

$$\theta = n\pi \qquad\qquad\quad \theta = (2n+1)\frac{\pi}{4}$$

Hence the general solution is $\theta = n\pi$ or $\theta = (2n+1)\dfrac{\pi}{4}$ where n is an integer.

Example 3   Find the general solution of the equation $\cos 4\theta = \sin 5\theta$ giving your answer in degrees.

$$\cos 4\theta = \sin 5\theta \Leftrightarrow \cos 4\theta = \cos(90° - 5\theta)$$

$\therefore$ either $4\theta = 360n° + 90° - 5\theta$ or $4\theta = 360n° - (90° - 5\theta)$
$\qquad\qquad\quad 9\theta = 360n° + 90° \qquad\qquad\quad -\theta = 360n° - 90°$
$\qquad\qquad\quad \theta = 40n° + 10° \qquad\qquad\quad \theta = 90° - 360n°.$

Since every solution of the form $90° - 360n°$ can also be expressed in the form $40n° + 10°$, the required general solution is simply $\theta = 40n° + 10°$, where n is an integer.

Example 4   Find the general solution of the equation $3\cos\theta + 4\sin\theta = 2$, giving your answer in degrees and minutes.

Let $3\cos\theta + 4\sin\theta \equiv r\cos(\theta - \alpha) \equiv r\cos\alpha\cos\theta + r\sin\alpha\sin\theta$

then $r\cos\alpha = 3$ and $r\sin\alpha = 4$.

$\therefore \quad r = \sqrt{(3^2 + 4^2)} = 5$ and $\tan\alpha = 4/3, \quad \alpha = 53\cdot13°$.

Hence $3\cos\theta + 4\sin\theta = 2 \Leftrightarrow 5\cos(\theta - 53\cdot13°) = 2$

$$\therefore \quad \cos(\theta - 53\cdot13°) = 0\cdot4 = \cos 66\cdot42°$$

$\therefore$ either $\theta - 53 \cdot 13° = 360n° + 66 \cdot 42°$
$$\theta = 360n° + 119 \cdot 6°$$
or $\theta - 53 \cdot 13° = 360n° - 66 \cdot 42°$
$$\theta = 360n° - 13 \cdot 3°.$$

Thus the general solution is $\theta = 360n° + 119 \cdot 6°$ or $\theta = 360n° - 13 \cdot 3°$ where n is an integer.

Exercise 14.2

Find, in radians, the general solutions of the following equations.

1. $\cos \theta = \cos \dfrac{2}{5}\pi.$

2. $\tan \theta = \tan \dfrac{1}{3}\pi.$

3. $\sin \theta = \sin \dfrac{\pi}{8}.$

4. $\cos \theta = \cos \pi.$

5. $\tan \theta = -1.$

6. $2 \sin \theta = 1.$

7. $\sin^2 \theta = 1.$

8. $2 \cos^2 \theta = 1.$

9. $\sin 2x = \sin x.$

10. $\cos 2x = \cos x.$

11. $2 \cos 2x = 1 - 4 \cos x.$

12. $\sec^2 x = 2 \tan x.$

13. $\tan \theta = \tan 4\theta.$

14. $\tan 2\theta = \tan (\tfrac{1}{2}\pi - 3\theta).$

15. $\tan 3\theta = \cot \theta.$

16. $\cos 3\theta = \sin 5\theta.$

Find, in degrees and minutes, the general solutions of the following equations.

17. $\cos \theta + 3 \sin \theta = 2.$

18. $16 \cos \theta + 30 \sin \theta = 17.$

14.3 Useful limits and approximations

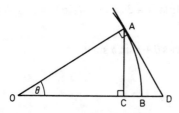

If $OA = OB = 1$ and θ is measured in radians, then

$$OC = \cos \theta. \quad \text{arc } AB = \theta$$
$$AC = \sin \theta, \quad AD = \tan \theta$$

When θ is small, $OC \approx OB$ and $AC \approx AB \approx AD$.
Hence $\cos \theta \approx 1$ and $\sin \theta \approx \theta \approx \tan \theta$.

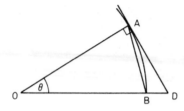

In a more formal approach to small values of θ, are considered.

$$
\begin{array}{ccccc}
\triangle AOB & < & \text{sector } AOB & < & \triangle AOD \\
\Rightarrow \tfrac{1}{2}r^2 \sin\theta & < & \tfrac{1}{2}r^2\theta & < & \tfrac{1}{2}r^2\tan\theta \\
\Rightarrow \quad \sin\theta & < & \theta & < & \tan\theta \\
\Rightarrow \quad \dfrac{\sin\theta}{\sin\theta} & < & \dfrac{\theta}{\sin\theta} & < & \dfrac{\tan\theta}{\sin\theta} \\
\Rightarrow \quad 1 & < & \dfrac{\theta}{\sin\theta} & < & \dfrac{1}{\cos\theta}
\end{array}
$$

As $\quad \theta \to 0, \quad \cos\theta \to 1 \qquad \therefore \quad \dfrac{1}{\cos\theta} \to 1$

Hence, as $\quad \theta \to 0, \qquad \dfrac{\theta}{\sin\theta} \to 1$.

Thus
$$
\lim_{\theta \to 0} \frac{\theta}{\sin\theta} = 1 \quad \text{and} \quad \lim_{\theta \to 0} \frac{\sin\theta}{\theta} = 1.
$$

As before, we see that for small values of θ, $\sin\theta \approx \theta$.
Similarly $\sin\tfrac{1}{2}\theta \approx \tfrac{1}{2}\theta \quad \therefore \cos\theta = 1 - 2\sin^2\tfrac{1}{2}\theta \approx 1 - 2(\tfrac{1}{2}\theta)^2 \approx 1 - \tfrac{1}{2}\theta^2$.

$\therefore \quad$ for small values of θ in radians, we write

$$\sin\theta \approx \theta, \quad \tan\theta \approx \theta \quad \text{and} \quad \cos\theta \approx 1 - \tfrac{1}{2}\theta^2$$

[or $\cos\theta \approx 1$ when less accuracy is required].

These limits and approximations have many applications especially in calculus. Some of their uses are simply illustrated in the following examples.

Example 1   Find an approximate value for $\sin 2°$ given $\pi = 3\cdot1416$.

$$2° = \frac{2}{360} \times 2\pi = \frac{\pi}{90} \approx \frac{3\cdot1416}{90} \approx 0\cdot0349$$

$\therefore \quad$ an approximate value for $\sin 2°$ is $0\cdot0349$.

[The true value of $\sin 2°$ correct to 6 significant figures is $0\cdot0348995$.]

Example 2   Find an approximation for the expression $\dfrac{\sin 2\theta \tan 2\theta}{1 - \cos\theta}$ when θ is small.

$$\sin 2\theta \approx 2\theta, \quad \tan 2\theta \approx 2\theta, \quad \cos \theta \approx 1 - \tfrac{1}{2}\theta^2$$

$$\therefore \quad \frac{\sin 2\theta \tan 2\theta}{1 - \cos \theta} \approx \frac{2\theta \cdot 2\theta}{1 - (1 - \tfrac{1}{2}\theta^2)} = \frac{4\theta^2}{\tfrac{1}{2}\theta^2} = 8.$$

Hence when θ is small $\dfrac{\sin 2\theta \tan 2\theta}{1 - \cos \theta} \approx 8.$

Example 3   Evaluate $\lim\limits_{\theta \to 0} \dfrac{\sin 2\theta + \sin 4\theta}{\theta}$

$$\frac{\sin 2\theta + \sin 4\theta}{\theta} = \frac{\sin 2\theta}{\theta} + \frac{\sin 4\theta}{\theta} = 2 \cdot \frac{\sin 2\theta}{2\theta} + 4 \cdot \frac{\sin 4\theta}{4\theta}$$

$$\therefore \quad \lim_{\theta \to 0} \frac{\sin 2\theta + \sin 4\theta}{\theta} = 2\left(\lim_{\theta \to 0} \frac{\sin 2\theta}{2\theta} \right) + 4\left(\lim_{\theta \to 0} \frac{\sin 4\theta}{4\theta} \right)$$
$$= 2 \cdot 1 + 4 \cdot 1 = 6.$$

Exercise 14.3

1.   Find approximate values for the following expressions without using tables or calculators, given that $\pi = 3 \cdot 1416$.
(a) $\sin 1 \cdot 2°$ (b) $\tan 0 \cdot 7°$ (c) $\sin 0 \cdot 015°$.

2.   Find approximate values of the following expressions when θ is small,
(a) $\dfrac{\theta \sin \theta}{1 - \cos \theta}$,   (b) $\dfrac{\tan 3\theta}{2\theta}$,   (c) $\dfrac{\sin 2\theta \tan \theta}{1 - \cos 3\theta}$.

3.   Find an expression of the form $a + b\theta + c\theta^2$ which is approximately equal to the following when θ is small,
(a) $\sin (\theta + \tfrac{1}{3}\pi)$,   (b) $\cos (\theta - \tfrac{1}{4}\pi)$,   (c) $\cos \theta \cos 2\theta$.

4.   Find the limit as $\theta \to 0$ of
(a) $\dfrac{\sin \theta}{2\theta}$,   (b) $\dfrac{\sin 3\theta}{\theta}$,   (c) $\dfrac{\sin 5\theta}{\sin 4\theta}$.

5.   Find the limit as $x \to 0$ of
(a) $\dfrac{1 - \cos 2x}{x^2}$,   (b) $\dfrac{\cos 3x - \cos x}{\cos 4x - \cos 2x}$.

6.   Use areas to prove that as $\theta \to 0$, $\dfrac{\tan \theta}{\theta} \to 1$.

14.4 Derivatives of trigonometric functions

The function $\sin x$ is differentiated from first principles as follows:

$$y = \sin x \quad \text{(where } x \text{ is in radians)}$$
$$\Rightarrow y + \delta y = \sin(x + \delta x)$$
$$\Rightarrow \quad \delta y = \sin(x + \delta x) - \sin x = 2 \cos(x + \tfrac{1}{2}\delta x) \sin \tfrac{1}{2} \delta x$$
$$\Rightarrow \quad \frac{\delta y}{\delta x} = \frac{2 \cos(x + \tfrac{1}{2}\delta x) \sin \tfrac{1}{2} \delta x}{\delta x} = \cos(x + \tfrac{1}{2}\delta x) . \frac{\sin \tfrac{1}{2} \delta x}{\tfrac{1}{2}\delta x}.$$

As $\quad \delta x \to 0, \quad \cos(x + \tfrac{1}{2}\delta x) \to \cos x \quad$ and $\quad \dfrac{\sin \tfrac{1}{2}\delta x}{\tfrac{1}{2}\delta x} \to 1$

$$\therefore \quad \frac{dy}{dx} = \lim_{\delta x \to 0} \frac{\delta y}{\delta x} = \cos x . 1.$$

Hence $\quad \dfrac{d}{dx}(\sin x) = \cos x.$

Similarly it can be shown that $\dfrac{d}{dx}(\cos x) = -\sin x.$

Example 1 Differentiate (a) $\sin^3 x$, (b) $\cos(4 - 3x)$, (c) $\sin x°$.

(a) Let $\quad y = \sin^3 x = u^3 \quad$ where $\quad u = \sin x$

$$\frac{dy}{du} = 3u^2 \qquad\qquad \frac{du}{dx} = \cos x$$

$$\therefore \quad \frac{dy}{dx} = \frac{dy}{du}.\frac{du}{dx} = 3u^2 . \cos x = 3 \sin^2 x \cos x.$$

Hence $\quad \dfrac{d}{dx}(\sin^3 x) = 3 \sin^2 x \cos x.$

(b) Let $y = \cos(4 - 3x) = \cos u \quad$ where $\quad u = 4 - 3x$

$$\frac{dy}{du} = -\sin u \qquad\qquad \frac{du}{dx} = -3$$

$$\therefore \quad \frac{dy}{dx} = \frac{dy}{du}.\frac{du}{dx} = -\sin u . (-3) = 3 \sin u.$$

Hence $\quad \dfrac{d}{dx}(\cos(4 - 3x)) = 3 \sin(4 - 3x).$

(c) Let $y = \sin x°$, then expressing $x°$ in radians $y = \sin \dfrac{\pi}{180} x$

$$\therefore \quad \frac{dy}{dx} = \cos \frac{\pi}{180} x . \frac{\pi}{180} = \frac{\pi}{180} \cos x°.$$

Hence $\quad \dfrac{d}{dx}(\sin x°) = \dfrac{\pi}{180} \cos x°.$

Example 2   Find the maximum and minimum values of the function
$f(x) = 2\sin x + \cos 2x$ for $0 < x < \pi$.

Differentiating, $f'(x) = 2\cos x - 2\sin 2x = 2\cos x - 4\sin x \cos x$
$$= 2\cos x(1 - 2\sin x)$$

$f'(x) = 0 \Leftrightarrow \cos x(1 - 2\sin x) = 0$
$$\Leftrightarrow \cos x = 0 \quad \text{or} \quad \sin x = \tfrac{1}{2}$$

∴   for $0 < x < \pi, f'(x) = 0$ when $x = \dfrac{1}{6}\pi, \dfrac{1}{2}\pi$ or $\dfrac{5}{6}\pi$.

	$0 < x < \dfrac{1}{6}\pi$	$\dfrac{1}{6}\pi < x < \dfrac{1}{2}\pi$	$\dfrac{1}{2}\pi < x < \dfrac{5}{6}\pi$	$\dfrac{5}{6}\pi < x < \pi$
$\cos x$	$+$	$+$	$-$	$-$
$1 - 2\sin x$	$+$	$-$	$-$	$+$
$f'(x)$	$+$	$-$	$+$	$-$

∴   at $x = \dfrac{1}{6}\pi, f(x)$ takes the maximum value $f\left(\dfrac{1}{6}\pi\right) = 1\tfrac{1}{2}$,

at $x = \dfrac{1}{2}\pi, f(x)$ takes the minimum value $f\left(\dfrac{1}{2}\pi\right) = 1$.

at $x = \dfrac{5}{6}\pi, f(x)$ takes the maximum value $f\left(\dfrac{5}{6}\pi\right) = 1\tfrac{1}{2}$.

The function $\tan x$ is differentiated by expressing it as the quotient $\dfrac{\sin x}{\cos x}$ and

using the formula $\dfrac{d}{dx}\left(\dfrac{u}{v}\right) = \left(v\dfrac{du}{dx} - u\dfrac{dv}{dx}\right)\bigg/ v^2$ derived in §13.3.

If $y = \tan x = \dfrac{\sin x}{\cos x} = \dfrac{u}{v}$ where $u = \sin x, \quad v = \cos x$

then $\dfrac{du}{dx} = \cos x, \quad \dfrac{dv}{dx} = -\sin x$

∴ $\dfrac{dy}{dx} = \dfrac{\cos x \cdot \cos x - \sin x \cdot (-\sin x)}{\cos^2 x} = \dfrac{\cos^2 x + \sin^2 x}{\cos^2 x} = \dfrac{1}{\cos^2 x} = \sec^2 x$

∴ $\dfrac{d}{dx}(\tan x) = \sec^2 x.$

Similarly it can be shown that $\dfrac{d}{dx}(\cot x) = -\operatorname{cosec}^2 x.$

To differentiate $\sec x$ we write:

$$y = \sec x = (\cos x)^{-1} = u^{-1} \quad \text{where} \quad u = \cos x$$

∴ $\dfrac{dy}{du} = -\dfrac{1}{u^2}$ and $\dfrac{du}{dx} = -\sin x.$

Hence $\dfrac{dy}{dx} = \dfrac{dy}{du} \cdot \dfrac{du}{dx} = -\dfrac{1}{u^2}(-\sin x) = \dfrac{\sin x}{\cos^2 x} = \dfrac{1}{\cos x} \times \dfrac{\sin x}{\cos x}$

$$\therefore \quad \frac{d}{dx}(\sec x) = \sec x \tan x.$$

Similarly it can be shown that $\dfrac{d}{dx}(\operatorname{cosec} x) = -\operatorname{cosec} x \cot x$.

Example 3 If $y^2 = \tan 2x + \sec 2x$, show that when $y \neq 0$, $\dfrac{dy}{dx} = y \sec 2x$.

Differentiating with respect to x,

$$y^2 = \tan 2x + \sec 2x \Leftrightarrow 2y\frac{dy}{dx} = 2\sec^2 2x + 2\sec 2x \tan 2x$$

$$\therefore \quad y\frac{dy}{dx} = \sec 2x(\sec 2x + \tan 2x) = y^2 \sec 2x.$$

Hence, when $y \neq 0$, $\dfrac{dy}{dx} = y \sec 2x$.

The table below shows the results obtained in this section.

y	$\sin x$	$\cos x$	$\tan x$	$\cot x$	$\sec x$	$\operatorname{cosec} x$
$\dfrac{dy}{dx}$	$\cos x$	$-\sin x$	$\sec^2 x$	$-\operatorname{cosec}^2 x$	$\sec x \tan x$	$-\operatorname{cosec} x \cot x$

Exercise 14.4

Differentiate the following with respect to x, simplifying your answers where possible.

1. (a) $\sin 2x$, (b) $3\cos 4x$, (c) $4\tan \frac{1}{2}x$.

2. (a) $\cos^4 x$, (b) $\sqrt{(\sin x)}$, (c) $\sin^2 5x$.

3. (a) $(x + \sin x)^3$, (b) $\sin(x^2)$, (c) $\tan(6x°)$.

4. (a) $x^4 \cos x$, (b) $(4x^2 + 1)\tan x$, (c) $x^2 \sin 3x$.

5. (a) $\operatorname{cosec}(x + 1)$, (b) $2\cot(1 - 2x)$, (c) $\sec(3x - 4)$.

6. (a) $\tan(\cos x)$, (b) $\sec(1 + \sqrt{x})$, (c) $\cot(1/x)$.

7. (a) $\sin x \cos 2x$, (b) $\cos^3 x \sin 3x$, (c) $\cos x \tan x$.

8. (a) $\dfrac{\sin x}{x^2}$,   (b) $\dfrac{1}{\cos x - \sin x}$,   (c) $\dfrac{\cos 3x}{1 + \sin 3x}$.

In questions 9 to 12 find the coordinates of any stationary points on the given curve for $0 \leqslant x \leqslant 2\pi$. Hence sketch the curve in that interval.

9. $y = 4 \sin x - \cos 2x$.

10. $y = x - 2 \sin x$.

11. $y = \tan x (\tan x + 2)$.

12. $y = \sin^3 x \cos x$.

13. A particle P moves in a straight line such that after t seconds its displacement from a fixed point O is s metres, where $s = 10 \sin 3t$. Find (a) the time at which P first returns to O, (b) the maximum distance from O reached by P, (c) the maximum speed attained by the particle and its position each time this speed is reached.

14. A right circular cone has semi-vertical angle θ and height $15\,\text{cm}$. If θ is increasing at the rate of k rad/s, where k is constant, find the rate of change of the volume of the cone when $\theta = \frac{1}{4}\pi$.

15. Points A and B lie on the circumference of a circle with centre O and radius $10\,\text{cm}$. If $\angle AOB = \theta$ and θ is increasing at the rate of $0 \cdot 1$ rad/s, find the rate of increase when $\theta = \frac{1}{3}\pi$ of (a) the area of $\triangle AOB$, (b) the perimeter of the region enclosed by the chord AB and the arc AB.

16. Given that $\sqrt{3} \approx 1 \cdot 7321$ and $0 \cdot 1° \approx 0 \cdot 00175$ rad, find approximate values for
(a) $\cos 30 \cdot 7°$,   (b) $\tan 45 \cdot 3°$,   (c) $\sin 60 \cdot 5°$.

17. In a trapezium $ABCD$ the sides AB, DC are parallel and $\angle ABC$ is a right angle. If $AD = DB = 10\,\text{cm}$ and $\angle DAB = \theta$, find the value of $\tan \theta$ for which the perimeter, p cm, is a maximum. Hence find the range of values taken by p for $0 < \theta < \frac{1}{2}\pi$.

18. A particle moves in a straight line such that after t seconds its displacement from a fixed point is s metres, where $s = 4 \cos t - \cos 2t$. If the particle first comes to rest after T seconds, where $T > 0$, find (a) its acceleration at time T, (b) the maximum speed it attains for $0 < t < T$.

14.5 Integration of trigonometric functions

From the differentiation results obtained so far:

$$\int \sin x \, dx = -\cos x + c \qquad\qquad \int \cos x \, dx = \sin x + c$$

$$\int \sec^2 x \, dx = \tan x + c \qquad\qquad \int \text{cosec}^2 x \, dx = -\cot x + c$$

$$\int \sec x \tan x \, dx = \sec x + c \qquad\qquad \int \operatorname{cosec} x \cot x \, dx = -\operatorname{cosec} x + c.$$

These results can be extended as shown in the following examples.

Example 1   Integrate   (a) $\cos 2x$,   (b) $\sin^4 x \cos x$.

(a) Since $\dfrac{d}{dx}(\sin 2x) = 2\cos 2x$, $\displaystyle\int \cos 2x \, dx = \tfrac{1}{2}\sin 2x + c.$

(b) Since $\dfrac{d}{dx}(\sin^5 x) = 5\sin^4 x \cos x$, $\displaystyle\int \sin^4 x \cos x \, dx = \frac{1}{5}\sin^5 x + c.$

In some cases trigonometric identities are used to change the form of the integrand (i.e. the function to be integrated).

Example 2   Integrate   (a) $\tan^2 x$,   (b) $\sin^2 x$.

(a) $\displaystyle\int \tan^2 x \, dx = \int (\sec^2 x - 1) \, dx = \tan x - x + c.$

(b) $\displaystyle\int \sin^2 x \, dx = \int \tfrac{1}{2}(1 - \cos 2x) \, dx = \tfrac{1}{2}(x - \tfrac{1}{2}\sin 2x) + c = \tfrac{1}{2}x - \tfrac{1}{4}\sin 2x + c.$

Example 3   Find $\displaystyle\int \sin 5x \cos x \, dx.$

$$2\sin 5x \cos x = \sin 6x + \sin 4x$$

$$\therefore \quad \int \sin 5x \cos x \, dx = \int (\tfrac{1}{2}\sin 6x + \tfrac{1}{2}\sin 4x) \, dx$$

$$= \frac{1}{2}\cdot\frac{1}{6}(-\cos 6x) + \frac{1}{2}\cdot\frac{1}{4}(-\cos 4x) + c$$

$$= -\frac{1}{12}\cos 6x - \frac{1}{8}\cos 4x + c.$$

Example 4   Evaluate $\displaystyle\int_0^{\pi/6} \cos^5 x \, dx.$

$$\cos^5 x = (\cos^2 x)^2 \cos x = (1 - \sin^2 x)^2 \cos x = (1 - 2\sin^2 x + \sin^4 x)\cos x,$$

$$\therefore \quad \int_0^{\pi/6} \cos^5 x \, dx = \int_0^{\pi/6} (\cos x - 2\sin^2 x \cos x + \sin^4 x \cos x) \, dx$$

$$= \left[\sin x - \frac{2}{3}\sin^3 x + \frac{1}{5}\sin^5 x \right]_0^{\pi/6}$$

$$= \left(\frac{1}{2} - \frac{2}{3}\left(\frac{1}{2}\right)^3 + \frac{1}{5}\left(\frac{1}{2}\right)^5 \right) - 0 = \frac{1}{2} - \frac{1}{12} + \frac{1}{160} = \frac{203}{480}.$$

Exercise 14.5

Find the following indefinite integrals

1. (a) $\int \sin 3x\, dx,$ (b) $\int \cos \frac{1}{2}x\, dx,$ (c) $\int 4\sec^2 2x\, dx.$

2. (a) $\int \cos^2 x\, dx,$ (b) $\int \cot^2 x\, dx,$ (c) $\int \sin^2 x \cos x\, dx.$

3. (a) $\int \sin x \cos^3 x\, dx,$ (b) $\int \sin x \cos 3x\, dx,$ (c) $\int \cos 2x \cos x\, dx.$

Evaluate the following definite integrals

4. (a) $\int_0^{\pi/4} \cos 2x\, dx,$ (b) $\int_0^{\pi/12} \sin^2 3x\, dx,$ (c) $\int_0^{\pi/2} \sin x \cos^2 x\, dx.$

5. (a) $\int_0^{\pi/3} \tan^2 x\, dx,$ (b) $\int_0^{\pi/3} \sin 3x \cos 2x\, dx,$ (c) $\int_0^{\pi/4} \sin 3x \sin x\, dx.$

6. (a) $\int_0^{\pi/3} \frac{\sin x}{\cos^2 x}\, dx,$ (b) $\int_{\pi/4}^{\pi/2} \frac{\cos 2x}{\sin^2 x}\, dx,$ (c) $\int_0^{\pi/2} \cos^3 x\, dx.$

7. Differentiate $\tan^3 x$ with respect to x and use your result to find $\int \tan^4 x\, dx.$

8. Show that   (a) $\int 4\sin x \cos x\, dx = 2\sin^2 x + \text{constant},$

    (b) $\int 4\sin x \cos x\, dx = -\cos 2x + \text{constant}.$

Write down a third expression for the same integral and explain how all three expressions can be correct.

9. Express $\cos^4 x$ in terms of $\cos 2x$ and $\cos 4x$. Hence evaluate $\int_0^{\pi/4} \cos^4 x\, dx.$

10. Find the area of the region bounded by the curve $y = 2/(1+\cos 2x)$, the x-axis, the y-axis and the line $x = \frac{1}{4}\pi.$

11. Find the volume of the solid formed when the region bounded by the curve $y = 3 - 2\cos x$, the x-axis, the y-axis and the line $x = \pi$ is rotated completely about the x-axis.

12. Find the mean value of $\sin^3 x$ over the interval $0 \leqslant x \leqslant \pi.$

14.6 Inverse trigonometric functions

The inverse trigonometric functions $\sin^{-1} x$, $\cos^{-1} x$ and $\tan^{-1} x$ were introduced in §7.2. At that stage it was convenient to use angles measured in degrees. However, in more advanced work, these functions are usually evaluated in radians.

To differentiate the function $\sin^{-1} x$,

let $y = \sin^{-1} x$,   then $\sin y = x$

Differentiating $\cos y \dfrac{dy}{dx} = 1$

$$\therefore \quad \frac{dy}{dx} = \frac{1}{\cos y}$$

But $\cos^2 y + \sin^2 y = 1$ $\therefore$ $\cos^2 y = 1 - \sin^2 y$

Since $\cos y$ is positive for $-\frac{1}{2}\pi < y < \frac{1}{2}\pi$,

$$\cos y = \sqrt{(1 - \sin^2 y)} = \sqrt{(1 - x^2)}.$$

Hence $\dfrac{dy}{dx} = \dfrac{1}{\sqrt{(1-x^2)}}$,   i.e. $\dfrac{d}{dx}(\sin^{-1} x) = \dfrac{1}{\sqrt{(1-x^2)}}$.

By a similar method, it can be shown that

$$\frac{d}{dx}(\cos^{-1} x) = -\frac{1}{\sqrt{(1-x^2)}}$$

[This result can also be obtained by using the fact that $\sin^{-1} x + \cos^{-1} x = \frac{1}{2}\pi$.]

To differentiate the function $\tan^{-1} x$,

let $y = \tan^{-1} x$,   then $\tan y = x$

Differentiating $\sec^2 y \dfrac{dy}{dx} = 1$

$$\therefore \quad \frac{dy}{dx} = \frac{1}{\sec^2 y}.$$

But $\sec^2 y = 1 + \tan^2 y = 1 + x^2$

$\therefore$ $\dfrac{dy}{dx} = \dfrac{1}{1+x^2}$,   i.e. $\dfrac{d}{dx}(\tan^{-1} x) = \dfrac{1}{1+x^2}$.

Summarising:

$$\frac{d}{dx}(\sin^{-1} x) = \frac{1}{\sqrt{(1-x^2)}} \quad \text{and} \quad \int \frac{dx}{\sqrt{(1-x^2)}} = \sin^{-1} x + c$$

$$\frac{d}{dx}(\tan^{-1} x) = \frac{1}{1+x^2} \quad \text{and} \quad \int \frac{dx}{1+x^2} = \tan^{-1} x + c.$$

Example 1   Find $\dfrac{dy}{dx}$ if $y = \sin^{-1}(3x-1)$.

Let $y = \sin^{-1} u$ where $u = 3x - 1$

then $\dfrac{dy}{du} = \dfrac{1}{\sqrt{(1-u^2)}}$ $\dfrac{du}{dx} = 3$

$\therefore \quad \dfrac{dy}{dx} = \dfrac{dy}{du} \cdot \dfrac{du}{dx} = \dfrac{3}{\sqrt{(1-u^2)}} = \dfrac{3}{\sqrt{(1-\{3x-1\}^2)}} = \dfrac{3}{\sqrt{(6x-9x^2)}}.$

To obtain more general integration results, we consider the functions $\sin^{-1}\dfrac{x}{a}$ and $\tan^{-1}\dfrac{x}{a}$.

Let $y = \sin^{-1} u$ where $u = \dfrac{x}{a}$

$\dfrac{dy}{du} = \dfrac{1}{\sqrt{(1-u^2)}}$ $\dfrac{du}{dx} = \dfrac{1}{a}$

$\therefore \quad \dfrac{dy}{dx} = \dfrac{dy}{du} \cdot \dfrac{du}{dx} = \dfrac{1}{\sqrt{(1-u^2)}} \cdot \dfrac{1}{a} = \dfrac{1}{a\sqrt{\left(1-\dfrac{x^2}{a^2}\right)}} = \dfrac{1}{\sqrt{(a^2-x^2)}}.$

Hence $\dfrac{d}{dx}\left(\sin^{-1}\dfrac{x}{a}\right) = \dfrac{1}{\sqrt{(a^2-x^2)}}.$

Similarly, $\dfrac{d}{dx}\left(\tan^{-1}\dfrac{x}{a}\right) = \dfrac{1}{\left(1+\dfrac{x^2}{a^2}\right)} \cdot \dfrac{1}{a} = \dfrac{a}{a^2\left(1+\dfrac{x^2}{a^2}\right)} = \dfrac{a}{a^2+x^2}.$

Thus $\boxed{\displaystyle\int \dfrac{dx}{\sqrt{(a^2-x^2)}} = \sin^{-1}\dfrac{x}{a} + c, \quad \int \dfrac{dx}{a^2+x^2} = \dfrac{1}{a}\tan^{-1}\dfrac{x}{a} + c.}$

Example 2   Evaluate $\displaystyle\int_0^3 \dfrac{dx}{\sqrt{(9-x^2)}}.$

$\displaystyle\int_0^3 \dfrac{dx}{\sqrt{(9-x^2)}} = \left[\sin^{-1}\dfrac{x}{3}\right]_0^3 = \sin^{-1}1 - \sin^{-1}0 = \tfrac{1}{2}\pi.$

Example 3   Find   (a) $\displaystyle\int \dfrac{dx}{1+4x^2}$,   (b) $\displaystyle\int \dfrac{dx}{\sqrt{(4-9x^2)}}.$

(a) [The fact that $4x^2 = (2x)^2$ suggests that the answer may involve $\tan^{-1} 2x$.]

$$\frac{d}{dx}(\tan^{-1} 2x) = \frac{1}{1+(2x)^2} \cdot 2 = \frac{2}{1+4x^2}$$

$$\therefore \quad \int \frac{dx}{1+4x^2} = \tfrac{1}{2}\tan^{-1} 2x + c.$$

(b) [The fact that $4 - 9x^2 = 2^2 - (3x)^2$ suggests that the answer may involve $\sin^{-1}\frac{3x}{2}$.]

$$\frac{d}{dx}\left(\sin^{-1}\frac{3x}{2}\right) = \frac{1}{\sqrt{\left(1-\frac{9x^2}{4}\right)}} \cdot \frac{3}{2} = \frac{3}{\sqrt{\left\{4\left(1-\frac{9x^2}{4}\right)\right\}}} = \frac{3}{\sqrt{(4-9x^2)}}$$

$$\therefore \quad \int \frac{dx}{\sqrt{(4-9x^2)}} = \frac{1}{3}\sin^{-1}\frac{3x}{2} + c.$$

Exercise 14.6

Differentiate with respect to x:

1. $\sin^{-1}\tfrac{1}{2}x$.

2. $\tan^{-1} 3x$.

3. $\cos^{-1}(2x+1)$

4. $x\tan^{-1} x$.

5. $\sin^{-1}(\cos x)$.

6. $(1-x^2)^{1/2}\sin^{-1} x$.

7. $\sin^{-1}\left(\dfrac{1}{x}\right)$.

8. $\dfrac{\tan^{-1} 2x}{1+4x^2}$.

9. $\tan^{-1}\left(\dfrac{1+x}{1-x}\right)$.

Integrate the following functions:

10. $\dfrac{1}{\sqrt{(25-x^2)}}$.

11. $\dfrac{1}{9+x^2}$.

12. $\dfrac{1}{\sqrt{(1-4x^2)}}$.

13. $\dfrac{1}{1+9x^2}$.

14. $\dfrac{1}{\sqrt{(16-9x^2)}}$.

15. $\dfrac{1}{9+4x^2}$.

Evaluate the following definite integrals:

16. $\displaystyle\int_0^1 \frac{dx}{\sqrt{(4-x^2)}}$.

17. $\displaystyle\int_0^2 \frac{dx}{4+x^2}$.

18. $\displaystyle\int_{1/6}^{1/3} \frac{dx}{\sqrt{(1-9x^2)}}$.

19. $\displaystyle\int_0^{\sqrt{3}/5} \frac{dx}{1+25x^2}$.

20. $\displaystyle\int_0^{\sqrt{3}/4} \frac{dx}{16x^2+9}$.

21. $\displaystyle\int_{-3/4}^0 \frac{dx}{\sqrt{(9-16x^2)}}$.

Exercise 14.7   (*miscellaneous*)

1. Two circles with centres A and B intersect at points P and Q, such that $\angle APB$ is a right angle. If $AB = x$ cm and $\angle PAQ = \frac{1}{3}\pi$ radians, find in terms of x the length of the perimeter and the area of the region common to the two circles.

2. The points A, B and C lie on a circle and $\angle ACB = \frac{1}{3}\pi$ radians. Show that AB divides the circle into major and minor segments whose areas are in the ratio $(8\pi + 3\sqrt{3}) : (4\pi - 3\sqrt{3})$.

3. Two parallel chords of lengths x cm and y cm lie on the same side of the centre of a circle with radius r cm. Construct a flow chart for a procedure to determine the perimeter and the area of the part of the circle between the two chords.

4. Solve, for $0 \leqslant \theta \leqslant 2\pi$, the equations
(a) $\cos 3\theta = \cos \theta$,   (b) $\sin \theta + \sin 2\theta = \sin 3\theta$.

5. Find, in radians, the general solutions of the equations
(a) $\cos 2x = \sin x$,   (b) $\cos x + \cos 3x + \cos 5x = 0$.

6. Find, in degrees, the general solution of the equation $6 \cos \theta - 4 \sin \theta = 5$.

7. Find the limit as $x \to 0$ of $\dfrac{\sin(x+\alpha) - \sin \alpha}{\sin 2x}$.

8. Differentiate the function $\cos x$ from first principles.

9. Write down the derivatives of $\sin x$ and $\cos x$. Hence find the derivatives of $\cot x$ and $\operatorname{cosec} x$.

10. Differentiate with respect to x
(a) $\sin^2 (2x - 5)$,   (b) $x^4 \tan 4x$,
(c) $\dfrac{\sec x + \tan x}{\sec x - \tan x}$,   (d) $\dfrac{\sin x}{\sqrt{(\cos 2x)}}$.

11. Sketch the following curves for $0 \leqslant x \leqslant 2\pi$, showing clearly the positions of any stationary points
(a) $y = 2 \cos x + \sin 2x$,   (b) $y = 3 \sin x - \sin 3x$.

12. Find the following indefinite integrals
(a) $\displaystyle\int \cos (2x - 1) \, dx$,   (b) $\displaystyle\int \sin^2 \tfrac{1}{2}x \, dx$,

(c) $\displaystyle\int \tan x \sec^2 x \, dx$,   (d) $\displaystyle\int \sin 4x \cos x \, dx$.

13. Evaluate (a) $\displaystyle\int_{\pi/6}^{\pi/2} \sin^5 x \cos x \, dx$,   (b) $\displaystyle\int_0^{\pi/4} \cos x \cos 2x \cos 3x \, dx$.

14. Differentiate $x \sin x$ with respect to x and use your result to find $\displaystyle\int x \cos x \, dx$.

By a similar method find $\displaystyle\int x^2 \sin x \, dx$.

15. A particle moves in a straight line such that after t seconds its acceleration is $a \, \text{ms}^{-2}$, where $a = 8 \sin 2t$. If the velocity of the particle after $\pi/3$ seconds is $5 \, \text{ms}^{-1}$, find its displacement from its initial position at that instant.

16. Given that R is the region in the first quadrant enclosed by the curves $y = 2 \cos x$, $y = \sin 2x$ and the y-axis, find the volume of the solid formed by rotating R completely about the x-axis.

17. Use the identity $1 - \sin^2 x = \cos^2 x$ to find the integral $\displaystyle\int \frac{dx}{1 - \sin x}$. Hence find the mean value of the function $1/(1 - \sin x)$ over the interval $-\pi/3 \leqslant x \leqslant \pi/6$.

18. Differentiate with respect to x
(a) $x/\sin^{-1} x$,   (b) $\tan^{-1}(\sec x)$,   (c) $\cos(\sin^{-1} x)$.

19. Find (a) $\displaystyle\int \frac{dx}{\sqrt{(9 - x^2)}}$,   (b) $\displaystyle\int \frac{dx}{9x^2 + 25}$.

20. Evaluate (a) $\displaystyle\int_0^1 \frac{dx}{1 + 3x^2}$,   (b) $\displaystyle\int_{\sqrt{2}}^2 \frac{dx}{\sqrt{(8 - x^2)}}$.

21. Given that p and q are solutions of the equation $x \tan x = 1$,
(i) show that $\displaystyle\int_0^1 \cos px \cos qx \, dx = 0$ when $p \neq q$,

(ii) find an expression for $\displaystyle\int_0^1 \cos^2 px \, dx$ entirely in terms of p, not involving any trigonometric functions.   (AEB 1976)

22. The equal sides AB and AC of an isosceles triangle have length a, and the angle at A is denoted by 2θ. Show that the radius of the inscribed circle of the triangle is given by $r = \dfrac{a \sin \theta \cos \theta}{1 + \sin \theta}$. Show that, if a is constant and θ varies between $0°$ and $90°$, the maximum value of r occurs when $\sin \theta = (\sqrt{5} - 1)/2$. [You need not verify that this value of θ gives a maximum rather than a minimum.]   (W)

23. The points A and B are on the same horizontal level, and at a distance b apart. A particle P falls vertically from rest at B, so that, at time t, its depth below

B is kt^2, where k is constant. At this time, the angle of depression of P from A is θ.

Prove that $\dfrac{d\theta}{dt} = \dfrac{2bkt}{b^2 + k^2 t^4}$. Show that $\dfrac{d\theta}{dt}$ is greatest (and not least) when $\theta = \dfrac{1}{6}\pi$.

(C)

24. Differentiate $\tan^{-1}(x+a)$ with respect to x and use your result to find

(a) $\displaystyle\int \dfrac{dx}{x^2 + 4x + 5}$,

(b) $\displaystyle\int \dfrac{dx}{x^2 + 2x + 5}$.

25. A rectangular sheet of paper $ABCD$ is folded about the line joining points P on AB and Q on AD so that the new position of A is on CD. If $AB = a$ and $AD = b$, where $a \geqslant 2b/\sqrt{3}$, show that the least possible area of the triangle APQ is obtained when the angle AQP is equal to $\pi/3$. What is the significance of the condition $a \geqslant 2b/\sqrt{3}$?

(O)

15 Probability

15.1 Permutations

A *permutation* is an arrangement of objects chosen from a given set. In this section we discuss ways of finding the total number of permutations of a set of objects under various conditions. For example, consider the number of ways in which we could arrange four guests in a row of four numbered chairs at a concert.

Any one of the 4 people may sit in the first chair. When the first person is seated, there are 3 people to be considered for the second place. This means that for each of the 4 ways of choosing the first person, there are 3 ways of choosing the second person to be seated. Similarly, in each of these $4 \times 3 = 12$ cases, we are left with 2 ways of choosing the occupant of the third chair. When 3 people are seated, the fourth chair must be filled by the 1 remaining person. Hence the total number of ways of seating the 4 people is $4.3.2.1$, i.e. 24.

Using the letters A, B, C, D to represent the 4 people, these 24 arrangements are as follows:

A B C D	B A C D	C A B D	D A B C
A B D C	B A D C	C A D B	D A C B
A C B D	B C A D	C B A D	D B A C
A C D B	B C D A	C B D A	D B C A
A D B C	B D A C	C D A B	D C A B
A D C B	B D C A	C D B A	D C B A

Since products of the type $4.3.2.1$ frequently arise in this type of work, the *factorial* notation is used. We write $4.3.2.1 = 4!$ which is read 'four factorial' or 'factorial four'. Thus for any positive integer n, $n! = n(n-1)(n-2)...3.2.1$.

The result of the above seating problem can now be generalised.

The number of permutations or arrangements of n different objects is $n!$.

Example 1 In how many different ways can 4 people chosen from a set of 6 be seated in a row of four chairs?

There are 6 possible choices for the first chair, then 5 choices for the second, 4 for the third and 3 for the fourth. Hence the total number of ways of seating 4 people out of a set of six is $6.5.4.3$, i.e. 360.
[In factorial notation this number is $6!/2!$]

In general, the number of permutations or arrangements of r different objects chosen from a set of n objects, written $_nP_r$, is given by $_nP_r = \dfrac{n!}{(n-r)!}$.

When $r = n$ this expression becomes $_nP_n = n!/0!$, but as shown above $_nP_n = n!$. Hence, it is convenient to define $0!$ to be 1. This is equivalent to assuming that there is exactly one way of arranging a set containing no objects.

These methods can be extended to other types of problem.

Example 2 How many different three-digit numbers can be formed using the digits 0, 1, 2, 3, 4 (excluding numbers which begin with 0) if (a) no digit may be repeated, (b) repetitions are allowed?

(a) Since the first digit must not be 0, the number of ways of choosing the first digit is 4. As no digit may be repeated, there are 4 digits to choose from for the second position, then 3 for the third. Hence the total number of ways of forming the three-digit number is $4.4.3$, i.e. 48.
(b) If repetitions are allowed, any of the 5 digits may be used in the second and third positions. Hence the total number of three-digit numbers is $4.5.5$, i.e. 100.

Difficulties arise when some of the objects to be arranged are identical. For instance, let us consider the number of ways of arranging the letters of the word TOTTER.

If we label the Ts with suffixes, then for the word $T_1OT_2T_3ER$ there are $6!$, i.e. 120 different arrangements. However, many of these would be indistinguishable with the suffixes removed, e.g. the set

$$RT_1ET_2T_3O, \qquad RT_2ET_1T_3O, \qquad RT_3ET_1T_2O,$$
$$RT_1ET_3T_2O, \qquad RT_2ET_3T_1O, \qquad RT_3ET_2T_1O.$$

Since T_1, T_2, T_3 can be arranged in $3!$, i.e. 6 different ways, all 120 arrangements can be grouped into sets of 6, like the one already listed. Hence, with no suffixes, the number of different arrangements will be $6!/3! = 120/6 = 20$.

In general, the number of permutations of n objects, r of which are identical, is $n!/r!$.

Example 3   In how many ways can the letters of the word NECESSITIES be arranged?

The total number of letters in the word is 11.
The letters E and S each occur 3 times. The letter I occurs twice.
Hence the total number of arrangements $= \dfrac{11!}{3!3!2!} = 554\,400.$

Exercise 15.1

1. Without using a calculator, find the values of

(a) 7!,   (b) 6! − 5!,   (c) $\dfrac{8!}{4!}$,   (d) $\dfrac{10!}{3!7!}$,   (e) $_4P_1$,   (f) $_8P_5$.

2. Express in factorial notation
(a) 5.4.3.2,   (b) 10.9.8.7,   (c) $n(n-1)(n-2)$.

3. Find the number of ways of arranging 6 different books on a shelf.

4. Find the number of arrangements of 8 items on a shopping list.

5. A lady has 8 house plants. In how many ways can she arrange 6 of them in a line on a window sill?

6. In a competition 6 household products chosen from 10 are to be listed in order of preference. In how many ways can this be done?

7. There are 25 entrants in a gymnastics competition. In how many different ways can the gold, silver and bronze medals be awarded?

8. Find the number of arrangements of 4 different letters chosen from the word PROBLEM which (a) begin with a vowel, (b) end with a consonant.

9. How many different four-digit numbers greater than 6000 can be formed using the digits 1, 2, 4, 5, 6, 8, if (a) no digit can be repeated, (b) repetitions are allowed?

10. In how many ways can 3 books be distributed among 10 people if (a) each person can receive any number of books, (b) nobody can be given more than 1 book, (c) nobody can be given more than 2 books?

11. Find the number of different ways in which the letters of the following words can be arranged
(a) NUMBER,   (b) POSSIBLE,   (c) PEPPER,   (d) STATISTICS.

12. How many different six-digit numbers can be formed using the digits 2, 3, 3, 3, 4, 4? How many of these are even?

13. In how many different ways can 8 books be arranged on a shelf if 2 particular books must be placed next to each other?

14. Three girls and 4 boys are to sit in a row of 7 chairs. If the girls wish to sit in adjacent chairs, how many different arrangements are possible?

15. In how many ways can 6 children form a circle to play a game? If Joan refuses to stand next to Tom, how many possible arrangements are there?

16. Find the number of permutations of the word PARABOLA. In how many of these permutations are (a) all three As together, (b) no two As together?

17. How many numbers less than 3500 can be formed using one or more of the digits 1, 3, 5, 7, if (a) no digit can be repeated, (b) repetitions are allowed?

18. Find the total number of possible arrangements of 3 letters chosen from the word CALCULUS.

19. A set of 10 flags, 5 red, 3 blue and 2 yellow, are to be arranged in a line along a balcony. If flags of the same colour are indistinguishable, find the number of arrangements in which (a) the three blue flags are together, (b) the yellow flags are not together, (c) the red flags occupy alternate positions in the line. If there is room for only 9 of the flags, find the total number of possible arrangements.

20. On a shelf there are 4 saucers of different colours and 4 matching cups. In how many ways can the cups be arranged on the saucers, so that no cup is on a matching saucer?

15.2 Combinations

In some problems the order in which objects are arranged is not important. The number of *combinations* of r different objects out of a set of n is the number of different selections, irrespective of order, and is denoted by $_nC_r$, or sometimes $\binom{n}{r}$.

Let us suppose that a set of 3 cards is to be dealt from a pack of 52 playing cards. The number of permutations of 3 cards chosen from $52 = 52!/49! = 52.51.50$. However, since any set of 3 cards can be arranged in 3! different ways, each set of 3 cards will appear 6 times in the list of all possible arrangements. For instance, the set containing cards X, Y and Z will appear as XYZ, XZY, YXZ, YZX, ZXY and ZYX. Hence the total number of ways of selecting 3 cards $= \dfrac{52!}{3!49!} = \dfrac{52.51.50}{3.2.1} = 22\,100$.

> In general, the number of ways of selecting r objects from n unlike objects $= {_nC_r} = \dfrac{n!}{r!(n-r)!}$.

Example 1 A committee of 5 is to be formed from 12 men and 8 women. In how many ways can the committee be chosen so that there are 3 men and 2 women on it?

Number of ways of choosing 3 men from 12

$$= {}_{12}C_3 = \frac{12!}{3!9!} = \frac{12.11.10}{3.2.1} = 220.$$

Number of ways of choosing 2 women from 8

$$= {}_8C_2 = \frac{8!}{2!6!} = \frac{8.7}{2.1} = 28.$$

$\therefore$ the total number of ways of forming the committee

$$= {}_{12}C_3 \times {}_8C_2 = 220 \times 28 = 6160.$$

There is no general formula for dealing with selections made from sets containing objects which are not all different, as there was for permutations of such sets.

Example 2 How many ways are there of selecting 4 letters from the letters in the word TOTTER?

Number of selections containing one T and three other letters $= 1$.
Number of selections containing two Ts and two letters from the remaining three $= {}_3C_2 = 3$.
Number of selections containing three Ts and one letter from the remaining three $= {}_3C_1 = 3$.
Hence the total number of ways of selecting 4 letters from TOTTER is 7.

When we select r objects from n, we are dividing the set containing n objects into 2 sets containing r and $n-r$ objects respectively. It is not, therefore, surprising to find that the number of ways of selecting $n-r$ objects from n unlike objects is the same as the number of ways of selecting r objects.

$${}_nC_{n-r} = \frac{n!}{(n-r)!(n-\{n-r\})!} = \frac{n!}{(n-r)!r!} = {}_nC_r.$$

Another identity involving the quantity ${}_nC_r$ is proved in the next example.

Example 3 Prove that ${}_nC_r + {}_nC_{r-1} = {}_{n+1}C_r$.

$$\begin{aligned}
{}_nC_r + {}_nC_{r-1} &= \frac{n!}{r!(n-r)!} + \frac{n!}{(r-1)!(n-\{r-1\})!} \\
&= \frac{n!}{r!(n-r+1)!}\{(n-r+1)+r\} \\
&= \frac{(n+1)!}{r!(n+1-r)!} = {}_{n+1}C_r.
\end{aligned}$$

Exercise 15.2

1. Evaluate (a) $_7C_4$, (b) $_5C_4$, (c) $_6C_1$, (d) $_8C_0$.

2. Verify that $_nC_{n-r} = {}_nC_r$ for the cases
(a) $n = 8, r = 5$;   (b) $n = 10, r = 2$.

3. In how many ways can
(a) 4 photographs be chosen from 10 proofs,
(b) 3 representatives be chosen from 20 students,
(c) a hand of 5 cards be dealt from a set of 13,
(d) 11 players be selected from 12 cricketers?

4. Nine people are to go on a journey in cars which can take 2, 3 and 4 passengers respectively. In how many different ways can the party travel, assuming that the seating arrangements inside the cars are not important?

5. In how many ways can a set of 12 unlike objects be divided into (a) 2 sets of 6, (b) 3 sets of 4, (c) 6 sets of 2?

6. A team of 5 students, including a captain and a reserve, is to be selected for a general knowledge contest. In how many ways can the team be chosen from a short list of 12?

7. Find the number of different selections of two letters which can be made from the letters of the word **PROBABILITY**. How many of these selections do not contain a vowel?

8. A chess team of 5 players is to be selected from 15 boys. In how many ways can the team be chosen if (a) no more than one of the three best players is to be included, (b) at least one of the four youngest players is to be included?

9. Find the number of ways in which 8 books can be distributed to 2 boys, if each boy is to receive at least 2 books.

10. Find the number of different selections of three letters which can be made from the letters of the word **PARALLELOGRAM**. How many of these selections contain the letter **P**?

11. A tennis team of 6 players consists of a 1st pair, a 2nd pair and a 3rd pair. In how many ways can the team be selected, if there are (a) only 6 players available, (b) 9 players available?

12. A committee of 6 is to be formed from 13 men and 7 women. In how many ways can the committee be selected given that (a) it must consist of 4 men and 2 women, (b) it must have at least one member of each sex?

15.3 Further set language

The language of sets was introduced in §1.1. To develop these ideas further we now define two binary operations on sets.

The *union* of sets A and B, written $A \cup B$, is the set containing all elements which belong to A or B (or both). For instance, if $A = \{1,2,3,4\}$ and $B = \{1,3,5\}$, then $A \cup B = \{1,2,3,4,5\}$.

The *intersection* of sets A and B, written $A \cap B$, is the set containing all elements which belong to both A and B. For instance, if $A = \{2,4,6,8\}$ and $B = \{1,2,3,4\}$, then $A \cap B = \{2,4\}$.

Example 1   If $S = \{x \in \mathbb{R} : 3 \leqslant x \leqslant 5\}$ and $T = \{x \in \mathbb{R} : -1 < x < 4\}$, find $S \cup T$ and $S \cap T$.

Representing S and T on the real number line we find that
$S \cup T = \{x \in \mathbb{R} : -1 < x \leqslant 5\}$ and $S \cap T = \{x \in \mathbb{R} : 3 \leqslant x < 4\}$.

Venn† diagrams are used to represent relationships between sets. The universal set is shown as a rectangle and other sets as regions, often circles, within this rectangle.

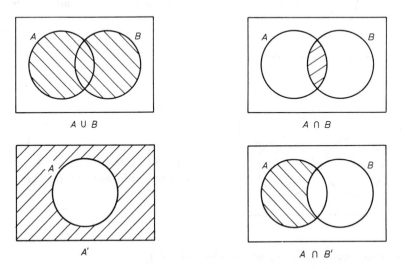

Venn diagrams can be used to demonstrate properties of union and intersection, but such verifications are not regarded as rigorous proofs of the results.

† *Venn, John* (1834–1923) English logician. He developed the theories of probability and mathematical logic. In his *Symbolic Logic* (1881) he improved the diagrammatic methods of earlier writers.

Example 2   Verify using Venn diagrams, the associative law

$$A \cap (B \cap C) = (A \cap B) \cap C.$$

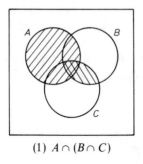

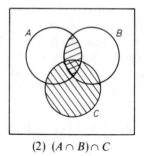

(1) $A \cap (B \cap C)$ (2) $(A \cap B) \cap C$

In (1) the sets A and $(B \cap C)$ are shaded in different ways, so that the region shaded twice represents $A \cap (B \cap C)$. Similarly in (2) the region shaded twice represents $(A \cap B) \cap C$. Hence, in these diagrams, $A \cap (B \cap C) = (A \cap B) \cap C$.

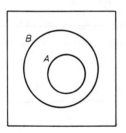

This diagram illustrates the fact that A is a subset of B, i.e. $A \subset B$. We see that in this case $A \cup B = B$ and $A \cap B = A$.

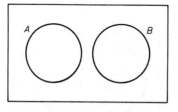

Here the sets A and B have no common elements, i.e. $A \cap B = \emptyset$. Such sets are said to be *disjoint*.

A', the complement of A, contains no elements of A,

∴ A and A' are disjoint and $A \cap A' = \emptyset$.

Since every element of the universal set $\mathscr{E}$ belongs to either A or A', for any set A, $A \cup A' = \mathscr{E}$.

Venn diagrams can also be used in problems concerning the numbers of elements in sets. The number of elements in a finite set A is called the *order* or *cardinal number* of the set and is denoted by $n(A)$.

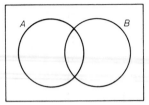

By considering the diagram we see that in the sum $n(A)+n(B)$, the elements of $A \cap B$ have been counted twice.

Hence
$$n(A \cup B) = n(A)+n(B)-n(A \cap B).$$

Example 3 On a shelf of 50 books, there are 37 historical novels and 29 romances, but no other types of book. What conclusion may be drawn from this?

Let H and R be the sets of historical novels and romances respectively, then $n(H) = 37, n(R) = 29$ and $n(H \cup R) = 50$.
$$n(H \cup R) = n(H)+n(R)-n(H \cap R)$$
$$50 = 37+29 - n(H \cap R)$$
$$\therefore \quad n(H \cap R) = 16.$$
It follows that there are 16 historical romances on the shelf.

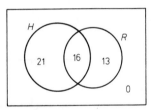

The diagram shows one way of illustrating this result.

Example 4 In a survey of 100 housewives it is found that 65 use brand X and 55 use brand Y washing powder. What can be said about the numbers of housewives using both X and Y.

Let X and Y be the sets of housewives using those brands,
then
$$n(X \cup Y) = n(X)+n(Y)-n(X \cap Y)$$
$$= 65+55-n(X \cap Y)$$
$$= 120-n(X \cap Y).$$
As there may be housewives who use neither X nor Y, we may assume only that $n(X \cup Y) \leqslant 100$,
$$\therefore \quad n(X \cap Y) \geqslant 20, \text{ i.e. at least 20 housewives use both } X \text{ and } Y.$$

Karnaugh† *maps* provide a useful alternative to Venn diagrams when dealing with intersecting sets. Karnaugh maps for one set A and for two sets A and B are given below.

† Karnaugh maps representing sets are developments of diagrams used to simplify logic circuits in computers and electronic control systems, as described by *M. Karnaugh* in his article *The Map Method for Synthesis of Combinational Logic Circuits* (1953).

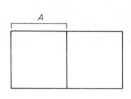

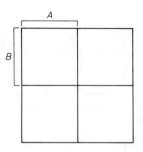

Any set can be represented on a Karnaugh map by shading the appropriate region.

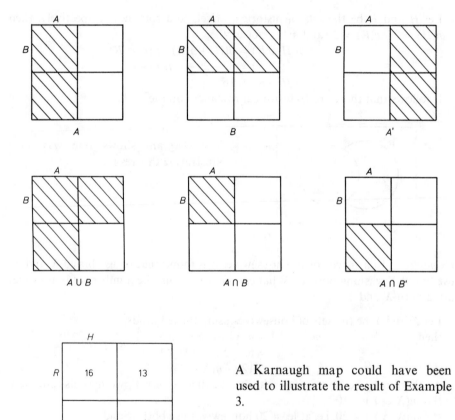

A Karnaugh map could have been used to illustrate the result of Example 3.

The Karnaugh system can be extended to larger numbers of sets. The maps for 3 sets and 4 sets are given here.

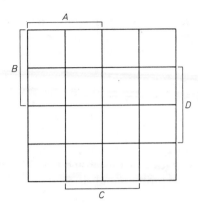

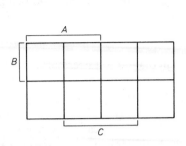

Exercise 15.3

1. If $A = \{x, y, z\}$, $B = \{x, p, q, r\}$, $C = \{x, y, p\}$, list the following sets:
(a) $A \cap B$,
(b) $B \cup C$,
(c) $(A \cup B) \cup C$,
(d) $(A \cup C) \cap B$,
(e) $(A \cap C) \cap B$,
(f) $B \cup (A \cap C)$.

2. Find $P \cup Q$ and $P \cap Q$ in each of the following cases
(a) $P = \{x \in \mathbb{R} : -2 < x < 4\}$, $Q = \{x \in \mathbb{R} : 0 < x < 7\}$,
(b) $P = \{x \in \mathbb{R} : 2 \leqslant x \leqslant 9\}$, $Q = \{x \in \mathbb{R} : x \leqslant 5\}$,
(c) $P = \{x \in \mathbb{R} : |x| < 10\}$, $Q = \{x \in \mathbb{R} : -4 < x < 3\}$,
(d) $P = \{x \in \mathbb{R} : -3 < x < 1\}$, $Q = \{x \in \mathbb{R} : 1 \leqslant x \leqslant 6\}$.

3. If $\mathscr{E} = \{1, 2, 3, 4, 5\}$, $A = \{1, 2, 3\}$, $B = \{5\}$, $C = \{3, 4\}$, list the following sets:
(a) $A \cap B$,
(b) $C \cap A'$,
(c) $(B \cup C) \cap \mathscr{E}$,
(d) $B' \cap C'$,
(e) $A \cup \varnothing$,
(f) $A' \cap (B \cup C)$.

4. If $\mathscr{E} = \{a, b, c, d\}$, $X = \{a, b\}$, $Y = \{b, c\}$, find expressions for the following sets using the symbols X, Y, X', Y', $\cup$ and $\cap$.
(a) $\{c, d\}$,
(b) $\{b\}$,
(c) $\{d\}$,
(d) $\{a, b, d\}$,
(e) $\{a, c, d\}$,
(f) $\{a, c\}$.

5. In each of the following cases draw a Venn diagram showing the universal set $\mathscr{E}$ and three sets A, B and C, then shade the region representing the given set.
(a) $A \cap B \cap C$,
(b) $A' \cap B$,
(c) $B' \cup C$,
(d) $(A \cup B) \cap C'$,
(e) $A \cup (B \cap C')$,
(f) $A' \cap B' \cap C'$.

6. Use Venn diagrams to verify the following:
(a) $A \cap (A \cup B) = A$,
(b) $(A \cap B)' = A' \cup B'$,
(c) $A \cap (B \cup C) = (A \cap B) \cup (A \cap C)$.

7. Draw a Venn diagram to show the relationships between the following sets:
$\mathscr{E} = \{\text{all triangles}\}$, $P = \{\text{isosceles triangles}\}$, $Q = \{\text{equilateral triangles}\}$, $R = \{\text{right-angled triangles}\}$, $S = \{\text{acute angled triangles}\}$.

8. Draw a Venn diagram to show the following sets and their non-empty intersections: $\mathscr{E} = \{\text{positive integers}\}$, $A = \{\text{even numbers}\}$, $B = \{\text{multiples of 3}\}$, $C = \{\text{multiples of 9}\}$, $D = \{\text{multiples of 12}\}$.

9. Simplify the following expressions, using Venn diagrams if necessary.
(a) $A \cup (A \cap B)$,
(b) $A \cap (A' \cap B)$,
(c) $(A \cap B) \cup (B' \cap A)$,
(d) $(A \cup B') \cap [A \cup (B \cap C)]$.

10. In a survey of 100 children, it is found that 47 have at least one brother, 58 have at least one sister and that 32 children have both. How many of the children have no brothers and sisters?

11. Find an expression for $n(A \cup B \cup C)$ in terms of the numbers of elements in the sets A, B, C and their intersections.

12. Out of 50 people who enrol for evening classes, 30 enrol for Mathematics, 15 for Further Mathematics, 22 for Physics and 8 enrol for none of these subjects. Denoting the sets of students who enrol for Mathematics, Further Mathematics and Physics by M, F and P respectively, what conclusions may be drawn about the values of $n(M \cap F)$, $n(M \cap P)$ and $n(M \cap F \cap P)$?

13. In each of the following cases draw a Karnaugh map for three sets A, B and C, then shade the region representing the given set.
(a) $A \cap B \cap C$,
(b) $A \cap C'$,
(c) $B' \cup C$,
(d) $A' \cap B \cap C'$,
(e) $A \cup (B' \cap C)$,
(f) $(A' \cup B') \cap C'$.

14. Using Karnaugh maps or otherwise, simplify the following expressions.
(a) $A \cap (A \cup B')$,
(b) $(B \cap C) \cup (B \cap C')$,
(c) $(A \cap B \cap C) \cup (A \cap B' \cap C)$,
(d) $(A \cup B \cup C) \cap (A' \cup B \cup C)$.

15.4 Elementary theory of probability

In mathematics probability is the numerical value assigned to the likelihood that a particular event will take place. For instance, if we throw an unbiased die, we have equal chances of scoring any of the numbers 1, 2, 3, 4, 5 and 6. Since there is one chance in six of throwing a 3 the probability of the event occurring is said to be 1/6. Similarly when tossing a coin the probability that it lands heads is considered to be 1/2. This does not, of course, mean that in any two tosses we expect the coin to fall heads once, only that in a long series of tosses the number of heads will be approximately half the total number of throws.

To arrive at a formal definition, consider the set S of all possible *outcomes* of an experiment or *trial*. A set of this kind is sometimes called a *possibility space* or *sample space*. Any *event* A can be represented by the subset A of S, which contains all the outcomes in which the event occurs. If S contains a finite number of *equally likely* outcomes, then the probability $P(A)$ that the event A will occur is given by

$$P(A) = \frac{n(A)}{n(S)}, \quad \text{i.e.} \quad \frac{\text{no. of favourable outcomes}}{\text{no. of possible outcomes}}.$$

If the event A is impossible, then $A = \emptyset$ and $P(A) = 0$. However, if the event A is certain to occur, then $A = S$ and $P(A) = 1$. Otherwise $P(A)$ will take some value between 0 and 1.

[When probabilities are calculated from experimental data, a definition such as the following may be used.

$$\text{Estimated probability} = \frac{\text{no. of successes}}{\text{no. of trials}}.$$

However, this value for the probability could be unreliable unless large numbers of observations are used.]

Example 1 What is the probability of drawing an ace at random from a pack of cards?

Since there are 4 aces in a pack of 52 cards, the probability of drawing an ace is 4/52, i.e. 1/13.

Example 2 If a letter is chosen at random from the word FACETIOUS, what is the probability that it is a vowel?

Since there are 5 vowels out of a total of 9 letters, the probability of choosing a vowel is 5/9.

Returning to the problem of throwing an unbiased die, the probability of throwing a 3 is 1/6, but the probability of not throwing a 3 is 5/6. We notice that the sum of these probabilities is 1. More generally, in a set S of equally likely possible outcomes, if A' denotes the subset of outcomes in which the event A does not occur, then A and A' are called *complementary* events and

$$P(A') = \frac{n(A')}{n(S)} = \frac{n(S) - n(A)}{n(S)} = 1 - \frac{n(A)}{n(S)}$$

$$\therefore \quad \boxed{P(A') = 1 - P(A).}$$

Hence, it seems reasonable to assume that in any sample space, if p is the probability that an event occurs and q is the probability that the event does not occur, then $p + q = 1$.

[Note that the event A' is sometimes referred to as 'not A' and may also be written $\bar{A}$.]

For experiments in which a trial is a throw of two or more dice, we can construct the sample space from the set of outcomes for a single die, $S = \{1, 2, 3, 4, 5, 6\}$. For instance, in the two-dice case, the possible outcomes can be expressed as ordered pairs such as $(1, 3)$, $(4, 6)$ and $(6, 4)$ formed from the elements of S.

A *sample space diagram* can be used to find probabilities as shown in the next example.

Example 3   If two dice are thrown together, what is the probability of the following events?

A: scoring a total of 2 B: scoring a total of 3
C: the same score on both dice D: 3 or more on each die.

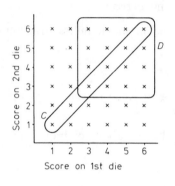

The diagram shows that when two dice are thrown, there are 36 possible outcomes. A score of 2 is obtained only when the score on each die is 1

$$\therefore \quad P(A) = \frac{1}{36}.$$

There are two outcomes $(1, 2)$ and $(2, 1)$ which produce a score of 3

$$\therefore \quad P(B) = \frac{2}{36} = \frac{1}{18}.$$

The set of outcomes in which both dice show the same score has 6 elements, as indicated in the diagram

$$\therefore \quad P(C) = \frac{6}{36} = \frac{1}{6}.$$

The diagram also shows that there are 16 outcomes in which both dice show 3 or more

$$\therefore \quad P(D) = \frac{16}{36} = \frac{4}{9}.$$

Example 4   If a coin is spun three times, what is the probability that it lands heads once and tails twice?

[In this case the elements of the sample space may be represented by the paths along a simple *tree diagram*.]

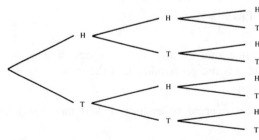

Since the given event occurs in 3 out of 8 equally likely outcomes, i.e. HTT, THT, TTH, the required probability is 3/8.

Let us now consider two different events A and B which may occur in an experiment. The set containing the outcomes in which A or B (or both) occur is the set $A \cup B$, so the probability of event A or event B (or both) is written $P(A \cup B)$. Similarly, the probability that both A and B occur is $P(A \cap B)$.

Example 5 An integer is chosen at random from the set $\{1, 2, 3, \ldots, 14, 15\}$. If A is the event of choosing an even number and B is the event of choosing a multiple of 3, find $P(A \cap B)$, $P(A \cup B)$ and $P(A' \cap B)$.

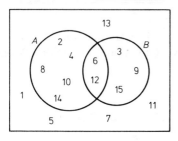

Using the possibility space diagram,

$$P(A \cap B) = \frac{2}{15}$$

$$P(A \cup B) = \frac{10}{15} = \frac{2}{3}$$

$$P(A' \cap B) = \frac{3}{15} = \frac{1}{5}$$

Some problems can be solved using the theory of permutations and combinations.

Example 6 If 4 cards are selected at random from a pack of 52, what is the probability that exactly 3 of them are diamonds?

The number of ways of selecting 3 diamonds from $13 = {}_{13}C_3$.
The number of ways of selecting 1 card from the 39 which are not diamonds $= 39$.
∴ the number of ways of selecting 4 cards, 3 of which are diamonds $= 39 \times {}_{13}C_3$.
The number of ways of selecting any 4 cards from $52 = {}_{52}C_4$.
Hence the probability that exactly 3 cards out of 4 are diamonds

$$= 39 \times {}_{13}C_3 \div {}_{52}C_4 = 39 \times \frac{13!}{3!10!} \times \frac{4!48!}{52!} \approx 0 \cdot 0412.$$

For a trial in which the possible outcomes are not equally likely, it is more difficult to arrive at a precise definition of the probability of an event. However, the subsets of a sample space S are regarded as events whatever the exact nature of the set S. In all cases $P(A)$, the probability of an event A, is a number or a weighting associated with the subset A. We can consider P to be a *probability mapping* which maps any subset A onto a real number p, where $0 \leqslant p \leqslant 1$. In any experiment the set of events (i.e. the set of subsets of the sample space) to be considered together with the appropriate probability mapping may be referred to as the *probability space*.

Exercise 15.4

In questions 1 to 4 find the probability of events A, B and C.

1. A die is thrown. A: scoring a four, B: scoring an odd number, C: scoring four or less.

2. A card is drawn from a pack. A: drawing a red card, B: drawing a seven, C: drawing a king, a queen or a jack.

3. Three coins are tossed. A: three heads, B: at least two tails, C: at least one of each.

4. It rained on exactly two days last week. A: it rained on Monday and Tuesday, B: it rained on two consecutive days, C: it rained on neither Monday nor Tuesday.

5. For the events in question 1, find
(a) $P(A \cup B)$,   (b) $P(A \cap B)$,   (c) $P(B \cup C)$,   (d) $P(B \cap C')$.

6. For the events in question 2, find
(a) $P(A \cup B)$,   (b) $P(A' \cap B)$,   (c) $P(B \cap C')$,   (d) $P(A \cap C')$.

7. An integer is chosen at random from the first 200 positive integers. Find the probability that it is (a) not divisible by 5, (b) a perfect square, (c) divisible by both 5 and 2, (d) divisible by neither 2 nor 7.

8. Two unbiased dice are thrown. Find the probability that the product of the scores is (a) odd, (b) a multiple of 3, (c) a multiple of 12.

9. A domino is drawn from a standard set of 28. Find the probability that the sum of its spots is (a) 2, (b) 8, (c) odd, (d) even.

10. In a cafeteria 80% of the customers order chips and 60% order peas. If 20% of those ordering peas do not want chips, find the probability that a customer chosen at random orders chips but not peas.

11. The points A, B, C, D and E are the vertices of a regular pentagon. All possible lines joining pairs of these points are drawn. If two of these lines are chosen at random, that is the probability that their point of intersection is (a) inside the pentagon, (b) one of the points A, B, C, D, E.

12. A box contains 4 discs numbered 1, 2, 2, 3. A disc is drawn from the box then replaced and a second disc is drawn. With the aid of a sample space diagram or otherwise, find the probability that (a) the total score is 6, (b) the total score is 4, (c) the numbers drawn are different, (d) the difference between the two numbers drawn is less than two.

13. A shopper with a keen eye for a bargain has bought a carton of assorted tins without labels. Assuming that the tins came from a large warehouse containing equal numbers of tins of soup, peaches and rice pudding, find the probability that (a) the first tin opened contains soup, (b) the first tin does not contain soup but the second does, (c) none of the first three tins contains soup. [You may find tree diagrams helpful.]

14. Two cards are selected at random without replacement from a set of five numbered 2, 3, 4, 4, 5. With the aid of a diagram or otherwise, find the probability that the numbers on the cards (a) are both even, (b) have a difference of 2, (c) have a sum of 7, (d) have a sum of 8.

15. A bag contains 3 blue beads and 2 red beads. A second bag contains 1 blue bead and 3 red beads. If one bead is drawn from each bag, find the probability that the beads are (a) both blue, (b) one blue and one red.

16. Write down in factorial form the number of different arrangements of the letters of the word EQUILIBRIUM. One of these arrangements is chosen at random. Find the probability that (a) the first two letters of the arrangement are consonants, (b) all the vowels are together.

17. There are 8 dusters in a drawer, 5 orange and 3 pink. If two of the dusters are chosen at random simultaneously, what is the probability that there will be one of each colour.

18. Four balls, two red, one blue and one white, are placed in a bag. The balls are then drawn at random from the bag one at a time and not replaced. Find the probability that (a) the first ball drawn is red, (b) the last ball drawn is white, (c) the red balls are drawn consecutively.

19. A hand of three cards is dealt from a well shuffled pack of 52. Find the probability that the hand contains (a) exactly one ace, (b) three cards of the same suit, (c) no two cards of the same suit.

20. On a plate of 12 assorted cakes, 3 are doughnuts. If 3 cakes are selected at random, find the probabilities that 0, 1, 2, 3 doughnuts are chosen.

21. A bag contains 25 clothes pegs, 15 plastic pegs and 10 wooden pegs. If 4 pegs are taken from the bag at random, find the probability that (a) all 4 are plastic, (b) 2 are plastic and 2 are wooden.

22. Four books are selected at random from a shelf containing 3 cookery books, 5 novels and 2 biographies. Find the probability that the four books (a) are all novels, (b) are 2 novels and 2 cookery books, (c) include at least one biography, (d) include at least one of each type.

15.5 Sum and product laws

If A and B are two different events which may occur in a sample space S, the set theory result

$$n(A \cup B) = n(A) + n(B) - n(A \cap B)$$

discussed in §15.3 can be used to find the probability that A or B (or both) will occur.

$$P(A \cup B) = \frac{n(A \cup B)}{n(S)} = \frac{n(A) + n(B) - n(A \cap B)}{n(S)}$$

$$= \frac{n(A)}{n(S)} + \frac{n(B)}{n(S)} - \frac{n(A \cap B)}{n(S)}$$

$$\therefore \quad \boxed{P(A \cup B) = P(A) + P(B) - P(A \cap B).}$$

This is the *addition law* of probability.

Example 1   A card is drawn at random from a pack. A is the event of drawing an ace and B is the event of drawing a diamond. Find $P(A)$, $P(B)$, $P(A \cap B)$ and deduce $P(A \cup B)$.

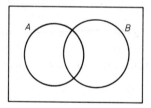

$$P(A) = \frac{4}{52} = \frac{1}{13}, \quad P(B) = \frac{13}{52} = \frac{1}{4}.$$

$P(A \cap B)$ is the probability of drawing the ace of diamonds, which is 1/52.

Hence using the addition law,

$$P(A \cup B) = P(A) + P(B) - P(A \cap B) = \frac{4}{52} + \frac{13}{52} - \frac{1}{52} = \frac{16}{52} = \frac{4}{13}.$$

Events A and B are said to be *mutually exclusive* if they cannot occur at the same time, i.e. if the sets A and B are disjoint. This means that $A \cap B = \varnothing$ and $P(A \cap B) = 0$, so that for mutually exclusive events the addition law reduces to $P(A \cup B) = P(A) + P(B)$. In particular, if $B = A'$, then $P(A \cup B) = P(A \cup A') = P(S) = 1$, giving the result $1 = P(A) + P(A')$, i.e. $P(A') = 1 - P(A)$, which was obtained in the previous section.

Example 2   A disc is drawn at random from a box containing 12 discs of which 6 are red and 2 are blue. If event A is the choice of a red disc and event B is the choice of a blue disc, find $P(A)$, $P(B)$ and deduce $P(A \cup B)$.

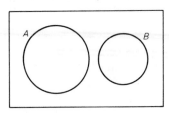

$$P(A) = \frac{6}{12} = \frac{1}{2}, \; P(B) = \frac{2}{12} = \frac{1}{6}$$

Since a disc cannot be both red and blue, the events A and B are mutually exclusive,

$$\therefore \quad P(A \cup B) = P(A) + P(B)$$

$$= \frac{6}{12} + \frac{2}{12} = \frac{8}{12} = \frac{2}{3}.$$

A set of events is said to be *exhaustive* if at least one of them is certain to occur in any trial, e.g. if the events A and B are exhaustive, then $P(A \cup B) = 1$ and the addition law becomes $P(A) + P(B) - P(A \cap B) = 1$. This will reduce to $P(A) + P(B) = 1$, if A and B are also mutually exclusive.

In general, for events $A_1, A_2, \ldots, A_n$ which are both mutually exclusive and exhaustive

$$P(A_1) + P(A_2) + \ldots + P(A_n) = 1.$$

Two events A and B are said to be *independent* if the occurrence of event A has no effect on the probability of event B. However, if the probability of event B changes when A is known to have occurred then A and B are *dependent* upon one another.

Let us consider a playing card drawn at random from a pack and events A, B and C defined as follows: A: drawing a king, B: drawing a club, C: drawing a black king. The probability of drawing a club from the complete pack is 1/4. If the card drawn is known to be a king, i.e. if event A has occurred, then the probability that the card is a club remains 1/4. Hence events A and B are independent. Since $P(A) = 1/13$, $P(B) = 1/4$ and $P(A \cap B) = 1/52$, we find that $P(A \cap B) = P(A).P(B)$. However, if the card drawn is known to be a black king, i.e. if event C has occurred, then the probability that the card is a club is now 1/2 rather than 1/4. Hence events B and C are not independent. Since $P(B) = 1/4$, $P(C) = 1/26$ and $P(B \cap C) = 1/52$, in this case $P(B \cap C) \neq P(B).P(C)$.

This example illustrates the fact that

two events A and B are independent if and only if

$$P(A \cap B) = P(A).P(B)$$

This is the simplest form of the *multiplication* or *product law*. [The extension to events which are not independent is dealt with in the next section.]

When two dice are thrown, the scores obtained are independent of each other, so the product law may be applied. This provides an alternative approach to Example 3, §15.4.

The probability of throwing 1 on a single die is $\frac{1}{6}$, so the probability of throwing two 1's, i.e. of scoring a total of 2, is $\frac{1}{6} \times \frac{1}{6} = \frac{1}{36}$.

The probabilities of the outcomes $(1,2)$ and $(2,1)$ are both equal to $\frac{1}{6} \times \frac{1}{6} = \frac{1}{36}$. Hence the probability of scoring a total of 3 is $\frac{1}{36} + \frac{1}{36} = \frac{2}{36} = \frac{1}{18}$.

For any score on the 1st die, the probability of the 2nd score being the same is $\frac{1}{6}$. Thus the probability of throwing the same score on both dice is $\frac{1}{6}$.

Since the probability of scoring 3 or more on a single die is $\frac{2}{3}$, the probability of throwing 3 or more on each die is $\frac{2}{3} \times \frac{2}{3} = \frac{4}{9}$.

Example 3   A boy must throw a 6 on a single die to start a game. Find the probability that he succeeds at his third attempt.

The probability of not throwing a 6 in his first turn $= \frac{5}{6}$.

The probability of not throwing a 6 in his second turn $= \frac{5}{6}$.

The probability of throwing a 6 in his third turn $= \frac{1}{6}$.

Hence the probability that he succeeds at his third attempt

$$= \frac{5}{6} \times \frac{5}{6} \times \frac{1}{6} = \frac{25}{216}.$$

Example 4   A card is drawn from a pack and then replaced, the suit being noted. Find the probability that in a sequence of four such trials exactly two diamonds are drawn.

The number of orders in which two diamonds and two other cards may be drawn in the four trials is $\frac{4!}{2!2!}$, i.e. 6.

In each of these 6 cases, the probability of drawing two diamonds and two other cards in the appropriate order is $\left(\frac{1}{4}\right)^2 \left(\frac{3}{4}\right)^2$.

Hence the probability that two diamonds are drawn

$$= 6 \times \frac{3 \cdot 3}{4 \cdot 4 \cdot 4 \cdot 4} = \frac{27}{128}.$$

Exercise 15.5

1. If A and B are random events such that $P(A) = 5/12$, $P(B) = 1/3$ and $P(A \cap B) = 1/4$, find $P(A \cup B)$.

2. In a certain school the probability that a pupil takes part in the annual musical evening is $0 \cdot 25$. The probability that a pupil helps with the school play is $0 \cdot 1$. If the probability of doing both is $0 \cdot 02$, find the probability that a pupil is involved in at least one of these activities.

3. A butcher observes that the probability that a certain customer buys beef is $\frac{1}{2}$ and the probability that she buys pork is $\frac{1}{3}$. If the probability that this customer buys neither pork nor beef is $\frac{1}{4}$, find the probability that she buys both beef and pork.

In questions 4 and 5 for the given events A, B and C
(i) write down any pairs of events which are (a) mutually exclusive, (b) exhaustive;
(ii) find (a) $P(A \cap B)$,   (b) $P(A \cup C)$,   (c) $P(B \cap C)$,
   (d) $P(A' \cup C)$,   (e) $P(B \cap C')$,   (f) $P(A' \cup B')$.

4. A card is drawn at random from a pack of 52. A: drawing a black card, B: drawing a red card, C: drawing a heart.

5. Two unbiased dice are thrown and the total score recorded. A: a score less than 8, B: a score divisible by 5, C: a score greater than 4.

6. The events A and B are mutually exclusive. The events A, B and C are exhaustive. If $P(B) = 0 \cdot 6$ and $P(C) = 0 \cdot 3$, find the range of possible values of $P(A')$.

In questions 7 and 8, for the given events A, B and C, find $P(A)$, $P(B)$, $P(C)$, $P(A \cap B)$, $P(B \cap C)$, $P(A \cap C)$. Write down any pairs of events which are independent.

7. A card is drawn at random from a pack of 52. A: drawing a black card, B: drawing the jack of hearts, C: drawing a court card (i.e. jack, queen or king).

8. A disc is drawn at random from a bag containing discs numbered 1 to 48. A: drawing a multiple of 3, B: drawing a multiple of 4, C: drawing a multiple of 6.

9. If A and B are two independent events such that $P(A) = 1/4$ and $P(B) = 3/5$, find
(a) $P(A \cap B)$,   (b) $P(A \cup B)$,   (c) $P(A \cap B')$,   (d) $P(A' \cup B')$.

10. A and B are two independent events, such that $P(A \cap B) = 1/3$ and $P(A \cup B) = 9/10$. Given that $P(A) > P(B)$, find $P(A)$ and $P(B)$.

11. If A, B and C are independent random events such that $P(A) = 2/5$, $P(B) = 1/3$, $P(C) = 3/4$, calculate $P(A \cup B \cup C)$ and $P(A' \cup B' \cup C')$.

12. A box contains 5 red balls and 3 blue balls. A second box contains 2 red balls and 3 blue balls. Find the probability that (a) if a ball is drawn at random from each box both will be blue, (b) if two balls are drawn at random from each box all four will be red.

13. A die is biased so that the probability of throwing a six is $\frac{1}{4}$. If the die is thrown twice find the probability of (a) two sixes, (b) at least one six.

14. A man estimates that the probability that he will be early for work is 1/3, the probability that he will be on time is 1/2 and the probability that he will be late is 1/6. Find the probability that in a particular 3-day period he will (a) arrive early every day, (b) never arrive on time, (c) be late at least once.

15. Three unbiased dice are thrown. Find the probability that (a) the score on each die is the same, (b) the total score is 5, (c) the total score is even.

16. Alan and Bob are playing a game which ends when either player is two points ahead of his opponent. If the probability that Alan wins any particular point is $\frac{2}{3}$, find the probabilities that the game ends when the number of points played is (a) 2, (b) 3, (c) 4.

17. Tom and Peter throw a single die in turn. The first player to throw a six wins. If Tom throws first, find the probability that Peter wins on his first, second or third turn. [Give your answer to 3 significant figures.]

18. Cards are drawn at random, with replacement, from a pack of 52. Find the probability that (a) the first two cards drawn are spades, (b) two out of the first three cards drawn are spades, (c) the third card drawn is a club, (d) the fourth card is the first black card to be drawn.

19. A certain committee has 6 members. The probability that any member will attend a particular meeting is 4/5. Find the probability that the chairman and at least 4 of the 5 other members will attend.

20. An unbiased coin is tossed 5 times. Find the probabilities of (a) 4 heads, (b) 3 heads, (c) at least 2 heads.

21. A bag contains 3 red discs, 4 blue discs and 5 green discs. A trial consists of selecting a disc at random, noting its colour then replacing it. In three such trials, what is the probability of selecting (a) three red discs, (b) two blue discs, (c) one green and two blue discs?

15.6 Conditional probability

The probability of an event B, given that an event A has occurred, is called the *conditional probability* of B upon A, or the probability of B given A, and is written $P(B|A)$. This means that $P(B|A)$ is the probability that B occurs, considering A as sample space.

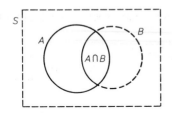

Since the subset of A in which B occurs is $A \cap B$,

$$P(B|A) = \frac{n(A \cap B)}{n(A)} = \frac{n(A \cap B)}{n(S)} \Big/ \frac{n(A)}{n(S)}$$

$$\therefore \quad \boxed{P(B|A) = \frac{P(A \cap B)}{P(A)}.}$$

The general statement of the product law is obtained by rearranging this result:

$$P(A \cap B) = P(A) \cdot P(B|A).$$

Similarly

$$P(A \cap B) = P(B) \cdot P(A|B).$$

In the case of independent events the probability of B is unaffected by the occurrence of A, so $P(B|A) = P(B)$

$$\therefore \quad P(A \cap B) = P(A) \cdot P(B).$$

Example 1   Two cards are drawn in succession from a pack. What is the probability that they are both diamonds if (a) the first card is replaced before the second is drawn, (b) the first card is not replaced?

(a) Since both cards are drawn from a full pack, in each case the probability of drawing a diamond is $\dfrac{13}{52}$, i.e. $\dfrac{1}{4}$

    $\therefore$ with replacement the probability of two diamonds is

$$\frac{1}{4} \times \frac{1}{4} = \frac{1}{16}.$$

(b) The probability that the second card is a diamond given that a diamond has already been drawn is $\dfrac{12}{51} = \dfrac{4}{17}$,

    $\therefore$ without replacement, the probability of two diamonds is

$$\frac{1}{4} \times \frac{4}{17} = \frac{1}{17}.$$

Consider now a trial in which two dice are thrown, where event C is throwing the same score on both dice and event D is throwing 3 or more on each of the dice.

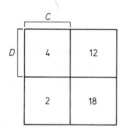

The 36 possible outcomes in the sample space for this trial can be represented in a Karnaugh map for events C and D.

Relationships between probabilities associated with such a trial are often displayed in a *probability tree diagram*.

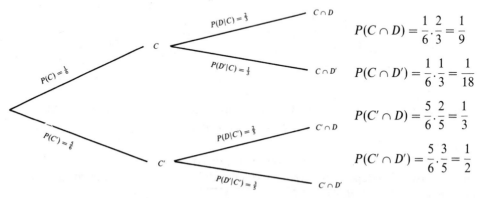

$$P(C \cap D) = \frac{1}{6} \cdot \frac{2}{3} = \frac{1}{9}$$

$$P(C \cap D') = \frac{1}{6} \cdot \frac{1}{3} = \frac{1}{18}$$

$$P(C' \cap D) = \frac{5}{6} \cdot \frac{2}{5} = \frac{1}{3}$$

$$P(C' \cap D') = \frac{5}{6} \cdot \frac{3}{5} = \frac{1}{2}$$

Probabilities can also be recorded in a *contingency table* as shown below. The diagram on the left shows which probabilities should be inserted in each part of the table. The completed table is on the right.

	D	D'	
C	$P(C \cap D)$	$P(C \cap D')$	$P(C)$
C'	$P(C' \cap D)$	$P(C' \cap D')$	$P(C')$
	$P(D)$	$P(D')$	

	D	D'	
C	$\frac{1}{9}$	$\frac{1}{18}$	$\frac{1}{6}$
C'	$\frac{1}{3}$	$\frac{1}{2}$	$\frac{5}{6}$
	$\frac{4}{9}$	$\frac{5}{9}$	1

Both the tree diagram and the contingency table demonstrate that

$$P(C \cap D) + P(C \cap D') + P(C' \cap D) + P(C' \cap D') = P(C) + P(C') = 1.$$

Example 2   A motorist plans a journey and event X is arrival at his destination in less than three hours. He estimates the probabilities of dry weather D, rain R or snow S to be $\frac{1}{3}$, $\frac{1}{2}$ and $\frac{1}{6}$ respectively. The probabilities of event X in these conditions are $\frac{3}{4}$, $\frac{2}{5}$ and $\frac{1}{10}$ respectively. What is the probability that (a) the motorist completes his journey in under 3 hours, (b) if he fails to arrive in less than 3 hours, there was a fall of snow?

(a) On the assumption that events D, R and S are mutually exclusive, the following tree diagram can be constructed.

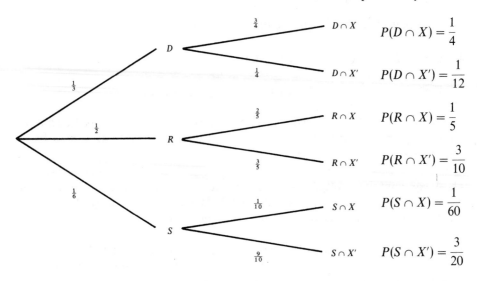

$$P(X) = P(D \cap X) + P(R \cap X) + P(S \cap X) = \frac{1}{4} + \frac{1}{5} + \frac{1}{60} = \frac{7}{15}$$

∴   the probability that the journey is completed in less than 3 hours is 7/15.

(b) $P(S|X') = \dfrac{P(S \cap X')}{P(X')} = \dfrac{P(S \cap X')}{1 - P(X)} = \dfrac{\frac{3}{20}}{1 - \frac{7}{15}} = \dfrac{3}{20} \cdot \dfrac{15}{8} = \dfrac{9}{32}$.

∴   given that the motorist failed to arrive in less than three hours, the probability that it snowed is 9/32.

	D	R	S
X	$\frac{1}{4}$	$\frac{1}{5}$	$\frac{1}{60}$
X'	$\frac{1}{12}$	$\frac{3}{10}$	$\frac{3}{20}$

It is sometimes helpful to display probabilities on a Karnaugh map. We give here the probability map for Example 2. This shows clearly that

$$P(X) = \frac{1}{4} + \frac{1}{5} + \frac{1}{60} = \frac{7}{15}.$$

When finding $P(S|X')$ we consider X' as sample space. Thus, using the lower half of the diagram only,

$$P(S|X') = \frac{3}{20} \Big/ \left(\frac{1}{12} + \frac{3}{10} + \frac{3}{20} \right) = \frac{3}{20} \Big/ \frac{32}{60} = \frac{9}{32}.$$

Exercise 15.6

Questions 1 to 5 concern a trial in which a domino is drawn from a complete set of 28. Event A: it has at least one six on it. Event B: it is a double, i.e. has two identical numbers on it. Event C: the sum of the numbers is six or less.

1.   Find $P(A)$, $P(B)$, $P(C)$.

2.   Find $P(B|A)$, $P(C|A)$, $P(A \cap B)$, $P(A \cap C)$.

3. Find $P(B|C)$, $P(C|B)$, $P(B|C')$, $P(C|B')$.

4. Find $P(B \cap C)$, $P(B \cap C')$, $P(B' \cap C)$.

5. Find $P(A \cap C')$, $P(A' \cap B')$, $P(A' \cap C')$.

Questions 6 to 10 Repeat questions 1 to 5 for a trial in which a card is drawn from a normal pack. Event A: drawing a king or a queen. Event B: drawing a black card. Event C: drawing a heart.

11. A bag contains 4 black discs and 2 white ones. If discs are drawn from the bag one at a time, find the probability that 2 black discs followed by 2 white discs are drawn (a) with replacement, (b) without replacement.

12. A bag contains 10 cards bearing the letters of the word STATISTICS. Find the probability that the first four letters drawn at random from the bag are S, I, T, S in that order (a) with replacement, (b) without replacement.

13. The probability that it will be foggy on a November morning is $\frac{1}{3}$. The probability that Mr. Jones will be late for work when it is foggy is $\frac{1}{2}$. The probability that he will be late if it is not foggy is $\frac{1}{8}$. If on a particular November morning Mr. Jones is late, find the probability that it is foggy.

14. On a day on the summer holidays the probability that Jane will go swimming is 0·6. The probability that both Jane and Karen will go swimming is 0·45. If Jane is seen in the swimming pool what is the probability that Karen also swims that day?

15. The probability that a light bulb has a life of more than 1000 hours is $\frac{1}{2}$ and the probability of a life of more than 2000 hours is $\frac{1}{10}$. What is the probability that a bulb that has already been in use for 1000 hours will last a further 1000 hours?

16. A football star is having injury problems. When he is playing the probability that his team will win is $\frac{3}{4}$, but otherwise it is only $\frac{1}{2}$. The probability that the player will be fit this Saturday is $\frac{1}{3}$. Find the probability that his team will win the match.

17. The probability that it will be sunny tomorrow is 0·25. The probability that Mrs. Brown will go shopping tomorrow is 0·2. The probability that both these events occur is 0·15. Find the probability that (a) neither event occurs, (b) Mrs. Brown goes shopping, given that it is sunny, (c) it is sunny given that Mrs. Brown goes shopping.

18. A bag contains 5 black beads and 3 white beads. A second bag contains 3 black beads and 5 white beads. A bead is drawn at random from the first bag and placed in the second bag. A bead is then drawn at random from the second bag

and placed in the first. Find the probability that each bag now contains (a) 4 black and 4 white beads, (b) the same numbers of each colour as it did initially.

19. A box *A* contains 3 red balls and 4 black balls. A box *B* contains 3 red balls and 2 black balls. One box is selected at random and then from that box one ball is selected at random. Find (a) the probability that the ball is red, (b) the probability that the ball came from box *A*, given that it is red.

20. A member of a slimming club claims that the probability that she will have a chocolate biscuit with her morning coffee is $\frac{1}{6}$ and that on the days that she resists the temptation, the probability that she exceeds her daily calorie allowance is $\frac{1}{5}$. If the probability that she consumes too many calories on any day is $\frac{1}{4}$, find the probability that she does so on a day when she has had a chocolate biscuit. Given that yesterday she kept within her calorie allowance, find the probability that she ate a chocolate biscuit.

Exercise 15.7 (miscellaneous)

1.   How many even five-digit numbers can be formed using the digits 0, 1, 2, 3, 4, 5, 6 (excluding numbers which start with 0) if (a) no digit may be repeated, (b) repetition is allowed?

2.   Find the number of different ways in which the letters of the word ISOSCELES can be arranged. How many of these arrangements (a) begin and end with E, (b) begin with S, (c) end with a vowel?

3.   A man has found 6 records that he likes in a shop, but cannot decide which of them to buy. Assuming that he buys at least one, find the number of different selections he could make.

4.   Six lines are drawn, no two of which are parallel. If no more than two of the lines pass through any one point, find the number of triangles formed.

5.   In how many ways can a group of 9 people attending a conference be split into three sets of 3 people for discussion? Later the same people form new sets of 3. In how many ways can this be done if no two people remain together?

6.   Find the number of ways in which 5 books can be distributed between three people *A*, *B* and *C* if the books are (a) all different, (b) indistinguishable.

7.   Draw a Venn diagram to show the following sets and their non-empty intersections: $\mathscr{E} = \{$all quadrilaterals$\}$, $T = \{$trapeziums$\}$, $R = \{$rectangles$\}$, $P = \{$parallelograms$\}$, $K = \{$kites, i.e. two pairs of adjacent sides equal$\}$. Describe the elements of the sets $K \cap R$ and $K \cap P$.

8.   Decide whether the following statements are true or false for sets *A*, *B* and *C*, giving brief reasons for your answers.

(a) $A \cap B = A \cup B \Leftrightarrow A = B$,
(b) $A \cup B = A \cup C \Leftrightarrow B = C$,
(c) $A \cap B = A \cap C \Leftrightarrow A = B \cap C$,
(d) $A \cap (B' \cup C') = \varnothing \Leftrightarrow A \subset (B \cup C)$.

9.   In a certain school half the pupils under 16 years of age are girls, two thirds of the girls in the school are under 16 and a quarter of the pupils are boys of 16 and over. If there are 540 girls in the school, find the total number of pupils.

10. A drawer contains 4 different pairs of socks. Find the probability that (a) if 2 socks are selected at random they will form a pair, (b) if 4 socks are selected at random they will form two pairs.

11. (a) Two dice are thrown together, and the scores added. What is the probability that (i) the total score exceeds 8, (ii) the total score is 9, or the individual scores differ by 1, or both?
(b) A bag contains 3 red balls and 4 black ones. 3 balls are picked out, one at a time and not replaced. What is the probability that there will be 2 red and 1 black in the sample?
(c) A committee of 4 is to be chosen from 6 men and 5 women. One particular man and one particular woman refuse to serve if the other person is on the committee. How many different committees may be formed?   (SU)

12. An analysis of the other subjects taken by A-level Mathematics candidates in a certain year showed that 20% of them took Further Mathematics, 50% took Physics and 5% took both Further Mathematics and Physics. A candidate is chosen at random from those who took A-level Mathematics. (i) Calculate the probability that the chosen candidate took neither Further Mathematics nor Physics. (ii) Given that the chosen candidate took at least one of Further Mathematics and Physics, calculate the probability that the candidate took Further Mathematics.   (JMB)

13. Two cards are drawn simultaneously from a pack of 52. What is the probability that both are spades if one card is known to be (a) black, (b) red, (c) a spade, (d) a king?

14. An experiment consists of tossing two fair coins one after the other. Let X be the set of events in which the first coin is a head, Y be the set of events in which the second coin is a head and Z be the set of events in which both coins are heads or both coins are tails. Calculate the following probabilities: $p(X)$, $p(Y)$, $p(Z)$, $p(X \cap Z)$, $p(X \cap Y \cap Z)$.   (AEB 1977)

15. Three cards are drawn at random without replacement from a pack of ten cards which are numbered from 1 to 10, respectively. Calculate (i) the probability that the numbers drawn consist of two even numbers and one odd number, (ii) the probability that at least one of the numbers drawn is a perfect square greater than 1, (iii) the probability that the smallest number drawn is the 5.   (JMB)

16. A disc of diameter 5 cm is tossed at random on to a large sheet of paper ruled with parallel lines 8 cm apart. Assuming the lines to be of negligible thickness, find the probability that the disc lands on a line. If the paper is now ruled with 8 cm squares, find the probability that (a) the disc lands on a line, (b) the disc lands on the intersection of two lines.

17. A pack of 52 cards contains 4 suits each of 13 cards. If 13 cards are taken at random from the pack what is the probability that exactly 10 of them are spades? [You may take $\binom{52}{13} = 6\cdot35 \times 10^{11}$.]   (O&C)

18. A bag contains 5 red, 4 orange and 3 yellow sweets. One after another, three children select and eat one sweet each. When the bag contains n sweets, the probability of any one child choosing any particular sweet is $1/n$. What are the probabilities that (a) they all choose red sweets, (b) at least one orange sweet is chosen, (c) each chooses a different colour, (d) all choose the same colour? [Answers to this question may be left as fractions in their lowest terms.]   (O&C)

19. (a) Two independent events are such that there is a probability of $\frac{1}{6}$ that they will both occur and a probability of $\frac{1}{3}$ that neither will occur. Calculate their individual probabilities of occurring.
(b) A fair cubical die has three of its faces coloured red, two coloured blue and one coloured white. If the die is thrown six times calculate the probabilities that (i) a red face will be uppermost at least once, (ii) a red face will be uppermost exactly three times, (iii) each colour will be uppermost exactly twice.   (W)

20. Four cards are drawn at random from a pack, one at a time with replacement. Find the probability that (a) no heart is drawn, (b) four hearts are drawn, (c) two hearts and two diamonds are drawn (in any order), (d) one card from each suit is drawn.   (L)

21. Two events A and B are such that $P(A) = 0\cdot4$ and $P(A \cup B) = 0\cdot7$. (i) Find the value of $P(A' \cap B)$. (ii) Find the value of $P(B)$ if A and B are mutually exclusive. (iii) Find the value of $P(B)$ if A and B are independent.   (JMB)

22. The following are three of the classical problems in probability.
(a) Compare the probability of a total of 9 with the probability of a total of 10 when three fair dice are tossed once (Galileo and Duke of Tuscany).
(b) Compare the probability of at least one six in 4 tosses of a fair die with the probability of at least one double-six in 24 tosses of two fair dice (Chevalier de Méré).
(c) Compare the probability of at least one six when 6 dice are rolled with the probability of at least two sixes when 12 dice are rolled (Pepys to Newton). Solve each of these problems.   (AEB 1978)

23. (a) From an ordinary pack of 52 cards two are dealt face downwards on a table. What is the probability that (i) the first card dealt is a heart, (ii) the second card dealt is a heart, (iii) both cards are hearts, (iv) at least one card is a heart?

(b) Bag A contains 3 white counters and 2 black counters whilst bag B contains 2 white and 3 black. One counter is removed from bag A and placed in bag B without its colour being seen. What is the probability that a counter removed from bag B will be white?

(c) A box of 24 eggs is known to contain 4 old and 20 new. If 3 eggs are picked at random determine the probability that (i) 2 are new and the other old, (ii) they are all new.   (SU)

24. A hand of 5 cards is drawn from a pack of 52 playing cards. Find the probability of drawing (a) 5 cards of the same suit, (b) a pair plus a triple (e.g. 2 sevens and 3 aces). If a hand consists of 2 aces, a king, a jack and a 4 and a player keeps the pair of aces but discards the other 3 cards and draws 3 more cards from the remaining 47, find the probability of obtaining a pair plus a triple.   (L)

25. If A, B and C are exhaustive events, such that A cannot occur at the same time as either B or C, show that
(a) $P(A)+P(B)+P(C) = 1+P(B \cap C)$,   (b) $P(A') = P(B \cup C)$.

26. Two cards are drawn without replacement from a pack of playing cards. Using a tree diagram, or otherwise, calculate the probability (a) that both cards are aces, (b) that one (and only one) card is an ace, (c) that the two cards are of different suits. Given that at least one ace is drawn, find the probability that the two cards are of different suits.   (L)

27. (a) Two events A and B are such that $P(A) = \frac{1}{3}$ and $P(B) = \frac{1}{2}$. If A' denotes the complement of A, calculate $P(A' \cap B)$ in each of the cases when (i) $P(A \cap B) = \frac{1}{8}$, (ii) A and B are mutually exclusive, (iii) A is a subset of B.
(b) A Scottish court may give any one of the three verdicts: 'guilty', 'not guilty' and 'not proven'. Of all the cases tried by the court, 70% of the verdicts are 'guilty', 20% are 'not guilty' and 10% are 'not proven'. Suppose that when the court's verdict is 'guilty', 'not guilty' and 'not proven', the probabilities that the accused is really innocent are 0·05, 0·95 and 0·25, respectively. Calculate the probability that an innocent person will be found 'guilty' by the court.   (W)

28. Suppose that letters sent by first and second class post have probabilities of being delivered a given number of days after posting according to the following table (weekends are ignored).

Days to delivery	1	2	3
1st class	0·9	0·1	0
2nd class	0·1	0·6	0·3

The secretary of a committee posts a letter to a committee member who replies immediately using the same class of post. What is the probability that four or more days are taken from the secretary posting the letter to receiving the reply if (a) first class, (b) second class post is used? The secretary sends out four letters and each member replies immediately by the same class of post. Assuming the letters move independently, what is the probability that the secretary receives (a) all the replies within three days using first class post, (b) at least two replies within three days using second class post?   (O)

29. In a game, three cubical dice are thrown by a player who attempts to throw the same number on all three. What is the chance of the player (a) throwing the same number on all three, (b) throwing the same number on just two? If the first throw results in just two dice showing the same number, then the third is thrown again. If no two dice show the same number, then all are thrown again. The player then comes to the end of his turn. What is the chance of the player succeeding in throwing three identical numbers in a complete turn? What is the chance that all the numbers are different at the end of a turn?   (O&C)

30. An experiment is performed with a die and two packs of cards. The die is thrown, and if it shows 1, 2, 3 or 4 a card is drawn at random from the first pack, which contains the usual 52 cards; if the score on the die is 5 or 6 a card is drawn from the second pack, which contains only 39 cards, all the clubs having been removed. X denotes the event 'The first pack with 52 cards is used', and Y denotes the event 'The card drawn is a diamond'. Calculate the probabilities (i) $P(X)$, (ii) $P(X \cap Y)$, (iii) $P(Y)$, (iv) $P(Y|X)$, (v) $P(X|Y)$.   (C)

31. Each of two boxes contains ten discs. In one box four of the discs are red, two are white and four are blue; in the other box two are red, three are white and five are blue. One of these two boxes is chosen at random and three discs are drawn at random from it without replacement. Calculate (i) the probability that one disc of each colour will be drawn, (ii) the probability that no white disc will be drawn, (iii) the most probable number of white discs that will be drawn. Given that three blue discs were drawn, calculate the conditional probability that they came from the box that contained four blue discs.   (W)

32. Write down an equation, in terms of probabilities, corresponding to each of the statements (i) the events A and B are independent, (ii) the events A and C are mutually exclusive. The events A, B and C are such that A and B are independent and A and C are mutually exclusive. Given that $P(A) = 0.4$, $P(B) = 0.2$, $P(C) = 0.3$, $P(B \cap C) = 0.1$, calculate $P(A \cup B)$, $P(C|B)$, $P(B|A \cup C)$. Also calculate the probability that one and only one of the events B, C will occur.   (W)

33. A pack of sixteen playing cards consists of the ace, king, queen and jack of each of the four suits, spades, hearts, diamonds and clubs. A man is dealt four cards at random from the pack. (i) Calculate the probability that he has been dealt exactly two aces. (ii) The man is asked whether he has been dealt any aces. He truthfully replies that he has. Calculate the probability that he has been dealt exactly two aces. (iii) By accident the man displays one of his four cards, which is seen to be the ace of spades. Calculate the probability that he has been dealt exactly two aces. Give your answers correct to two significant figures.   (C)

16 Series and the binomial theorem

16.1 Sequences and series

Below are some sets of numbers given in a definite order. In each case the numbers are produced according to some simple rule.

(a) $2, 4, 6, 8, 10, \ldots$

(b) $1, 4, 9, 16, 25, \ldots$

(c) $1, -1, 1, -1, 1, \ldots$

(d) $8, -4, 2, -1, \frac{1}{2}, \ldots$

(e) $10, 11, 9, 12, 8, 13, \ldots$

(f) $1, 1, 2, 3, 5, 8, \ldots$

These sets of numbers are called *sequences*. Each member of a sequence is a *term*. For instance, in (a) 6 is the third term of the sequence.

One way of stating the rule by which a sequence is obtained is to write down a formula for the *nth term*, often denoted by u_n. For sequences (a), (b) and (c) given above we may write,

(a) $u_n = 2n$ and $u_6 = 12, u_7 = 14$,

(b) $u_n = n^2$ and $u_6 = 36, u_7 = 49$,

(c) $u_n = (-1)^{n+1}$ and $u_6 = -1, u_7 = 1$.

In some cases the formula for u_n is not obvious. In (d) to obtain each successive term we divide by 2, then change the sign. Thus, at each stage we are multiplying by $-\frac{1}{2}$, which means that

$$u_n = 8 . (-\tfrac{1}{2})^{n-1}, \text{ so that } u_6 = 8 . (-\tfrac{1}{2})^5 = -\frac{8}{2^5} = -\tfrac{1}{4}.$$

In (e) a general formula for u_n would be difficult to find, but we can give separate formulae for odd and even terms of the sequence.

$u_{2n+1} = 10 - n$ and for $n = 3, u_7 = 7$,

$u_{2n} \quad = 10 + n$ and for $n = 4, u_8 = 14$.

In (f) each term is the sum of the two previous terms. This can be expressed formally as a relation between u_n and u_{n-1}, u_{n-2}:

$$u_n = u_{n-1} + u_{n-2} \qquad (n \geqslant 3).$$

In this case more advanced techniques are needed to find a formula for u_n in terms of n.

If a sequence ends after a certain number of terms it is said to be *finite*. A sequence which continues indefinitely is said to be *infinite*. The sequence $1, 3, 5, \ldots, 2n-1$ is finite with n terms, whereas $1, 3, 5, \ldots, 2n-1, \ldots$ denotes an infinite sequence.

A sum of a sequence of numbers is called a *series*. $1+3+5+ \ldots +97+99$ is a finite series and $1+3+5+ \ldots$ indicates an infinite series. The nth term of both these series is $u_n = 2n-1$, where $1 \leqslant n \leqslant 50$ in the finite case.

S_n is used to denote the sum of the first n terms of a series, so that $S_n = u_1 + u_2 + \ldots + u_n$.

Example 1 Find S_1, S_5 and S_6 for the series $1 + \dfrac{1}{2} + \dfrac{1}{4} + \dfrac{1}{8} + \ldots$.

S_1 is simply the first term $\quad \therefore \quad S_1 = 1$

$$S_5 = 1 + \frac{1}{2} + \frac{1}{4} + \frac{1}{8} + \frac{1}{16} \quad \therefore \quad S_5 = 1\frac{15}{16}$$

$$S_6 = S_5 + \frac{1}{32} = 1\frac{15}{16} + \frac{1}{32} \quad \therefore \quad S_6 = 1\frac{31}{32}.$$

Example 2 The sum of the first n terms of a series is given by $S_n = n^2 + n$. Write down the first three terms of the series and find an expression for the nth term, u_n.

$$S_1 = 1^2 + 1 = 2, \quad S_2 = 2^2 + 2 = 6, \quad S_3 = 3^2 + 3 = 12,$$

$\therefore \quad$ the first three terms of the series must be $2 + 4 + 6 + \ldots$.

$$S_n = n^2 + n, \quad S_{n-1} = (n-1)^2 + (n-1) = n^2 - n$$

$$\therefore \quad u_n = S_n - S_{n-1} = n^2 + n - (n^2 - n) = 2n.$$

Hence the nth term of the series is $2n$.

Exercise 16.1

In questions 1 to 10 write down the next two terms of the given sequence and an expression for the nth term.

1. $5, 10, 15, 20, \ldots$

2. $4, 7, 10, 13, \ldots$

3. $1, \dfrac{1}{2}, \dfrac{1}{3}, \dfrac{1}{4}, \ldots$

4. $\dfrac{1}{2}, \dfrac{2}{3}, \dfrac{3}{4}, \dfrac{4}{5}, \ldots$

5. $1, 2, 4, 8, \ldots$

6. $0, 3, 8, 15, 24, \ldots$

7. $\dfrac{1}{12}, \dfrac{1}{6}, \dfrac{1}{3}, \dfrac{2}{3}, \ldots$

8. $\dfrac{1}{2}, \dfrac{1}{6}, \dfrac{1}{12}, \dfrac{1}{20}, \ldots$

9. $1, -2, 3, -4, \ldots$ 10. $11, 8, 13, 6, 15, \ldots$

In questions 11 to 16 find the first three terms and the nth term of the series with the given sum to n terms.

11. $S_n = 4 + 7n^2$.   12. $S_n = 3n - 7n^3$.

13. $S_n = 2^n$.   14. $S_n = \dfrac{1}{n}$.

15. $S_n = \dfrac{1}{4}n^2(n+1)^2$.   16. $S_n = \dfrac{1}{6}n(n+1)(2n+1)$.

17. Given that $1^3 + 2^3 + 3^3 + \ldots + n^3 = \dfrac{1}{4}n^2(n+1)^2$, find the sum of the first 20 terms of the series $2 + 16 + 54 + 128 + 250 + \ldots$.

18. Given that $1^2 + 2^2 + 3^2 + \ldots + n^2 = \dfrac{1}{6}n(n+1)(2n+1)$, find the sum of the series $2^2 + 4^2 + 6^2 + \ldots + 50^2$. Use your result to find the sum of the series $1^2 + 3^2 + 5^2 + \ldots + 49^2$.

16.2 Arithmetic progressions

An *arithmetic progression* is a series in which one term is obtained from the previous term by adding a fixed number.
For example: (a) $1 + 2 + 3 + 4 + \ldots + 98 + 99$,
                (b) $6 + 10 + 14 + 18 + \ldots + 46 + 50$,
                (c) $10 + 7 + 4 + 1 + \ldots - 47 - 50$.
This fixed number is called the *common difference*. In the above examples the common differences are 1, 4 and -3 respectively.

An arithmetic progression is completely defined when the first term a and the common difference d are given:

$$a + (a+d) + (a+2d) + \ldots .$$

Example 1   Write down the first three terms, the 10th term and the nth term of the A.P. (arithmetic progression) with first term -20 and common difference 3.

The first three terms are -20, -17 and -14.
The 10th term, $u_{10} = -20 + (3 \times 9) = 7$.
The nth term, $u_n = -20 + 3(n-1) = 3n - 23$.

More generally, the nth term of the A.P. with first term a and common difference d is $\boxed{u_n = a + (n-1)d.}$

To illustrate the general approach to the summation of arithmetic progressions, we consider the following:

$$1+ \quad 2+ \quad 3+ \ldots + \quad 98+ \quad 99$$
$$99+ \quad 98+ \quad 97+ \ldots + \quad 2+ \quad 1$$

$$\overline{}$$

$$100+100+100+\ldots+100+100$$

Since there are 99 columns,

$$2(1+2+3+\ldots+99) = 99 \times 100.$$

Hence $1+2+3+\ldots+99 = 4950.$

Applying this method to an A.P. with n terms, first term a, last term l and common difference d:

$$S_n = a+(a+d)+\ldots+(l-d)+l$$
$$S_n = l+(l-d)+\ldots+(a+d)+a$$
$$\therefore \quad 2S_n = (a+l)+(a+l)+\ldots+(a+l)+(a+l) = n(a+l)$$

$$\therefore \quad \text{the sum of the first } n \text{ terms,} \boxed{S_n = \tfrac{1}{2}n(a+l).}$$

As l is the nth term of the series, $l = a+(n-1)d$

$$\therefore \quad S_n = \tfrac{1}{2}n\{a+a+(n-1)d\} = \tfrac{1}{2}n\{2a+(n-1)d\}.$$

> Hence the sum of the first n terms of an arithmetic progression with first term a and common difference d is $\tfrac{1}{2}n\{2a+(n-1)d\}$.

Example 2   Find the sums of the series
(a) $6+10+14+\ldots+50$,   (b) $10+7+4+\ldots-50$.

(a) The series has first term 6, common difference 4

$$\therefore \quad \text{the number of terms} = \frac{50-6}{4}+1 = 12.$$

Hence the sum of the series $= \tfrac{1}{2}.12(6+50) = 336.$
(b) The series has first term 10, common difference -3

$$\therefore \quad \text{the number of terms} = \frac{-50-10}{-3}+1 = 21.$$

Hence the sum of the series $= \tfrac{1}{2}.21(10-50) = -420.$

Example 3   Find the sum of the first 20 terms of the A.P. with first term 3 and common difference $\tfrac{1}{2}$.

$$S_{20} = \tfrac{1}{2}.20(2.3+19.\tfrac{1}{2}) = 10\left(6 + \frac{19}{2}\right) = 60+95 = 155$$

$$\therefore \quad \text{the sum of the first twenty terms is 155.}$$

Example 4   In an A.P. the sum of the first 10 terms is 520 and the 7th term is double the 3rd term. Find the first term a and the common difference d.

$$S_{10} = \tfrac{1}{2} \cdot 10(2a+9d) = 520 \qquad \therefore \quad 2a+9d = 104. \tag{1}$$

The 7th term, $u_7 = a+6d$, and the 3rd term, $u_3 = a+2d$

$$\therefore \quad a+6d = 2(a+2d), \quad \text{i.e.} \quad 2d = a \tag{2}$$

Substituting in (1) $13d = 104 \qquad \therefore \quad d = 8$
Substituting in (2) $a = 16.$

Hence the first term of the A.P. is 16 and the common difference is 8.

If three numbers a, b and c are in arithmetic progression, then b is called the *arithmetic mean* of a and c.

The common difference $= b-a = c-b$, $\therefore$ $b = \tfrac{1}{2}(a+c)$. Thus the arithmetic mean of any two numbers, p and q, is the 'average', $\tfrac{1}{2}(p+q)$.

Exercise 16.2

1.   Write down the stated term and the nth term of the following A.P.s
(a) $7+11+15+ \ldots$ (7th),
(b) $18+11+4+ \ldots$ (6th),
(c) $-7-5-3- \ldots$ (23rd),
(d) $3+3\tfrac{1}{2}+4+ \ldots$ (16th).

2.   Find the sums of the following series
(a) $5+9+13+ \ldots +101.$
(b) $83+80+77+ \ldots +5,$
(c) $-17-12-7- \ldots +33,$
(d) $1+1\tfrac{1}{4}+1\tfrac{1}{2}+ \ldots +9\tfrac{3}{4}.$

3.   Find the sums of the following A.P.s
(a) $4+11+ \ldots$ to 16 terms,
(b) $3+8\tfrac{1}{2}+ \ldots$ to 20 terms,
(c) $19+13+ \ldots$ to 10 terms,
(d) $-9-1+ \ldots$ to 8 terms.

4.   Find the sum of the A.P. $-7-3+1+ \ldots$ from the seventh to the thirtieth term inclusive.

5.   Find the sum of all odd numbers between 0 and 500 which are divisible by 7.

6.   Show that the sum $1+3+5+ \ldots +(2n-1)$ is always a perfect square.

7.   The first and last terms of an A.P. with 25 terms are 29 and 179. Find the sum of the series and its common difference.

8.   The rth term of a series is $10-3r$. Find the first three terms of the series and the sum of the first 18 terms.

9.   Given that the first and third terms of an A.P. are 13 and 25 respectively, find the 100th term and the sum of the first 15 terms.

10. A piece of string of length 5 m is cut into n pieces in such a way that the lengths of the pieces are in arithmetic progression. If the lengths of the longest and the shortest pieces are 1 m and 25 cm respectively, calculate n.

11. The second and seventh terms of an A.P. are -5 and 10 respectively. Find the fifth term and the least number of terms that must be taken for their sum to exceed 200.

12. The tenth term of an A.P. is 10 and the sum of the first 10 terms is -35. Find the first term and the common difference of the progression.

13. The sum of the first four terms of an A.P. is twice the fifth term. Show that the common difference is equal to the first term.

14. In an A.P. the sum of the first 15 terms is 615 and the 13th term is six times the 2nd term. Find the first three terms.

15. Find the arithmetic mean of

(a) 3 and 27,   (b) 3 and -27,   (c) $\dfrac{1}{3}$ and $\dfrac{1}{27}$,   (d) lg 3 and lg 27.

16. Three numbers in A.P. have sum 33 and product 1232. Find the numbers.

17. The sum of three numbers in A.P. is 30 and the sum of their squares is 398. Find the numbers.

18. The sum of the first n terms of a certain series is $3n^2 + n$. Show that the series is an A.P. and find the first term and the common difference.

19. Show that the sum to 20 terms of the series

$$\log a + \log (ab) + \log (ab^2) + \log (ab^3) + \ldots$$

can be written in the form $\log (a^x b^y)$ and find the values of x and y.

20. In an A.P. the sum of the first $2n$ terms is equal to the sum of the next n terms. If the first term is 12 and the common difference is 3, find the non-zero value of n.

16.3 Geometric progressions

A *geometric progression* (G.P.) is a series in which any term is obtained from the previous term by multiplying by a fixed number.

For example: (a) $1 + 2 + 4 + 8 + \ldots + 128 + 256$,

(b) $27 - 9 + 3 - 1 + \ldots + \dfrac{1}{27} - \dfrac{1}{81}$.

This fixed number is called the *common ratio*. In the above examples the common ratios are 2 and $-\frac{1}{3}$ respectively.

A geometric progression is completely defined when the first term a and the common ratio r are given:

$$a + ar + ar^2 + ar^3 + \ldots.$$

The nth term of this G.P. is $\boxed{u_n = ar^{n-1}.}$

Example 1   Write down the formula for the nth term and find the number of terms in series (a) and (b) above.

(a) The nth term $= 1 \times 2^{n-1} = 2^{n-1}$.

   If $2^{n-1} = 256 = 2^8$, then $n = 9$

   $\therefore$ the series has 9 terms.

(b) The nth term $= 27 \times (-\frac{1}{3})^{n-1}$

   If $27 \times (-\frac{1}{3})^{n-1} = -\dfrac{1}{81}$, then $(-\frac{1}{3})^{n-1} = -\dfrac{1}{3^7} = (-\frac{1}{3})^7$

   $\therefore$ $n = 8$ and the series must have 8 terms.

We illustrate the method by which G.P.s are summed by considering the sum S of series (a).

$$S = 1 + 2 + 4 + \ldots + 256$$
$$2S = \quad 2 + 4 + 8 + \ldots \quad + 512.$$

Subtracting:
$$2S - S = 512 - 1$$
$$\therefore \quad S = 511.$$

Applying this method to a G.P. with n terms, first term a and common ratio r, we have

$$S_n = a + ar + ar^2 + \ldots + ar^{n-1}$$
$$rS_n = \quad ar + ar^2 + \ldots + ar^{n-1} + ar^n.$$

Subtracting:
$$S_n - rS_n = a - ar^n$$
$$S_n(1-r) = a(1-r^n)$$

$$\therefore \quad S_n = \frac{a(1-r^n)}{1-r} \qquad (r \neq 1)$$

$\therefore$ the sum of the first n terms of a geometric progression with first term a and common ratio r is given by

$$\boxed{S_n = \frac{a(1-r^n)}{1-r} \quad \text{or} \quad S_n = \frac{a(r^n-1)}{r-1}.}$$

[The second expression is more convenient if $r > 1$.]

Example 2   Find the sum of the first 6 terms of the series
(a) $2-6+18-\ldots$ (b) $14+7+3\frac{1}{2}+\ldots$

(a) The series is a G.P. with first term 2 and common ratio -3

$$\therefore \quad \text{the sum} = \frac{2\{1-(-3)^6\}}{1-(-3)} = \frac{2(1-729)}{4} = -364.$$

(b) The series is a G.P. with first term 14 and common ratio $\frac{1}{2}$

$$\therefore \quad \text{the sum} = \frac{14\{1-(\frac{1}{2})^6\}}{1-\frac{1}{2}} = \frac{14(1-1/64)}{\frac{1}{2}} = 14 \, . \, 2 \, . \frac{63}{64} = \frac{441}{16}.$$

Example 3   A G.P. has first term 10 and common ratio 1·5. How many terms of the series are needed to reach a sum greater than 200?

$$\text{The sum of } n \text{ terms} = \frac{10\{(1\cdot5)^n - 1\}}{1\cdot5 - 1} = 20\{(1\cdot5)^n - 1\}$$

$$
\begin{aligned}
20\{(1\cdot5)^n - 1\} > 200 &\Rightarrow (1\cdot5)^n - 1 > 10 \\
&\Rightarrow \quad\quad (1\cdot5)^n > 11 \\
&\Rightarrow \quad \lg(1\cdot5)^n > \lg 11 \\
&\Rightarrow \quad\quad n\lg 1\cdot5 > \lg 11 \\
&\Rightarrow \quad\quad\quad n > \frac{\lg 11}{\lg 1\cdot5} \approx 5\cdot9.
\end{aligned}
$$

Hence 6 terms are needed to reach a sum greater than 200.

If three positive numbers a, b and c are in geometric progression, then b is called the *geometric mean* of a and c.

$$\text{The common ratio} = \frac{b}{a} = \frac{c}{b} \quad \therefore \quad b^2 = ac.$$

Hence the geometric mean of two numbers, p and q, is $\sqrt{(pq)}$.

Exercise 16.3

1.   Write down the stated term and the *n*th term of the following G.P.s
(a) $\frac{1}{2}+1+2+\ldots$ (8th),   (b) $162+54+18+\ldots$ (6th),

(c) $200-50+12\frac{1}{2}-\ldots$ (5th),   (d) $-\frac{4}{9}-\frac{2}{3}-1-\ldots$ (7th).

2.   Find the number of terms in each of the following G.P.s and the sum of the series.
(a) $\frac{1}{4}+\frac{1}{2}+\ldots+64$,   (b) $\frac{1}{4}-\frac{1}{2}+\ldots+64$,

(c) $1000+200+\ldots+0\cdot32$,   (d) $2-3+\ldots+22\frac{25}{32}$.

3.   Find the sums of the following G.P.s
(a) $100+10+\ldots$ to 7 terms,   (b) $1-\frac{1}{3}+\ldots$ to 6 terms,
(c) $3-6+\ldots$ to n terms,   (d) $a^p+a^{p+3}+a^{p+6}+\ldots$ to k terms.

Put in Ex 16.4 here

4.   The first term of a G.P. with positive terms is 80. If the sum of the first three terms is 185, find the common ratio.

5.   Find two distinct numbers p and q such that p, q, 10 are in arithmetic progression and q, p, 10 are in geometric progression.

6.   Find the geometric mean of

(a) 3 and 27,   (b) $\dfrac{1}{3}$ and $\dfrac{1}{27}$,   (c) 10^3 and 10^{27}.

7.   Given that the geometric mean of the numbers $4x-3$ and $9x+4$ is $6x-1$, find the value of x.

8.   The second and fifth terms in a G.P. are 405 and -120 respectively. Find the seventh term and the sum of the first seven terms.

9.   In a G.P. the second term exceeds the first by 20 and the fourth term exceeds the second by 15. Find the two possible values of the first term.

10. If the sum of the first two terms of a G.P. is 162 and the sum of its first four terms is 180, find the sum of the first six terms. Find also the two possible values of the sixth term.

11. The sum of the first six terms of a G.P. is nine times the sum of the first three terms. Find the common ratio.

12. The sum of $(n+12)$ terms of the G.P. $2+4+8+\ldots$ is twice the sum of n terms of the G.P. $3+12+48+\ldots$. Calculate the value of n.

13. The sum of the first seven terms of a G.P. is 7 and the sum of the next seven terms is 896. Find the common ratio of the progression. If the kth term is the first term of the G.P. which is greater than 1, find k.

14. Find the sum of the first n terms of the G.P. $\dfrac{1}{12}+\dfrac{1}{4}+\dfrac{3}{4}+\ldots$. How many terms of the series are needed to reach a sum greater than 100?

15. A G.P. has first term 16 and common ratio $\frac{3}{4}$. If the sum of the first n terms is greater than 60, find the least possible value of n.

16.4 Infinite geometric series

Consider the infinite geometric progression (or geometric series)

$$1 + \frac{1}{2} + \frac{1}{4} + \frac{1}{8} + \ldots + \left(\frac{1}{2}\right)^{n-1} + \ldots .$$

The sum of the first n terms is

$$S_n = \frac{1\{1-(\frac{1}{2})^n\}}{1-\frac{1}{2}} = 2\{1-(\frac{1}{2})^n\} = 2 - \frac{1}{2^{n-1}}.$$

As n increases, $\dfrac{1}{2^{n-1}}$ approaches 0 and S_n takes values closer and closer to 2.

Thus as $n \to \infty$, $1/2^{n-1} \to 0$ and $S_n \to 2$.

Since S_n approaches a finite limit, as n increases, the infinite series is said to be *convergent* with sum 2.

More generally, for a geometric series with first term a and common ratio r,

$$S_n = \frac{a(1-r^n)}{1-r} = \frac{a}{1-r} - r^n \left(\frac{a}{1-r}\right).$$

The value of S_n, as n increases, depends on the value of r^n. If

$$|r| < 1, \quad \text{i.e. if} \quad -1 < r < 1, \quad \text{then as} \quad n \to \infty, r^n \to 0 \quad \text{and} \quad S_n \to a/(1-r).$$

In this case, we say that the series $a + ar + ar^2 + \ldots$ *converges* and its *sum to infinity*, denoted by S or S_∞, is $\dfrac{a}{1-r}$.

[If $r > 1$, then as $n \to \infty$, $r^n \to \infty$. If $r < -1$, then as $n \to \infty$, r^n oscillates between large positive and large negative values. If $r = 1$, then $S_n = na$. If $r = -1$, then either $S_n = a$ or $S_n = 0$. In none of these cases does S_n approach a finite limit and the series is said to be *divergent*.]

Hence, provided that $|r| < 1$, the sum to infinity of the geometric series $a + ar + ar^2 + \ldots$ is $\dfrac{a}{1-r}$.

Example 1   Find the sum to infinity of the series $18 - 6 + 2 - \ldots$.

This is a geometric series with first term 18 and common ratio $-\frac{1}{3}$

$$\therefore \quad \text{the sum to infinity} = \frac{18}{1-(-\frac{1}{3})} = 18 \Big/ \frac{4}{3} = 18 \cdot \frac{3}{4} = 13\frac{1}{2}.$$

Example 2   Express as a fraction in its lowest terms the recurring decimal $0\cdot 37\dot{0}$.

$$0\cdot 37\dot{0} = 0\cdot 370370370\ldots = 0\cdot 37 + 0\cdot 00037 + 0\cdot 00000037 + \ldots.$$

This is a geometric series with first term $0\cdot 37$ and common ratio $1/1000$, i.e. $0\cdot 001$

$$\therefore \quad 0\cdot 37\dot{0} = \frac{0\cdot 37}{1-0\cdot 001} = \frac{0\cdot 37}{0\cdot 999} = \frac{370}{999} = \frac{10}{27}.$$

Example 3   Find the sum of the infinite series $1+2x+4x^2+8x^3+\ldots$, stating for which values of x your result is valid.

This is a geometric series with first term 1 and common ratio $2x$.

$\therefore$ its sum is $1/(1-2x)$.

The result is valid for $|2x| < 1$, i.e. for $|x| < \frac{1}{2}$.

Exercise 16.4

1.   Find the sums to infinity of the following geometric series
(a) $6+2+\frac{2}{3}+\ldots$,   (b) $1-\frac{1}{2}+\frac{1}{4}-\ldots$,
(c) $10+1+0\cdot1+\ldots$,   (d) $45-30+20-\ldots$.

2.   Express as fractions in their lowest terms
(a) $0\cdot5\dot4$,   (b) $0\cdot0\dot7\dot2$,   (c) $0\cdot5\dot74\dot0$.

3.   A geometric series with common ratio $0\cdot8$ converges to the sum 250. Find the fourth term of the series.

4.   The first and fourth terms of a geometric series are 135 and -40 respectively. Find the common ratio of the series and its sum to infinity.

5.   The sum of the first n terms of a geometric series is $8-2^{3-2n}$. Find the first term of the series, its common ratio and its sum to infinity.

6.   Find the sums of the following infinite series, stating the values of x for which your results are valid.
(a) $1+3x+9x^2+\ldots$,   (b) $1-\frac{1}{2}x+\frac{1}{4}x^2-\ldots$,

(c) $2-4x+8x^2-\ldots$,   (d) $x+\frac{1}{3}x^2+\frac{1}{9}x^3+\ldots$.

7. A geometric series has first term 35 and common ratio 2^x. State the set of values of x for which the series is convergent. Find the value of x for which the sum to infinity of the series is 40.

8.   Find the sum to n terms of the geometric series $108+60+33\frac{1}{3}+\ldots$. If k is the least number which exceeds this sum for all values of n, find k. Find also the least value of n for which the sum exceeds 99% of k.

16.5 The Σ notation

As writing out a series can often be cumbersome, the Greek capital letter Σ (pronounced 'sigma') is used to mean 'the sum of'.

$\displaystyle\sum_{r=1}^{n} u_r \left(\text{or} \sum_{1}^{n} u_r\right)$ means 'the sum of all the terms u_r from $r = 1$ to $r = n$', i.e.

$$S_n = \sum_{r=1}^{n} u_r = u_1 + u_2 + u_3 + \ldots + u_n.$$

Example 1 Find $\displaystyle\sum_{r=3}^{6} r(r-1)$.

$$\sum_{r=3}^{6} r(r-1) = 3(3-1) + 4(4-1) + 5(5-1) + 6(6-1)$$

$$= 6 + 12 + 20 + 30 = 68.$$

Example 2 Write in Σ notation the series
(a) $2 + 5 + 10 + 17 + 26 + \ldots + 401$
(b) $3 + 2 + \dfrac{11}{7} + \dfrac{12}{9} + \dfrac{13}{11} + \ldots + \dfrac{20}{25}.$

(a) The rth term of the series, $u_r = r^2 + 1$.
  Since 401 is given by $r = 20$, the series has 20 terms.
  Hence in Σ notation the series is $\displaystyle\sum_{r=1}^{20} (r^2 + 1)$.

(b) The rth term, $u_r = \dfrac{r+8}{2r+1}$.

  The final term, $\dfrac{20}{25}$, is given by $r = 12$.

  Hence in Σ notation the series is $\displaystyle\sum_{r=1}^{12} \dfrac{r+8}{2r+1}.$

Some basic rules for manipulating expressions involving Σ can be established as follows:

$$\sum_{1}^{n} (ku_r) = ku_1 + ku_2 + \ldots + ku_n$$

$$= k(u_1 + u_2 + \ldots + u_n) = k\sum_{1}^{n} u_r.$$

In particular $\displaystyle\sum_{1}^{n} k = k + k + \ldots + k = kn.$

$$\sum_{1}^{n} (u_r + v_r) = (u_1 + v_1) + (u_2 + v_2) + \ldots + (u_n + v_n)$$

$$= (u_1 + u_2 + \ldots + u_n) + (v_1 + v_2 + \ldots + v_n)$$

$$= \sum_{1}^{n} u_r + \sum_{1}^{n} v_r.$$

Similarly
$$\sum_{1}^{n}(u_r - v_r) = \sum_{1}^{n}u_r - \sum_{1}^{n}v_r.$$

$$\left[\text{Note that, in general, } \sum_{1}^{n}u_rv_r = u_1v_1 + u_2v_2 + \ldots + u_nv_n \text{ is } \textit{not} \text{ equal to}\right.$$

$$\left.\left(\sum_{1}^{n}u_r\right)\left(\sum_{1}^{n}v_r\right).\right]$$

Exercise 16.5

1. Write in full and hence evaluate:

(a) $\sum_{r=1}^{4}(r^3 + 3r)$,

(b) $\sum_{r=10}^{12}(150 - r^2)$,

(c) $\sum_{r=2}^{7}(4r + 3)$,

(d) $\sum_{r=1}^{6}\dfrac{120}{r}$,

(e) $\sum_{r=1}^{6}\sin\dfrac{r\pi}{3}$,

(f) $\sum_{r=0}^{5}(-1)^r(1 + 2^{r+1})$.

2. Find expressions for the following series in the form $\sum_{1}^{n}f(r)$.

(a) $5 + 7 + 9 + \ldots + 27$,

(b) $1 + 8 + 27 + \ldots + 4096$

(c) $-2 + 3 - 4 + 5 - \ldots + 41$,

(d) $360 - 180 + 90 - \ldots + 5\dfrac{5}{8}$,

(e) $\dfrac{1}{2} + \dfrac{3}{4} + \dfrac{5}{6} + \ldots + \dfrac{19}{20}$,

(f) $\dfrac{1}{4} + \dfrac{4}{7} + \dfrac{9}{10} + \ldots + \dfrac{144}{37}$.

3. Evaluate the following without writing down the series in full.

(a) $\sum_{1}^{20}(4r + 5)$,

(b) $\sum_{1}^{16}(-1)^r(2r - 1)$,

(c) $\sum_{0}^{50}(25 - 2r)$

(d) $\sum_{1}^{9}5(-2)^{r-1}$,

(e) $\sum_{0}^{\infty}(\tfrac{1}{2})^r$,

(f) $\sum_{2}^{\infty}(0 \cdot 1)^r$.

4. By expressing r in terms of s, or otherwise, show that

(a) $\sum_{r=0}^{n-1}r(r+1) = \sum_{s=1}^{n}s(s-1)$,

(b) $\sum_{r=1}^{n}(3r+2) = \sum_{s=3}^{n+2}(3s-4)$.

5. Show that   (a) $\sum_{1}^{100}(2-r)(2+r) = 400 - \sum_{1}^{100}r^2$

(b) $\sum_{21}^{40}(r-5) = 300 + \sum_{1}^{20}r$.

6. Show that $1^3 + 3^3 + 5^3 + \ldots + 25^3 = \sum_{1}^{25}r^3 - 8\sum_{1}^{12}r^3$.

7.   Prove that $\sum_{1}^{n} (au_r + bv_r) = a\sum_{1}^{n} u_r + b\sum_{1}^{n} v_r.$

16.6 The binomial theorem

This section concerns powers of binomial (or 'two term') expressions such as $(a+b)$. By multiplication it is found that

$$
\begin{aligned}
(a+b)^0 &= 1\\
(a+b)^1 &= a+b\\
(a+b)^2 &= a^2 + 2ab + b^2\\
(a+b)^3 &= a^3 + 3a^2b + 3ab^2 + b^3\\
(a+b)^4 &= a^4 + 4a^3b + 6a^2b^2 + 4ab^3 + b^4\\
(a+b)^5 &= a^5 + 5a^4b + 10a^3b^2 + 10a^2b^3 + 5ab^4 + b^5
\end{aligned}
$$

These coefficients form an array known as Pascal's† triangle:

$$
\begin{array}{ccccccccccc}
&&&&&1\\
&&&&1&&1\\
&&&1&&2&&1\\
&&1&&3&&3&&1\\
&1&&4&&6&&4&&1\\
1&&5&&10&&10&&5&&1
\end{array}
$$

....

Each entry in the array is the sum of the two entries on either side of it in the previous line. This means that the triangle can easily be extended to provide the coefficients in the expansions of higher powers of $(a+b)$. To see why coefficients in successive lines are related in this way, we consider

$$(a+b)^6 = (a+b)(a^5 + 5a^4b + 10a^3b^2 + 10a^2b^3 + 5ab^4 + b^5)$$

In the product there are two terms in a^4b^2, $b.5a^4b = 5a^4b^2$ and $a.10a^3b^2 = 10a^4b^2$. Hence the coefficient of $a^4b^2 = 5 + 10 = 15$.

Although Pascal's triangle can always be used to obtain expansions of $(a+b)^n$, for large values of n the method becomes rather long. To find another approach, we consider the product

$$
\begin{aligned}
&(a_1 + b_1)(a_2 + b_2)(a_3 + b_3)\\
&= (a_1 + b_1)(a_2a_3 + a_2b_3 + b_2a_3 + b_2b_3)\\
&= a_1a_2a_3 + a_1a_2b_3 + a_1b_2a_3 + a_1b_2b_3 + b_1a_2a_3 + b_1a_2b_3 + b_1b_2a_3 + b_1b_2b_3.
\end{aligned}
$$

Writing $a_1 = a_2 = a_3 = a$ and $b_1 = b_2 = b_3 = b$, we see that the terms a^2b and ab^2 each appear three times, giving

$$(a+b)^3 = a^3 + 3a^2b + 3ab^2 + b^3.$$

† *Pascal, Blaise* (1623–1662) French theologian, mathematician and inventor. His *Traité du Triangle Arithmétique* is one of several essays on mathematical subjects. He also anticipated the invention of differential calculus and contributed to the foundation of the theory of probability.

Extending this method to the product

$$(a+b)^6 = (a+b)(a+b)(a+b)(a+b)(a+b)(a+b)$$

we find that each individual term in the expansion is the product of an 'a' or a 'b' from each of the six brackets. A term in a^4b^2 arises when 4 a's and 2 b's are selected. Since there are $_6C_2$ ways of selecting b's from 2 of the 6 brackets and a's from the rest, the term a^4b^2 must be produced $_6C_2$ times. (See §15.2.) Hence the coefficient of a^4b^2 is $_6C_2 = \dfrac{6!}{2!4!} = 15$, as shown above. This approach leads to the general result:

$$(a+b)^n = a^n + {_nC_1}a^{n-1}b + {_nC_2}a^{n-2}b^2 + \ldots + {_nC_{n-1}}ab^{n-1} + b^n$$

i.e. $(a+b)^n = a^n + na^{n-1}b + \dfrac{n(n-1)}{2!}a^{n-2}b^2 + \ldots + nab^{n-1} + b^n$. This is called the *binomial theorem*.

A general term in the expansion, such as the term in $a^{n-r}b^r$, takes the form:

$${_nC_r}a^{n-r}b^r = \frac{n!}{r!(n-r)!}a^{n-r}b^r = \frac{n(n-1)\ldots(n-r+1)}{r!}a^{n-r}b^r.$$

Example 1   Use the binomial theorem to expand $(1+x)^7$.

$$(1+x)^7 = 1 + {_7C_1}x + {_7C_2}x^2 + {_7C_3}x^3 + {_7C_4}x^4 + {_7C_5}x^5 + {_7C_6}x^6 + x^7$$

$$= 1 + \frac{7!}{1!6!}x + \frac{7!}{2!5!}x^2 + \frac{7!}{3!4!}x^3 + \frac{7!}{4!3!}x^4 + \frac{7!}{5!2!}x^5 + \frac{7!}{6!1!}x^6 + x^7$$

$$= 1 + 7x + \frac{7.6}{1.2}x^2 + \frac{7.6.5}{1.2.3}x^3 + \frac{7.6.5}{1.2.3}x^4 + \frac{7.6}{1.2}x^5 + 7x^6 + x^7$$

$$= 1 + 7x + 21x^2 + 35x^3 + 35x^4 + 21x^5 + 7x^6 + x^7.$$

Example 2   Use the binomial theorem to expand $(3p+2q)^4$.

$$(3p+2q)^4 = (3p)^4 + {_4C_1}(3p)^3 2q + {_4C_2}(3p)^2(2q)^2 + {_4C_3}3p(2q)^3 + (2q)^4$$

$$= 3^4p^4 + 4.3^3.2p^3q + 6.3^2.2^2p^2q^2 + 4.3.2^3pq^3 + 2^4q^4$$

$$\therefore \quad (3p+2q)^4 = 81p^4 + 216p^3q + 216p^2q^2 + 96pq^3 + 16q^4.$$

Example 3   Find the coefficient of x^4 in the expansion of $(2x-1)^{15}$.

By the binomial theorem the term in $x^4 = {_{15}C_{11}}(2x)^4(-1)^{11}$.

Hence the coefficient of $x^4 = -{_{15}C_{11}}.2^4 = -\dfrac{15.14.13.12}{4.3.2.1}.2^4$

$$= -21\,840.$$

Example 4   Write down the first three terms in the expansion of $(1-x)^{10}$. Hence find an approximate value for $(0\cdot99)^{10}$.

Using the binomial theorem

$$(1-x)^{10} = 1^{10} + {}_{10}C_1 1^9(-x) + {}_{10}C_2 1^8(-x)^2 + \ldots$$

$$= 1 - 10x + \frac{10.9}{2.1}x^2 - \ldots$$

$$\therefore \quad (1-x)^{10} = 1 - 10x + 45x^2 - \ldots$$

Substituting $x = 0\cdot01$, we find that

$$(0\cdot99)^{10} \approx 1 - 10(0\cdot01) + 45(0\cdot01)^2 = 1 - 0\cdot1 + 0\cdot0045$$

$$\therefore \quad (0\cdot99)^{10} \approx 0\cdot9045.$$

Example 5   Expand $(1 + x + 2x^2)^6$ as far as the term in x^3.

$$(1 + x + 2x^2)^6 = 1 + 6(x + 2x^2) + \frac{6.5}{2.1}(x + 2x^2)^2$$

$$+ \frac{6.5.4}{3.2.1}(x + 2x^2)^3 + \ldots.$$

Neglecting terms in x^4 and higher powers:

$$(1 + x + 2x^2)^6 = 1 + 6(x + 2x^2) + 15(x^2 + 4x^3) + 20x^3 + \ldots$$

$$= 1 + 6x + 27x^2 + 80x^3 + \ldots.$$

One important application of the binomial theorem is to the proof of the rule for differentiating powers of x introduced in §5.4, namely

$$\frac{d}{dx}(x^n) = nx^{n-1}, \quad \text{where } n \text{ is a positive integer.}$$

Let δx be a small increase in x and let δy be the corresponding small increase in y, then

$$y = x^n$$

$$\Rightarrow y + \delta y = (x + \delta x)^n$$

$$= x^n + {}_nC_1 x^{n-1}\delta x + {}_nC_2 x^{n-2}(\delta x)^2 + \ldots + (\delta x)^n$$

$$\Rightarrow \quad \delta y = nx^{n-1}\delta x + \frac{n(n-1)}{2}x^{n-2}(\delta x)^2 + \ldots + (\delta x)^n$$

$$\Rightarrow \quad \frac{\delta y}{\delta x} = nx^{n-1} + \frac{n(n-1)}{2}x^{n-2}\delta x + \ldots + (\delta x)^{n-1}$$

$$\therefore \quad \frac{dy}{dx} = \lim_{\delta x \to 0}\frac{\delta y}{\delta x} = nx^{n-1}.$$

Exercise 16.6

1. Use the binomial theorem to expand:
(a) $(x+y)^4$,
(b) $(a-b)^7$,
(c) $(2+p^2)^6$,
(d) $(2h-k)^5$,
(e) $\left(x + \dfrac{1}{x}\right)^3$,
(f) $\left(z - \dfrac{1}{2z}\right)^8$.

2. Find the given terms in the following expansions:
(a) $(1+x)^{10}$, 5th term,
(b) $(2-3x)^8$, term in x^2,
(c) $(2a+b)^{12}$, 10th term,
(d) $(p-3q^2)^7$, term in p^4q^6,
(e) $\left(x - \dfrac{1}{x}\right)^6$, constant term,
(f) $\left(x^2 + \dfrac{1}{x}\right)^9$, term in $\dfrac{1}{x^3}$.

3. Expand and simplify $\left(2x + \dfrac{1}{x^2}\right)^5 + \left(2x - \dfrac{1}{x^2}\right)^5$.

4. The coefficient of x^3 in the expansion of $(1+x)^n$ is four times the coefficient of x^2. Find the value of n.

5. In the binomial expansion of $(1+\frac{1}{3})^n$ the fourth and fifth terms are equal. Find the value of n.

6. The coefficient of x^5 in the expansion of $(1+5x)^8$ is equal to the coefficient of x^4 in the expansion of $(a+5x)^7$. Find the value of a.

7. If the first three terms of the expansion of $(1+ax)^n$ in ascending powers of x are $1-4x+7x^2$, find n and a.

8. Use the expansion of $(a+b)^4$ to evaluate $(1\cdot03)^4$ correct to 2 decimal places.

9. Use the expansion of $(2-x)^5$ to evaluate $(1\cdot98)^5$ correct to 5 decimal places.

10. Obtain the expansion in ascending powers of x of $(1+2x)^{15}$ as far as the term in x^3. Hence evaluate $(1\cdot002)^{15}$ correct to 5 decimal places.

11. Find the first four terms in the expansions, in ascending powers of x, of
(a) $(1+x)^7$, (b) $(1+x-x^2)^7$.

12. Expand $(1+2x+3x^2)^8$ in ascending powers of x as far as the term in x^3.

13. Find the first three terms in the expansion in ascending powers of x of $(1-3x)(1+2x)^6$.

14. Find the coefficient of the given power of x in the expansion of
(a) $(1+x^2)(2-3x)^7$, x^3,
(b) $(1-3x-2x^2)(1+x^2)^{10}$, x^{20},
(c) $x\left(x - \dfrac{2}{x^2}\right)^{12}$, x^4,
(d) $\left(x + \dfrac{1}{x}\right)^2(1-x)^5$, x^2.

16.7 Use of the binomial series

By the binomial theorem, if n is a positive integer

$$(1+x)^n = 1 + nx + \frac{n(n-1)}{2!}x^2 + \frac{n(n-1)(n-2)}{3!}x^3 + \ldots + x^n.$$

This result can be extended to other values of n by writing

$$(1+x)^n = 1 + nx + \frac{n(n-1)}{2!}x^2 + \frac{n(n-1)(n-2)}{3!}x^3 + \ldots$$

$$\ldots + \frac{n(n-1)\ldots(n-r+1)}{r!}x^r + \ldots.$$

When n is a positive integer this expansion is the same as that obtained using the binomial theorem. Otherwise the expansion produces an infinite series called the *binomial series*. It can be shown that this series is a valid expansion of $(1+x)^n$ when $|x| < 1$, i.e. when $-1 < x < 1$.

Example 1 Expand $1/(1-x)^2$ in ascending powers of x, giving the first four terms and the term in x^r.

$$\frac{1}{(1-x)^2} = (1-x)^{-2} = 1 + (-2)(-x) + \frac{(-2)(-3)}{2!}(-x)^2$$

$$+ \frac{(-2)(-3)(-4)}{3!}(-x)^3 + \ldots$$

$$+ \frac{(-2)(-3)\ldots(-2-r+1)}{r!}(-x)^r + \ldots$$

$$\therefore \quad \frac{1}{(1-x)^2} = 1 + 2x + 3x^2 + 4x^3 + \ldots + (r+1)x^r + \ldots$$

[Note: The coefficients of the terms in the expansions of $(1-x)^{-1}$, $(1-x)^{-2}$, $(1-x)^{-3}$,... form the diagonal lines in Pascal's triangle given in §16.6.]

Example 2 Find the expansion of $(1+2x)^{3/2}$ in ascending powers of x as far as the term in x^3. State the range of values of x for which the expansion is valid.

$$(1+2x)^{3/2} = 1 + \frac{3}{2} \cdot 2x + \frac{\frac{3}{2} \cdot \frac{1}{2}}{2!}(2x)^2 + \frac{\frac{3}{2} \cdot \frac{1}{2} \cdot (-\frac{1}{2})}{3!}(2x)^3 + \ldots$$

$$= 1 + 3x + \frac{3}{2}x^2 - \frac{1}{2}x^3 + \ldots$$

The expansion is valid for $|2x| < 1$, i.e. for $|x| < \frac{1}{2}$.

Example 3 Find the first three terms in the expansion of $(4+x)^{-1/2}$ in ascending powers of x. Deduce an approximate value of $1/\sqrt{(4 \cdot 16)}$.

[Note that the binomial series cannot be used to expand $(4+x)^{-1/2}$ directly. We must rearrange to create an expression of the form $(1+\ldots)^{-1/2}$.]

$$(4+x)^{-1/2} = \{4(1+\tfrac{1}{4}x)\}^{-1/2} = 4^{-1/2}(1+\tfrac{1}{4}x)^{-1/2} = \tfrac{1}{2}(1+\tfrac{1}{4}x)^{-1/2}$$

$$\therefore \quad (4+x)^{-1/2} = \tfrac{1}{2}\left\{1+(-\tfrac{1}{2})(\tfrac{1}{4}x) + \frac{(-\tfrac{1}{2})(-\tfrac{3}{2})}{2!}(\tfrac{1}{4}x)^2 + \ldots\right\}$$

$$= \frac{1}{2} - \frac{1}{16}x + \frac{3}{256}x^2 - \ldots$$

Substituting $x = 0\cdot 16$, we have

$$(4\cdot16)^{-1/2} = \frac{1}{2} - \frac{1}{16}(0\cdot16) + \frac{3}{256}(0\cdot16)^2 - \ldots = 0\cdot5 - 0\cdot01 + 0\cdot0003 - \ldots .$$

Hence $1/\sqrt{(4\cdot16)} \approx 0\cdot4903.$

Exercise 16.7

In questions 1 to 3 expand the given function in a series of ascending powers of x giving the first three terms and the term in x^r.

1. $(1+x)^{-2}$,   2. $(1-x)^{-1}$,   3. $1/(1+2x)^3$.

In questions 4 to 9 expand the given function in a series of ascending powers of x giving the first three terms. State the values of x for which each expansion is valid.

4. $(1+x)^{1/2}$.   5. $\sqrt[3]{(1-3x)}$.   6. $1/\sqrt{(1+\tfrac{1}{2}x)}$.

7. $1/(3+x)$.   8. $\sqrt{(4-x)}$.   9. $(9-4x)^{3/2}$.

10. Find the first four non-zero terms in the expansion of $(1+2x^2)^{-1/2}$ in ascending powers of x.

11. Find the first four terms in the expansion of $(1-x)^{1/2}$ in ascending powers of x. Deduce the value of $\sqrt{(0\cdot9)}$ correct to 4 decimal places.

12. Expand $(1+3x)^{-1/3} + (1-4x)^{-1/4}$ as a series in ascending powers of x, giving the first three non-zero terms.

13. Expand $(2-x)^{-2}$ as a series of ascending powers of x as far as the term in x^4. Deduce the value of $1/(1\cdot8)^2$ correct to 3 significant figures.

14. Expand $(1+2x)^{1/2}$ in ascending powers of x as far as the term in x^3. By substituting $x = 1/25$, find an approximate value for $\sqrt{3}$ giving your answer to 3 decimal places.

15. Expand $(1-3x)/(1+4x)$ in ascending powers of x as far as the term in x^3, stating the values of x for which the expansion is valid.

16. Expand the following in ascending powers of x as far as the term in x^3
(a) $1/(1+x-x^2)$,   (b) $\sqrt{(1+x+2x^2)}$.

Exercise 16.8   (miscellaneous)

1.   Write down an expression for the nth term of the sequence 0, 2, 6, 12, 20, ... Find which term of the sequence is equal to 210.

2.   Find the first three terms and the nth term of the series whose sums are given by

(a) $S_n = 4n - \dfrac{1}{n}$,   (b) $S_n = (n+1)!$

3.   Evaluate the following:

(a) $\displaystyle\sum_{r=1}^{20} (r+2)$,   (b) $\displaystyle\sum_{r=1}^{8} 2^r$,   (c) $\displaystyle\sum_{r=1}^{n} 2r$.

4.   Given that the arithmetic mean of $1/(a+b)$ and $1/(b+c)$ is $1/(a+c)$, find in terms of b the arithmetic mean of a^2 and c^2.

5.   Find the least value of n for which the nth term of the series $12+10\cdot7+9\cdot4$ $+8\cdot1+ \ldots$ is negative. Find also the least value of n for which the sum to n terms is negative.

6.   The sum of the first twenty terms of an arithmetic progression is 45, and the sum of the first forty terms is 290. Find the first term and the common difference. Find the number of terms in the progression which are less than 100.   (JMB)

7.   The second, fourth and ninth terms of an arithmetic progression are in geometric progression. Find the common ratio of the geometric progression.

8.   A rod one metre in length is divided into ten pieces whose lengths are in geometrical progression. The length of the longest piece is eight times the length of the shortest piece. Find, to the nearest millimetre, the length of the shortest piece.
(JMB)

9.   (i) The sum to infinity of a geometric progression is 3. When the terms of this geometric progression are squared a new geometric progression is obtained whose sum to infinity is 1·8. Find the first term and the common ratio of each series.
(ii) Express the recurring decimal $0\cdot3\dot{2}\dot{1}$ in the form p/q, where p and q are integers with no common factor.   (AEB 1978)

10. (i) Write down the first four terms of a geometric series with sum to infinity 4/3 and first term 1.
(ii) Find the values of r for which the series

$$r^2 + \frac{r^2}{1+r^2} + \frac{r^2}{(1+r^2)^2} + \ldots \text{ is convergent, and find its sum.} \qquad \text{(AEB 1975)}$$

11. (i) Given that the sum of the first and second terms of an arithmetical progression is x and that the sum of the $(n-1)$th and nth terms is y, prove that the sum of the first n terms is $\frac{1}{4}n(x+y)$.
(ii) The sum of the first four terms of a geometric series of positive terms is 15 and the sum to infinity of the series is 16. Show that the sum of the first eight terms of the series differs from the sum to infinity by 1/16. (L)

12. (a) Prove that the sum of all the integers between m and n inclusive $(m, n \in \mathbb{Z}_+, n > m)$ is $\frac{1}{2}(m+n)(n-m+1)$. Find the sum of all the integers between 1000 and 2000 which are not divisible by 5.
(b) A geometric series has first term 2 and common ratio 0·95. The sum of the first n terms of the series is denoted by S_n and the sum to infinity is denoted by S. Calculate the least value of n for which $S - S_n < 1$. (C)

13. The first term of a geometric series is 2 and the second term is x. State the set of values of x for which the series is convergent. Show that when convergent the series converges to a sum greater than 1. If $x = \frac{1}{2}$, find the smallest positive integer n such that the sum of the first n terms differs from the sum to infinity by less than 2^{-10}. (L)

14. An arithmetic series and a geometric series have r as the common difference and the common ratio respectively. The first term of the arithmetic series is 1 and the first term of the geometric series is 2. If the fourth term of the arithmetic series is equal to the sum of the third and fourth terms of the geometric series, find the three possible values of r. When $|r| < 1$ find, in the form $p + q\sqrt{2}$, (i) the sum to infinity of the geometric series, (ii) the sum of the first ten terms of the arithmetic series. (AEB 1978)

15. Mary and Joan play a game as follows. Mary cuts a pack of 52 playing cards, notes the card turned up and then shuffles the pack. The process is repeated by Joan and the game continues with the girls taking alternate cuts. What is the probability that Mary will turn up a spade before Joan? (AEB 1976)

16. A and B play a game in which the probability that A wins any particular point is p, and that B wins any particular point is q, where $p + q = 1$. The game ends when either player is two points ahead of his opponent. Determine the probabilities that (i) the game ends after two points have been played, (ii) A wins the game. (C)

17. If S is the sum of the series $1+3x+5x^2+ \ldots +(2n+1)x^n$, prove, by considering $(1-x)S$, or otherwise, that if $x \neq 1$,

$$S = \frac{1+x-(2n+3)x^{n+1}+(2n+1)x^{n+2}}{(1-x)^2}.$$
(O)

18. (i) Find the set of values of x for which the series $\displaystyle\sum_{n=0}^{\infty} \left(\frac{2x}{x+1} \right)^n$ is convergent.

If the sum to infinity is 3, find x.

(ii) Without using tables, find the value of

$$\left(2 + \frac{1}{\sqrt{2}} \right)^5 + \left(2 - \frac{1}{\sqrt{2}} \right)^5.$$
(L)

19. Obtain the first three terms in the expansion in ascending powers of x of
(a) $(1+2x-x^2)^6$,   (b) $(1-2x)^5(1+x)^7$.

20. Find the term independent of x in the expansion of

(a) $\left(x^2 - \frac{1}{3x} \right)^9$,   (b) $\left(x - \frac{1}{x} \right)^8 \left(x + \frac{1}{x^2} \right)^4$.

21. The coefficients of x^4 in the expansions of $(1+x)^{2n}$ and $(1+15x^2)^n$ are equal. Given that n is a positive integer, find its value.

22. By substituting $a = b = 1$ in the binomial expansion of $(a+b)^8$, find the sum of $_8C_1 + _8C_2 + _8C_3 + \ldots + _8C_8$. Using a similar method, evaluate $_8C_1 - _8C_2 + _8C_3 - \ldots - _8C_8$.

23. The first three terms in the expansion of $(1+ax)^n$ are $1+12x+81x^2$. Find the values of a and n. Find also the coefficient of x^3 in the expansion.

24. Show that $\dfrac{1}{1-2x} - \dfrac{2}{2+x} = \dfrac{5x}{(1-2x)(2+x)}$. Hence expand $\dfrac{5x}{(1-2x)(2+x)}$ in ascending powers of x as far as the term in x^3.

25. Write down the first five terms of the expansion of $(1+3x)^{1/3}$ in ascending powers of x and state the range of values of x for which the infinite expansion of this form is valid. Hence find an approximation to the cube root of 0.97, correct to five places of decimals.   (AEB 1976)

26. Write down the first four terms in the binomial expansion of $(1+2x)^{-3}$ in ascending powers of x. State the range of values of x for which it is valid. By putting $x = -0.01$, find a value for 7^{-6} correct to four significant figures.   (AEB 1978)

27. Write down the expansion in ascending powers of x up to the term in x^2 of (i) $(1+x)^{1/2}$, (ii) $(1-x)^{-1/2}$ and simplify the coefficients. Hence, or otherwise, expand

$\sqrt{\left(\dfrac{1+x}{1-x}\right)}$ in ascending powers of x up to the term in x^2. By using $x = 1/10$ obtain an estimate, to three decimal places, for $\sqrt{11}$.   (JMB)

28. (a) If $x - \dfrac{1}{x} = u$, express $x^3 - \dfrac{1}{x^3}$ and $x^5 - \dfrac{1}{x^5}$ in terms of u.

(b) Assuming that $(1 - 2kx + x^2)^{-1/2}$ may be expanded in a series of ascending powers of x, obtain the expansion as far as the term in x^3, simplifying the coefficients.   (C)

29. For each of the following series construct a flow diagram for a procedure to list the first 10 terms of the series and determine their sum.
(a) An arithmetic progression with the first term a and common difference d given as data.
(b) A geometric progression with the first term a and common ratio r given as data.
(c) The series $1 + 2x + 3x^2 + \ldots$ with the value of x given as data.

30. Construct a flow diagram to output the terms of the binomial expansion of $(\tfrac{1}{4} + \tfrac{3}{4})^n$, where n is a given positive integer. Test your procedure using a calculator and the value $n = 3$.

17 Newton's laws of motion

17.1 Newton's laws

The theory of elementary dynamics is based on three laws first stated by Newton† in the seventeenth century. Although slight discrepancies can be detected when comparing the predictions of Newtonian mechanics with the true motion of atomic particles travelling at high speed, and with certain astronomical observations (when results based on the theory of relativity are more reliable), Newton's laws can be used to describe the behaviour of moving objects with great accuracy.

> *Newton's first law*: Every body continues in its state of rest or of uniform motion in a straight line unless compelled to change that state by external forces.

This law implies that a force changes the velocity of the body it acts upon. However, if two or more forces act on a body there will be no change in velocity unless the forces have a non-zero resultant. Newton's first law also implies that no force is required to maintain uniform motion, but practical experience tells us that the thrust of an engine is needed to drive a car along a horizontal road with uniform velocity. The explanation of this apparent contradiction is that the resultant force on the car is zero, the thrust of the engine exactly balancing the various forces opposing motion.

> *Newton's second law*: The rate of change of momentum of a body is proportional to the applied force and has the direction of that force.

---

† *Newton, Sir Isaac* (1642–1727) English scientist and mathematician. He invented, a little before the German mathematician Leibnitz, differential calculus (1665) and integral calculus (1666). He did significant work on the nature of light and the construction of telescopes. However, Newton is chiefly remembered for his statement of the laws of dynamics and his theory of gravitation, which are explained in *Philosophiae Naturalis Principia Mathematica* (1687).

The *momentum* of a body is defined as the product of its mass and its velocity, i.e. the momentum of a particle of mass m moving with velocity v is the vector quantity mv. Thus for a particle of constant mass m,

$$\text{rate of change of momentum} = \frac{d}{dt}(mv) = m\frac{dv}{dt} = ma,$$

where a is the acceleration of the particle.

Hence Newton's second law states that, for a body of constant mass m, the quantity ma is proportional to the force $\mathbf{F}$ acting on the body, i.e. $\mathbf{F} = km\mathbf{a}$, where k is constant. The value of k depends on the units of measurement chosen for $\mathbf{F}$, m and $\mathbf{a}$. A newton is defined as the force which gives a mass of 1 kg an acceleration of $1\,\mathrm{m\,s^{-2}}$. Hence when using SI units, $k = 1$ and we obtain the fundamental equation of motion for a body of constant mass

$$\boxed{\mathbf{F} = m\mathbf{a}.}$$

Example 1   A particle of mass 5 kg is moving with an acceleration of $3\,\mathrm{m\,s^{-2}}$. Find the magnitude of the resultant force acting on the particle.

If the resultant force on the particle is F newtons, by Newton's second law, $F = 5 \times 3$.
Hence there is a force of 15 N acting on the particle.

In Chapter 10 various ways of finding the resultant of a set of forces were discussed. Any of these methods, together with Newton's second law, can be used when finding the acceleration of a body moving under the action of two or more forces.

Example 2   A body of mass 8 kg on a smooth horizontal plane is acted upon by horizontal forces of magnitudes 20, 12 and 16 newtons in the directions N 45° E, due south and S 30° E respectively. Find the magnitude and direction of the acceleration of the body.

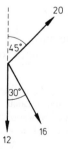

Let $\mathbf{F}$ be the resultant force on the body acting at an angle θ east of south and let $\mathbf{a}$ be the acceleration produced. If $\mathbf{P}$ and $\mathbf{Q}$ are the components of $\mathbf{F}$ in the directions south and east respectively, then resolving in these directions

$$\downarrow P = 12 + 16\cos 30° - 20\cos 45°$$
$$\rightarrow Q = 16\sin 30° + 20\sin 45°.$$

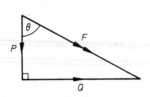

[By calculator $P = 11\cdot71426$, $Q = 22\cdot14214$.]

Using Newton's second law, $\mathbf{F} = 8\mathbf{a}$

$$\therefore \quad a = \frac{1}{8}F = \frac{1}{8}\sqrt{(P^2 + Q^2)} \approx 3\cdot13$$

and $\tan\theta = Q/P$, $\theta \approx 62\cdot1°$.

Hence the body moves with an acceleration of $3\cdot13\,\mathrm{m\,s^{-2}}$ in the direction S $62\cdot1°$ E.

When a body is moving in a straight line under the action of constant forces then its acceleration is constant. Hence the motion is governed by the constant acceleration equations derived in §12.2.

Example 3   A particle of mass 2 kg starting from rest is acted on by a force of 4 N for 5 seconds. Find its acceleration and the speed attained. The particle is then brought to rest in a further 8 seconds by a constant resistance of R N. Find the value of R and the total distance travelled by the particle.

Let $a_1\,\mathrm{m\,s^{-2}}$ and $a_2\,\mathrm{m\,s^{-2}}$ be the accelerations of the particle in the two parts of its motion and let $V\,\mathrm{m\,s^{-1}}$ be the speed attained after 5 seconds.

By Newton's 2nd law, $4 = 2a_1$ $\therefore$ $a_1 = 2$.

Using the formula $v = u + at$, substituting $u = 0$, $v = V$, $a = 2$, $t = 5$:
$V = 0 + 2 \times 5 = 10$.

Hence the initial acceleration of the particle is $2\,\mathrm{m\,s^{-2}}$ and the speed attained is $10\,\mathrm{m\,s^{-1}}$.

Using the formula $v = u + at$, substituting $u = 10$, $v = 0$, $a = a_2$, $t = 8$
$0 = 10 + 8a_2$ $\therefore$ $a_2 = -1\cdot25$.

By Newton's 2nd law $R = 2 \times 1\cdot25 = 2\cdot5$.

Hence the resistance is of magnitude $2\cdot5$ N.

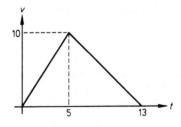

The total distance travelled is represented by the area under the velocity-time graph for the motion

$\therefore$ total distance travelled

$= \frac{1}{2} \times 10 \times 13\,\mathrm{m} = 65\,\mathrm{m}$.

Newton's third law: To every action there is an equal and opposite reaction.

This means that if a body A exerts a force on a body B, then B exerts an equal and opposite force on A. For instance, consider a body of weight W. The gravitational pull of the earth on the body is W, but Newton's third law tells us that the body also exerts a force of magnitude W on the earth. Similarly, when a body is in contact with a smooth surface, the surface exerts a force called the normal reaction on the body. As stated in Newton's third law, the body exerts an equal and opposite force on the surface. This law will be particularly useful when examining systems involving two or more objects in contact.

Exercise 17.1

[Vector quantities are measured in SI units.]

1. A particle of mass 12 kg is moving with an acceleration of $4\,\mathrm{m\,s^{-2}}$. Find the magnitude of the resultant force on the particle.

2. Find the acceleration of a mass of 45 kg acted upon by a single force of 9 N.

3. A resultant force of 36 N acting on a particle produces an acceleration of $3\,\mathrm{m\,s^{-2}}$. Find the mass of the particle.

4. A particle of mass 8 kg is pulled along a smooth horizontal plane by a horizontal string. Find the tension in the string when the acceleration of the particle is $5\,\mathrm{m\,s^{-2}}$.

5. A particle of mass 15 kg is acted upon by a resultant force represented by the vector $30\mathbf{i} - 75\mathbf{j}$. Find, in vector form, the acceleration of the particle.

6. A particle of mass 4 kg is moving with acceleration vector $3\mathbf{i} + \mathbf{j}$. Find a vector representing the resultant force on the particle.

7. Forces of magnitude 5 N and 9 N act at an angle of 50° on a particle of mass 2 kg. Find the magnitude and direction of the acceleration produced.

8. Two forces of magnitude 16 N and 24 N act on a particle of mass 5 kg. Given that an acceleration of $6\,\mathrm{m\,s^{-2}}$ is produced, find the angle between the lines of action of the two forces.

9. Forces represented by the vectors $3\mathbf{i} + \mathbf{j}$, $-5\mathbf{i} + 2\mathbf{j}$, $4\mathbf{i} - 5\mathbf{j}$ and $2\mathbf{i} - \mathbf{j}$ act on a particle of mass 0·8 kg moving on a smooth horizontal plane. Given that $\mathbf{i}$ and $\mathbf{j}$ are unit vectors in the directions due east and due north respectively, find the magnitude and direction of the acceleration of the particle.

10. A body of mass 2 kg on a smooth horizontal plane is acted upon by horizontal forces of magnitudes 9, 4 and 7 newtons in the directions due west, N 20° W and S 75° E respectively. Find the magnitude and direction of the acceleration of the body.

11. A train of mass $300\,000\,\text{kg}$ travelling at $0.3\,\text{m s}^{-1}$ is brought to rest at a station by buffers exerting an average force of $50\,000\,\text{N}$. Find the time taken to bring the train to rest.

12. A particle of mass $20\,\text{kg}$ is brought to rest in a distance of $600\,\text{m}$ from a speed of $30\,\text{m s}^{-1}$ by a constant retarding force. Find the magnitude of this force.

13. A body is moving in a straight line with a speed of $40\,\text{m s}^{-1}$. A retarding force of $25\,\text{N}$ reduces its speed to $15\,\text{m s}^{-1}$ in 10 seconds. Find the mass of the body.

14. A car of mass $800\,\text{kg}$ is moving along a straight horizontal road under the action of a constant resultant force of $1000\,\text{N}$. Find its speed when it has moved $40\,\text{m}$ from rest.

15. A particle of mass $5\,\text{kg}$ starting from rest is acted on by a force of $12\,\text{N}$ for 10 seconds. The particle continues to move with constant velocity for a further 10 seconds. It is then brought to rest by a constant resistance of $R\,\text{N}$ in a further time of 15 seconds. Find the value of R and the total distance travelled.

16. A particle of mass $4\,\text{kg}$ starts from rest and moves under the action of a constant force of F newtons for 12 seconds. It is then brought to rest in a further 8 seconds by a constant retarding force of R newtons. If the total distance travelled is $200\,\text{m}$, find the maximum speed attained and the values of F and R.

17.2 Further applications

As stated in §12.2, an object falling freely under gravity has a constant downward acceleration denoted by g. Thus if a body of mass m kg has a weight of WN, then by Newton's second law,

$$W = mg.$$

Hence the force due to the gravitational pull of the earth on a body of mass m kg is of magnitude mg N.

[In this and later chapters the value of g will be taken as $9.8\,\text{m s}^{-2}$ unless otherwise stated.]

Example 1　A package of mass $5\,\text{kg}$ is being lifted by means of a vertical cable. If the tension in the cable is $60\,\text{N}$, find the acceleration of the package.

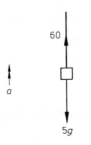

Let the acceleration of the package be $a\,\text{m s}^{-2}$ vertically upwards.
Applying Newton's 2nd law vertically

$$\uparrow 60 - 5g = 5a$$

$$\therefore \quad a = 12 - g = 12 - 9.8 = 2.2.$$

Hence the acceleration of the package is $2.2\,\text{m s}^{-2}$.

[Note that to avoid confusion in diagrams showing both forces and accelerations, the forces are denoted by single-headed arrows and the accelerations by double-headed arrows.]

Example 2   A lift descends with an acceleration of $1\cdot5\,\mathrm{m\,s^{-2}}$, then moves with constant speed until it is retarded at $1\,\mathrm{m\,s^{-2}}$. A suitcase of mass $20\,\mathrm{kg}$ stands on the floor of the lift during the journey. Find the magnitude of the force it exerts on the floor of the lift at each stage of the journey.

Let R_1, R_2 and R_3 newtons be the reactions of the floor on the suitcase.

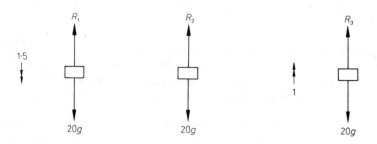

Applying Newton's 2nd law vertically in each case

$\downarrow 20g - R_1 = 20 \times 1\cdot5 \qquad \therefore \quad R_1 = 20g - 30 = 196 - 30 = 166$
$\uparrow R_2 - 20g = 0 \qquad\qquad \therefore \quad R_2 = 20g = 196$
$\uparrow R_3 - 20g = 20 \times 1 \qquad \therefore \quad R_3 = 20g + 20 = 196 + 20 = 216.$

By Newton's third law, the force exerted by the suitcase on the floor is equal in magnitude to the reaction of the floor on the suitcase. Hence the forces exerted by the suitcase on the floor of the lift are 166, 196 then 216 newtons.

Example 3   A particle of mass $4\,\mathrm{kg}$ is being pulled across a smooth horizontal plane by a string inclined at $40°$ to the horizontal. If the acceleration of the particle is $5\,\mathrm{m\,s^{-2}}$, find the tension in the string and the reaction between the particle and the plane.

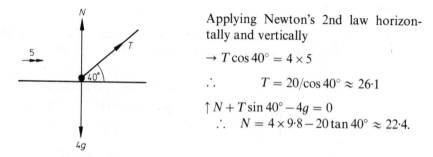

Applying Newton's 2nd law horizontally and vertically

$\rightarrow T\cos 40° = 4 \times 5$

$\therefore \qquad\qquad T = 20/\cos 40° \approx 26\cdot1$

$\uparrow N + T\sin 40° - 4g = 0$
$\therefore \quad N = 4 \times 9\cdot8 - 20\tan 40° \approx 22\cdot4.$

Hence the tension in the string is $26\cdot1\,\mathrm{N}$ and the reaction between the particle and the plane is $22\cdot4\,\mathrm{N}$.

Example 4   A particle of mass 10 kg slides down a smooth plane inclined at 30° to the horizontal. If the particle starts from rest, find its speed after sliding a distance of 5 m.

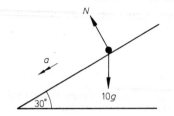

Let the acceleration of the particle down the plane be a m s^{-2}.
Applying Newton's 2nd law down the plane

$$10g \sin 30° = 10a$$
$$\therefore \quad a = g \sin 30° = 4 \cdot 9$$

Using the formula $v^2 = u^2 + 2as$, where $u = 0$, $s = 5$, $a = 4 \cdot 9$,

$$v^2 = 0 + 2 \times 4 \cdot 9 \times 5 = 49 \quad \therefore \quad v = 7$$

Hence the particle attains a speed of 7 m s^{-1}.

Exercise 17.2

1. Find the weight in newtons of an object with mass
(a) 10 kg,   (b) 1·5 kg,   (c) 250 grammes.

2. A particle of mass 2 kg is being lifted by a vertical force of 32 N. Find its acceleration.

3. A package of mass 8 kg is being lowered by means of a vertical cable with a downward acceleration of 2 m s^{-2}. Find the tension in the cable.

4. A boy of mass 40 kg is ascending in a lift. Find the force exerted by the floor of the lift on the boy when the lift is (a) accelerating at 1·2 m s^{-2}, (b) moving with constant speed, (c) decelerating at 0·8 m s^{-2}.

5. A man descends in a lift carrying a parcel of mass 3 kg. Describe the motion of the lift when the parcel has an apparent weight of (a) 33 N, (b) 27 N, (c) 29·4 N.

6. A box of mass 30 kg is being towed across a smooth horizontal surface by means of a light chain inclined at an angle of 25° to the horizontal. If the box is accelerating at 1 m s^{-2}, find the tension in the chain and the reaction between the box and the surface.

7. A particle of mass 6 kg is being pushed across a smooth horizontal plane by a force of 30 N acting downwards at 20° to the vertical. Find the acceleration of the particle and the force the particle exerts on the plane.

8. A particle of mass 0·5 kg is suspended from a string. If the acceleration of the particle is horizontal and of magnitude 2 m s^{-2}, find the tension in the string and its angle of inclination to the vertical.

9. A particle of mass 5 kg slides down a smooth plane inclined at an angle of 60° to the horizontal. Find the acceleration of the particle down the plane and the reaction between the particle and the plane.

10. A body of mass 12 kg slides down a smooth plane inclined at an angle of 20° to the horizontal. If it starts from rest, find the time it takes to slide a distance of 10 m.

11. A particle of mass 2 kg is being pulled up a smooth plane inclined at 25° to the horizontal by a string parallel to the plane. Find the tension in the string if the particle is (a) moving with uniform velocity, (b) accelerating at $3\,\mathrm{m\,s^{-2}}$.

12. An object of mass 4 kg is being pushed up a smooth plane inclined at an angle α to the horizontal by a horizontal force of 40 N. If $\sin\alpha = 0\cdot6$, find the acceleration of the particle and the reaction between the particle and the plane.

17.3 Connected particles

When considering a set of objects which are connected in some way, it is necessary to apply Newton's third law. Any force exerted by one of the objects upon another is matched by an equal and opposite reaction. For instance, in the case of a railway engine pulling a truck, the truck exerts a pull on the engine equal in magnitude to the pull of the engine on the truck. In problems involving such systems it is usually helpful to examine the forces on one or more of the objects separately.

Example 1   An engine of mass 50 tonnes is pulling several trucks of total mass 200 tonnes along a horizontal track. The resistance to the motion of the engine is 60 N per tonne and the resistance to the motion of the trucks is 35 N per tonne. If the tractive force exerted by the engine is 60 kN, find (a) the acceleration, (b) the tension in the coupling between the engine and the trucks.

$$[1\text{ tonne} = 1000\,\mathrm{kg},\ 1\,\mathrm{kN} = 1000\,\mathrm{N}.]$$

Let $a\,\mathrm{m\,s^{-2}}$ be the acceleration and let $T\,\mathrm{N}$ be the tension in the coupling.
(a) Considering the forces on the whole train, the total resistance to motion is
$(50 \times 60) + (200 \times 35)\,\mathrm{N}$, i.e. $10\,000\,\mathrm{N}$.

Applying Newton's 2nd law

$\leftarrow 60\,000 - 10\,000 = 250\,000a$

$$\therefore\quad a = \frac{50\,000}{250\,000} = 0\cdot2.$$

Hence the acceleration of the train is $0\cdot2\,\mathrm{m\,s^{-2}}$.

(b) Considering the forces on the trucks, the resistance to motion is $(200 \times 35)\,\text{N}$, i.e. $7000\,\text{N}$.

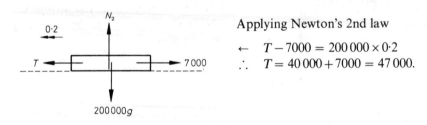

Applying Newton's 2nd law

$$\leftarrow \quad T - 7000 = 200\,000 \times 0\cdot2$$
$$\therefore \quad T = 40\,000 + 7000 = 47\,000.$$

Hence the tension in the coupling is $47\,\text{kN}$.

Each of the remaining examples in this section involves two particles connected by a light inextensible string passing over a smooth fixed light pulley. Under these conditions we can assume that the tension in the string is constant throughout its length.

Example 2 Two particles of mass $3\,\text{kg}$ and $4\,\text{kg}$ are connected by a light inextensible string passing over a smooth fixed pulley. The particles are released from rest with the strings taut and vertical. Find the acceleration of the particles and the tension in the string.

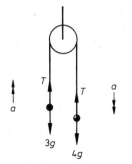

Let the acceleration be $a\,\text{m s}^{-2}$ and the tension $T\,\text{N}$.
Using Newton's 2nd law

| for the $3\,\text{kg}$ mass: | $T - 3g = 3a$ | (1) |
| for the $4\,\text{kg}$ mass: | $4g - T = 4a$ | (2) |

Adding (1) to (2) $\qquad g = 7a \qquad \therefore \quad a = \dfrac{1}{7}g = 1\cdot4$

Substituting in (1) $\qquad T - 3g = \dfrac{3}{7}g \qquad \therefore \quad T = 33\cdot6.$

Hence the acceleration of the system is $1\cdot4\,\text{m s}^{-2}$ and the tension in the string is $33\cdot6\,\text{N}$.

Example 3 A particle of mass $15\,\text{kg}$ rests on a smooth horizontal table. It is connected by a light inextensible string passing over a smooth pulley fixed at the edge of the table to a particle of mass $10\,\text{kg}$ which hangs freely. Find the acceleration of the system when it is released from rest. Find also the force exerted by the string on the pulley.

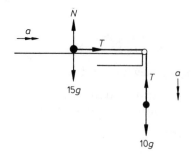

Let $a\,\mathrm{m\,s^{-2}}$ be the acceleration and let $T\,\mathrm{N}$ be the tension in the string. Let $F\,\mathrm{N}$ be the force exerted by the string on the pulley.

Using Newton's 2nd law,

horizontally for the 15 kg mass: $\qquad T = 15a$ $\qquad\qquad$ (1)

vertically for the 10 kg mass: $\quad 10g - T = 10a$ $\qquad\qquad$ (2)

Adding (1) to (2), $\quad 10g = 25a \qquad \therefore \quad a = \dfrac{2}{5}g = 3 \cdot 92.$

Hence the acceleration of the system is $3 \cdot 92\,\mathrm{m\,s^{-2}}$.

Substituting in (1) $\quad T = 15 . \dfrac{2}{5}g = 6g.$

The string exerts on the pulley a force T horizontally and a force T vertically.

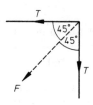

Since these forces are equal in magnitude, their resultant F acts along the bisector of the angle between the strings.

Resolving along this bisector:

$$F = 2T \cos 45° = 12g \cos 45° \approx 83 \cdot 2$$

Hence the string exerts a force on the pulley of $83 \cdot 2\,\mathrm{N}$ at $45°$ to the horizontal.

Example 4   A particle of mass m rests on a smooth plane inclined at an angle α to the horizontal where $\sin \alpha = \dfrac{3}{5}$. It is connected by a light inextensible string passing over a smooth pulley fixed at the top of the plane to a particle of mass $3\,\mathrm{m}$ which hangs freely. Find the acceleration of the system when it is released from rest and the tension in the string.

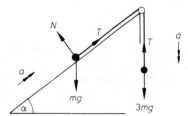

Let a be the acceleration and let T be the tension in the string.

Using Newton's 2nd law,

parallel to the plane for the mass m: $T - mg\sin\alpha = ma$ (1)
vertically for the mass $3m$: $3mg - T = 3ma$ (2)

Adding (1) to (2) $3mg - mg\sin\alpha = 4ma$

$$\therefore \quad a = \frac{1}{4}g\left(3 - \frac{3}{5}\right) = \frac{3}{5}g.$$

Substituting in (2) $3mg - T = 3m \cdot \frac{3}{5}g$

$$\therefore \quad T = 3mg\left(1 - \frac{3}{5}\right) = \frac{6}{5}mg.$$

Hence the acceleration of the system is $\frac{3}{5}g$ and the tension in the string is $\frac{6}{5}mg$.

Example 5 Two particles of mass 2 kg and 3 kg are connected by a light inextensible string passing over a fixed smooth pulley. Initially the system is at rest with the strings taut and vertical with both particles at a height of 2 m above the ground. When the system is released, find the time which elapses before the 3 kg mass hits the ground and the maximum height reached by the 2 kg mass.

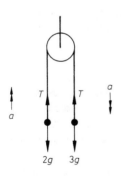

Let $a\,\mathrm{m\,s^{-2}}$ be the acceleration and let $T\,\mathrm{N}$ be the tension in the string.
Using Newton's 2nd law,

for the 2 kg mass: $T - 2g = 2a$ (1)
for the 3 kg mass: $3g - T = 3a$ (2)

$(1) + (2) \quad g = 5a \quad \therefore \quad a = \frac{1}{5}g.$

Using the formula $s = ut + \frac{1}{2}at^2$

where $s = 2, u = 0, a = \frac{1}{5}g$

$$2 = 0 + \frac{1}{2} \cdot \frac{1}{5}gt^2$$

$$\therefore \quad t = \sqrt{\left(\frac{20}{g}\right)} = \sqrt{\left(\frac{20}{9 \cdot 8}\right)} = \sqrt{\left(\frac{100}{49}\right)}$$

$$= \frac{10}{7}.$$

2 m

initial position

Hence the 3 kg mass hits the ground after $1\frac{3}{7}$ seconds.

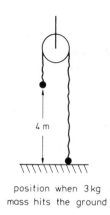

4 m

position when 3 kg
mass hits the ground

Using the formula $v^2 = u^2 + 2as$,

where $u = 0, a = \dfrac{1}{5}g, s = 2$,

$$v^2 = 0 + 2 \cdot \dfrac{1}{5}g \cdot 2 = \dfrac{4}{5}g.$$

Hence when the 3 kg mass hits the ground both particles are travelling at $\sqrt{(4g/5)}\,\mathrm{m\,s}^{-1}$. Since the string is no longer taut, the 2 kg mass now moves freely under gravity with a downward acceleration of $g\,\mathrm{m\,s}^{-2}$.

Using the formula $v^2 = u^2 + 2as$, where $u = \sqrt{(4g/5)}, v = 0, a = -g$:

$$0 = \dfrac{4}{5}g - 2gs \qquad \therefore \quad s = \dfrac{2}{5}.$$

Hence the 2 kg mass rises a further 0·4 m, reaching a maximum height above the ground of 4·4 m.

Exercise 17.3

1.   A motor-boat of mass 1000 kg is towing a water-skier of mass 75 kg by means of a horizontal cable. The resistance to the motion of the boat is 650 N and the resistance to the motion of the skier is 125 N. If the thrust of the boat's engine is 4000 N, find (a) the acceleration, (b) the tension in the cable.

2.   A car of mass 800 kg tows a trailer of mass 400 kg. The resistance to motion for both car and trailer is 0·2 N per kg. Find the tractive force exerted by the car engine and the tension in the coupling between the car and the trailer when they are travelling (a) with uniform velocity, (b) with an acceleration of 2 m s^{-2}.

3.   A railway engine of mass 60 tonnes is pulling two trucks each of mass 12 tonnes along a horizontal track. The resistance to motion is 80 N per tonne for the engine and 50 N per tonne for the trucks. Given that the train is travelling at constant speed, find the tractive force exerted by the engine, the tension in the coupling between the engine and the first truck and the tension in the coupling between the two trucks. If the tractive force is increased to 27 kN, find the acceleration and the new tensions in the couplings.

4.   Two particles of mass 2 kg and 5 kg are connected by a light inextensible string passing over a smooth fixed pulley. If the system is moving freely with the strings taut and vertical, find the acceleration of the particles and the tension in the string.

5.   Repeat question 4 for particles of mass 9 kg and 3 kg.

6. Repeat question 4 for particles of mass m and M, where $m < M$.

7. Two particles of mass $2m$ and $3m$ are connected by a light inextensible string passing over a smooth fixed pulley. If the system is moving freely with the strings vertical, find the force exerted by the string on the pulley.

8. A particle of mass 3·6 kg rests on a smooth horizontal table. It is connected by a light inextensible string passing over a smooth pulley fixed at the edge of the table to a particle of mass 1·2 kg which hangs freely. Find the acceleration of the system when it is released from rest and the tension in the string.

9. A particle of mass m rests on a smooth horizontal table. It is connected by a light inextensible string passing over a smooth pulley fixed at the edge of the table to a second particle of mass m which hangs freely. If the system is released from rest find the distance travelled in the first 0·4 seconds of the subsequent motion. Find also the force exerted by the string on the pulley.

10. A particle of mass 2 kg rests on a smooth plane inclined at an angle of 30° to the horizontal. It is connected by a light inextensible string passing over a smooth pulley fixed at the top of the plane to a particle of mass 3 kg which hangs freely. Find the acceleration of the system when it is released from rest and the tension in the string. Find also the force exerted by the string on the pulley.

11. A particle of mass m rests on a smooth plane inclined at an angle α to the horizontal. It is connected by a light inextensible string passing over a smooth pulley fixed at the top of the plane to a particle of mass M which hangs freely. Find the condition that the mass m should slide down the plane when the system is released from rest.

12. Two particles of mass $3m$ and $5m$ are connected by a light inextensible string passing over a fixed smooth pulley. The system is released from rest with the strings taut and vertical. After 2 seconds the $5m$ mass hits the ground. Find the further time which elapses before the $3m$ mass reaches its greatest height.

13. Two particles of mass 0·5 kg and 0·7 kg are connected by a light inextensible string passing over a fixed smooth pulley. Initially both parts of the string are taut and vertical, and the 0·5 kg mass is moving vertically downwards with a speed of 3·5 m s^{-1}. Find the distance it travels before coming instantaneously to rest.

14.

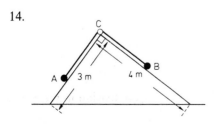

Two particles A and B, of mass 5 kg, rest on the smooth inclined faces of a fixed wedge as shown. They are connected by a light inextensible string passing over a smooth pulley C which lies in the same vertical plane as the particles. Find the acceleration of the system when moving freely and the tension in the string. If the system is released from rest with the string taut and both particles 2 m from the pulley, find the speed of the particles when *A* hits the horizontal plane. Find also the distance that B has travelled when it first comes to rest.

17.4 Related accelerations

Some problems involve bodies moving with different but related accelerations. A system including both fixed and moving pulleys can be dealt with by the methods of the previous section when the relationships between the accelerations have been established.

Example 1   A light inextensible string attached to the ceiling passes under a smooth movable pulley of mass 2 kg and then over a smooth fixed pulley. A particle of mass 3 kg hangs freely from the end of the string. All parts of the string not touching the pulleys are vertical. If the system is released from rest, find the acceleration of the particle and the tension in the string.

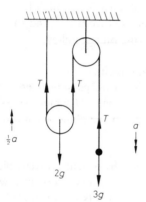

If the particle moves a distance s m, then the movable pulley moves a distance $\frac{1}{2}s$ m in the opposite direction. Hence if the acceleration of the particle is $a\,\mathrm{m\,s^{-2}}$ downwards, then the acceleration of the movable pulley is $\frac{1}{2}a\,\mathrm{m\,s^{-2}}$ upwards.

Using Newton's 2nd law,

for the particle	$3g - T = 3a$	(1)
for the pulley	$2T - 2g = 2 \cdot \frac{1}{2}a$	(2)
$2 \times (1) + (2)$	$4g = 7a$	

$$\therefore \quad a = \frac{4}{7}g = \frac{4}{7} \times 9\cdot8 = 5\cdot6.$$

Substituting in (1) $3g - T = 3 \cdot \frac{4}{7}g$

$$\therefore \quad T = \frac{9}{7}g = \frac{9}{7} \times 9\cdot8 = 12\cdot6.$$

Hence the acceleration of the particle is $5\cdot6\,\mathrm{m\,s^{-2}}$ and the tension in the string is $12\cdot6\,\mathrm{N}$.

Example 2   A light inextensible string which passes over a smooth fixed pulley P carries at one end a particle A of mass $2\,\text{kg}$ and at the other end a smooth light pulley Q. A light inextensible string passes over pulley Q and carries at its ends a particle B of mass $1\,\text{kg}$ and a particle C of mass $2\,\text{kg}$. Find the acceleration of particle A when the system is moving freely and the tensions in the two strings.

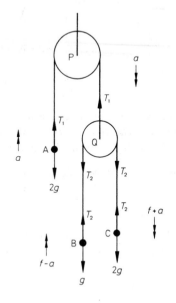

The tensions and accelerations in the different parts of the system are shown in the diagram, f being the acceleration of particles B and C relative to pulley Q.

Applying Newton's 2nd law, to pulley Q (which is assumed to have zero mass):

$$2T_2 - T_1 = 0 \qquad\qquad (1)$$
to A: $\qquad\qquad T_1 - 2g = 2a \qquad\qquad (2)$
to B: $\qquad\qquad T_2 - g = f - a \qquad\qquad (3)$
to C: $\qquad\qquad 2g - T_2 = 2(f+a) \qquad (4)$

$(1)+(2)$ $\qquad\qquad\qquad 2T_2 - 2g = 2a \qquad\qquad \therefore \quad T_2 = a+g.$

Substituting in (3) $\qquad a+g-g = f-a \qquad \therefore \quad f = 2a$

Substituting in (4) $\qquad 2g-(a+g) = 2(2a+a)$

$$g - a = 6a$$

$$\therefore \quad a = \frac{1}{7}g = 1\cdot 4.$$

Hence the acceleration of particle A is $1\cdot 4\,\text{m}\,\text{s}^{-2}$ upwards.

Thus $\quad T_2 = a+g = 1\cdot 4+9\cdot 8 = 11\cdot 2$

and $\quad\;\; T_1 = 2T_2 = 22\cdot 4.$

Hence the tensions in the strings over pulleys P and Q are $22\cdot 4\,\text{N}$ and $11\cdot 2\,\text{N}$ respectively.

The last example concerns two bodies in contact, which are free to move with different accelerations. In problems of this kind it is helpful to consider the acceleration of one body relative to the other.

Example 3   A particle of mass m rests on a smooth face of a wedge which stands on a smooth horizontal plane. The face of the wedge is inclined at an angle of $45°$ to the horizontal and the mass of the wedge is $10m$. Find the acceleration of the wedge when the system is released.

Motion of the particle   Motion of the wedge

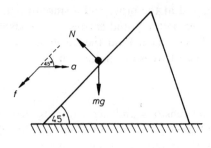
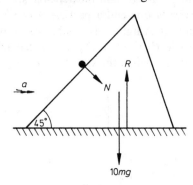

Let a be the acceleration of the wedge and let f be the acceleration of the particle relative to the wedge. [This means that the true acceleration of the particle has components of magnitudes a and f as shown in the diagram.]

Applying Newton's second law to the motion of the particle in a direction perpendicular to the wedge:

$$\searrow mg\cos 45° - N = ma\sin 45°.$$

Applying Newton's second law horizontally to the motion of the wedge:
$$\to N\sin 45° = 10ma$$

Hence $\qquad\qquad mg \cdot \dfrac{1}{\sqrt{2}} - N = ma \cdot \dfrac{1}{\sqrt{2}}$ $\qquad\qquad$ (1)

and $\qquad\qquad N \cdot \dfrac{1}{\sqrt{2}} = 10ma$ $\qquad\qquad$ (2)

$\dfrac{1}{\sqrt{2}} \times (1) + (2)$ $\qquad\qquad \tfrac{1}{2}mg = \tfrac{1}{2}ma + 10ma$

$$\therefore \quad g = a + 20a$$

i.e. $\quad a = \dfrac{g}{21} = \dfrac{9\cdot 8}{21} = \dfrac{1\cdot 4}{3} = \dfrac{7}{15}.$

Hence the acceleration of the wedge is $7/15\,\mathrm{m\,s^{-2}}$ horizontally.

Exercise 17.4

1. A light inextensible string attached to the ceiling passes under a smooth movable pulley of mass 6 kg and then over a smooth fixed pulley. A particle of mass 1 kg hangs freely from the end of the string. All parts of the string not touching the pulleys are vertical. If the system is released from rest, find the acceleration of the particle and state its direction. Find also the tension in the string.

2. Repeat question 1 for a movable pulley of mass 5 kg and a particle of mass 4 kg.

3. A light inextensible string which passes over a smooth fixed pulley P carries at one end a particle A of mass 2 kg and at the other end a smooth light pulley Q. Particles B and C of masses 4 kg and 5 kg respectively are connected by a light inextensible string passing over pulley Q. Find the accelerations of particles A and B when the system is moving freely. Find also the tensions in the two strings.

4. Repeat question 3 for particles A, B and C of masses 4 kg, 1 kg and 2 kg.

5. Repeat question 3 for particles A, B and C of masses 5 kg, 2 kg and 1 kg. Assume in this case that pulley Q has a non-zero mass of 1 kg.

6. A particle of mass 2 kg is in contact with a smooth face of a wedge which stands on a smooth horizontal plane. The face of the wedge is inclined at 30° to the horizontal and the mass of the wedge is 4 kg. Find the acceleration of the wedge when the system is released.

7. A particle of mass m is placed on a smooth face of a wedge which stands on a smooth horizontal plane. The face of the wedge is inclined at an angle α to the horizontal and the mass of the wedge is $4m$. If the system is released from rest, find the speed of the particle relative to the wedge after 1 second.

8. A particle of mass 1 kg is placed on the smooth horizontal upper face of a wedge, which is itself standing on a smooth plane inclined at an angle α to the horizontal. The mass of the wedge is 3 kg and $\sin \alpha = \frac{1}{4}$. Find the accelerations of the particle and the wedge when the system is released.

Exercise 17.5   (miscellaneous)

1. Two forces of magnitude 10 N and 7 N act on a particle producing an acceleration of $4\,\mathrm{m\,s^{-2}}$. If the forces act at an angle of 60°, find the mass of the particle.

2. A body of mass 5 kg rests on a smooth horizontal plane. Horizontal forces of magnitudes 6, 12, 5 and F newtons act on the body in the directions 015°, 130°, 180° and 270° respectively. Given that the acceleration of the body is $3\,\mathrm{m\,s^{-2}}$, find the value of F and the direction of the acceleration.

3. A vehicle of mass 2000 kg is travelling along a straight horizontal road at 90 km/h. It is brought to rest by a constant force of F newtons in a distance of 500 m. Find F and the time taken for the car to come to rest.

4. A particle A of mass 2 kg is acted upon by a constant force of magnitude 10 N in the direction of the vector $3\mathbf{i} - 4\mathbf{j}$ and a constant force of magnitude 50 N in the direction of the vector $7\mathbf{i} + 24\mathbf{j}$. Find the acceleration of A. Initially the particle is at the origin moving with speed $39\,\mathrm{m\,s^{-1}}$ in the direction of the vector $12\mathbf{i} - 5\mathbf{j}$. Find the position vector of the particle 2 s later.   (O & C)

5. A particle of mass 3 kg is being pulled across a smooth horizontal surface by a string inclined at an angle of 35° to the horizontal. If the particle started from rest and moved a distance of 5 m in the first 2 seconds of its motion, find the constant tension in the string.

6. A man of mass 70 kg is travelling in a lift. Find the force between the man and the floor of the lift when the lift is (a) ascending with an acceleration of $1·5\,\mathrm{m\,s^{-2}}$, (b) descending with an acceleration of $0·5\,\mathrm{m\,s^{-2}}$.

7. A load of 200 kg is lifted by means of a vertical cable through a distance of 50 m. The load accelerates uniformly from rest for 20 seconds, then continues with constant speed for 10 seconds. It then decelerates uniformly to rest in a further 10 seconds. Find the maximum speed attained and the tensions in the cable at every stage of the motion.

8. A package of mass 80 kg is suspended by a rope from a helicopter. Find the tension in the rope and its angle of inclination to the vertical when the helicopter is flying (a) with uniform velocity, (b) with a constant acceleration of $2·5\,\mathrm{m\,s^{-2}}$ horizontally.

9. A particle of mass m slides down a smooth plane inclined at an angle α to the horizontal. Find the speed attained by the particle when it has travelled a distance x from rest.

10. A particle of mass 5 kg rests on a smooth plane inclined at an angle α to the horizontal, where $\sin \alpha = 0·2$. If the particle is projected up the plane with an initial speed of $7\,\mathrm{m\,s^{-1}}$, find the distance up the plane that it travels.

11. A ring of mass m kg slides down a smooth wire inclined at an angle α to the horizontal. The ring is released 4·9 m vertically above a horizontal plane and hits the plane t seconds later. Find (a) the value of t if $\alpha = 30°$, (b) the value of α if $t = 1·2$.

12. A car of mass 1200 kg is pulling a trailer of mass 500 kg up a road inclined at an angle α to the horizontal where $\sin \alpha = 0·1$. The resistance to motion for both car and trailer is 0·15 N per kg. Find the tractive force exerted by the car engine and the tension in the coupling between the car and the trailer if they are decelerating at $0·5\,\mathrm{m\,s^{-2}}$.

13. Two particles of mass 1·5 kg and 2 kg are connected by a light inextensible string passing over a smooth fixed pulley. If the system is released from rest with the strings taut and vertical, find the distance moved by the 1·5 kg mass relative to the 2 kg mass in the first $1\frac{1}{2}$ seconds of the subsequent motion.

14. A particle of mass 4 kg rests on a smooth horizontal platform at a distance of 2·5 m from the edge. It is connected by a light inextensible string passing over a smooth pulley at the edge of the platform to a particle of mass 0·5 kg which is

hanging freely. If the particles are released from rest, find the speed of the 4 kg mass when it reaches the edge of the platform given that initially the 0·5 kg mass is (a) 3 m, (b) 1·6 m above the ground.

15. A particle of mass 10 kg rests on a smooth plane inclined at an angle of 40° to the horizontal. It is connected by a light inextensible string passing over a smooth pulley fixed at the top of the plane to a particle of mass 4 kg which is hanging freely. Find the acceleration of the system when it is released from rest and the force exerted by the string on the pulley.

16. A particle *A* of mass 4 kg is connected by a light inextensible string passing over a smooth fixed pulley to a particle *B* of mass 3 kg. Particle *B* is connected by another light inextensible string to a particle *C* of mass 3 kg. The system is released from rest with the strings taut and vertical. Find the tensions in the strings. Given that *B* and *C* are initially 2 m and 1 m respectively above the ground, find the distance travelled by *A* before it first comes to rest.

17. A particle *A* of mass 2*m* is initially at rest on a smooth plane inclined at an angle α to the horizontal. It is supported by a light inextensible string which passes over a smooth light pulley *P* at the top edge of the plane. The other end of the string supports a particle *B*, of mass *m*, which hangs freely. Given that the system is in equilibrium, find α, and the magnitude and direction of the resultant force exerted by the string on the pulley.

A further particle of mass *m* is now attached to *B* and the system is released. Find, for the ensuing motion, the tension in the string, the acceleration of *B* and the magnitude and direction of the resultant force exerted by the string on the pulley. (L)

18. A light inextensible string has one end attached to a ceiling. The string passes under a smooth movable pulley of mass *m* and then over a smooth fixed pulley. A particle of mass 2*m* is attached at the free end of the string. The sections of the string not in contact with the pulleys are vertical. If the system is released from rest and moves in a vertical plane, show that the acceleration of the particle is 2*g*/3 and find the tension in the string. (AEB 1977)

19.

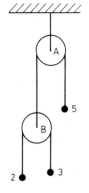

The figure shows a light inextensible string which passes over a smooth light fixed pulley *A* and carries at one end a mass 5 kg and at the other end a smooth light pulley *B*. A light inextensible string passes over the pulley *B* and carries masses 3 kg and 2 kg at its ends. The system is in motion with the masses moving vertically. Find the tensions in the two moving strings stating the units. (C)

20. A crate of mass 50 kg is suspended by a light cable from a helicopter. The helicopter has a horizontal acceleration of $2\,\text{m}\,\text{s}^{-2}$. The crate is being winched aboard the helicopter with an acceleration relative to the helicopter of $0.5\,\text{m}\,\text{s}^{-2}$. Find the tension in the cable and its angle of inclination to the vertical.

18 Friction; Hooke's law

18.1 The laws of friction

When two objects are in contact each exerts a force upon the other. As stated in Newton's third law, these forces are equal and opposite. The force exerted by a smooth surface on a body in contact with it is perpendicular to the surface. Thus there is no resistance to sliding motion across the surface. However, the force exerted by a rough surface on a body has a component perpendicular to the surface, but may also have a component which acts along or tangentially to the surface and which opposes sliding or a tendency to slide. These components are referred to as the *normal reaction* **N** and the *frictional force* **F**. Their resultant is called the *total reaction* **R**.

It can be shown experimentally that the following laws provide a fairly good approximation to the behaviour of frictional forces.
(1) When two rough surfaces are in contact friction acts to oppose the sliding of one surface relative to the other.
(2) When there is no sliding the frictional force is just sufficient to prevent the relative motion of the surfaces. However, the friction between two surfaces cannot exceed a maximum value called *limiting friction.*
(3) The limiting value of the frictional force F is proportional to the normal reaction N. Thus the maximum value of F is μN, where μ is a constant called the *coefficient of friction.*
(4) The value of μ depends only on the nature of the two surfaces involved.
(5) When sliding occurs the frictional force takes its limiting value μN and acts in the direction that opposes the relative motion of the surfaces.

[In practice, the magnitude of the frictional force that acts during sliding is found to be slightly less than the limiting value reached just before sliding begins. Thus for any pair of surfaces the coefficient of sliding friction is slightly less than the coefficient of static friction. However, in elementary work the difference is assumed to be negligible and a single value of μ is used.]

Example 1　A block of weight $50\,\text{N}$ rests in equilibrium on a rough horizontal plane. It is attached to a string inclined at $30°$ to the horizontal and the tension in the string is $30\,\text{N}$. Find the magnitude of the frictional force acting on the block.

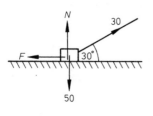

Resolving horizontally

$$\leftarrow F - 30\cos 30° = 0$$

$$\therefore \quad F = 30 \times \frac{\sqrt{3}}{2} = 15\sqrt{3}.$$

Hence the magnitude of the frictional force is $15\sqrt{3}\,\text{N}$.

Example 2　A block of weight $50\,\text{N}$ rests on a horizontal plane. The coefficient of friction between the block and the plane is $0\cdot4$. A horizontal force of $P\,\text{N}$ is applied to the block. Find the value of P given that the block is just about to slide.

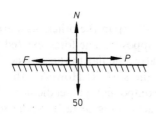

Resolving horizontally and vertically,

$$\leftarrow F - P = 0, \quad \text{i.e.} \quad F = P$$
$$\uparrow N - 50 = 0, \quad \text{i.e.} \quad N = 50.$$

When the block is just about to slide friction is limiting, i.e. $F = \mu N$

$$\therefore \quad P = 0\cdot4 \times 50 = 20.$$

Example 3　A block of mass $5\,\text{kg}$ is sliding down a plane inclined at $35°$ to the horizontal. If the coefficient of friction between the block and the plane is $0\cdot3$, find the acceleration of the block.

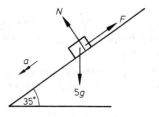

Let the acceleration of the block be $a\,\text{m}\,\text{s}^{-2}$ down the plane.

Applying Newton's second law parallel and perpendicular to the plane

$$\diagup \quad 5g\sin 35° - F = 5a \qquad (1)$$
$$\diagdown \quad N - 5g\cos 35° = 0 \qquad (2)$$

Since the block is sliding, $F = \mu N$

$\therefore$　using (2), $F = 0\cdot3 \times 5g\cos 35°$.

Substituting in (1)　$5g\sin 35° - 0\cdot3 \times 5g\cos 35° = 5a$

$$\therefore \quad a = g(\sin 35° - 0\cdot3\cos 35°) \approx 3\cdot21.$$

Hence the acceleration of the block down the plane is $3\cdot21\,\text{m}\,\text{s}^{-2}$.

Exercise 18.1

1. A block of weight 20 N rests in equilibrium on a rough horizontal plane under the action of a force of 10 N. Find the magnitude of the frictional force on the block, given that the external force acts (a) horizontally, (b) vertically downwards, (c) downwards at 30° to the vertical.

2. A block of weight 20 N rests in equilibrium on a rough plane inclined at 30° to the horizontal. Find the magnitude and direction of the frictional force on the block, given that there is a force on the block acting up the plane, which has magnitude (a) 15 N, (b) 10 N, (c) 5 N.

3. A block of weight 12 N rests on a horizontal plane. The coefficient of friction between the block and the plane is $\frac{1}{4}$. A force of P N is applied to the block, such that the block is just about to slide. Find the value of P, given that this force acts (a) horizontally, (b) upwards at 45° to the horizontal, (c) downwards at 45° to the horizontal.

4. A block of weight W rests on a rough plane inclined at 25° to the horizontal. Given that the block is just about to slide down the plane, find the coefficient of friction.

5. A particle of weight 5 N is at rest on a plane inclined at 40° to the horizontal under the action of a force of P N up the plane. The coefficient of friction is 0·2. Find the value of P if the particle is just about to slide (a) up the plane, (b) down the plane.

6. A particle of mass 2 kg is being pulled across a horizontal plane by a horizontal force of magnitude 8 N. Find the acceleration of the particle if the coefficient of friction is $\frac{1}{3}$.

7. A particle of mass 3 kg is being pulled across a rough horizontal plane by a string inclined at 35° to the horizontal. The coefficient of friction between the particle and the plane is 0·6. Find the acceleration of the particle when the tension in the string is 30 N.

8. A particle of mass 10 kg is sliding down a plane inclined at 30° to the horizontal. If the coefficient of friction is 0·2, find the acceleration of the particle.

9. A particle of mass 4 kg is sliding down a plane inclined at 20° to the horizontal. If its acceleration is $2 \, \mathrm{m \, s^{-2}}$, find the coefficient of friction between the particle and the plane.

10. A particle of mass 5 kg is being pulled up a plane inclined at 45° to the horizontal by a string parallel to the plane. If the coefficient of friction is $\frac{1}{4}$ and the tension in the string is 50 N, find the acceleration of the particle.

18.2 Motion on a rough surface

When a particle is in motion on a rough surface the frictional force F and the normal reaction N are connected by the equation $F = \mu N$. Many problems can be solved using this relationship together with the methods of Chapter 17.

Example 1   A particle of mass 4 kg is being pulled across a rough horizontal plane by a string inclined at an angle of $20°$ above the horizontal. When the tension in the string is 25 N the particle accelerates at $4\,\mathrm{m\,s^{-2}}$. Find the coefficient of friction between the particle and the plane.

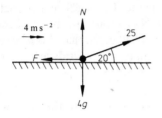

Applying Newton's second law horizontally and vertically

$$\rightarrow 25\cos 20° - F = 4 \times 4$$
$$\uparrow N + 25\sin 20° - 4g = 0$$
$$\therefore F = 25\cos 20° - 16$$
$$N = 4g - 25\sin 20°.$$

Since the particle is in motion, $F = \mu N$

$$\therefore \quad \mu = \frac{F}{N} = \frac{25\cos 20° - 16}{4 \times 9\cdot 8 - 25\sin 20°} \approx 0\cdot 244.$$

Hence the coefficient of friction is $0\cdot 244$.

Example 2   A particle of mass 6 kg rests on a rough horizontal plane, the coefficient of friction between the particle and the plane being $0\cdot 4$. The particle is connected by a light inextensible string passing over a smooth pulley at the edge of the plane to a second particle of mass 3 kg which hangs freely. If the system is released from rest, find the distance moved by each of the particles in the first 3 seconds of the subsequent motion.

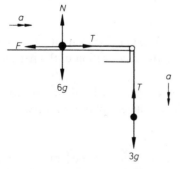

Let the tension in the string be T N and the acceleration of the particles $a\,\mathrm{m\,s^{-2}}$.

Using Newton's second law for the 6 kg mass:

$$\rightarrow \quad T - F = 6a \qquad (1)$$
$$\uparrow N - 6g = 0 \qquad (2)$$

for the 3 kg mass: $\downarrow 3g - T = 3a \qquad (3)$

When the 6 kg mass is sliding $\qquad F = \mu N$

$\therefore$ using (2) $\qquad F = 0.4N = 0.4 \times 6g = 2.4g$

Substituting in (1) $\qquad T - 2.4g = 6a$

Adding (3) $\qquad 0.6g = 9a$

$$\therefore \quad a = \frac{1}{9} \times \frac{3}{5}g = \frac{1}{15}g.$$

Using the formula $s = ut + \frac{1}{2}at^2$, where $u = 0$, $t = 3$, $a = \dfrac{g}{15}$

$$s = 0 + \frac{1}{2} \times \frac{g}{15} \times 9 = \frac{3g}{10} = 2.94.$$

Hence the distance moved by the particles in the first 3 seconds is 2·94 m.

Example 3   A particle A of mass 3 kg rests on a rough plane inclined at $30°$ to the horizontal, the coefficient of friction between the particle and the plane being $1/10$. It is connected by a light inextensible string passing over a smooth pulley at the top edge of the plane to a particle B of mass 2 kg which hangs freely. Find the acceleration of the system when it is released from rest and the tension in the string.

The component of the weight of particle A acting down the plane is $3g \sin 30°$, i.e. $\dfrac{3}{2}g$. Since this is less than the weight of particle B, A will have a tendency to move up the plane. Thus the force of friction will act down the plane.

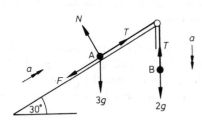

Using Newton's second law, for particle A:

$\nearrow \quad T - F - 3g \sin 30° = 3a$ $\qquad$ (1)

$\searrow \quad N - 3g \cos 30° = 0$ $\qquad$ (2)

for particle B:

$\downarrow 2g - T = 2a.$ $\qquad$ (3)

When particle A is sliding $\qquad F = \mu N$

$\therefore$ using (2) $\qquad F = \dfrac{1}{10}N = \dfrac{3}{10}g \cdot \dfrac{\sqrt{3}}{2} = \dfrac{3\sqrt{3}}{20}g.$

Substituting in (1) $\qquad T - \dfrac{3\sqrt{3}}{20}g - 3g \cdot \dfrac{1}{2} = 3a.$

Adding (3) $\qquad 2g - \dfrac{3\sqrt{3}}{20}g - \dfrac{3}{2}g = 5a.$

$$\therefore \quad a = \frac{1}{5}g\left(\frac{1}{2} - \frac{3\sqrt{3}}{20}\right) = \frac{g}{100}(10 - 3\sqrt{3}) \approx 0\cdot47.$$

Rearranging (3) $T = 2(g-a) \approx 18\cdot7.$

Hence the acceleration of the particles is approximately $0\cdot47\,\text{m s}^{-2}$ and the tension in the string $18\cdot7\,\text{N}$.

Exercise 18.2

1.   A particle of mass $4\,\text{kg}$ is projected across a horizontal plane with a speed of $7\,\text{m s}^{-1}$. If it comes to rest in a distance of $10\,\text{m}$, find the magnitude of the frictional force acting on the particle and the coefficient of friction between the particle and the plane.

2.   A body of mass $5\,\text{kg}$ is at rest on a horizontal plane, the coefficient of friction being $0\cdot3$. A horizontal force of $21\,\text{N}$ is applied to the body. If after $3\frac{1}{2}$ seconds this force ceases to act, find the further time that elapses before the particle comes to rest.

3.   A block of mass $2\,\text{kg}$ rests on a horizontal plane. The coefficient of friction is $\frac{1}{2}$. Find the acceleration of the block when a force of $20\,\text{N}$ acts on it (a) upwards at an angle of $20°$ to the horizontal, (b) downwards at an angle of $20°$ to the horizontal.

4.   A particle slides down a plane inclined at an angle of $30°$ to the horizontal. The coefficient of friction is $1/7$. Find the distance the particle slides in 2 seconds from rest.

5.   A particle is sliding down a plane inclined at an angle α to the horizontal. Find an expression for the coefficient of friction, given that the particle moves with (a) uniform velocity, (b) constant acceleration a.

6.   A block of mass $2\,\text{kg}$ rests on a rough plane inclined at an angle α to the horizontal, where $\sin\alpha = 7/25$. The coefficient of friction is $\frac{1}{2}$. Find the acceleration of the block when a force of $50\,\text{N}$ acts on it down the plane. Find the magnitude of the force acting parallel to the plane that would give the block an equal acceleration up the plane.

7.   A body of mass $8\,\text{kg}$ is projected down a plane inclined at an angle of $15°$ to the horizontal with a velocity of $10\,\text{m s}^{-1}$. If the body comes to rest after travelling a distance of $25\,\text{m}$, find the magnitude of the frictional force acting on the body during its motion and the coefficient of friction.

8.   A particle is placed at a point O on a plane inclined at $45°$ to the horizontal. The particle is projected up the plane along a line of greatest slope with velocity $15\,\text{m s}^{-1}$. Find the condition that must be satisfied by the coefficient of friction μ if the particle is to return through 0 after it comes to rest. Given that the particle does return through 0 with a velocity of $10\,\text{m s}^{-1}$, find the value of μ.

9. A particle of mass 3 kg rests on a rough horizontal table, the coefficient of friction between the particle and the table being $\frac{1}{3}$. The particle is connected by a light inextensible string passing over a smooth pulley at the edge of the table to a second particle of mass 4 kg which hangs freely. If the system is released from rest, find the acceleration of the particles and the tension in the string.

10. A particle A of mass 6 kg rests on a rough horizontal plane. It is connected by a light inextensible string passing over a smooth pulley at the edge of the plane to a particle B of mass 2 kg which hangs vertically. If particle B is projected vertically downwards with a velocity of $3 \cdot 5 \, \text{m s}^{-1}$, it comes to rest after travelling a distance of $2 \cdot 5 \, \text{m}$. Find the coefficient of friction between particle A and the plane. Find also the force exerted by the string on the pulley during the motion.

11. A particle of mass 5 kg rests on a rough plane inclined at 25° to the horizontal, the coefficient of friction between the particle and the plane being 0·2. The particle is connected by a light inextensible string passing over a smooth pulley at the top of the plane to a second particle of mass 5 kg which hangs freely. Find the acceleration of the system and the tension in the string.

12. A particle of mass 5 kg rests on a rough plane inclined at an angle α to the horizontal, the coefficient of friction between the particle and the plane being 1/8. The particle is connected by a light inextensible string passing over a smooth pulley at the top of the plane to a particle of mass 3 kg which hangs freely. Find the acceleration of the system when it is released from rest if (a) $\alpha = 10°$, (b) $\alpha = 60°$.

18.3 Equilibrium on a rough surface

When a particle is in equilibrium on a rough surface the frictional force F and the normal reaction N satisfy the inequality $F \leqslant \mu N$. When the particle is on the point of sliding $F = \mu N$. This state is sometimes called limiting equilibrium.

Example 1 A particle is placed on a rough plane inclined at an angle α to the horizontal. The coefficient of friction between the particle and the plane is μ. Find the condition that the particle should be in equilibrium.

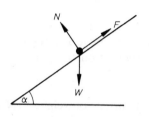

If the particle is in equilibrium, resolving along and perpendicular to the plane,

$$\nearrow \quad F - W \sin \alpha = 0$$
$$\nwarrow \quad N - W \cos \alpha = 0$$

i.e. $F = W \sin \alpha$ and $N = W \cos \alpha$.

Since $F \leqslant \mu N$, $\qquad W \sin \alpha \leqslant \mu W \cos \alpha \qquad \therefore \quad \tan \alpha \leqslant \mu$.

Hence equilibrium will be possible if $\mu \geqslant \tan \alpha$.

In some problems it is necessary to find the least force required to maintain the equilibrium of a particle on a rough surface. As this least force just prevents motion, when it is applied the frictional force F takes its limiting value μN.

Example 2   A particle of weight 6 N is placed on a rough plane inclined at 45° to the horizontal, the coefficient of friction being $\frac{1}{2}$. Find the magnitude of the least horizontal force required to maintain equilibrium.

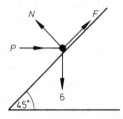

Consider a horizontal force of P N acting on the particle to maintain equilibrium.

Resolving along and perpendicular to the plane

$$\nearrow \quad P\cos 45° + F - 6\sin 45° = 0$$
$$\nwarrow \quad N - P\sin 45° - 6\cos 45° = 0$$

$$\therefore \quad F = 6\sin 45° - P\cos 45° = \frac{1}{\sqrt{2}}(6 - P)$$

$$N = 6\cos 45° + P\sin 45° = \frac{1}{\sqrt{2}}(6 + P).$$

The value of P is least when friction is limiting, i.e. when

$$F = \mu N$$

$$\therefore \quad \text{since} \quad \mu = \frac{1}{2}, \quad \frac{1}{\sqrt{2}}(6 - P) = \frac{1}{2} \cdot \frac{1}{\sqrt{2}}(6 + P)$$

i.e. $12 - 2P = 6 + P$ $\therefore \quad P = 2$

Hence the least horizontal force required to maintain equilibrium is 2 N.

[A similar method can be used to find the least force required to move a particle across a rough surface.]

Exercise 18.3

1.   A particle is at rest on a rough plane inclined at an angle α to the horizontal. The coefficient of friction is $\frac{1}{4}$. Given that the particle is on the point of sliding down the plane, find α.

2.   A horizontal force of magnitude P N acts on a particle of weight 10 N at rest on a plane inclined at 40° to the horizontal. The coefficient of friction between the particle and the plane is 0·2. Find the value of P if the particle is just about to slide (a) up the plane, (b) down the plane.

3. A particle of weight 3 N rests on a rough horizontal plane. The coefficient of friction is μ. A force of 1 N acts on the particle. Find the condition that the particle should be in equilibrium if this force acts (a) horizontally, (b) upwards at 30° to the horizontal, (c) downwards at 30° to the horizontal.

4. A body of weight 4 N is placed on a rough plane inclined at an angle α to the horizontal. The coefficient of friction between the body and the plane is $\frac{1}{2}$. A force of 1·5 N acting downwards parallel to the plane is applied to the particle. Prove that the condition that the particle should be in equilibrium is $3 \sec \alpha + 8 \tan \alpha \leqslant 4$.

5. A body of weight 6 N is kept in equilibrium on a rough plane inclined at 30° to the horizontal by a force of P N acting up the plane. If the least value of P for which equilibrium is maintained is 1·5, find the coefficient of friction.

6. A block of weight 5 N is on a plane inclined at 50° to the horizontal. The coefficient of friction is 0·3. The block is kept in equilibrium by a force of P N acting up the plane. Find the range of possible values of P.

7. A particle of weight 20 N rests on a rough plane inclined at an angle α to the horizontal, where $\tan \alpha = \frac{3}{4}$. The coefficient of friction is $\frac{1}{4}$. The particle moves up the plane under the action of a force of P N. Find the least value of P, given that this force acts (a) up the plane, (b) horizontally.

8. A particle of mass M rests on a rough horizontal plane, the coefficient of friction being $\frac{1}{3}$. The particle is connected by a light inextensible string passing over a smooth pulley at the edge of the plane to a particle of mass m which hangs vertically. Find the condition that the system should remain at rest when released.

18.4 Angle of friction

The total reaction **R** of a surface on a body in contact with it is the resultant of the normal reaction **N** and the frictional force **F**.

If **R** acts at an angle θ to the normal to the surface then

$$F = R \sin \theta \quad \text{and} \quad N = R \cos \theta.$$

Since $F \leqslant \mu N$, $R \sin \theta \leqslant \mu R \cos \theta$, i.e. $\tan \theta \leqslant \mu$.

If λ is the acute angle such that $\tan \lambda = \mu$, then $\tan \theta \leqslant \tan \lambda$ and hence $\theta \leqslant \lambda$. The angle λ is called the *angle of friction*. It is the maximum angle which the total reaction can make with the normal to the surface. Thus the total reaction can lie anywhere on or within a 'cone of friction' with the normal to the surface as axis and λ as semi-vertical angle.

Some problems are simplified by using the total reaction $\mathbf{R}$ instead of its components $\mathbf{F}$ and $\mathbf{N}$.

Example 1   A particle of weight $\mathbf{W}$ is suspended by a light inextensible string from a light ring. The ring can slide along a rough horizontal rod, the coefficient of friction being $\frac{1}{3}$. A horizontal force $\mathbf{P}$ is applied to the particle. Given that the system is at rest with the ring on the point of sliding along the rod, find P in terms of W.

Since the only forces acting on the ring are the tension in the string and the reaction of the rod on the ring, these forces must be equal and opposite. If the reaction makes an angle θ with the vertical, then since the ring is about to slide, $\theta = \lambda$.

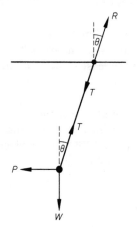

Resolving the forces on the particle horizontally and vertically

$$\leftarrow \quad P - T \sin \lambda = 0$$
$$\downarrow \quad W - T \cos \lambda = 0$$
$$\therefore \quad P = T \sin \lambda, \quad W = T \cos \lambda,$$

Hence $\dfrac{P}{W} = \tan \lambda = \frac{1}{3}$ so that

$$P = \tfrac{1}{3}W.$$

Example 2   A particle of weight $\mathbf{W}$ is placed on a rough plane inclined at an angle α to the horizontal. The coefficient of friction is μ, where $\mu < \tan \alpha$. Find the magnitude and direction of the least force required to prevent motion down the plane.

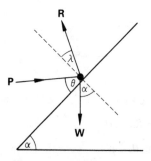

Suppose that sliding is prevented by a force of magnitude P acting at an angle θ to the plane. When P takes its least value, friction will be limiting and the reaction $\mathbf{R}$ will make an angle λ with the normal to the plane, where $\tan \lambda = \mu$.

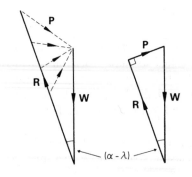

Assuming that the particle is on the point of sliding, the value of P depends on the value chosen for θ. By considering the triangle of forces as θ varies, we see that the magnitude of **P** is least when its direction is perpendicular to that of **R**, i.e. when $\theta = \lambda$. Hence the least force required to prevent motion down the plane is of magnitude $W \sin(\alpha - \lambda)$ and acts at an angle λ to the plane where $\tan \lambda = \mu$.

Exercise 18.4

1. A particle of weight $12\,\text{N}$ is suspended by a light inextensible string from a light ring. The ring can slide along a rough horizontal rod, the coefficient of friction being $\frac{1}{2}$. A force of $P\,\text{N}$ is acting on the particle. The system is in equilibrium with the ring on the point of sliding along the rod. Find P given that the force is acting (a) horizontally, (b) upwards at $45°$ to the horizontal, (c) downwards at $30°$ to the horizontal.

2. One end of a light inextensible string is attached to a fixed point on a horizontal rod. The other end is fastened to a light ring which can slide on the rod. The coefficient of friction is $\frac{3}{4}$. A particle of weight $20\,\text{N}$ is suspended from the mid-point of the string. Given that the system is in equilibrium with the ring on the point of sliding along the rod, find the tension in each part of the string and the angle of inclination to the vertical.

3. A block of weight $8\,\text{N}$ rests on a horizontal plane. The coefficient of friction between the block and the plane is $\frac{1}{4}$. Find the magnitude and direction of the least force required to move the block.

4. A particle of weight W rests on a horizontal plane. If the angle of friction is λ, find the magnitude and direction of the least force required to move the block.

5. A particle of weight $6\,\text{N}$ is placed on a rough plane inclined at $30°$ to the horizontal. The coefficient of friction is $\frac{1}{3}$. Find the magnitude and direction of the least force required (a) to prevent motion down the plane, (b) to move the particle up the plane.

6. A particle of weight W is placed on a rough plane inclined at an angle α to the horizontal. The angle of friction is λ, where $\lambda > \alpha$. Find the magnitude and direction of the least force required to move the particle (a) down the plane, (b) up the plane.

18.5 Hooke's† law

In this section we examine the behaviour of springs and elastic strings. The length of an elastic string when it is unstretched is called its *natural length*. When the string is stretched by forces exerted on its ends, its increase in length is called the *extension* of the string. It can be shown experimentally that as an elastic string is stretched the tension in the string is proportional to the extension. This result is called *Hooke's law* after the man who discovered it.

When a string is stretched beyond a certain point it ceases to obey Hooke's law. However, in elementary work with elastic strings it is assumed that conditions are such that Hooke's law can be used.

For a string of natural length l, tension T and extension x, the law is written in the form

$$T = \lambda \frac{x}{l}$$

where λ is a constant called the *modulus of elasticity*. The value of λ depends on the nature of the string in use. Since x/l is a ratio of two lengths it has a purely numerical value. Hence the modulus λ must be measured in the same units as T, i.e. in units of force. By writing $x = l$, we see that λ is equal to the tension required to double the length of an elastic string.

Example 1   An elastic string of natural length $1 \cdot 5$ m is stretched to 2 m. If its modulus of elasticity is 12 N, find the tension in the string.

By Hooke's law $$T = \frac{\lambda x}{l}.$$

Putting $\lambda = 12, l = 1 \cdot 5, x = 0 \cdot 5$:

$$T = \frac{12 \times 0 \cdot 5}{1 \cdot 5} = 4$$

∴   the tension in the string is 4 N.

Example 2   A light elastic string of natural length 1 m is fixed at one end and a particle of weight 4 N is attached to the other end. When the particle hangs freely

† *Hooke, Robert* (1635–1703) English experimental physicist. He contributed to many of the important inventions of his time including the microscope. He worked on planetary motion and proposed a law of gravitation similar to Newton's. A pamphlet containing his theory of elasticity was published in 1678.

in equilibrium the length of the string is 1·4 m. The string is now held at an angle θ to the vertical by a horizontal force of 3 N acting on the particle. Find the value of θ and the new length of the string.

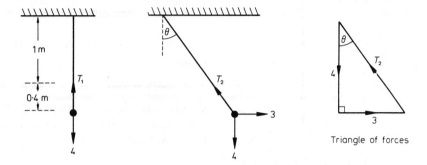

Triangle of forces

In the first equilibrium position $T_1 = 4$.

Using Hooke's law, $T = \dfrac{\lambda x}{l}$, where $T = 4$, $x = 0·4$, $l = 1$,

$$4 = \frac{\lambda \times 0·4}{1} \qquad \therefore \quad \lambda = 10.$$

Hence the modulus of elasticity for the string is 10 N.

Using the triangle of forces for the second equilibrium position

$$\tan \theta = \tfrac{3}{4} \qquad \therefore \quad \theta = 36° \, 52'$$
$$T_2 = \sqrt{(3^2 + 4^2)} = 5.$$

Using Hooke's law for $T = 5$, $\lambda = 10$, $l = 1$

$$5 = \frac{10x}{1} \qquad \therefore \quad x = 0·5.$$

Hence the value of θ is 36° 52' and the new length of the string is 1·5 m.

Hooke's law also applies to the stretching of springs.

Example 3 A spring AB of natural length 0·8 m and modulus 16 N is fixed at A. The other end is joined to a second spring BC of natural length 0·5 m and modulus 20 N. A particle of weight W N is then attached at the point C. When the system is hanging freely in equilibrium, the point C is a distance of 2·5 m vertically below A. Find the value of W.

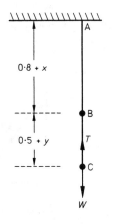

Let the extensions in springs AB and BC be x m and y m respectively.

Applying Hooke's law to both springs

$$T = \frac{16x}{0\cdot8} = 20x \quad \text{and} \quad T = \frac{20y}{0\cdot5} = 40y$$

Since the tensions in the springs are equal,

$$20x = 40y, \quad \text{i.e.} \quad x = 2y \qquad (1)$$

Considering the two springs together, the natural length is $1\cdot3$ m and the stretched length is $2\cdot5$ m.

Hence $\qquad x+y = 2\cdot5 - 1\cdot3 = 1\cdot2$

Using (1) $\qquad 3y = 1\cdot2, \qquad \text{i.e.} \quad y = 0\cdot4$

$$\therefore \quad T = 40y = 40 \times 0\cdot4 = 16.$$

Since the particle is in equilibrium, $\quad W = T = 16$.

Thus the value of W is 16.

A spring also obeys Hooke's law when compressed. We again write $T = \lambda x/l$, where in this case T represents a thrust and x a compression, i.e. a reduction in length.

Exercise 18.5

1. An elastic string of natural length 2 m is stretched to $2\cdot5$ m. If its modulus of elasticity is 20 N, find the tension in the string.

2. A spring of natural length 60 cm is stretched to a length of 1 m. If the tension in the spring is 6 N, find its modulus of elasticity.

3. A spring of natural length 10 cm is compressed to a length of 8 cm. If its modulus of elasticity is 12 N, find the thrust in the spring.

4. When the tension in an elastic string is 8 N, its length is $1\cdot6$ m. If the modulus of elasticity of the string is 24 N, find its natural length.

5. An elastic string of natural length $3a$ is fixed at one end and a particle of weight W is attached to the other end. When the particle hangs freely in equilibrium the length of the string is $5a$. If the string is held at an angle θ to the vertical by a horizontal force of magnitude $\frac{1}{2}W$, find the value of θ and the new length of the string.

6. A particle of mass 3 kg is suspended from an elastic string of natural length $0\cdot5$ m and modulus 48 N. If the particle is pulled vertically downwards and then

released when the length of the string is 1 m, find its acceleration at the instant that it is released.

7. An elastic string AB of natural length 1·2 m and modulus 10 N lies along a line of greatest slope of a smooth plane inclined at 30° to the horizontal. The end A is fixed and to the end B is attached a particle of weight 10 N. Find the length of the string when the particle at B rests in equilibrium on the plane.

8. The ends of an elastic string of natural length $4a$ are fixed to points A and B on the same horizontal level, where $AB = 3a$. A particle P of weight W is attached to the mid-point of the string and hangs in equilibrium at a depth of $2a$ below the level of AB. Find the modulus of elasticity of the string in terms of W.

9. A particle P of weight 5·2 N is hanging in equilibrium attached to two strings PA and PB, where A and B are fixed points on the same horizontal level. The string PA is inextensible, but the string PB is elastic with modulus of elasticity 6 N. Given that $AB = 1·3$ m, $PA = 0·5$ m and $\angle APB = 90°$, find the tension in each string and the natural length of the elastic string.

10. A spring PQ of natural length 1·5 m and modulus λ N is fixed at P. The other end is joined to a second spring QR of natural length 1 m and modulus 2λ N. A particle of weight 15 N is then attached to the end R of the second spring. When the system is hanging freely in equilibrium, the distance PR is 4 m. Find the value of λ.

11. Three light springs of natural lengths 80 cm, 100 cm and 120 cm whose moduli of elasticity are 320 N, 240 N and 180 N respectively are attached end to end and confined in a horizontal tube of length 276 cm. Calculate the thrust on the ends of the tube and the amount by which each spring is compressed.

12. A body of weight 50 N is suspended from a ceiling by a spring of natural length 1 m and modulus of elasticity 120 N. It is supported by another spring of natural length 2 m and modulus 160 N attached to the floor. The springs are vertical, and the distance from floor to ceiling is 3 m. Find the lengths of the springs.

Exercise 18.6 (*miscellaneous*)

1. A box of weight 400 N will just move across a rough horizontal floor when a horizontal force of 100 N is applied to it. Find the least force required to move the box if it acts (a) upwards at 60° to the horizontal, (b) downwards at 30° to the horizontal.

2. A particle of weight W rests on a horizontal plane, the coefficient of friction between the particle and the plane being $\frac{1}{3}$. When a horizontal force of magnitude P is applied to the particle its acceleration is a. When a horizontal force of magnitude $2P$ is applied, the acceleration is $4a$. Find P in terms of W.

3.   A particle of mass 2 kg is placed on a rough plane inclined at an angle α to the horizontal, where $\sin \alpha = 3/5$. The coefficient of friction between the particle and the plane is $1/4$. Initially the particle is at rest under the action of a force P N acting upwards parallel to the plane. Find the value of P given that the particle is about to move (a) up the plane, (b) down the plane. If the force is then removed, find the acceleration of the particle down the plane.

4.   A block of weight W is placed on a rough plane inclined at an angle α to the horizontal, where $\tan \alpha = 5/12$. The block is on the point of sliding when a force of 5 N acts directly down the plane and when a force of 55 N acts directly up the plane. Find the value of W and the coefficient of friction between the block and the plane.

5.   A particle of weight W rests on a rough plane inclined to the horizontal at an angle whose tangent is 2μ, where μ is the coefficient of friction between the particle and the plane. The particle is acted on by a force of magnitude P. (i) Given that the force acts horizontally in a vertical plane through a line of greatest slope, and that the particle is on the point of sliding down the plane, find the magnitude of P. (ii) Given that the force acts along a line of greatest slope and that the particle is on the point of sliding up the plane, find the magnitude of P.   (C)

6.   Two small light rings which are connected by a light inextensible string of length a can slide on a rough horizontal rod. A weight is attached to the mid-point of the string and the system hangs in equilibrium. If the coefficient of friction between the rings and the rod is μ, show that the greatest possible distance between the rings is $\mu a/(1+\mu^2)^{1/2}$.   (O)

7.   A mass of 5 kg is moved along a rough horizontal table by means of a light inextensible string which passes over a smooth light pulley at the edge of the table and is attached to a mass of 1 kg hanging over the edge of the table. The system is released from rest with the string taut, each portion being at right angles to the edge of the table. If the masses take twice the time to acquire the same velocity from rest that they would have done had the table been smooth, prove that the coefficient of friction is $1/10$. After falling a distance $1/5$ m, the 1 kg mass reaches the floor and comes to rest. Prove that the 5 kg mass then moves a further $1/6$ m before coming to rest. [Assume throughout that the 5 kg mass does not leave the table.]   (O & C)

8.   Two particles A and B, of masses 5 kg and 10 kg respectively, lie with B above A on a rough slope inclined at an angle of 30° to the horizontal. The coefficient of friction between the particles and the slope is 0·2. A light inextensible string along a line of greatest slope joins A to B and a force of P N acts on B up the slope. If the system is in equilibrium with both particles just about to move up the slope, find the value of P and the tension in the string.

9.   Three particles, A, B, C, are of masses 4, 4, 2 kg respectively. They lie at rest on a horizontal table in a straight line, with particle B attached to the mid-point of a light inextensible string. The string has particle A attached at one end and

particle C at the other, and is taut. A force of 60 N is applied to A in the direction CA produced, and a force of 15 N is applied to C in the opposite direction. Find the acceleration of the particles and the tension in each part of the string, (a) if the table is smooth, (b) if the coefficient of friction between each particle and the table is $\frac{1}{4}$. [Take g as $10\,\text{m/s}^2$.] (L)

10. A particle of mass m moves on a line of greatest slope of a plane inclined to the horizontal at an angle θ. The coefficient of friction between the particle and the plane is μ. The particle is released from rest at a point A and it slides down to B, where the distance $AB = d$. Prove that $\mu < \tan\theta$, and that V, the particle's speed at B, is given by $V^2 = 2gd(\sin\theta - \mu\cos\theta)$. The particle is projected up the plane from B with speed U, comes instantaneously to rest at A and then returns to B. If $U = 2V$, prove that (i) $\mu = \frac{3}{5}\tan\theta$, (ii) $V^2 = \frac{4}{5}gd\sin\theta$. (C)

11. A particle of mass $2m$ rests on a rough plane inclined at an angle α to the horizontal. It is connected by a light inextensible string passing over a smooth pulley at the top edge of the plane to a second particle of mass m which hangs freely. Given that when the system is released from rest it begins to move, find the condition that must be satisfied by μ, the coefficient of friction if (a) $\tan\alpha = 3/4$, (b) $\tan\alpha = 5/12$.

12.

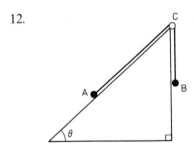

The diagram shows a fixed wedge whose sloping face is inclined at an angle θ to the horizontal. Two particles A and B are connected by a light inextensible string which passes over a smooth pulley C at the top of the wedge.

The particle A rests on the sloping face of the wedge, the string between A and C being parallel to a line of greatest slope, and B hangs freely. Particle A has mass M, particle B has mass m and the coefficient of friction between A and the sloping plane is μ. The system is released from rest.

(i) Show that, if $\sin\theta - \mu\cos\theta \leqslant \dfrac{m}{M} \leqslant \sin\theta + \mu\cos\theta$, the masses remain at rest.

(ii) If $M = m$, $\theta = 30°$ and $\mu = \frac{1}{2}$, find expressions for the acceleration of the system and the tension in the string. (C)

13. The inclined faces of a fixed triangular wedge make angles α and β with the horizontal. Two particles of equal weight, connected by a light inextensible string which passes without friction over the vertex of the wedge, rest on the two faces of the wedge. The system rests in equilibrium with the plane containing the string perpendicular to both faces of the wedge. If both particles are in limiting equilibrium, show that $\mu = \tan\frac{1}{2}|\alpha - \beta|$, where μ is the coefficient of friction between each particle and the face of the wedge. (O)

14. A body of weight W N is being pulled across a rough horizontal plane by a string inclined at an angle θ to the horizontal. The tension in the string is $\frac{1}{2}W$N and the coefficient of friction between the body and the plane is $\frac{1}{4}$. Find the value of θ for which the acceleration of the body is greatest and the magnitude of the maximum acceleration.

15. A particle of weight 4 N is placed on a rough plane inclined at an angle of 10° to the horizontal. The coefficient of friction between the particle and the plane is 0·2. Find the magnitude and direction of the least force required to move the particle (a) down the plane, (b) up the plane.

16.

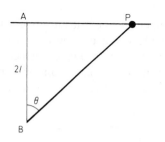

A small rough ring P of weight W is threaded on to a fixed horizontal straight wire which passes through a point A, as shown in the diagram. The ring is joined by a light elastic string of natural length l and modulus λ to a fixed point B which is at a distance $2l$ vertically below A. The ring is in equilibrium with the angle $ABP = \theta$. Find the tension in the string and the vertical and horizontal components of the force exerted by the wire on the ring. Show that

$$W \geqslant \lambda(2-\cos\theta)\left(\frac{\tan\theta}{\mu} - 1\right),$$

where μ is the coefficient of friction between the ring and the wire.   (JMB)

17. A particle of mass $3m$ is tied to the end C and a particle of mass m is tied at the mid-point B of a light unstretched elastic string ABC. The end A of the string

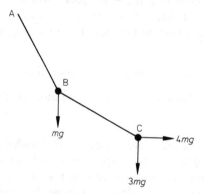

is fixed and a horizontal force of magnitude $4mg$ is applied to the particle at C so that the system hangs in equilibrium as shown. Calculate (i) the tensions in BC and AB, (ii) the inclinations of BC and AB to the vertical. Given that the modulus of elasticity of the string is $6mg$, show that, for this position of equilibrium, $AB:BC = (6+4\sqrt{2}):11$.   (AEB 1978)

18. A light elastic string is of natural length a, and is such that when it is stretched by a length a, the tension in it is mg.
(i) The string is fastened to two points, $2a$ apart, which are in the same

horizontal line. A particle of mass m is attached to the mid-point, and gently lowered until it comes to rest. Show that, in its equilibrium position, each of the parts of the string is inclined at θ to the horizontal, where θ is given by $2 \tan \theta - \sin \theta = \frac{1}{2}$.

(ii) The same string is now fastened to two points, $2a$ apart, which are in the same vertical line. A particle of mass m is attached to the mid-point and gently released until it comes to rest. Show that, when the system comes to rest, the particle is $\frac{1}{4}a$ below its original position.   (SU)

19. A wedge of mass M is resting on a smooth horizontal table. A particle of mass m is placed gently on a rough face of the wedge which is inclined at an angle θ to the horizontal, and slides down this face. Given that the coefficient of friction between the particle and the face is μ, where $\mu < \tan \theta$, prove that the acceleration of the wedge is $\dfrac{mg \cos \theta(\sin \theta - \mu \cos \theta)}{M + m \sin \theta(\sin \theta - \mu \cos \theta)}$.   (O)

19 Matrices

19.1 Introduction to matrices

A *matrix* is a rectangular array of numbers, called *elements* or *entries*.

For example, $\begin{pmatrix} 4 & 2 & 0 \\ 1 & -3 & 4 \end{pmatrix}$ is a 2×3 matrix, or a matrix of *order* 2×3, because it has 2 *rows* and 3 *columns*.

Similarly, $\begin{pmatrix} 5 & -2 \\ 3 & -4 \end{pmatrix}$ is a 2×2 matrix, but may also be referred to as a *square matrix* of order 2.

Two matrices are *equal* if they are of the same order and their corresponding elements are equal.

The most elementary type of matrix is simply a collection of numerical data in table form. For example, the table below shows the number of packs of butter, cheese and eggs that a wholesaler supplies every week to regular customers I, II, III and IV. As illustrated, this information can be given in matrix form.

	I	II	III	IV
Butter	3	1	0	1
Cheese	1	1	1	0
Eggs	0	1	2	2

$\begin{pmatrix} 3 & 1 & 0 & 1 \\ 1 & 1 & 1 & 0 \\ 0 & 1 & 2 & 2 \end{pmatrix}$.

If a second matrix is used to show the extra provisions required by these customers one holiday week-end, the calculation of the total orders can be written as follows:

$$\begin{pmatrix} 3 & 1 & 0 & 1 \\ 1 & 1 & 1 & 0 \\ 0 & 1 & 2 & 2 \end{pmatrix} + \begin{pmatrix} 2 & 0 & 0 & 0 \\ 1 & 1 & 0 & 2 \\ 1 & 0 & 1 & 1 \end{pmatrix} = \begin{pmatrix} 5 & 1 & 0 & 1 \\ 2 & 2 & 1 & 2 \\ 1 & 1 & 3 & 3 \end{pmatrix}.$$

This is an example of *matrix addition*. In general, if **A** and **B** denote two matrices of the same order, their sum **A**+**B** is found by adding corresponding elements. Since the addition of real numbers is commutative and associative, matrix addition must be commutative and associative, so that for matrices of the same order

$$\mathbf{A}+\mathbf{B} = \mathbf{B}+\mathbf{A} \quad \text{and} \quad (\mathbf{A}+\mathbf{B})+\mathbf{C} = \mathbf{A}+(\mathbf{B}+\mathbf{C}).$$

It is also possible to define the operation of subtraction for matrices. The difference **A**−**B** is obtained by subtracting from the elements of **A** the corresponding elements of **B**.

Any matrix whose elements are all zero is called a *zero matrix* (or null matrix) and written **0**. Clearly

$$\begin{pmatrix} a & b \\ c & d \end{pmatrix} + \begin{pmatrix} 0 & 0 \\ 0 & 0 \end{pmatrix} = \begin{pmatrix} 0 & 0 \\ 0 & 0 \end{pmatrix} + \begin{pmatrix} a & b \\ c & d \end{pmatrix} = \begin{pmatrix} a & b \\ c & d \end{pmatrix}.$$

Thus for any matrix **A** and the zero matrix **0** of the same order

$$\mathbf{A}+\mathbf{0} = \mathbf{0}+\mathbf{A} = \mathbf{A} \quad \text{and} \quad \mathbf{A}-\mathbf{A} = \mathbf{0}.$$

The multiplication of a matrix **A** by a scalar k is performed by multiplying each element of **A** by k, e.g.

$$\text{if} \quad \mathbf{A} = \begin{pmatrix} a & b \\ c & d \end{pmatrix}, \quad \text{then} \quad k\mathbf{A} = k\begin{pmatrix} a & b \\ c & d \end{pmatrix} = \begin{pmatrix} ka & kb \\ kc & kd \end{pmatrix}.$$

Returning to our earlier illustration, we see that the matrix

$$4\begin{pmatrix} 3 & 1 & 0 & 1 \\ 1 & 1 & 1 & 0 \\ 0 & 1 & 2 & 2 \end{pmatrix} = \begin{pmatrix} 12 & 4 & 0 & 4 \\ 4 & 4 & 4 & 0 \\ 0 & 4 & 8 & 8 \end{pmatrix}$$

might represent the provisions supplied to regular customers over a four-week period.

The prices of the items supplied are given in the table below.

	Butter	Cheese	Eggs
Price in £	15	20	6

The cost of the four customers' regular orders can now be calculated. For example, the cost in £ of the order supplied to customer $I = 15 \times 3 + 20 \times 1 + 6 \times 0 = 65$.

This illustrates the process by which matrix products are formed and may be written:

$$(15 \quad 20 \quad 6)\begin{pmatrix} 3 \\ 1 \\ 0 \end{pmatrix} = (15 \times 3 + 20 \times 1 + 6 \times 0) = (65).$$

Using a similar method to calculate the cost of the other customers' orders we obtain:

$$(15 \quad 20 \quad 6)\begin{pmatrix} 3 & 1 & 0 & 1 \\ 1 & 1 & 1 & 0 \\ 0 & 1 & 2 & 2 \end{pmatrix} = (65 \quad 41 \quad 32 \quad 27).$$

After a rise in prices a new calculation is required.

$$(18 \quad 21 \quad 8)\begin{pmatrix} 3 & 1 & 0 & 1 \\ 1 & 1 & 1 & 0 \\ 0 & 1 & 2 & 2 \end{pmatrix} = (75 \quad 47 \quad 37 \quad 34).$$

These results can be combined as follows:

$$\begin{pmatrix} 15 & 20 & 6 \\ 18 & 21 & 8 \end{pmatrix}\begin{pmatrix} 3 & 1 & 0 & 1 \\ 1 & 1 & 1 & 0 \\ 0 & 1 & 2 & 2 \end{pmatrix} = \begin{pmatrix} 65 & 41 & 32 & 27 \\ 75 & 47 & 37 & 34 \end{pmatrix}.$$

We see that the cost at second row prices of the third column provisions, i.e. $18.0 + 21.1 + 8.2 = 37$, appears in the second row and third column of the product matrix.

[It would clearly be unrealistic to suggest that a man running a small business should use matrix multiplication when handling customers' orders. However, since addition and multiplication of matrices are within the capabilities of a computer, procedures based on the matrix operations already outlined can be used by larger companies.]

In general, if the product $\mathbf{AB}$ of two matrices $\mathbf{A}$ and $\mathbf{B}$ exists, its elements are found by combining each row of $\mathbf{A}$ term by term with each column of $\mathbf{B}$, so that the element in the pth row and qth column of $\mathbf{AB}$ is calculated using the pth row of $\mathbf{A}$ and the qth column of $\mathbf{B}$, e.g.

$$\begin{pmatrix} a & b \\ c & d \end{pmatrix}\begin{pmatrix} r & s \\ t & u \end{pmatrix} = \begin{pmatrix} ar+bt & as+bu \\ cr+dt & cs+du \end{pmatrix}.$$

Example 1   If $\mathbf{P} = \begin{pmatrix} 2 & -1 \\ 0 & 1 \end{pmatrix}$ and $\mathbf{Q} = \begin{pmatrix} 4 & -2 & 0 \\ 0 & 1 & -3 \end{pmatrix}$, find if possible $\mathbf{PQ}$ and $\mathbf{QP}$.

$$\mathbf{PQ} = \begin{pmatrix} 2 & -1 \\ 0 & 1 \end{pmatrix}\begin{pmatrix} 4 & -2 & 0 \\ 0 & 1 & -3 \end{pmatrix}$$

$$= \begin{pmatrix} 2.4+(-1)0 & 2(-2)+(-1)1 & 2.0+(-1)(-3) \\ 0.4+1.0 & 0(-2)+1.1 & 0.0+1(-3) \end{pmatrix}$$

$$\therefore \quad \mathbf{PQ} = \begin{pmatrix} 8 & -5 & 3 \\ 0 & 1 & -3 \end{pmatrix}.$$

To find the element in the first row and first column of $\mathbf{QP}$ we would need to combine the row $(4 \quad -2 \quad 0)$ with the column $\begin{pmatrix} 2 \\ 0 \end{pmatrix}$. Since these have different numbers of elements, their product cannot be found. Thus the product $\mathbf{QP}$ does not exist.

This example demonstrates that it is possible to find a matrix product **AB** if and only if the number of columns of **A** is equal to the number of rows of **B**. The matrices **A** and **B** are then said to be *conformable* for multiplication. For an $m \times n$ matrix **A** and an $n \times r$ matrix **B** the product matrix **AB** is of order $m \times r$.

Example 2   If $\mathbf{S} = \begin{pmatrix} 5 & -3 \\ 2 & 1 \end{pmatrix}$ and $\mathbf{T} = \begin{pmatrix} 3 \\ -2 \end{pmatrix}$, find if possible **ST** and **TS**.

$$\mathbf{ST} = \begin{pmatrix} 5 & -3 \\ 2 & 1 \end{pmatrix}\begin{pmatrix} 3 \\ -2 \end{pmatrix} = \begin{pmatrix} 21 \\ 4 \end{pmatrix} \quad \text{and } \mathbf{TS} \text{ does not exist.}$$

Example 3   If $\mathbf{A} = (1 \quad -3)$ and $\mathbf{B} = \begin{pmatrix} -2 \\ 5 \end{pmatrix}$, find **AB** and **BA**.

$$\mathbf{AB} = (1 \quad -3)\begin{pmatrix} -2 \\ 5 \end{pmatrix} = (-17), \quad \mathbf{BA} = \begin{pmatrix} -2 \\ 5 \end{pmatrix}(1 \quad -3) = \begin{pmatrix} -2 & 6 \\ 5 & -15 \end{pmatrix}.$$

[We see that although both **AB** and **BA** exist they are of different orders.]

Example 4   If $\mathbf{A} = \begin{pmatrix} 1 & -2 \\ 0 & 1 \end{pmatrix}$ and $\mathbf{B} = \begin{pmatrix} 2 & 0 \\ 3 & -1 \end{pmatrix}$, find **AB** and **BA**.

$$\mathbf{AB} = \begin{pmatrix} 1 & -2 \\ 0 & 1 \end{pmatrix}\begin{pmatrix} 2 & 0 \\ 3 & -1 \end{pmatrix} = \begin{pmatrix} -4 & 2 \\ 3 & -1 \end{pmatrix}, \quad \mathbf{BA} = \begin{pmatrix} 2 & 0 \\ 3 & -1 \end{pmatrix}\begin{pmatrix} 1 & -2 \\ 0 & 1 \end{pmatrix}$$

$$= \begin{pmatrix} 2 & -4 \\ 3 & -7 \end{pmatrix}.$$

[In this case the product matrices are of the same order, but $\mathbf{AB} \neq \mathbf{BA}$.]

Examples 3 and 4 show that, in general, matrix multiplication is *not commutative*. However, it is sometimes possible to find a pair of square matrices which do commute.

Example 5   If $\mathbf{P} = \begin{pmatrix} 1 & 2 \\ -3 & 0 \end{pmatrix}$ and $\mathbf{Q} = \begin{pmatrix} x & -2 \\ y & 5 \end{pmatrix}$, find the values of x and y, such that $\mathbf{PQ} = \mathbf{QP}$.

$$\mathbf{PQ} = \begin{pmatrix} 1 & 2 \\ -3 & 0 \end{pmatrix}\begin{pmatrix} x & -2 \\ y & 5 \end{pmatrix} = \begin{pmatrix} x+2y & 8 \\ -3x & 6 \end{pmatrix}, \quad \mathbf{QP} = \begin{pmatrix} x & -2 \\ y & 5 \end{pmatrix}\begin{pmatrix} 1 & 2 \\ -3 & 0 \end{pmatrix}$$

$$= \begin{pmatrix} x+6 & 2x \\ y-15 & 2y \end{pmatrix}.$$

Since $\mathbf{PQ} = \mathbf{QP}$, we may equate corresponding elements

$$x+2y = x+6 \qquad\qquad 8 = 2x$$
$$-3x = y-15 \qquad\qquad 6 = 2y.$$

The values of x and y which satisfy these equations are $x = 4$, $y = 3$.

Hence $\mathbf{PQ} = \mathbf{QP} = \begin{pmatrix} 10 & 8 \\ -12 & 6 \end{pmatrix}$ when $x = 4$, $y = 3$.

We now consider the properties of the matrix $\mathbf{I} = \begin{pmatrix} 1 & 0 \\ 0 & 1 \end{pmatrix}$. If $\mathbf{A} = \begin{pmatrix} a & b \\ c & d \end{pmatrix}$, then

$\mathbf{AI} = \begin{pmatrix} a & b \\ c & d \end{pmatrix}\begin{pmatrix} 1 & 0 \\ 0 & 1 \end{pmatrix} = \begin{pmatrix} a & b \\ c & d \end{pmatrix} = \mathbf{A}$ and similarly $\mathbf{IA} = \mathbf{A}$. Since it acts as an identity element for matrix multiplication, $\mathbf{I}$ is called an *identity matrix* (or unit matrix).

Example 6   If $\mathbf{A} = \begin{pmatrix} 3 & -5 \\ 1 & -2 \end{pmatrix}$, show that $\mathbf{A}^2 - \mathbf{A} = \mathbf{I}$, where $\mathbf{I}$ is the 2×2 identity matrix.

$$\mathbf{A}^2 - \mathbf{A} = \begin{pmatrix} 3 & -5 \\ 1 & -2 \end{pmatrix}\begin{pmatrix} 3 & -5 \\ 1 & -2 \end{pmatrix} - \begin{pmatrix} 3 & -5 \\ 1 & -2 \end{pmatrix} = \begin{pmatrix} 4 & -5 \\ 1 & -1 \end{pmatrix} - \begin{pmatrix} 3 & -5 \\ 1 & -2 \end{pmatrix} = \begin{pmatrix} 1 & 0 \\ 0 & 1 \end{pmatrix} = \mathbf{I}.$$

It has already been demonstrated that in general matrix multiplication is not commutative. However, matrix multiplication is associative and it is distributive over matrix addition. Thus for matrices $\mathbf{A}$, $\mathbf{B}$ and $\mathbf{C}$ of appropriate orders:

$$(\mathbf{AB})\mathbf{C} = \mathbf{A}(\mathbf{BC}) \qquad \text{associative law}$$
$$\mathbf{A}(\mathbf{B}+\mathbf{C}) = \mathbf{AB}+\mathbf{AC}$$
$$(\mathbf{A}+\mathbf{B})\mathbf{C} = \mathbf{AC}+\mathbf{BC} \qquad \text{distributive laws.}$$

[For verification of these laws and proofs in the case of 2×2 matrices see Exercise 19.1.]

Example 7   Given that $\mathbf{M}$ is a 2×2 matrix such that $\mathbf{M}^2 = \mathbf{M} - \mathbf{I}$, show that $\mathbf{M}^4 = -\mathbf{M}$.

$$\mathbf{M}^4 = (\mathbf{M}^2)^2 = (\mathbf{M}-\mathbf{I})(\mathbf{M}-\mathbf{I}) = \mathbf{M}^2 - \mathbf{MI} - \mathbf{IM} + \mathbf{I}^2$$
$$= (\mathbf{M}-\mathbf{I}) - \mathbf{M} - \mathbf{M} + \mathbf{I} = -\mathbf{M}.$$

The *transpose* of a matrix $\mathbf{A}$, written $\mathbf{A}^T$, is obtained by interchanging the rows and columns of $\mathbf{A}$, e.g.

if $\mathbf{A} = \begin{pmatrix} 3 & -1 \\ 2 & 4 \end{pmatrix}$, then $\mathbf{A}^T = \begin{pmatrix} 3 & 2 \\ -1 & 4 \end{pmatrix}$

and if $\mathbf{M} = \begin{pmatrix} 1 & -3 \\ -5 & 2 \\ 0 & 1 \end{pmatrix}$, then $\mathbf{M}^T = \begin{pmatrix} 1 & -5 & 0 \\ -3 & 2 & 1 \end{pmatrix}$.

It can be shown that for matrices $\mathbf{A}$ and $\mathbf{B}$, which are conformable for multiplication $(\mathbf{AB})^T = \mathbf{B}^T\mathbf{A}^T$.

Exercise 19.1

1. If $A = \begin{pmatrix} 1 & 4 & 0 & 3 \\ -2 & 0 & 2 & -1 \end{pmatrix}$ and $B = \begin{pmatrix} 0 & -1 & -3 & 1 \\ 3 & 5 & 1 & -2 \end{pmatrix}$ find $3A, -B, A+B$, $A-B$ and $2A-3B$.

2. If $A = \begin{pmatrix} 2 & 1 \\ -1 & 3 \end{pmatrix}$, $B = \begin{pmatrix} 1 & -1 & 3 \\ 0 & 4 & 2 \end{pmatrix}$, $C = \begin{pmatrix} -3 & 1 & -2 \\ 0 & 4 & -1 \\ 2 & 0 & 1 \end{pmatrix}$, $D = \begin{pmatrix} 1 & -4 \\ 2 & 0 \\ -3 & 1 \end{pmatrix}$

calculate where possible the products A^2, AB, AC, AD, BA, B^2, BC, BD, CA, CB, C^2, CD, DA, DB, DC and D^2. Verify that $(AB)C = A(BC)$ and that $(BC)D = B(CD)$.

3. Find all possible sums and products that can be formed using two of the following matrices,

$$A = (1 \quad -2 \quad 3), \quad B = \begin{pmatrix} 5 \\ 1 \\ -1 \end{pmatrix}, \quad C = (0 \quad 4 \quad -3), \quad D = \begin{pmatrix} 3 \\ -1 \\ 4 \end{pmatrix}.$$

Verify that $(AB)C = A(BC)$ and that $(BC)D = B(CD)$.

4. If $A = \begin{pmatrix} -2 & 0 \\ 3 & 4 \end{pmatrix}$, $B = \begin{pmatrix} 2 & -1 \\ 1 & 3 \end{pmatrix}$, $C = \begin{pmatrix} 4 & 1 \\ -1 & 3 \end{pmatrix}$, find AB, BA, BC, CB, AC and CA. Verify that $(A+B)C = AC+BC$ and that $A(B+C) = AB+AC$.

5. Repeat question 4 for $A = \begin{pmatrix} 4 & 2 \\ 5 & -1 \end{pmatrix}$, $B = \begin{pmatrix} 5 & 4 \\ -2 & 3 \end{pmatrix}$, $C = \begin{pmatrix} 2 & -4 \\ -3 & 6 \end{pmatrix}$.

6. If $X = \begin{pmatrix} 2 & -1 \\ 3 & 2 \end{pmatrix}$ and $Y = \begin{pmatrix} a & 1 \\ b & 1 \end{pmatrix}$, find the values of a and b such that $XY = YX$.

7. Given that $A = \begin{pmatrix} 2 & 3 \\ -4 & -1 \end{pmatrix}$, show that $A - A^2 = 10I$.

8. Given that $M = \begin{pmatrix} 2 & -1 \\ -3 & 4 \end{pmatrix}$ and that $M^2 - 6M + kI = 0$, find k.

9. Given that M is a 2×2 matrix such that $M^2 = M+I$, show that $M^4 - 3M = 2I$.

10. Given that A is a 2×2 matrix such that $A^2 = A - 2I$, show that $A^4 + 3A^2 + 4I = 0$.

11. By letting $A = \begin{pmatrix} a & b \\ c & d \end{pmatrix}$, $B = \begin{pmatrix} p & q \\ r & s \end{pmatrix}$, $C = \begin{pmatrix} w & x \\ y & z \end{pmatrix}$, prove that for all 2×2 matrices A, B and C
(a) $(AB)C = A(BC)$,   (b) $A(B+C) = AB+AC$
(c) $(A+B)C = AC+BC$ (d) $(AB)^T = B^T A^T$.

19.2 The inverse of a 2 × 2 matrix

The *inverse* of a 2 × 2 matrix $\mathbf{A}$ is the matrix $\mathbf{A}^{-1}$ which has the property that

$$\mathbf{AA}^{-1} = \mathbf{A}^{-1}\mathbf{A} = \mathbf{I}$$

where $\mathbf{I}$ is the 2 × 2 identity matrix.

Let us assume that $\mathbf{A} = \begin{pmatrix} a & b \\ c & d \end{pmatrix}$ and $\mathbf{A}^{-1} = \begin{pmatrix} p & q \\ r & s \end{pmatrix}$, then

$$\mathbf{AA}^{-1} = \begin{pmatrix} a & b \\ c & d \end{pmatrix}\begin{pmatrix} p & q \\ r & s \end{pmatrix} = \begin{pmatrix} ap+br & aq+bs \\ cp+dr & cq+ds \end{pmatrix} = \begin{pmatrix} 1 & 0 \\ 0 & 1 \end{pmatrix}.$$

$\therefore$ $ap+br = 1$ (1) $aq+bs = 0$ (3)
   $cp+dr = 0$ (2) $cq+ds = 1$ (4)

Multiplying (1) by d $adp+bdr = d$
Multiplying (2) by b $bcp+bdr = 0$.
Subtracting $(ad-bc)p = d$

$\therefore$ provided that $ad-bc \neq 0$, $p = d/(ad-bc)$.

Using similar methods and letting $ad-bc = \Delta$,   we obtain

$$p = d/\Delta, \quad q = -b/\Delta, \quad r = -c/\Delta, \quad s = a/\Delta.$$

$\therefore$ if $\mathbf{A}^{-1}$ exists, it is of the form $\begin{pmatrix} d/\Delta & -b/\Delta \\ -c/\Delta & a/\Delta \end{pmatrix}$,   i.e. $\dfrac{1}{\Delta}\begin{pmatrix} d & -b \\ -c & a \end{pmatrix}$.

It is readily verified that

$$\begin{pmatrix} a & b \\ c & d \end{pmatrix} \times \frac{1}{\Delta}\begin{pmatrix} d & -b \\ -c & a \end{pmatrix} = \frac{1}{\Delta}\begin{pmatrix} d & -b \\ -c & a \end{pmatrix} \times \begin{pmatrix} a & b \\ c & d \end{pmatrix} = \begin{pmatrix} 1 & 0 \\ 0 & 1 \end{pmatrix}$$

Hence if $\mathbf{A} = \begin{pmatrix} a & b \\ c & d \end{pmatrix}$ and $\Delta \neq 0$, $\mathbf{A}^{-1} = \dfrac{1}{\Delta}\begin{pmatrix} d & -b \\ -c & a \end{pmatrix}$.

Thus to obtain the inverse of a matrix $\begin{pmatrix} a & b \\ c & d \end{pmatrix}$ we must

(i)   interchange the elements a and d in the leading diagonal,
(ii) change the signs of the remaining elements b and c,
(iii) divide each element by $\Delta = ad-bc$.

Example 1   Find the inverse of the matrix $\begin{pmatrix} 5 & -2 \\ 3 & -1 \end{pmatrix}$. Hence solve the

simultaneous equations $\begin{matrix} 5x-2y = 2 \\ 3x-\ y = 3 \end{matrix}$.

For the given matrix, $\Delta = 5(-1)-(-2)3 = 1$

$\therefore$ the inverse of the matrix $\begin{pmatrix} 5 & -2 \\ 3 & -1 \end{pmatrix}$ is $\begin{pmatrix} -1 & 2 \\ -3 & 5 \end{pmatrix}$.

In matrix form the simultaneous equations become

$$\begin{pmatrix} 5 & -2 \\ 3 & -1 \end{pmatrix}\begin{pmatrix} x \\ y \end{pmatrix} = \begin{pmatrix} 2 \\ 3 \end{pmatrix}$$

$$\therefore \quad \begin{pmatrix} -1 & 2 \\ -3 & 5 \end{pmatrix}\begin{pmatrix} 5 & -2 \\ 3 & -1 \end{pmatrix}\begin{pmatrix} x \\ y \end{pmatrix} = \begin{pmatrix} -1 & 2 \\ -3 & 5 \end{pmatrix}\begin{pmatrix} 2 \\ 3 \end{pmatrix}$$

$$\therefore \quad \begin{pmatrix} 1 & 0 \\ 0 & 1 \end{pmatrix}\begin{pmatrix} x \\ y \end{pmatrix} = \begin{pmatrix} 4 \\ 9 \end{pmatrix}$$

$$\therefore \quad \begin{pmatrix} x \\ y \end{pmatrix} = \begin{pmatrix} 4 \\ 9 \end{pmatrix}, \quad \text{i.e.} \quad \begin{matrix} x = 4 \\ y = 9. \end{matrix}$$

Example 2 Show that the matrix $\mathbf{A} = \begin{pmatrix} 3 & -2 \\ 4 & -3 \end{pmatrix}$ is its own inverse.

$$\mathbf{A}^2 = \begin{pmatrix} 3 & -2 \\ 4 & -3 \end{pmatrix}\begin{pmatrix} 3 & -2 \\ 4 & -3 \end{pmatrix} = \begin{pmatrix} 1 & 0 \\ 0 & 1 \end{pmatrix} = \mathbf{I}.$$

Hence $\mathbf{A}$ must be its own inverse.

For the matrix $\mathbf{A} = \begin{pmatrix} a & b \\ c & d \end{pmatrix}$, the quantity $\Delta = ad - bc$ is called the *determinant* of $\mathbf{A}$, written det $\mathbf{A}$, $|\mathbf{A}|$ or $\begin{vmatrix} a & b \\ c & d \end{vmatrix}$. If the determinant of $\mathbf{A}$ is zero, i.e. if $\Delta = 0$, then $\mathbf{A}$ has no inverse and is said to be *singular*. If $\mathbf{A}$ has a non-zero determinant, i.e. if $\Delta \neq 0$, then $\mathbf{A}^{-1}$ exists and $\mathbf{A}$ is *non-singular*.

In the next example, we see one type of problem in which singular matrices sometimes arise.

Example 3 Find the points of intersection, if any, of the following pairs of straight lines

(a) $x + y = 1$ (b) $2x - 3y = 4$ (c) $3x - 6y = 9$
      $3x + 5y = 7$ $4x - 6y = 1$ $2x - 4y = 6.$

(a) $\begin{matrix} x + y = 1 \\ 3x + 5y = 7 \end{matrix} \Leftrightarrow \begin{pmatrix} 1 & 1 \\ 3 & 5 \end{pmatrix}\begin{pmatrix} x \\ y \end{pmatrix} = \begin{pmatrix} 1 \\ 7 \end{pmatrix}$

$$\begin{vmatrix} 1 & 1 \\ 3 & 5 \end{vmatrix} = 1.5 - 1.3 = 5 - 3 = 2$$

$\therefore$ the inverse of $\begin{pmatrix} 1 & 1 \\ 3 & 5 \end{pmatrix}$ is $\dfrac{1}{2}\begin{pmatrix} 5 & -1 \\ -3 & 1 \end{pmatrix}$

Hence $\begin{pmatrix} x \\ y \end{pmatrix} = \frac{1}{2}\begin{pmatrix} 5 & -1 \\ -3 & 1 \end{pmatrix}\begin{pmatrix} 1 \\ 7 \end{pmatrix} = \frac{1}{2}\begin{pmatrix} -2 \\ 4 \end{pmatrix} = \begin{pmatrix} -1 \\ 2 \end{pmatrix}$

∴   the point of intersection is $(-1, 2)$.

(b) $\begin{array}{l} 2x - 3y = 4 \\ 4x - 6y = 1 \end{array} \Leftrightarrow \begin{pmatrix} 2 & -3 \\ 4 & -6 \end{pmatrix}\begin{pmatrix} x \\ y \end{pmatrix} = \begin{pmatrix} 4 \\ 1 \end{pmatrix}$

$$\begin{vmatrix} 2 & -3 \\ 4 & -6 \end{vmatrix} = 2(-6) - (-3)4 = -12 + 12 = 0$$

∴ $\begin{pmatrix} 2 & -3 \\ 4 & -6 \end{pmatrix}$ has no inverse.

Examining the original equations, we find that there is no point whose coordinates satisfy both equations. The equations represent a pair of parallel lines with no point of intersection.

(c) $\begin{array}{l} 3x - 6y = 9 \\ 2x - 4y = 6 \end{array} \Leftrightarrow \begin{pmatrix} 3 & -6 \\ 2 & -4 \end{pmatrix}\begin{pmatrix} x \\ y \end{pmatrix} = \begin{pmatrix} 9 \\ 6 \end{pmatrix}$

$$\begin{vmatrix} 3 & -6 \\ 2 & -4 \end{vmatrix} = 3(-4) - (-6)2 = -12 + 12 = 0$$

∴ $\begin{pmatrix} 3 & -6 \\ 2 & -4 \end{pmatrix}$ has no inverse.

We find that the original equations both reduce to $x - 2y = 3$, i.e. they both represent the same line. Hence the coordinates of any point on the line satisfy both equations.

In general, if $\begin{vmatrix} a & b \\ c & d \end{vmatrix} = 0$, the equation $\begin{pmatrix} a & b \\ c & d \end{pmatrix}\begin{pmatrix} x \\ y \end{pmatrix} = \begin{pmatrix} p \\ q \end{pmatrix}$ either has no solution or an infinite set of solutions.

For real numbers a and b, the statement $ab = 0$ implies that either $a = 0$ or $b = 0$. However, it is possible to find matrices **A** and **B** with non-zero entries such that $\mathbf{AB} = \mathbf{0}$, e.g.

$$\begin{pmatrix} 2 & -3 \\ 4 & -6 \end{pmatrix}\begin{pmatrix} 3 & -6 \\ 2 & -4 \end{pmatrix} = \begin{pmatrix} 0 & 0 \\ 0 & 0 \end{pmatrix}.$$

If we assume that **A** has an inverse $\mathbf{A}^{-1}$, then

$$\mathbf{AB} = \mathbf{0} \Rightarrow \mathbf{A}^{-1}(\mathbf{AB}) = \mathbf{A}^{-1}\mathbf{0} \Rightarrow (\mathbf{A}^{-1}\mathbf{A})\mathbf{B} = \mathbf{0} \Rightarrow \mathbf{IB} = \mathbf{0} \Rightarrow \mathbf{B} = \mathbf{0}$$

Hence, if $\mathbf{AB} = \mathbf{0}$, then $\mathbf{A} = \mathbf{0}$ or $\mathbf{B} = \mathbf{0}$ or both **A** and **B** are singular matrices.

Exercise 19.2

1.   Write down, if possible, the inverses of

(a) $\begin{pmatrix} 3 & 4 \\ 2 & 3 \end{pmatrix}$,   (b) $\begin{pmatrix} 5 & -2 \\ 7 & -3 \end{pmatrix}$,   (c) $\begin{pmatrix} 3 & -6 \\ -2 & 4 \end{pmatrix}$,   (d) $\begin{pmatrix} 2 & 1 \\ -1 & 1 \end{pmatrix}$,

(e) $\begin{pmatrix} -2 & -1 \\ 3 & 2 \end{pmatrix}$, (f) $\begin{pmatrix} 4 & -2 \\ -2 & 1 \end{pmatrix}$, (g) $\begin{pmatrix} 3 & 0 \\ 0 & -2 \end{pmatrix}$, (h) $\begin{pmatrix} 1 & 5 \\ 2 & 3 \end{pmatrix}$.

Verify that for each matrix A and its inverse A^{-1}, $AA^{-1} = A^{-1}A = I$.

2. Use matrix methods to solve the following pairs of simultaneous equations:

(a) $2x - y = 1$
     $5x - 2y = 2,$

(b) $3p - 5q = 7$
     $2p - 4q = 6,$

(c) $a + b = 11$
     $4a - b = 9.$

3. Use matrix methods, where possible, to find the points of intersection of the following pairs of lines:

(a) $3x + y + 1 = 0$
     $4x + 3y - 2 = 0,$

(b) $4x - 10y = 7$
     $5y - 2x = 3,$

(c) $y = 2x + 3$
     $x - 2y + 3 = 0.$

4. Given that A, B and X are 2×2 matrices and that A has an inverse A^{-1}, find expressions for X in terms of A, B and A^{-1} if

(a) $AX = B,$ (b) $XA = B,$ (c) $AX = AB,$ (d) $AX = BA.$

5. If $A = \begin{pmatrix} 5 & 1 \\ 4 & 2 \end{pmatrix}$ and $B = \begin{pmatrix} 1 & -1 \\ 2 & 4 \end{pmatrix}$, find X given that

(a) $AX = B,$ (b) $A = XB,$ (c) $A^{-1}XA = B.$

6. Use matrix methods to solve the given pairs of simultaneous equations, stating any restriction on the value of k in each case. Describe the relationship between the pair of lines represented by the equations when k takes each restricted value.

(a) $(2k+1)x - y = 1$
     $(k-1)x + y = 2,$

(b) $x - ky = 1$
     $kx - 4y = 2.$

19.3 Transformations of the x,y plane

A mapping of the set of points in the x,y plane into itself is called a *transformation* of the plane. Since any point can be represented by an ordered pair (x,y), i.e. by an element of $\mathbb{R}^2$, a transformation of the x,y plane may be represented as a mapping from $\mathbb{R}^2$ to $\mathbb{R}^2$.

When working with transformations it is often convenient to use matrix notation. The position vector of a typical point $P(x, y)$ with respect to the origin O can be given in the form of a 2×1 matrix or *column vector* $\begin{pmatrix} x \\ y \end{pmatrix}$. Thus for a transformation T under which the image of P is the point $P'(x', y')$, we write

$$T: \begin{pmatrix} x \\ y \end{pmatrix} \rightarrow \begin{pmatrix} x' \\ y' \end{pmatrix}.$$

A *translation* is a transformation in which the points of the plane move a fixed distance in a given direction. This means that if a translation maps A to A', B to

B', ... then **AA'**, **BB'**, ... are all equal to a fixed vector which can be used to represent the translation.

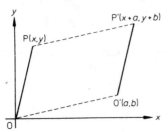

Consider a translation which maps $P(x, y)$ to $P'(x', y')$. If it maps $O(0,0)$ to $O'(a, b)$, then $x' = x+a$, $y' = y+b$.

In matrix notation

$$\begin{pmatrix} x' \\ y' \end{pmatrix} = \begin{pmatrix} x \\ y \end{pmatrix} + \begin{pmatrix} a \\ b \end{pmatrix}.$$

Hence the translation can be completely defined using the column vector $\begin{pmatrix} a \\ b \end{pmatrix}$.

Example 1   Find the image of the triangle with vertices

$$A(1, 2), B(5, -1), C(-4, -2)$$

under the translation with column vector $\begin{pmatrix} 2 \\ 1 \end{pmatrix}$.

$$\begin{pmatrix} 1 \\ 2 \end{pmatrix} + \begin{pmatrix} 2 \\ 1 \end{pmatrix} = \begin{pmatrix} 3 \\ 3 \end{pmatrix}, \quad \begin{pmatrix} 5 \\ -1 \end{pmatrix} + \begin{pmatrix} 2 \\ 1 \end{pmatrix} = \begin{pmatrix} 7 \\ 0 \end{pmatrix}, \quad \begin{pmatrix} -4 \\ -2 \end{pmatrix} + \begin{pmatrix} 2 \\ 1 \end{pmatrix} = \begin{pmatrix} -2 \\ -1 \end{pmatrix}.$$

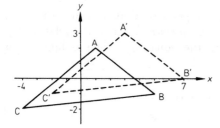

The image of the triangle ABC is the triangle with vertices $A'(3, 3)$, $B'(7, 0)$ and $C'(-2, -1)$.

Example 2   Find the column vector representing the translation which maps the point $(1, -3)$ onto the point $(-2, 4)$.

Let $\begin{pmatrix} -2 \\ 4 \end{pmatrix} = \begin{pmatrix} 1 \\ -3 \end{pmatrix} + \begin{pmatrix} a \\ b \end{pmatrix}$,   then $\begin{pmatrix} a \\ b \end{pmatrix} = \begin{pmatrix} -3 \\ 7 \end{pmatrix}$.

Thus the column vector of the given translation is $\begin{pmatrix} -3 \\ 7 \end{pmatrix}$.

Example 3   Find the image under the translation with column vector $\begin{pmatrix} 3 \\ -2 \end{pmatrix}$ of the line $y = 2x + 3$.

If the translation maps $P(x, y)$ to $P'(x', y')$, then

$$\begin{aligned} x' &= x+3 \\ y' &= y-2 \end{aligned} \quad \text{and} \quad \begin{aligned} x &= x'-3 \\ y &= y'+2. \end{aligned}$$

Hence if P lies on the line $y = 2x + 3$,

$$(y' + 2) = 2(x' - 3) + 3, \quad \text{i.e.} \quad y' = 2x' - 5$$

∴ P' lies on the line $y = 2x - 5$.

Hence the image of the line $y = 2x + 3$ is the line $y = 2x - 5$.

A *linear transformation* of the x, y plane is a mapping under which *linear* relationships between position vectors remain unchanged. Consider three points A, B, C with position vectors $\mathbf{a}$, $\mathbf{b}$, $\mathbf{c}$ and their images A', B', C' with position vectors $\mathbf{a}'$, $\mathbf{b}'$, $\mathbf{c}'$. If $\mathbf{c} = \lambda\mathbf{a} + \mu\mathbf{b}$, then under a linear transformation this relationship is preserved and $\mathbf{c}' = \lambda\mathbf{a}' + \mu\mathbf{b}'$. In particular if $OACB$ is a parallelogram, then $\mathbf{c} = \mathbf{a} + \mathbf{b}$. Under a linear transformation $\mathbf{c}' = \mathbf{a}' + \mathbf{b}'$ and hence $OA'C'B'$ is also a parallelogram.

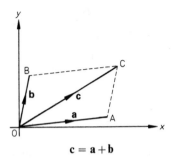

$\mathbf{c} = \mathbf{a} + \mathbf{b}$

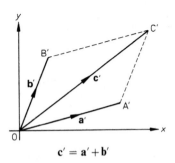

$\mathbf{c}' = \mathbf{a}' + \mathbf{b}'$

The diagrams below show that when a translation is performed, $OA'C'B'$ is not a parallelogram. Hence a translation is *not* a linear transformation.

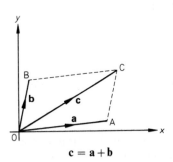

$\mathbf{c} = \mathbf{a} + \mathbf{b}$

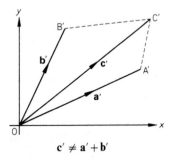

$\mathbf{c}' \neq \mathbf{a}' + \mathbf{b}'$

Let us consider now the points $A(1, 0)$, $B(0, 1)$, with position vectors $\mathbf{i}$, $\mathbf{j}$, and their images under a linear transformation A', B', with position vectors $\mathbf{u}$, $\mathbf{v}$. Since the transformation preserves linear relationships, the image of a point $P(x, y)$ with position vector $x\mathbf{i} + y\mathbf{j}$ will be the point $P'(x', y')$ with position vector $x\mathbf{u} + y\mathbf{v}$.

Thus if $\mathbf{u} = \begin{pmatrix} a \\ c \end{pmatrix}$ and $\mathbf{v} = \begin{pmatrix} b \\ d \end{pmatrix}$, then

$$\begin{pmatrix} x' \\ y' \end{pmatrix} = x\begin{pmatrix} a \\ c \end{pmatrix} + y\begin{pmatrix} b \\ d \end{pmatrix} = \begin{pmatrix} ax \\ cx \end{pmatrix} + \begin{pmatrix} by \\ dy \end{pmatrix} = \begin{pmatrix} ax + by \\ cx + dy \end{pmatrix} = \begin{pmatrix} a & b \\ c & d \end{pmatrix}\begin{pmatrix} x \\ y \end{pmatrix}.$$

Hence any linear transformation of the x, y plane is associated with a 2×2 matrix $\begin{pmatrix} a & b \\ c & d \end{pmatrix}$ whose columns $\begin{pmatrix} a \\ c \end{pmatrix}$ and $\begin{pmatrix} b \\ d \end{pmatrix}$ represent the images of the points with position vectors $\begin{pmatrix} 1 \\ 0 \end{pmatrix}$ and $\begin{pmatrix} 0 \\ 1 \end{pmatrix}$.

The diagrams below show the effect of a linear transformation on a system of lines parallel to the x- and y-axes.

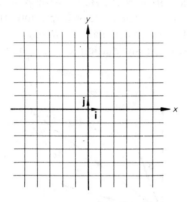 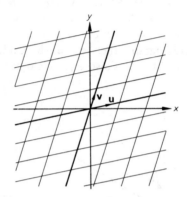

Example 4   Find the images of the points $(2, 0)$, $(1, -3)$ and $(-2, 1)$ under the linear transformation with matrix $\begin{pmatrix} 1 & -2 \\ -1 & 3 \end{pmatrix}$.

$$\begin{pmatrix} 1 & -2 \\ -1 & 3 \end{pmatrix} \begin{pmatrix} 2 \\ 0 \end{pmatrix} = \begin{pmatrix} 2 \\ -2 \end{pmatrix}, \quad \begin{pmatrix} 1 & -2 \\ -1 & 3 \end{pmatrix} \begin{pmatrix} 1 \\ -3 \end{pmatrix} = \begin{pmatrix} 7 \\ -10 \end{pmatrix},$$

$$\begin{pmatrix} 1 & -2 \\ -1 & 3 \end{pmatrix} \begin{pmatrix} -2 \\ 1 \end{pmatrix} = \begin{pmatrix} -4 \\ 5 \end{pmatrix}$$

∴   the images of the given points are $(2, -2)$, $(7, -10)$ and $(-4, 5)$ respectively.

Example 5   Find the matrix of the linear transformation which maps the points $(1, 1)$ and $(0, 3)$ to the points $(2, 3)$ and $(0, 6)$ respectively.

If $\begin{pmatrix} a & b \\ c & d \end{pmatrix} \begin{pmatrix} 1 \\ 1 \end{pmatrix} = \begin{pmatrix} 2 \\ 3 \end{pmatrix}$ and $\begin{pmatrix} a & b \\ c & d \end{pmatrix} \begin{pmatrix} 0 \\ 3 \end{pmatrix} = \begin{pmatrix} 0 \\ 6 \end{pmatrix}$

then $\begin{matrix} a+b = 2 \\ c+d = 3 \end{matrix}$ and $\begin{matrix} 3b = 0 \\ 3d = 6 \end{matrix}$

∴ $b = 0, \quad d = 2, \quad a = 2, \quad c = 1$.

Hence the matrix of the transformation is $\begin{pmatrix} 2 & 0 \\ 1 & 2 \end{pmatrix}$.

Example 6   Find the image of the line $y = 3x$ under the linear transformation with matrix $\begin{pmatrix} 4 & -2 \\ 5 & 1 \end{pmatrix}$.

The line $y = 3x$ is the set of points with coordinates of the form $(k, 3k)$.

$$\begin{pmatrix} 4 & -2 \\ 5 & 1 \end{pmatrix}\begin{pmatrix} k \\ 3k \end{pmatrix} = \begin{pmatrix} 4k - 6k \\ 5k + 3k \end{pmatrix} = \begin{pmatrix} -2k \\ 8k \end{pmatrix}.$$

Hence the image of the line $y = 3x$ is the set of points $(-2k, 8k)$, i.e. the line $y = -4x$.

If under a transformation T of the x, y plane each point $P'(x', y')$ is the image of exactly one point $P(x, y)$, then an *inverse transformation* T^{-1} can be defined such that $T^{-1} : (x', y') \rightarrow (x, y)$.

For instance, if a translation with column vector $\begin{pmatrix} a \\ b \end{pmatrix}$ maps (x, y) to (x', y') then

$$\begin{pmatrix} x' \\ y' \end{pmatrix} = \begin{pmatrix} x \\ y \end{pmatrix} + \begin{pmatrix} a \\ b \end{pmatrix} \quad \text{and} \quad \begin{pmatrix} x \\ y \end{pmatrix} = \begin{pmatrix} x' \\ y' \end{pmatrix} + \begin{pmatrix} -a \\ -b \end{pmatrix}.$$

Hence the inverse translation has column vector $\begin{pmatrix} -a \\ -b \end{pmatrix}$.

Consider now a linear transformation with non-singular matrix **M** which maps (x, y) to (x', y'). Using the properties of the inverse matrix $\mathbf{M}^{-1}$, as defined in §19.2, since

$$\begin{pmatrix} x' \\ y' \end{pmatrix} = \mathbf{M}\begin{pmatrix} x \\ y \end{pmatrix}, \quad \text{it follows that} \quad \begin{pmatrix} x \\ y \end{pmatrix} = \mathbf{M}^{-1}\begin{pmatrix} x' \\ y' \end{pmatrix}.$$

Hence the inverse of the linear transformation with non-singular matrix **M** is the linear transformation with matrix $\mathbf{M}^{-1}$.

Exercise 19.3

1. Find the images under the translation with column vector $\begin{pmatrix} -2 \\ 3 \end{pmatrix}$ of the points $P(1, -1), Q(7, 11), R(-3, 0)$.

2. Find the images under the translation with column vector $\begin{pmatrix} 4 \\ -3 \end{pmatrix}$ of the points $X(0, 1,), Y(-3, 2), Z(6, -2)$.

3. Find the column vectors representing the translations which map
(a) $(1, 1)$ to $(-7, 3)$,   (b) $(-1, 6)$ to $(3, 0)$,
(c) $(1, -2)$ to $(1, -2)$,   (d) $(3, -4)$ to $(4, -3)$.

4. Find the images of the following straight lines under the translations with given column vectors

(a) $y = 2x, \begin{pmatrix} 1 \\ 2 \end{pmatrix}$,

(b) $y = x - 1, \begin{pmatrix} -3 \\ 4 \end{pmatrix}$,

(c) $y = 3 - 4x, \begin{pmatrix} 2 \\ 0 \end{pmatrix}$,

(d) $2x + 3y = 5, \begin{pmatrix} 2 \\ -1 \end{pmatrix}$.

5. Find the images of the points $(1, 1)$, $(0, 3)$ and $(-4, -1)$ under the linear transformations with matrices

(a) $\begin{pmatrix} 0 & -1 \\ 1 & 0 \end{pmatrix}$;

(b) $\begin{pmatrix} 1 & 0 \\ 0 & 1 \end{pmatrix}$,

(c) $\begin{pmatrix} 2 & -1 \\ 1 & 3 \end{pmatrix}$,

(d) $\begin{pmatrix} 1 & 0 \\ -2 & 1 \end{pmatrix}$.

6. Write down the matrix of the linear transformation which maps the points $(1, 0)$ and $(0, 1)$ to the points $(4, -5)$ and $(-1, 7)$ respectively.

7. Find the matrix of the linear transformation which maps the points $(2, 0)$ and $(1, -1)$ to the points $(6, -2)$ and $(5, -1)$ respectively.

8. Find the images of the following straight lines under the linear transformations with the given matrices

(a) $y = x, \begin{pmatrix} 3 & 2 \\ 4 & 6 \end{pmatrix}$,

(b) $y = 2x, \begin{pmatrix} 1 & -2 \\ 4 & 1 \end{pmatrix}$,

(c) $y = -x, \begin{pmatrix} -3 & 0 \\ 2 & -1 \end{pmatrix}$,

(d) $y = x + 1, \begin{pmatrix} 2 & -1 \\ 0 & 1 \end{pmatrix}$.

9. Write down the column vector of the inverse T^{-1} of the translation T represented by $\begin{pmatrix} -3 \\ 2 \end{pmatrix}$. Hence find the coordinates of the points whose images under T are $(4, 1)$, $(-3, 2)$, $(-2, 3)$ and $(-5, 0)$.

10. Find the inverse of the matrix $\mathbf{M} = \begin{pmatrix} -2 & 3 \\ -3 & 4 \end{pmatrix}$. Hence find the coordinates of the points whose images under the linear transformation with matrix $\mathbf{M}$ are the points $(0, -1)$, $(1, 1)$ and $(7, 10)$.

11. Find the coordinates of the points whose images under the linear transformation with matrix $\begin{pmatrix} 2 & 1 \\ 1 & -1 \end{pmatrix}$ are the points $(4, -1)$, $(0, 0)$ and $(1, 5)$.

19.4 Geometrical properties of transformations

In this section we examine in more detail the properties of linear transformations of the x, y plane, i.e. transformations with equations of the form

$$\begin{matrix} x' = ax + by \\ y' = cx + dy \end{matrix} \quad \text{or in matrix notation} \quad \begin{pmatrix} x' \\ y' \end{pmatrix} = \begin{pmatrix} a & b \\ c & d \end{pmatrix} \begin{pmatrix} x \\ y \end{pmatrix}.$$

The nature of a particular transformation can often be determined by finding the image of the *unit square* $OACB$ with vertices $O(0,0)$, $A(1,0)$, $B(0,1)$, $C(1,1)$. Clearly under any linear transformation the point O is mapped onto itself, since

$$\begin{pmatrix} a & b \\ c & d \end{pmatrix}\begin{pmatrix} 0 \\ 0 \end{pmatrix} = \begin{pmatrix} 0 \\ 0 \end{pmatrix}.$$

Example 1   Describe the effect on the points of the x,y plane of the transformation with matrix $\begin{pmatrix} -1 & 0 \\ 0 & -1 \end{pmatrix}$.

$$\begin{pmatrix} -1 & 0 \\ 0 & -1 \end{pmatrix}\begin{pmatrix} 1 \\ 0 \end{pmatrix} = \begin{pmatrix} -1 \\ 0 \end{pmatrix}, \quad \begin{pmatrix} -1 & 0 \\ 0 & -1 \end{pmatrix}\begin{pmatrix} 0 \\ 1 \end{pmatrix} = \begin{pmatrix} 0 \\ -1 \end{pmatrix},$$

$$\begin{pmatrix} -1 & 0 \\ 0 & -1 \end{pmatrix}\begin{pmatrix} 1 \\ 1 \end{pmatrix} = \begin{pmatrix} -1 \\ -1 \end{pmatrix},$$

$\therefore$ the images of the points $A(1,0)$, $B(0,1)$, $C(1,1)$ are $A'(-1,0)$, $B'(0,-1)$ and $C'(-1,-1)$ respectively.

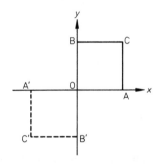

The diagram shows that the matrix $\begin{pmatrix} -1 & 0 \\ 0 & -1 \end{pmatrix}$ represents a rotation through $180°$ about the origin.

The matrix of a given transformation may be determined using the fact that its columns represent the images of the points $A(1,0)$ and $B(0,1)$. We use this method to write down the matrices of some basic transformations.

(1)   *Reflection* in the *x*-axis.

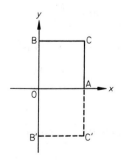

$$\begin{pmatrix} 1 \\ 0 \end{pmatrix} \rightarrow \begin{pmatrix} 1 \\ 0 \end{pmatrix} \quad \text{and} \quad \begin{pmatrix} 0 \\ 1 \end{pmatrix} \rightarrow \begin{pmatrix} 0 \\ -1 \end{pmatrix}$$

$\therefore$ the matrix of the transformation

is $\begin{pmatrix} 1 & 0 \\ 0 & -1 \end{pmatrix}$.

(2)   *Rotation* about O through $90°$ anti-clockwise.

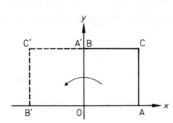

$$\begin{pmatrix} 1 \\ 0 \end{pmatrix} \rightarrow \begin{pmatrix} 0 \\ 1 \end{pmatrix} \quad \text{and} \quad \begin{pmatrix} 0 \\ 1 \end{pmatrix} \rightarrow \begin{pmatrix} -1 \\ 0 \end{pmatrix}$$

$\therefore$ the matrix of the transformation

is $\begin{pmatrix} 0 & -1 \\ 1 & 0 \end{pmatrix}$.

(3)   *Enlargement* (or dilatation), scale factor k.

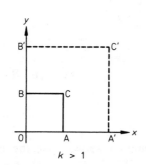

$$\begin{pmatrix} 1 \\ 0 \end{pmatrix} \rightarrow \begin{pmatrix} k \\ 0 \end{pmatrix} \quad \text{and} \quad \begin{pmatrix} 0 \\ 1 \end{pmatrix} \rightarrow \begin{pmatrix} 0 \\ k \end{pmatrix}$$

$\therefore$ the matrix of the transformation

is $\begin{pmatrix} k & 0 \\ 0 & k \end{pmatrix}$.

(4)   *Shear* with the x-axis fixed.

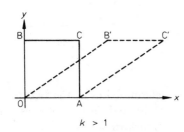

$$\begin{pmatrix} 1 \\ 0 \end{pmatrix} \rightarrow \begin{pmatrix} 1 \\ 0 \end{pmatrix} \quad \text{and} \quad \begin{pmatrix} 0 \\ 1 \end{pmatrix} \rightarrow \begin{pmatrix} k \\ 1 \end{pmatrix}$$

$\therefore$ the matrix of the transformation

is $\begin{pmatrix} 1 & k \\ 0 & 1 \end{pmatrix}$.

(5)   *One way stretch* in the direction of the x-axis, by a factor of k.

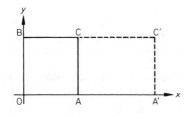

$$\begin{pmatrix} 1 \\ 0 \end{pmatrix} \rightarrow \begin{pmatrix} k \\ 0 \end{pmatrix} \quad \text{and} \quad \begin{pmatrix} 0 \\ 1 \end{pmatrix} \rightarrow \begin{pmatrix} 0 \\ 1 \end{pmatrix}$$

$\therefore$ the matrix of the transformation

is $\begin{pmatrix} k & 0 \\ 0 & 1 \end{pmatrix}$.

[The term 'stretch' has no generally accepted mathematical definition. It should therefore be used with care.]

Consider again a general linear transformation with matrix $\mathbf{M} = \begin{pmatrix} a & b \\ c & d \end{pmatrix}$ which maps the unit square $OACB$ onto a parallelogram $OA'C'B'$.

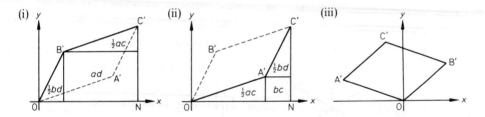

For the case illustrated in diagrams (i) and (ii) comparison of the areas $OB'C'N$ and $OA'C'N$ shows that the area of $OA'C'B'$ is $ad - bc$, i.e. $\det \mathbf{M}$. In other cases where there is a change in orientation, as in (iii), it is found that the area of $OA'C'B'$ is $bc - ad$, i.e. $-\det \mathbf{M}$. Hence $\det \mathbf{M}$ is the scale factor for change in area under the transformation represented by $\mathbf{M}$.

For a singular matrix with zero determinant the area of $OA'C'B'$ is zero and hence the points O, A', B' and C' must lie on a straight line. For instance, under the transformation with matrix $\begin{pmatrix} 2 & 3 \\ 4 & 6 \end{pmatrix}$, which has zero determinant, the images of the points $A(1,0)$, $B(0,1)$ are the points $A'(2,4)$, $B'(3,6)$. Hence the transformation maps every point of the plane onto the line $y = 2x$.

Considering the effect of this transformation in more detail,

$$\begin{pmatrix} 2 & 3 \\ 4 & 6 \end{pmatrix}\begin{pmatrix} x \\ y \end{pmatrix} = \begin{pmatrix} 2x+3y \\ 4x+6y \end{pmatrix} = (2x+3y)\begin{pmatrix} 1 \\ 2 \end{pmatrix}.$$

Thus if $2x + 3y = k$, then $\begin{pmatrix} 2 & 3 \\ 4 & 6 \end{pmatrix}\begin{pmatrix} x \\ y \end{pmatrix} = k\begin{pmatrix} 1 \\ 2 \end{pmatrix} = \begin{pmatrix} k \\ 2k \end{pmatrix}.$

Hence every point on the line $2x + 3y = k$ is mapped onto the single point $(k, 2k)$. For instance, the point $(2, 4)$ is the image of not only the point $A(1,0)$ but of every point on the line $2x + 3y = 2$.

Of particular interest are linear transformations in which both shape and area are preserved, namely rotations and reflections.

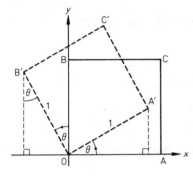

As shown in the diagram, for an anti-clockwise rotation about O through an angle θ

$$\begin{pmatrix} 1 \\ 0 \end{pmatrix} \rightarrow \begin{pmatrix} \cos\theta \\ \sin\theta \end{pmatrix} \quad \text{and} \quad \begin{pmatrix} 0 \\ 1 \end{pmatrix} \rightarrow \begin{pmatrix} -\sin\theta \\ \cos\theta \end{pmatrix}$$

$\therefore$ the matrix of the transformation

is $\mathbf{R} = \begin{pmatrix} \cos\theta & -\sin\theta \\ \sin\theta & \cos\theta \end{pmatrix}$, where

$\det \mathbf{R} = \cos^2\theta + \sin^2\theta = 1.$

Hence any rotation of the x, y plane about the origin is represented by a matrix of the form $\begin{pmatrix} a & -b \\ b & a \end{pmatrix}$ where $a^2 + b^2 = 1$.

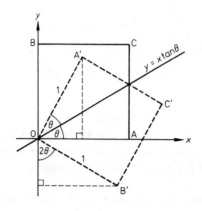

For a reflection in the line

$y = mx$, where $m = \tan \theta$,

$$\begin{pmatrix} 1 \\ 0 \end{pmatrix} \rightarrow \begin{pmatrix} \cos 2\theta \\ \sin 2\theta \end{pmatrix}, \quad \begin{pmatrix} 0 \\ 1 \end{pmatrix} \rightarrow \begin{pmatrix} \sin 2\theta \\ -\cos 2\theta \end{pmatrix}$$

$\therefore$ the matrix of the transformation is

$$\mathbf{M} = \begin{pmatrix} \cos 2\theta & \sin 2\theta \\ \sin 2\theta & -\cos 2\theta \end{pmatrix}, \quad \text{where}$$

$\det \mathbf{M} = -\cos^2 2\theta - \sin^2 2\theta = -1$.

Hence any reflection of the x, y plane in a line through the origin is represented by a matrix of the form $\begin{pmatrix} a & b \\ b & -a \end{pmatrix}$ where $a^2 + b^2 = 1$.

A point which is mapped onto itself under a transformation is called an *invariant point*. Similarly a line which is mapped onto itself is an *invariant line*. The origin is invariant for all linear transformations, but many linear transformations such as rotations and enlargements have no other invariant points. However, in an enlargement every line through the origin is invariant. The invariant lines in a reflection are the 'mirror' line and all lines perpendicular to it. The mirror line may be described as 'pointwise' invariant as every point on it is invariant. A shear has a set of parallel invariant lines, one of which is 'pointwise' invariant.

For any given linear transformation invariant lines through the origin can be found by either of the following methods.

Example 2   Find the invariant lines through the origin of the linear transformation with matrix $\begin{pmatrix} 4 & -2 \\ 5 & -3 \end{pmatrix}$.

Method A   Consider the image of a typical point (k, mk) on the line $y = mx$.

$$\begin{pmatrix} 4 & -2 \\ 5 & -3 \end{pmatrix} \begin{pmatrix} k \\ mk \end{pmatrix} = \begin{pmatrix} 4k - 2mk \\ 5k - 3mk \end{pmatrix} = \begin{pmatrix} \{4 - 2m\}k \\ \{5 - 3m\}k \end{pmatrix}.$$

Hence the image of the line $y = mx$ is the set of points $(\{4 - 2m\}k, \{5 - 3m\}k)$, i.e. the line $y = \dfrac{5 - 3m}{4 - 2m} x$.

Thus the line $y = mx$ is invariant if $\dfrac{5-3m}{4-2m} = m$

i.e.   if $5 - 3m = (4 - 2m)m$
$$2m^2 - 7m + 5 = 0$$
$$(2m - 5)(m - 1) = 0$$

i.e.   if either $m = 5/2$ or $m = 1$.

Hence the invariant lines of the transformation are $y = x$ and $y = \dfrac{5}{2}x$.

Method B   If the point (x, y) lies on an invariant line through the origin, then its image must lie on the same line and have coordinates of the form $(\lambda x, \lambda y)$.

$$\begin{pmatrix} 4 & -2 \\ 5 & -3 \end{pmatrix}\begin{pmatrix} x \\ y \end{pmatrix} = \begin{pmatrix} \lambda x \\ \lambda y \end{pmatrix} \Rightarrow \begin{matrix} 4x - 2y = \lambda x \\ 5x - 3y = \lambda y \end{matrix}$$

$$\Rightarrow \begin{matrix} (4 - \lambda)x + (-2)y = 0 \\ 5x + (-3 - \lambda)y = 0. \end{matrix}$$

These equations will hold for non-zero values of x and y only if

$$\begin{vmatrix} 4 - \lambda & -2 \\ 5 & -3 - \lambda \end{vmatrix} = 0, \quad \text{i.e.} \quad (4 - \lambda)(-3 - \lambda) - (-2)5 = 0$$

$$\therefore \qquad \lambda^2 - \lambda - 2 = 0$$
$$(\lambda - 2)(\lambda + 1) = 0.$$

Hence either $\lambda = 2$ or $\lambda = -1$.

For $\lambda = 2$ both equations reduce to $x - y = 0$.

For $\lambda = -1$ both equations reduce to $5x - 2y = 0$.

Hence the invariant lines of the transformation are $y = x$ and $y = \dfrac{5}{2}x$.

In general, if values of λ and column vectors $\begin{pmatrix} x \\ y \end{pmatrix}$ can be found such that $\begin{pmatrix} a & b \\ c & d \end{pmatrix}\begin{pmatrix} x \\ y \end{pmatrix} = \lambda \begin{pmatrix} x \\ y \end{pmatrix}$, then these are called the *eigenvalues* and *eigenvectors* of the matrix $\begin{pmatrix} a & b \\ c & d \end{pmatrix}$.

Exercise 19.4

1.   Write down the matrices representing the following transformations:
(a) reflection in the y-axis,
(b) enlargement, scale factor 2,
(c) one way stretch parallel to the y-axis, $\times 3$,
(d) rotation about O through $45°$ clockwise,

(e) reflection in the line $y = x$,

(f) shear parallel to the y-axis, $(1, 1) \rightarrow (1, 3)$.

2. Describe geometrically the transformations whose matrices are:

(a) $\begin{pmatrix} 3 & 0 \\ 0 & 3 \end{pmatrix}$,
(b) $\begin{pmatrix} 0 & 1 \\ -1 & 0 \end{pmatrix}$,
(c) $\begin{pmatrix} 1 & 0 \\ -2 & 1 \end{pmatrix}$,
(d) $\begin{pmatrix} 1 & 0 \\ 0 & 1 \end{pmatrix}$,

(e) $\begin{pmatrix} 2 & 0 \\ 0 & 1 \end{pmatrix}$,
(f) $\begin{pmatrix} 0 & 0 \\ 0 & 0 \end{pmatrix}$,
(g) $\begin{pmatrix} 1 & 1 \\ 1 & 1 \end{pmatrix}$,
(h) $\begin{pmatrix} 2 & 0 \\ 0 & 3 \end{pmatrix}$.

3. Describe geometrically the transformations whose matrices are:

(a) $\begin{pmatrix} 0\cdot6 & -0\cdot8 \\ 0\cdot8 & 0\cdot6 \end{pmatrix}$,
(b) $\begin{pmatrix} 0\cdot8 & 0\cdot6 \\ 0\cdot6 & -0\cdot8 \end{pmatrix}$,
(c) $\frac{1}{\sqrt{2}}\begin{pmatrix} 1 & 1 \\ 1 & -1 \end{pmatrix}$,

(d) $\begin{pmatrix} 1 & 1 \\ -1 & 1 \end{pmatrix}$,
(e) $\frac{1}{13}\begin{pmatrix} -12 & 5 \\ 5 & 12 \end{pmatrix}$,
(f) $\begin{pmatrix} 4 & -3 \\ 3 & 4 \end{pmatrix}$.

4. Find the matrix of the linear transformation which maps the points $P(1,0)$, $Q(0, 1)$ to the points $P'(5, 3)$, $Q'(1, 7)$. Use your answer to find the area of the triangle $OP'Q'$.

5. Use the method of question 4 to obtain an expression for the area of the triangle with vertices $(0,0)$, (x_1, y_1) and (x_2, y_2).

6. Show that the transformation with matrix $\begin{pmatrix} 1 & -2 \\ -2 & 4 \end{pmatrix}$ maps every point of the x, y plane onto the same straight line and give the equation of this line. Describe the set of points which are mapped onto the origin.

7. Repeat question 6 for the matrices (a) $\begin{pmatrix} 1 & -2 \\ 3 & -6 \end{pmatrix}$, (b) $\begin{pmatrix} 3 & 2 \\ 9 & 6 \end{pmatrix}$.

8. A transformation in which points (x, y) are mapped into points (x', y') in the same plane is given by

$$x' = \tfrac{1}{2}(x + \sqrt{3}y), \quad y' = \tfrac{1}{2}(\sqrt{3}x - y).$$

Describe the transformation fully, giving the equations of any invariant lines through the origin. Find the equation of the line which is mapped onto the x-axis.

9. Repeat question 8 for the transformation with equations

$$x' = y, \quad y' = 2y - x.$$

10. Find all possible vectors $\begin{pmatrix} x \\ y \end{pmatrix}$ such that

(a) $\begin{pmatrix} -5 & -3 \\ 2 & 2 \end{pmatrix}\begin{pmatrix} x \\ y \end{pmatrix} = \begin{pmatrix} x \\ y \end{pmatrix}$,
(b) $\begin{pmatrix} 4 & -6 \\ 1 & -3 \end{pmatrix}\begin{pmatrix} x \\ y \end{pmatrix} = 3\begin{pmatrix} x \\ y \end{pmatrix}$.

11. Find the equations of the invariant lines through the origin of the linear transformations with matrices

(a) $\begin{pmatrix} 2 & -1 \\ 5 & -4 \end{pmatrix}$,

(b) $\begin{pmatrix} 2 & 2 \\ 3 & 1 \end{pmatrix}$.

12. Find the equations of the invariant lines through the origin of the linear transformation with matrix $\begin{pmatrix} 5 & 2 \\ 2 & 2 \end{pmatrix}$. Describe geometrically the effect of the transformation.

19.5 Composite transformations

If two linear transformations with matrices **P** and **Q** are performed in succession, using first **P** then **Q**, the image of a point with position vector **r** under this composite transformation is given by

$$\mathbf{r}' = \mathbf{Q(Pr)} = \mathbf{(QP)r}.$$

Hence the composite transformation has matrix **QP**.

For instance, if **P** represents a reflection in the x-axis and **Q** an anti-clockwise rotation about the origin through 90°, then $\mathbf{P} = \begin{pmatrix} 1 & 0 \\ 0 & -1 \end{pmatrix}$ and $\mathbf{Q} = \begin{pmatrix} 0 & -1 \\ 1 & 0 \end{pmatrix}$.

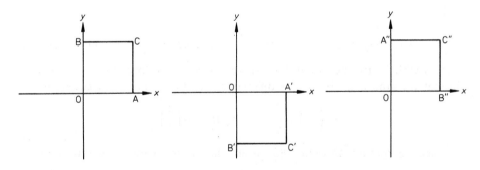

Using **P** then **Q** should produce a reflection in the line $y = x$, as shown in the diagrams, which has matrix $\begin{pmatrix} 0 & 1 \\ 1 & 0 \end{pmatrix}$. As expected $\mathbf{QP} = \begin{pmatrix} 0 & -1 \\ 1 & 0 \end{pmatrix}\begin{pmatrix} 1 & 0 \\ 0 & -1 \end{pmatrix}$

$= \begin{pmatrix} 0 & 1 \\ 1 & 0 \end{pmatrix}$. Similarly, using **Q** then **P** produces a reflection in the line $y = -x$,

which has matrix $\mathbf{PQ} = \begin{pmatrix} 1 & 0 \\ 0 & -1 \end{pmatrix}\begin{pmatrix} 0 & -1 \\ 1 & 0 \end{pmatrix} = \begin{pmatrix} 0 & -1 \\ -1 & 0 \end{pmatrix}$.

Let $\mathbf{R}_1$ and $\mathbf{R}_2$ represent anticlockwise rotations about the origin through angles θ and ϕ respectively. The composite transformation using $\mathbf{R}_1$ then $\mathbf{R}_2$ has matrix

$$R_2 R_1 = \begin{pmatrix} \cos\phi & -\sin\phi \\ \sin\phi & \cos\phi \end{pmatrix} \begin{pmatrix} \cos\theta & -\sin\theta \\ \sin\theta & \cos\theta \end{pmatrix}$$

$$= \begin{pmatrix} \cos\theta\cos\phi - \sin\theta\sin\phi & -\{\sin\theta\cos\phi + \cos\theta\sin\phi\} \\ \sin\theta\cos\phi + \cos\theta\sin\phi & \cos\theta\cos\phi - \sin\theta\sin\phi \end{pmatrix}.$$

However, since this transformation is equivalent to a single rotation through an angle $(\theta + \phi)$, its matrix may also be expressed in the form

$$\begin{pmatrix} \cos\{\theta+\phi\} & -\sin\{\theta+\phi\} \\ \sin\{\theta+\phi\} & \cos\{\theta+\phi\} \end{pmatrix}.$$

Hence we may deduce two trigonometric identities proved by other methods in §7.4:

$$\cos\{\theta+\phi\} = \cos\theta\cos\phi - \sin\theta\sin\phi$$
$$\sin\{\theta+\phi\} = \sin\theta\cos\phi + \cos\theta\sin\phi.$$

Finally we consider the set of non-linear transformations of the x, y plane with equations of the form

$$\begin{aligned} x' &= ax + by + p \\ y' &= cx + dy + q, \end{aligned} \quad \text{i.e.} \quad \begin{pmatrix} x' \\ y' \end{pmatrix} = \begin{pmatrix} a & b \\ c & d \end{pmatrix} \begin{pmatrix} x \\ y \end{pmatrix} + \begin{pmatrix} p \\ q \end{pmatrix}.$$

These can be considered as composite transformations consisting of a linear transformation with matrix $\begin{pmatrix} a & b \\ c & d \end{pmatrix}$ followed by a translation with column vector $\begin{pmatrix} p \\ q \end{pmatrix}$.

It can be shown that if the matrix $\begin{pmatrix} a & b \\ c & d \end{pmatrix}$ represents a rotation through an angle θ with centre the origin, then the composite transformation is a rotation through the same angle but with a different centre. For instance, the equation

$$\begin{pmatrix} x' \\ y' \end{pmatrix} = \begin{pmatrix} -1 & 0 \\ 0 & -1 \end{pmatrix} \begin{pmatrix} x \\ y \end{pmatrix} + \begin{pmatrix} 4 \\ 2 \end{pmatrix}$$

represents a rotation through $180°$ about the origin followed by a translation.

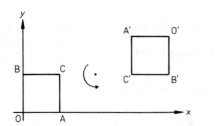

The diagram indicates that the composite transformation is a rotation through $180°$ about the point $(2, 1)$.

A reflection followed by a translation may produce either a reflection in a different line or a *glide-reflection*, which is a reflection with a translation along the 'mirror' line.

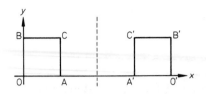

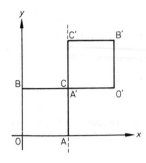

For example, the equation

$$\begin{pmatrix} x' \\ y' \end{pmatrix} = \begin{pmatrix} -1 & 0 \\ 0 & 1 \end{pmatrix}\begin{pmatrix} x \\ y \end{pmatrix} + \begin{pmatrix} 4 \\ 0 \end{pmatrix}.$$

represents a reflection in the y-axis followed by a translation. As shown in the diagram this is equivalent to a reflection in the line $x = 2$.

However, the equation

$$\begin{pmatrix} x' \\ y' \end{pmatrix} = \begin{pmatrix} -1 & 0 \\ 0 & 1 \end{pmatrix}\begin{pmatrix} x \\ y \end{pmatrix} + \begin{pmatrix} 2 \\ 1 \end{pmatrix}$$

represents a glide-reflection along the line $x = 1$.

Example 1 Describe fully the transformation of the x, y plane with equations

$$x' = -y+4, \quad y' = x.$$

The matrix equation of the transformation is

$$\begin{pmatrix} x' \\ y' \end{pmatrix} = \begin{pmatrix} 0 & -1 \\ 1 & 0 \end{pmatrix}\begin{pmatrix} x \\ y \end{pmatrix} + \begin{pmatrix} 4 \\ 0 \end{pmatrix}$$

The matrix $\begin{pmatrix} 0 & -1 \\ 1 & 0 \end{pmatrix}$ represents an anti-clockwise rotation of $90°$ about the origin and the column vector $\begin{pmatrix} 4 \\ 0 \end{pmatrix}$ represents a translation. Since a rotation followed by a translation is equivalent to a rotation through the same angle but with a different centre, the given transformation must be an anti-clockwise rotation through $90°$ about some point (h, k).

As the point (h, k) is invariant under the transformation,

$$h = -k+4, \quad k = h$$

$$\therefore \quad h = k = 2.$$

Hence the transformation is a rotation anti-clockwise through $90°$ about the point $(2, 2)$.

Exercise 19.5

1. Describe geometrically the transformations with matrices **A**, **B**, **BA** and **AB** where

(a) $\mathbf{A} = \begin{pmatrix} 2 & 0 \\ 0 & 1 \end{pmatrix}$, $\mathbf{B} = \begin{pmatrix} \frac{1}{2} & 0 \\ 0 & \frac{1}{2} \end{pmatrix}$, (b) $\mathbf{A} = \begin{pmatrix} 0 & 1 \\ 1 & 0 \end{pmatrix}$, $\mathbf{B} = \begin{pmatrix} -1 & 0 \\ 0 & 1 \end{pmatrix}$,

(c) $A = \dfrac{1}{2}\begin{pmatrix} \sqrt{3} & -1 \\ 1 & \sqrt{3} \end{pmatrix}$, $\quad B = \dfrac{1}{2}\begin{pmatrix} 1 & -\sqrt{3} \\ \sqrt{3} & 1 \end{pmatrix}$,

(d) $A = \begin{pmatrix} -1 & 0 \\ 0 & -1 \end{pmatrix}$, $\quad B = \dfrac{1}{5}\begin{pmatrix} 4 & 3 \\ 3 & -4 \end{pmatrix}$.

In each case verify that **BA** has the same effect as **A** followed by **B** and that **AB** is equivalent to **B** then **A**.

2.   Describe geometrically the transformations with matrices **A**, A^2 and A^{-1} where

(a) $A = \dfrac{1}{\sqrt{2}}\begin{pmatrix} 1 & -1 \\ 1 & 1 \end{pmatrix}$,   (b) $A = \begin{pmatrix} 3 & 0 \\ 0 & 3 \end{pmatrix}$,   (c) $A = \begin{pmatrix} 0 & -1 \\ -1 & 0 \end{pmatrix}$.

3.   By considering the matrix representation of two successive rotations about the origin through an angle θ, show that $\cos 2\theta = \cos^2 \theta - \sin^2 \theta$ and $\sin 2\theta = 2 \sin \theta \cos \theta$.

4.   If $P = \begin{pmatrix} 2 & -3 \\ 1 & -1 \end{pmatrix}$ and $Q = \begin{pmatrix} 3 & 1 \\ 4 & 2 \end{pmatrix}$, find the point whose image is $(5, 6)$ under the transformation with matrix
(a) **P**,   (b) **Q**,   (c) **QP**,   (d) **PQ**.

5.   Write down the matrix **R** representing an anti-clockwise rotation about the origin through an angle θ; the matrix **S** representing a shear that fixes the x-axis and maps the point $(1, 1)$ to the point $(2, 1)$; the matrix **T** representing a reflection in the x-axis. Describe the transformations represented by the matrices
(a) RTR^{-1},   (b) **STS**,   (c) $R^{-1}SR$.

6.   Describe geometrically the transformations represented by the following matrix equations:

(a) $\begin{pmatrix} x' \\ y' \end{pmatrix} = \begin{pmatrix} 1 & 0 \\ 0 & -1 \end{pmatrix}\begin{pmatrix} x \\ y \end{pmatrix} + \begin{pmatrix} 0 \\ 2 \end{pmatrix}$,   (b) $\begin{pmatrix} x' \\ y' \end{pmatrix} = \begin{pmatrix} -1 & 0 \\ 0 & -1 \end{pmatrix}\begin{pmatrix} x \\ y \end{pmatrix} + \begin{pmatrix} 2 \\ 2 \end{pmatrix}$,

(c) $\begin{pmatrix} x' \\ y' \end{pmatrix} = \begin{pmatrix} 3 & 0 \\ 0 & 3 \end{pmatrix}\begin{pmatrix} x \\ y \end{pmatrix} + \begin{pmatrix} -1 \\ -1 \end{pmatrix}$,   (d) $\begin{pmatrix} x' \\ y' \end{pmatrix} = \begin{pmatrix} 0 & 1 \\ 1 & 0 \end{pmatrix}\begin{pmatrix} x \\ y \end{pmatrix} + \begin{pmatrix} 2 \\ 0 \end{pmatrix}$.

7.   A transformation of the plane is given by the equations

$$x' = 5 - 2y, \quad y' = 4 - 2x.$$

Show that there is one invariant point and find it. Find also the invariant lines of the transformation. Give a full geometrical description of the transformation. (O)

8.   Find the matrix equations representing the following transformations:
(a) a clockwise rotation of $90°$ about the point $(-2, 1)$,
(b) a reflection in the line $x + y = 2$,
(c) an enlargement, centre $(-1, -1)$, scale factor 2,
(d) a glide-reflection along the line $y = 2$ such that the image of $(0, 2)$ is $(2, 2)$.

Exercise 19.6   *(miscellaneous)*

1.   Obtain all possible sums and products using two of the following matrices:

$$A = \begin{pmatrix} 1 & 3 \\ -2 & 0 \\ 2 & -1 \end{pmatrix}, \quad B = \begin{pmatrix} 2 & 0 & -1 \\ 0 & 1 & 3 \\ -3 & 1 & 2 \end{pmatrix}, \quad C = \begin{pmatrix} 3 \\ -1 \\ 0 \end{pmatrix}, \quad D = \begin{pmatrix} 2 & -1 \\ 0 & 1 \\ 1 & -2 \end{pmatrix},$$

$$E = \begin{pmatrix} 1 \\ 2 \end{pmatrix}, \quad F = (1 \quad 0 \quad 1), \quad G = \begin{pmatrix} -2 \\ 1 \end{pmatrix}, \quad H = (-1 \quad 0).$$

2.   If $P = \begin{pmatrix} 2 & -1 \\ 1 & 3 \end{pmatrix}$ and $P^2 - aP + bI = 0$, where a and b are scalars, find the values of a and b.

3.   If $X = \begin{pmatrix} 1 & 3 \\ 2 & 1 \end{pmatrix}$, evaluate $X^2 - 2X$ and hence express X^{-1} in terms of X.

4.   Given that A and B are square matrices of the same order, simplify
(a) $(A+B)(A-B) - (A-B)(A+B)$
(b) $(A+B)^3 + (A-B)^3 - 2A(A^2 + B^2)$.

5. A is an $m \times n$ matrix and B is a $p \times q$ matrix. Write down the conditions that m, n, p and q must satisfy if it is possible to calculate
(a) $A+B$,   (b) AB,   (c) BA,   (d) A^2,   (e) AB^{-1}.

6.   If $A = \begin{pmatrix} a & b \\ c & d \end{pmatrix}$ and $A \neq \pm I$, find the condition that A should be its own inverse.

7.   Prove that for all 2×2 matrices A and B, .
(a) $\det(kA) = k^2 \det A$, where k is a scalar,
(b) $\det(AB) = \det A \det B$,
(c) if A is singular, then AB is also singular.

8.   Find the conditions that the equations
$$3x + hy = p, \quad 2x + ky = q$$
should have (a) a unique solution, (b) infinitely many solutions, (c) no solution.

9.   Decide whether the following statement is true or false: provided that $A \neq 0$, then $AX = AY \Rightarrow X = Y$. Justify your answer either by proving the statement or by giving an example which disproves it.

10.   Describe completely the transformations with matrices

(a) $\begin{pmatrix} 0 & 2 \\ 2 & 0 \end{pmatrix}$,   (b) $\begin{pmatrix} 0 & -2 \\ 1 & 0 \end{pmatrix}$,   (c) $\begin{pmatrix} 1 & 0 \\ 0 & 0 \end{pmatrix}$,   (d) $\begin{pmatrix} 2 & 1 \\ -1 & 0 \end{pmatrix}$.

11. The matrix $\begin{pmatrix} 2 & 3 \\ -4 & -6 \end{pmatrix}$ represents a mapping of the plane to itself. On a diagram, draw the set of all points that map to $(1, -2)$. On another diagram draw the set of all possible images under this mapping.   (AEB 1975)

12. Give a full geometrical description of the transformations of the plane given by the matrices A and B, where

$$A = \frac{1}{\sqrt{2}}\begin{pmatrix} 1 & 1 \\ 1 & -1 \end{pmatrix}, \quad B = \frac{1}{\sqrt{2}}\begin{pmatrix} 1 & -1 \\ 1 & 1 \end{pmatrix}.$$

Describe also the transformations given by the matrices AB and BA.   (O)

13. Find the images under the transformation with matrix $\begin{pmatrix} 7 & 1 \\ 9 & 7 \end{pmatrix}$ of the lines (a) $y = x$, (b) $y = -x$, (c) $y = mx$. Find the values of m for which the line $y = mx$ is mapped onto itself.

14. When column vectors, representing points in a plane with the usual axes, are premultiplied by the matrix $\mathbf{M} = \begin{pmatrix} 3 & -2 \\ 6 & k \end{pmatrix}$, a transformation of the plane results.
(i) If $k = 3$, find (a) the point whose image is $(8, 9)$, (b) the image of the line $y = x$.
(ii) If $k = -4$, find (a) the set of points which map onto the origin, (b) the image of the line $y = \frac{3}{2}x - \frac{1}{2}$.   (C)

15. Given the matrix $M = \frac{1}{13}\begin{pmatrix} 5 & 12 \\ 12 & -5 \end{pmatrix}$, evaluate M^2 and the determinant of M.
Find a set of vectors $\begin{pmatrix} x \\ y \end{pmatrix}$ such that $M\begin{pmatrix} x \\ y \end{pmatrix} = \begin{pmatrix} x \\ y \end{pmatrix}$ and also a set of vectors $\begin{pmatrix} u \\ v \end{pmatrix}$ such that $M\begin{pmatrix} u \\ v \end{pmatrix} = \begin{pmatrix} -u \\ -v \end{pmatrix}$. Describe in geometrical terms the transformation represented by the matrix M.   (JMB)

16. A linear transformation is defined by the matrix $A = \begin{pmatrix} 1 & 1 \\ 1 & -1 \end{pmatrix}$. Describe, in geometrical terms, its effect upon the plane. Obtain the equations of the lines through the origin which map onto themselves under A.   (O & C)

17. A linear transformation T_1 of the plane takes the points $(1, 2)$, $(3, -2)$ to the points $(15, 27)$, $(-11, -7)$ respectively. Find the matrix of this transformation.
   A linear transformation T_2 of the plane is made up of the following sequence of elementary geometric transformations in the order given:
(a) a shear in which each point is moved parallel to the x-axis through a distance twice its distance from the x-axis,
(b) a rotation through $\frac{1}{4}\pi$ anti-clockwise about the origin,

(c) a stretch by a factor of 2 parallel to the x-axis and by a factor of 3 parallel to the y-axis,

(d) a further rotation through $\frac{1}{4}\pi$ anti-clockwise about the origin,

(e) a reflection in the y-axis.

Find the matrix of the transformation. State the geometric relation between the transformations T_1 and T_2. (O & C)

18. Obtain the matrix representing a reflection in the line $y = 2x$. If the image of a curve C under this transformation is the curve C' with equation $16x^2 + 24xy + 9y^2 + 15x - 20y = 0$, find the equation of C. Use your result to make a sketch of the curve C'.

19. Find and simplify the matrix representing a reflection in the line $y = x\tan\theta$ followed by a reflection in the line $y = x\tan\phi$. Describe geometrically the nature of this composite transformation.

20. Ox, Oy are rectangular axes in a plane. State the matrix representations of the following transformations of the plane into itself:

(a) reflection in a line through O making an angle θ with Ox (θ being measured in the anti-clockwise direction), (b) anti-clockwise rotation about O through an angle θ.

Linear transformations T_1 and T_2 are reflections of the plane in the lines $y = x$ and $y = x\tan(\pi/3)$ respectively. Write down the matrix representation of T_1 and T_2, find the matrix representation of the combined transformation $T_2 T_1$ and interpret the combined transformation geometrically.

A transformation T_3 is a magnification from the origin by a factor 2; T_4 is a translation with vector $\begin{pmatrix} -\sqrt{3} \\ -1 \end{pmatrix}$. Find the matrix of the transformation $T_3 T_2 T_1$ and the image of the point $(0, 1)$ under the transformation $T_4 T_3 T_2 T_1$. (L)

21. (a) A linear transformation T_1 of the plane has matrix $\begin{pmatrix} 4 & 2 \\ -1 & 1 \end{pmatrix}$. State the ratio of the area of the image under T_1 of a finite region R to the area of R. Write down the matrix of the inverse transformation T_1^{-1} and find the coordinates of the point which is mapped onto the point $(1, -\frac{1}{2})$ by T_1. Find also the equations of the two lines through the origin each of which is mapped onto itself by T_1.

(b) A second transformation T_2 has matrix $\begin{pmatrix} 4 & 2 \\ -2 & -1 \end{pmatrix}$. Explain why T_2 has no inverse transformation and show that T_2 maps all points of a certain line, whose equation should be given, onto $(1, -\frac{1}{2})$. Show also that T_2 maps all points of the plane onto a line, and give the equation of this line. (L)

22. Give a full geometrical description of the following transformations of the plane:

(i) $x' = -2y + 3$
    $y' = 2x - 1$

(ii) $5x' = 3x - 4y + 4$
     $5y' = -4x - 3y + 8$. (O)

20 Graphs and parameters

20.1 Odd and even functions

The diagrams below show the graphs of $y = x^n$ for various integral values of n: in (1) odd values of n and in (2) even values.

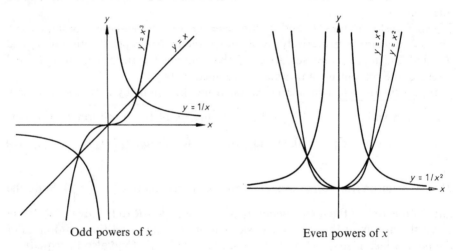

Odd powers of x Even powers of x

In diagram (1) each of the curves is symmetrical about the origin, i.e. each curve remains unchanged when rotated through $180°$ about the origin. In (2) the graphs of even powers of x are symmetrical about the y-axis, i.e. the curves remain unchanged when reflected in the y-axis. These symmetries arise because for odd powers $(-x)^n = -x^n$ and for even powers $(-x)^n = x^n$.

The words 'odd' and 'even' are used to describe functions with the same symmetries as the odd and even powers of x. Thus $f(x)$ is called an *odd function* if $f(-x) = -f(x)$. The graph is symmetrical about the origin, so that if (a, b) lies on the graph so does $(-a, -b)$. Similarly $f(x)$ is called an *even function* if $f(-x) = f(x)$. In this case the graph is symmetrical about the y-axis, so that if (a, b) lies on the graph so does $(-a, b)$.

Example 1   Decide whether the following functions are odd, even or neither.
(a) $f(x) = x^3 + x$,   (b) $g(x) = x^3 + 1$,   (c) $h(x) = \cos 3x$.

(a) $f(-x) = (-x)^3 + (-x) = -x^3 - x = -(x^3 + x) = -f(x)$
   $\therefore$ $f(x)$ is an odd function.
(b) $g(-x) = (-x)^3 + 1 = -x^3 + 1$
   $\therefore$ $g(-x) \neq g(x)$ and $g(-x) \neq -g(x)$.
   Hence $g(x)$ is neither odd nor even.
(c) $h(-x) = \cos 3(-x) = \cos(-3x) = \cos 3x = h(x)$
   $\therefore$ $h(x)$ is an even function.

When sketching the graphs of odd and even functions, it is only necessary to consider positive values of x. The sketch is then completed using the symmetry of the function.

Exercise 20.1

1.   State which of the following graphs represent odd or even functions.

(a)
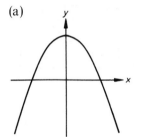

(b)
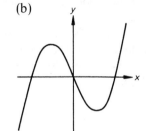

(c)
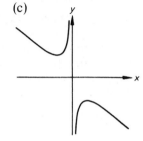

(d)
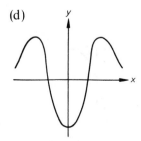

(e)
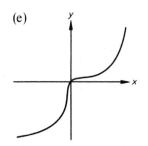

(f)
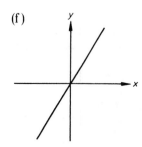

2.   Copy and complete each of the following sketches so that it represents (i) an odd function, (ii) an even function.

(a)
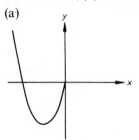

(b)
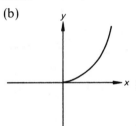

(c)
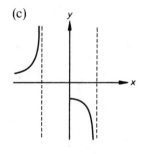

3. Decide whether the following functions are odd, even or neither.
(a) $3x^2 + 5$,   (b) $2x - x^3$,   (c) $(x-2)(x+4)$,
(d) $\dfrac{x^3 + 1}{x}$,   (e) $\dfrac{x^3 + x}{x^2}$,   (f) $\dfrac{x}{x^3 - x}$.

4. Decide whether the following functions are odd, even or neither.
(a) $\sin 2x$,   (b) $\cos 5x$,   (c) $\tan \pi x$,
(d) $\sin^2 x$,   (e) $\cos(x + \tfrac{1}{4}\pi)$,   (f) $1 + \cos x$.

20.2 Composite functions

When investigating the properties of a function it is sometimes helpful to think of it as a composite function. This approach was used in §13.1 when differentiating various types of function. It can also be used when sketching graphs.

To illustrate the method let us suppose that $y = f(x)$ is the equation of a graph already familiar to us or relatively easy to sketch. We will now consider how our knowledge can be used to sketch composite functions of the form $f(x) + k$, $f(x - k)$, $1/f(x)$ and $|f(x)|$.

The graph of $f(x) + k$ is produced from the graph of $f(x)$ by performing a translation of k units in the direction of the y-axis. Similarly a translation of k units in the direction of the x-axis gives the graph of $f(x - k)$. This is readily verified by considering some examples.

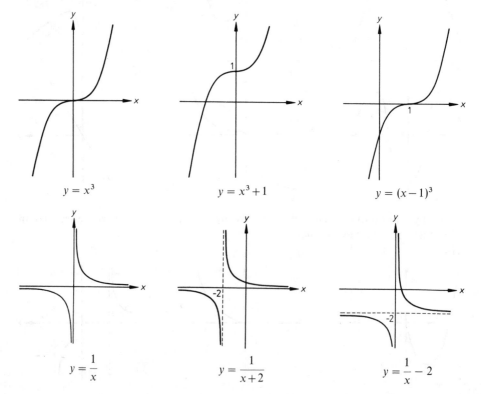

$y = x^3$ $y = x^3 + 1$ $y = (x - 1)^3$

$y = \dfrac{1}{x}$ $y = \dfrac{1}{x + 2}$ $y = \dfrac{1}{x} - 2$

When comparing the graphs of $f(x)$ and $1/f(x)$ it is useful to note the values of x for which $f(x) = 0$. If $f(a) = 0$ then the curve $y = f(x)$ cuts the x-axis at $(a, 0)$. However as $x \to a$, $1/f(x) \to \pm\infty$, which means that the line $x = a$ is an asymptote to the curve $y = 1/f(x)$.

Since as $f(x)$ increases, $1/f(x)$ must decrease, a non-zero maximum value of $f(x)$ corresponds to a minimum value of $1/f(x)$. Similarly when there is a minimum point on the curve $y = f(x)$, the corresponding point on the curve $y = 1/f(x)$, if defined, will be a maximum point.

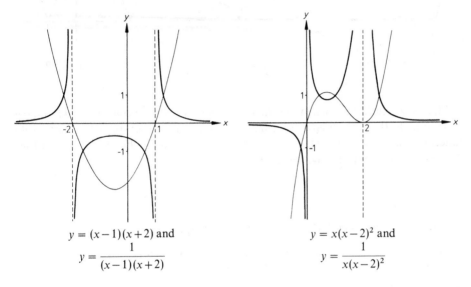

$y = (x-1)(x+2)$ and
$$y = \frac{1}{(x-1)(x+2)}$$

$y = x(x-2)^2$ and
$$y = \frac{1}{x(x-2)^2}$$

The sketches show the graphs of $y = f(x)$ and $y = 1/f(x)$ for $f(x) = (x-1)(x+2)$ and $f(x) = x(x-2)^2$. Since $f(x)$ and $1/f(x)$ always have the same sign, their graphs always lie on the same side of the x-axis. There are points of intersection where $f(x) = \pm 1$.

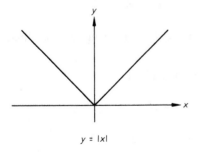

$y = |x|$

The modulus of a real number x, written $|x|$, was defined in §1.4 as the magnitude of x. Thus if $y = |x|$ then when $x \geqslant 0$, $y = x$ and when $x < 0$, $y = -x$.

Similarly the graph of $|f(x)|$ is the same as the graph of $f(x)$ when $f(x)$ is positive or zero. However, when $f(x)$ is negative, $|f(x)|$ takes the corresponding positive value. Hence when sketching the graph of $y = |f(x)|$ we must replace the parts of the curve $y = f(x)$ which lie below the x-axis by their reflection in the x-axis. For instance, using the graphs of $y = x^3 + 1$, $y = \dfrac{1}{x} - 2$ and

$y = (x-1)(x+2)$ which appear earlier in this section, the graphs of $y = |x^3+1|$,
$y = \left|\dfrac{1}{x} - 2\right|$ and $y = |(x-1)(x+2)|$ can be produced.

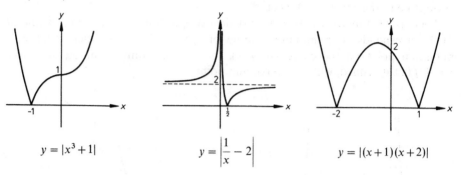

| $y = |x^3+1|$ | $y = \left|\dfrac{1}{x} - 2\right|$ | $y = |(x+1)(x+2)|$ |

Finally we note that when the turning points on the graph of $y = f(x)$ have been found, the coordinates of turning points on the graph of a composite function, such as $y = 1/f(x)$ or $y = |f(x)|$, can usually be deduced. Further differentiation should not be necessary.

Exercise 20.2

In questions 1 to 4 sketch the graphs with given equations

1. (a) $y = x^2$,   (b) $y = x^2+3$,   (c) $y = (x-3)^2$.

2. (a) $y = \dfrac{1}{x^2}$,   (b) $y = \dfrac{1}{x^2} + 4$,   (c) $y = \dfrac{1}{(x+4)^2}$.

3. (a) $y = \sin x$,   (b) $y = 1+\sin x$,   (c) $y = \sin(x-\tfrac{1}{3}\pi)$.

4. (a) $y = |x|$,   (b) $y = |x|-1$,   (c) $y = |x-1|$.

In questions 5 to 12 sketch the graphs of $f(x)$ and $1/f(x)$, giving the coordinates of any stationary points.

5. $f(x) = x^2-1$.   6. $f(x) = x^2+1$.

7. $f(x) = x(x-2)$.   8. $f(x) = (x+2)^2$.

9. $f(x) = x^2-4x+5$.   10. $f(x) = 2+x-x^2$.

11. $f(x) = x^2(x+3)$.   12. $f(x) = x^3-12x$.

In questions 13 to 18 sketch the graphs of $f(x)$ and $|f(x)|$, giving the coordinates of any stationary points.

13. $f(x) = x+2$.   14. $f(x) = 5-2x$.

15. $f(x) = x^2 - 2x - 3.$

16. $f(x) = 6x - x^2 - 9.$

17. $f(x) = 3 - \dfrac{1}{x}.$

18. $f(x) = \dfrac{1}{x^2} - 1.$

19. Discuss the graphs of composite functions of the forms $[f(x)]$ and $f([x])$, such as $[x^2]$ and $([x])^2$, where $[x]$ is the integral part of x as defined in §5.1.

20. Discuss the graphs of functions of the form $\sin\{f(x)\}$ or $\cos\{f(x)\}$, such as $\sin(x^2)$.

20.3 Addition of functions

When a function $f(x)$ can be expressed as the sum of two simpler functions $g(x)$ and $h(x)$, it is sometimes possible to sketch the graph of $f(x)$ by combining the graphs of $g(x)$ and $h(x)$. This approach can be particularly useful when dealing with a function such as $y = (x^2 + 3x + 2)/2x$. Rearranging we find that (i) $y = \dfrac{(x+1)(x+2)}{2x}$ and (ii) $y = \dfrac{1}{2}x + \dfrac{3}{2} + \dfrac{1}{x}$. The equation in form (i) tells us that the curve cuts the x-axis at the points $(-1,0)$ and $(-2,0)$ and that the y-axis is an asymptote to the curve. Form (ii) shows that the graph can be sketched by combining the graphs $y = \dfrac{1}{2}x + \dfrac{3}{2}$ and $y = \dfrac{1}{x}$.

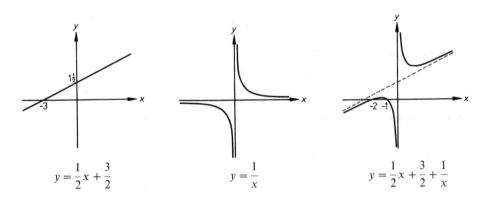

$$y = \frac{1}{2}x + \frac{3}{2} \qquad\qquad y = \frac{1}{x} \qquad\qquad y = \frac{1}{2}x + \frac{3}{2} + \frac{1}{x}$$

The sketches show that the line $y = \dfrac{1}{2}x + \dfrac{3}{2}$ is an asymptote to the curve. The coordinates of the two turning points can be obtained using dy/dx in the usual way.

The next set of sketches show the same method applied to the curve $y = (x^3 - 1)/x$, i.e. $y = x^2 - \dfrac{1}{x}.$

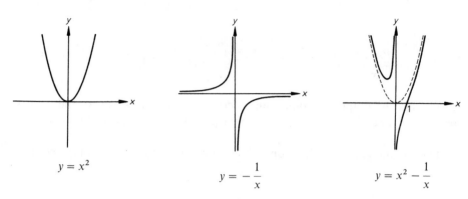

$$y = x^2$$

$$y = -\frac{1}{x}$$

$$y = x^2 - \frac{1}{x}$$

The sums of modulus functions need a different approach. For instance, consider the function $f(x) = |x+1| + |x-2|$.

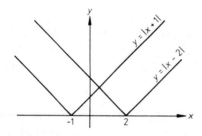

The sketch of the graphs of $|x+1|$ and $|x-2|$ shows that the behaviour of $f(x)$ is likely to change when $x = -1$ and again when $x = 2$.

When $x < -1$, $x+1 < 0$ and $x-2 < 0$
         $\therefore$ $|x+1| = -(x+1)$ and $|x-2| = -(x-2)$
$\therefore$ $f(x) = -(x+1)-(x-2) = -2x+1$.
When $-1 < x < 2$, $x+1 > 0$ and $x-2 < 0$
         $\therefore$ $|x+1| = x+1$ and $|x-2| = -(x-2)$
$\therefore$ $f(x) = (x+1)-(x-2) = 3$.
When $x > 2$, $x+1 > 0$ and $x-2 > 0$
         $\therefore$ $|x+1| = x+1$ and $|x-2| = x-2$
$\therefore$ $f(x) = (x+1)+(x-2) = 2x-1$.
Thus the graph of $f(x) = |x+1| + |x-2|$ can now be sketched.

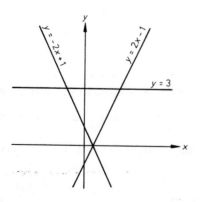

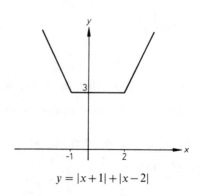

$$y = |x+1| + |x-2|$$

[Note that it is not generally advisable to apply the methods of this section to the sums of trigonometric functions. These methods would not, for instance, suggest that the function $a\cos x+b\sin x$ is of the form $r\cos(x\pm\alpha)$ as shown in §7.7.]

Exercise 20.3

Sketch the graphs with the following equations.

1. $y = 1 + \dfrac{1}{x}$.

2. $y = x - \dfrac{1}{x}$.

3. $y = x^2 + \dfrac{1}{x}$.

4. $y = 1 - \dfrac{1}{x^2}$.

5. $y = x + \dfrac{1}{x^2}$.

6. $y = x^2 + \dfrac{1}{x^2}$.

7. $y = \dfrac{x^2 + x + 1}{x}$.

8. $y = \dfrac{2x^2 - x - 1}{x}$.

9. $y = \dfrac{(x^2 - 1)^2}{x^2}$.

10. $y = \dfrac{(x+1)^2(2x-1)}{x^2}$.

11. $y = x + \dfrac{1}{x-1}$.

12. $y = |x| + \dfrac{1}{x}$.

13. $y = |x| + |x - 4|$.

14. $y = |x - 1| + |2x - 1|$.

20.4 The function $(ax+b)/(cx+d)$

We consider first a typical function of this type, namely $y = \dfrac{x+1}{x-1}$, i.e.

$$y = 1 + \frac{2}{x-1}.$$

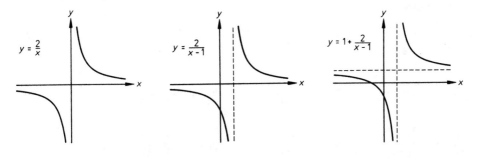

As shown in the sketches the graph of this function can be obtained from the graph $y = 2/x$ by means of translations parallel to the x- and y-axes. In general the function $y = (ax+b)/(cx+d)$ has a graph related to the graph $y = 1/x$ in a similar way. Thus to sketch such a graph it is usually sufficient to determine the points of intersection with the coordinate axes and the equations of the asymptotes.

Example 1   Sketch the curve $y = \dfrac{2x-9}{x+3}$.

$$x = 0 \Rightarrow y = -\frac{9}{3} = -3$$

∴   the curve cuts the y-axis at the point $(0, -3)$

$$y = 0 \Rightarrow 2x-9 = 0 \Rightarrow x = 4\tfrac{1}{2}$$

∴   the curve cuts the x-axis at the point $(4\tfrac{1}{2}, 0)$.

As $x \to -3$, $y \to \pm\infty$ ∴   the line $x = -3$ is an asymptote.

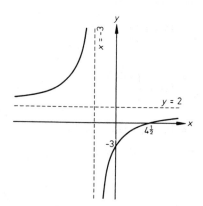

$$y = \frac{2x-9}{x+3} = \frac{2-9/x}{1+3/x}$$

∴   as $x \to \pm\infty$, $y \to 2$.
Hence the line $y = 2$ is an asymptote.

Exercise 20.4

In questions 1 to 9 sketch the given curve, stating the equations of the asymptotes.

1. $y = \dfrac{1}{x+2}$.

2. $y = \dfrac{3}{x-4}$.

3. $y = \dfrac{2}{2x-3}$.

4. $y = \dfrac{x}{x+2}$.

5. $y = \dfrac{x-2}{x+2}$.

6. $y = \dfrac{2x+3}{2x-3}$.

7. $y = \dfrac{2x-5}{x+2}$.

8. $y = \dfrac{1-3x}{3-x}$.

9. $y = \dfrac{5+2x}{2-5x}$.

10. Sketch the curve $y = x/(x-1)$ and write down the equation of its reflection in the y-axis.

11. Sketch the curve $y = (x-1)/(x+1)$ and write down the equation of its reflection in the x-axis.

12. If a, b, c and d are all positive, sketch the curve $y = \dfrac{ax+b}{cx+d}$ when
(i) $ad - bc < 0$, (ii) $ad - bc = 0$, (iii) $ad - bc > 0$.

20.5 Parametric equations

In this and earlier chapters we have considered lines and curves defined by equations connecting the coordinates (x, y) of a typical point. It is sometimes more convenient to express x and y in terms of a third variable called a *parameter*. For instance, the equations $x = 1-t$, $y = t^2 - 4$ are the *parametric equations* of a curve. The same curve may also be referred to as the *locus* or path of the point $(1-t, t^2-4)$ as the parameter t varies.

The procedure used when making a sketch of a curve defined by parametric equations is similar to that used when the Cartesian equation is known.

(1) Find the points of intersection with the x- and y-axes by letting $y = 0$ and then $x = 0$.
(2) Note any restrictions on the values that x and y can take.
(3) Decide whether the curve is symmetrical in any way.
(4) If necessary plot a few points on the curve.

Example 1   Sketch the curve $x = 1-t$, $y = t^2 - 4$.

$x = 0 \Rightarrow 1-t = 0 \Rightarrow t = 1 \Rightarrow y = -3$

$\therefore$ the curve cuts the y-axis at the point $(0, -3)$.
$y = 0 \Rightarrow t^2 - 4 = 0 \Rightarrow$ either $t = -2, x = 3$ or $t = 2, x = -1$
$\therefore$ the curve cuts the x-axis at $(-1, 0)$ and $(3, 0)$.

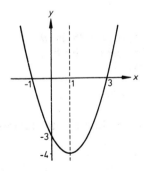

As t varies x may take any value, but since t^2 is never negative the minimum value of y is -4. For any given value of y greater than -4, there are two possible values of x of the form $1 \pm k$. Hence the curve is symmetrical about the line $x = 1$.

It is sometimes easier to work with the Cartesian equation of a curve. This is obtained from the parametric equations by eliminating the parameter.

Example 2   Find the Cartesian equation of the curve $x = 1-t$, $y = t^2 - 4$.

$x = 1-t \Leftrightarrow t = 1-x$
$\therefore \quad y = t^2 - 4 = (1-x)^2 - 4 = 1 - 2x + x^2 - 4$.
Hence the Cartesian equation is $y = x^2 - 2x - 3$.

Example 3   Find the Cartesian equation of the locus of the point $(2t^2, 1-t^2)$ as t varies. Sketch the locus.

The parametric equations of the locus are
$$x = 2t^2 \tag{1}$$
$$y = 1-t^2 \tag{2}$$
Adding (1) to $2 \times$ (2) $x + 2y = 2$.
However, since t^2 can never be negative, $x \geqslant 0$.
Hence the Cartesian equation of the locus is $x + 2y = 2$, where $x \geqslant 0$.

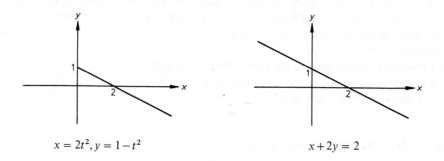

$x = 2t^2, y = 1-t^2$ $x + 2y = 2$

Example 4   Find the Cartesian equation of the curve $x = 2\cos\theta + \sin\theta$, $y = \cos\theta - 2\sin\theta$.

Eliminating $\sin\theta$, $2x + y = 5\cos\theta$
Eliminating $\cos\theta$, $x - 2y = 5\sin\theta$.
Squaring and adding these equations

$$(2x+y)^2 + (x-2y)^2 = 25\cos^2\theta + 25\sin^2\theta$$
$$4x^2 + 4xy + y^2 + x^2 - 4xy + 4y^2 = 25(\cos^2\theta + \sin^2\theta)$$
$$\therefore \quad 5x^2 + 5y^2 = 25.$$

Hence the required Cartesian equation is $x^2 + y^2 = 5$.

The parametric equations of a curve may be used to find points of intersection with other lines or curves.

Example 5   Find the points of intersection of the line $3x - 2y = 2$ with the curve $x = t-1$, $y = 1/t$.

Substituting for x and y in the equation of the line,

$$3x - 2y = 2 \Rightarrow 3(t-1) - 2.\frac{1}{t} = 2$$

$$\begin{aligned} \Rightarrow 3t(t-1) - 2 &= 2t \\ \Rightarrow 3t^2 - 5t - 2 &= 0 \\ \Rightarrow (3t+1)(t-2) &= 0 \\ \Rightarrow t = -\tfrac{1}{3} \text{ or } t &= 2 \end{aligned}$$

Hence the points of intersection are $(-1\frac{1}{3}, -3)$ and $(1, \frac{1}{2})$.

Exercise 20.5
1. Sketch the following graphs
(a) $x = 4t, y = 3 - t$,   (b) $x = t - 1, y = t^2 + 1$,
(c) $x = 2t^2, y = 4(t-1)$,   (d) $x = t^2 - 1, y = t^2 + 1$,
(e) $x = t + 2, y = 1/t$,   (f) $x = t^2 - 1, y = t^4 + 1$.

2. Find the Cartesian equations of the graphs in question 1.

3. Find the Cartesian equations of the following:
(a) $x = 3 \sin \theta$ (b) $x = \sec \theta$ (c) $x = 1 + \cos \theta$
     $y = 2 \cos \theta$, $y = 5 \tan \theta$, $y = 1 - 2 \sin \theta$,
(d) $x = \cos \theta + \sin \theta$ (e) $x = \cos(\theta + \frac{1}{4}\pi)$
     $y = 2 \cos \theta + \sin \theta$, $y = \sqrt{2} \sin \theta$.

4. Find the points of intersection of the following:
(a) $x = t^2, y = t^3; y = 3x$,
(b) $x = 3t, y = 3/t; y = 2x + 3$,
(c) $x = 4t^2, y = 8t; 3y + 16 = 4x$,
(d) $x = 2t + 1, y = t^2; 2x - 3y + 13 = 0$,
(e) $x = \dfrac{1}{1+t^2}, y = \dfrac{t}{1+t^2}; x + y = 1$,
(f) $x = \dfrac{t}{t+1}, y = \dfrac{1}{t+1}; x^2 + y^2 = 25$.

5. Draw sketches of the following curves on graph paper. Plot sufficient points to give a good indication of the general shape of the curve.
(a) The semi-cubical parabola $x = t^2, y = t^3$.
(b) The ellipse $x = 3 \cos \theta, y = 2 \sin \theta$.
(c) The hyperbola $x = t - \dfrac{1}{t}, y = 2\left(t + \dfrac{1}{t}\right)$.
(d) The astroid $x = \cos^3 t, y = \sin^3 t$.
(e) The cycloid $x = \theta - \sin \theta, y = 1 - \cos \theta$.

20.6 Differentiation and parameters

When a curve is defined by parametric equations of the form $x = f(t), y = g(t)$, $\dfrac{dy}{dx}$ can be obtained as a function of t by writing

$$\frac{dy}{dx} = \frac{dy}{dt} \cdot \frac{dt}{dx} = \frac{dy}{dt} \cdot \left(1 \bigg/ \frac{dx}{dt}\right) = \frac{dy}{dt} \bigg/ \frac{dx}{dt}$$

Example 1   Find $\dfrac{dy}{dx}$ in terms of t if $x = 3t - 2$, $y = t^3 + 1$.

$$\frac{dx}{dt} = 3 \qquad\qquad \frac{dy}{dt} = 3t^2$$

$$\therefore \quad \frac{dy}{dx} = \frac{dy}{dt} \bigg/ \frac{dx}{dt} = \frac{3t^2}{3} = t^2.$$

When expressed as a function of t, dy/dx represents the gradient of a curve at the point with parameter t.

Example 2   Find the equation of the tangent to the curve $y^2 = x - 5$ at the point $(4t^2 + 5, 2t)$.

This curve can be represented by the parametric equations

$$x = 4t^2 + 5 \qquad\qquad y = 2t$$

$$\frac{dx}{dt} = 8t \qquad\qquad \frac{dy}{dt} = 2$$

$$\therefore \quad \frac{dy}{dx} = \frac{dy}{dt} \bigg/ \frac{dx}{dt} = \frac{2}{8t} = \frac{1}{4t}.$$

Hence the gradient of the tangent to the curve at the point $(4t^2 + 5, 2t)$ is $1/4t$.

The equation of the tangent is $y - 2t = \dfrac{1}{4t}\left(x - \left\{4t^2 + 5\right\}\right)$,

i.e. $4ty - 8t^2 = x - 4t^2 - 5$.

Thus the required equation is $x - 4ty + 4t^2 - 5 = 0$.

Remembering that second derivatives do *not* behave like fractions, we now obtain d^2y/dx^2 for the equations of Example 1, $x = 3t - 2$, $y = t^3 + 1$. We look again at the method used to find dy/dx, then use a similar method to find d^2y/dx^2.

$$\frac{dy}{dx} = \frac{d}{dx}\left(t^3 + 1\right) = \frac{d}{dt}\left(t^3 + 1\right) \cdot \frac{dt}{dx} = \frac{d}{dt}\left(t^3 + 1\right) \bigg/ \frac{dx}{dt} = \frac{3t^2}{3} = t^2.$$

Similarly $\dfrac{d^2y}{dx^2} = \dfrac{d}{dx}\left(\dfrac{dy}{dx}\right) = \dfrac{d}{dx}\left(t^2\right) = \dfrac{d}{dt}\left(t^2\right) \cdot \dfrac{dt}{dx} = \dfrac{d}{dt}\left(t^2\right) \bigg/ \dfrac{dx}{dt} = \dfrac{2t}{3}.$

Example 3   If $x = 1 + \dfrac{1}{t}$ and $y = t + \dfrac{1}{t}$, find $\dfrac{dy}{dx}$ and $\dfrac{d^2y}{dx^2}$ in terms of t.

$$\frac{dx}{dt} = -\frac{1}{t^2} \qquad\qquad \frac{dy}{dt} = 1 - \frac{1}{t^2}$$

$$\therefore \quad \frac{dy}{dx} = \frac{dy}{dt} \Big/ \frac{dx}{dt} = \left(1 - \frac{1}{t^2}\right) \Big/ \left(-\frac{1}{t^2}\right) = 1 - t^2.$$

$$\frac{d^2y}{dx^2} = \frac{d}{dx}\left(1 - t^2\right) = \frac{d}{dt}\left(1 - t^2\right) \Big/ \frac{dx}{dt} = -2t \Big/ \left(-\frac{1}{t^2}\right) = 2t^3.$$

Exercise 20.6

Find $\dfrac{dy}{dx}$ in terms of t if:

1. $x = 2t + 1,\ y = t^2 - 1.$ 2. $x = t^3,\ y = 3t^2 + 2.$

3. $x = 4t,\ y = 1 - \dfrac{1}{t}.$ 4. $x = (1 - 2t)^3,\ y = t^2 - t.$

5. $x = 4\cos t,\ y = 3\sin t.$ 6. $x = 2\cos^3 t,\ y = 2\sin^3 t.$

Find $\dfrac{dy}{dx}$ and $\dfrac{d^2y}{dx^2}$ in terms of t if:

7. $x = \dfrac{1}{t^2},\ y = 1 + t.$ 8. $x = 6t^2,\ y = 12t - 3t^4.$

9. $x = t^3,\ y = t^2 + t.$ 10. $x = (t + 1)^2,\ y = t^2 - 1.$

11. $x = 2\sin t,\ y = \cos 2t.$ 12. $x = \cos^2 t,\ y = \sin 2t.$

13. Given that $x = t^3 - 2t,\ y = 5t^2 + \dfrac{1}{t}$, find the value of $\dfrac{dy}{dx}$ when $t = 1$.

14. The parametric equations of a curve are $x = t(t^2 + 1),\ y = t^2 + 1$. Find, in its simplest form, the equation of the tangent to the curve at the point with parameter t.

15. Find the equation of the tangent to the curve $x^2 - y^2 = 1$ at the point $(\sec\theta,\ \tan\theta)$.

16. The parametric equations of a curve are $x = (5 - 3t)^2,\ y = 6t - t^2$. Find dy/dx in terms of t and the coordinates of the stationary point on the curve.

17. Find the coordinates of the stationary points and the point of inflexion on the curve with parametric equations $x = t^2 + 1,\ y = t(t - 3)^2$.

18. Given that $x = \theta - \sin\theta,\ y = 1 - \cos\theta$, show that $y^2 \dfrac{d^2y}{dx^2} + 1 = 0$.

Exercise 20.7 (*miscellaneous*)

1. Decide whether the following functions are odd, even or neither
(a) $2x^5 - 3x^3 + 1$,   (b) $\sin 2x - \sin x$,   (c) $|x|$,

(d) $\dfrac{x}{2x^5 - 3x^3}$,   (e) $x^3 \tan x$,   (f) $[x + \frac{1}{2}]$.

2. Given that $f(x) = ax^3 + bx^2 + cx + d$, state what can be deduced about the values of a, b, c, d in each of the following cases
(a) $f(x)$ is an even function,
(b) $f(x)$ is an odd function,
(c) $f(x) = |f(x)|$ for all values of x,
(d) $f(x) = -|f(x)|$ for all values of x.

3. Let $f(x) = \dfrac{ax+b}{x+c}$ where x, a, b, c are real and $x \neq \pm c$. Show that if f is an even function then $ac = b$. Deduce that if f is an even function then $f(x)$ must reduce to the form $f(x) = k$, where k is constant. Find all odd functions of the form $\dfrac{ax+b}{x+c}$.   (JMB)

4. Sketch on the same diagram the graphs of $f(x)$ and $1/f(x)$ where (a) $f(x) = 4x - x^2$, (b) $f(x) = (x-1)^2$, (c) $f(x) = x^2 + 2x + 3$.

5. Sketch the graphs with the following equations
(a) $y = |2x - 3|$,   (b) $y = |x^2 + 3x|$,

(c) $y = \left|x + 3 + \dfrac{2}{x}\right|$,   (d) $y = \left|\dfrac{x-2}{x+2}\right|$.

6. Find the coordinates of the point of intersection of the curve $y = \dfrac{x^3 + 1}{x}$ and the x-axis. Find also the gradients when $x = 1$ and $x = -1$, the value of x at the stationary point, and the nature of this stationary point. Sketch this curve and also the curve $y = \dfrac{x}{x^3 + 1}$.   (L)

7. Find the local maxima and minima, and the points of inflexion, if any, of the functions f and g defined by $f(x) = \dfrac{x^4}{4(x^2 - 1)}$, $g(x) = \dfrac{4(x^2 - 1)}{x^4}$, and sketch the graphs of these two functions on one diagram.   (W)

8. Sketch the following curves
(a) $y = \dfrac{2x+1}{x-1}$,   (b) $y = \dfrac{x-1}{2x+1}$,

(c) $y = \left|\dfrac{2x+1}{x-1}\right|$,   (d) $y = \dfrac{2|x|+1}{|x|-1}$.

9. Below are listed five functions, numbered (1)–(5), and five graphs, lettered (A)–(E). *Four* of the graphs correspond to *four* of the functions.

$$\text{(1) } y = \frac{(x+1)^2}{x} \qquad \text{(2) } y = \frac{x}{(x+1)^2} \qquad \text{(3) } y = \frac{x}{x+1}$$

$$\text{(4) } y = \left| \frac{x}{x+1} \right| \qquad \text{(5) } y = \frac{x^2}{x+1}.$$

(A)

(B)

(C)

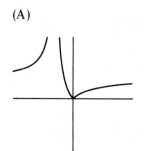

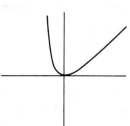

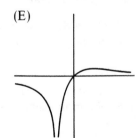

(D)

(E)

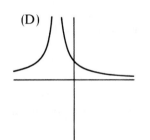

(a) Pair off the four corresponding functions and graphs. (Give your answers as four ordered pairs (n, X), n being a number and X a letter.)
(b) Sketch the graph of the fifth remaining function.
(c) Suggest a suitable function which could fit the fifth remaining graph.   (O&C)

10. Sketch the following graphs
(a) $y = |x| + |2x - 3|$,   (b) $y = |x+3| - |x-2|$.

11. Show that the equation $|x+2| + |x| + |x-1| + |x-3| = 6$ has an infinite number of real solutions.   (C)

12. The function f is defined by $f : x \to |x-2| + a/x$ where $x \in \mathbb{R}$, $x \neq 0$, and $a \in \mathbb{R}$, $a \neq \pm 4$ or 0. Determine the number of turning points of f for different sets of values of a. (A turning point exists for $x = c$ if $f'(c) = 0$ and $f'(x)$ changes sign as x passes through the value c.)

Sketch a graph of f (i) for $a = 3$ and (ii) for $a = -1$, showing carefully the nature of the curve near the point at which $x = 2$, and also for large values of $|x|$.

(JMB)

13. Find the Cartesian equations of the following curves

(a) $x = 2\left(t + \dfrac{1}{t}\right)$, $y = 3\left(t - \dfrac{1}{t}\right)$,

(b) $x = \dfrac{1+t^2}{1-t^2}$, $y = \dfrac{2t}{1-t^2}$,

(c) $x = \cos\theta - 2\sin\theta$
    $y = 3\cos\theta + \sin\theta$,

(d) $x = 3\sin(\theta + \tfrac{1}{3}\pi)$
    $y = 2\sin(\theta + \tfrac{1}{6}\pi)$.

14. The parametric equations of a curve are $x = 4 - t^2$, $y = 4t - t^3$. Find dy/dx in terms of t and hence find the coordinates of any turning points on the curve. Sketch the curve.

15. A curve is given parametrically by the equations $x = (1+t)^2$, $y = (1-t)^2$. Find dy/dx in terms of t, and hence find the point on the curve at which the gradient is zero. Find also the equation of the tangent to the curve at the point where $x = y$.

(C)

16. A locus is defined parametrically by the equations $x = \dfrac{3}{t+1}$, $y = \dfrac{1-5t}{t+1}$. Find the points of intersection between this locus and the curve $xy = 3$. Find dy/dx for the locus and hence its Cartesian equation.

17. Given that $x + y = \sin(\theta + 75°)$, $x - y = \sin(\theta + 15°)$, express x and y each in terms of the sine or cosine of $\theta + 45°$. If x and y are the coordinates of a point P and θ varies, obtain in a simplified form the Cartesian equation of the locus of P. Sketch this locus.

(JMB)

18. Find $\dfrac{dy}{dx}$ and $\dfrac{d^2y}{dx^2}$ in terms of t, given that
(a) $x = 3t(3 - 4t^2)$, $y = (3 - 2t)(1 - 2t)^2$,
(b) $x = 2\cos t + \cos 2t$, $y = 2\sin t + \sin 2t$.

19. A curve joining the points $(0, 1)$ and $(0, -1)$ is represented parametrically by the equations
$$x = \sin\theta, \quad y = (1 + \sin\theta)\cos\theta, \quad \text{where } 0 \leqslant \theta \leqslant \pi.$$

Find dy/dx in terms of θ, and determine the x, y coordinates of the points on the curve at which the tangents are parallel to the x-axis and of the point at which the tangent is perpendicular to the x-axis. Sketch the curve.

The region in the quadrant $x \geqslant 0$, $y \geqslant 0$ bounded by the curve and the coordinate axes is rotated about the x-axis through an angle of 2π. Show that the volume swept out is given by $V = \pi \displaystyle\int_0^1 (1+x)^2 (1-x^2)\,dx$. Evaluate V, leaving your result in terms of π.

(JMB)

21 Coordinate geometry

21.1 The locus of a point

A *locus* is the set of all points satisfying some condition. The Cartesian equation of a locus is obtained by expressing this condition as a relationship between the x- and y-coordinates of a typical point.

Example 1 Find the equation of the locus of points at a distance of 2 units from the point $C(-1, 2)$.

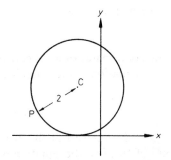

Let $P(x, y)$ be a point on the locus then

$$PC^2 = 2^2$$

$$\therefore \quad (x+1)^2 + (y-2)^2 = 2^2$$

$$x^2 + 2x + 1 + y^2 - 4y + 4 = 4$$

$\therefore$ the equation of the locus is

$$x^2 + y^2 + 2x - 4y + 1 = 0.$$

Clearly this equation must represent a circle with centre $(-1, 2)$ and radius 2.

Example 2 Find the equation of the locus of points 5 units above the x-axis.

For any point $P(x, y)$ on the locus, $y = 5$ and there is no restriction on the value of x. Hence the equation of the locus is simply $y = 5$.

Example 3 Find the equation of the locus of points equidistant from the x- and y-axes.

The distance of a point $P(x, y)$ from the x-axis is $|y|$. Similarly the distance of P from the y-axis is $|x|$. Since these distances are equal,

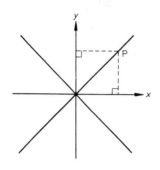

$$|x| = |y|.$$

Squaring to avoid the use of moduli, the equation of the locus may be written

$$x^2 = y^2 \quad \text{or} \quad x^2 - y^2 = 0.$$

This equation represents the pair of lines which bisect the angles between the x- and y-axes.

Example 4   Find the equation of the locus of a point which moves so that it is equidistant from the points $A(1, 0)$ and $B(0, 2)$.

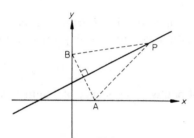

If $P(x, y)$ is any point on the locus then

$$PA^2 = PB^2$$
$$\therefore \quad (x-1)^2 + y^2 = x^2 + (y-2)^2$$
$$x^2 - 2x + 1 + y^2 = x^2 + y^2 - 4y + 4$$
$$\therefore \quad \text{the equation of the locus is}$$

$$2x - 4y + 3 = 0.$$

This equation must represent the perpendicular bisector of AB.

In some problems it is necessary to express the coordinates of a general point on a locus in terms of a parameter. The Cartesian equation is then obtained by eliminating the parameter.

Example 5   The line $y = mx$ and the curve $y = x^2 - 2x$ intersect at the origin O and meet again at a point A. If P is the mid-point of OA, find the equation of the locus of P as m varies.

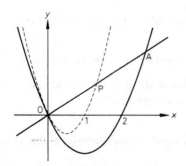

At the points of intersection

$$x^2 - 2x = mx$$
$$x(x - m - 2) = 0$$
$$\therefore \quad x = 0 \quad \text{or} \quad x = m + 2$$
$$\therefore \quad \text{the coordinates of } A \text{ are}$$

$(m+2, m\{m+2\})$ and the coordinates of P are $(\frac{1}{2}\{m+2\}, \frac{1}{2}m\{m+2\})$.

Hence in parametric form the equations of the locus of P are

$$x = \tfrac{1}{2}(m+2) \tag{1}$$

$$y = \tfrac{1}{2}m(m+2) \tag{2}$$

Rearranging (1) $m = 2x - 2$

Substituting in (2) $y = (x-1)2x$.

Hence the equation of the locus of P is $y = 2x(x-1)$.

Exercise 21.1

In questions 1 to 10 find the equation of the locus of a point P as it moves subject to the stated condition. In each case draw a sketch to illustrate your answer.

1.   The distance of P from the point $(-3, 4)$ is 5 units.

2. P is equidistant from the points $(1,0)$ and $(4,3)$.

3.   The perpendicular distance of P from the y-axis is 2 units.

4. A and B are the points $(-3,1)$ and $(5, -1)$ respectively and $\angle PAB$ is a right angle.

5. P is equidistant from the lines $x = 1$ and $y = 1$.

6.   The distance of P from the point $(0,2)$ is equal to its distance from the x-axis.

7.   The area of the rectangle with sides parallel to the axes and with diagonal OP is 9 square units.

8. A and B are the points $(-1,2)$ and $(3,2)$ respectively and $\angle APB$ is a right angle.

9.   The distance of P from point $(3,0)$ is equal to its perpendicular distance from the line $x + 3 = 0$.

10. The distance of P from the point $(-2,1)$ is twice its distance from the point $(4, 1)$.

11. Find the equation of the circle centre $(-1,1)$ and radius $\sqrt{2}$. Show that this circle passes through the origin.

12. Find the equation of the perpendicular bisector of the line joining the points $A(2,1)$ and $B(-4, -1)$. Hence find the point equidistant from A and B whose x-coordinate is -3.

13. The line $y = mx$ meets the curve $y^2 = 4x$ at the origin O and at a point A. Find the equation of the locus of the mid-point of OA as m varies.

14. If P is the point $(t, t^2 - 4)$ and Q is the point $(2, 0)$, find the coordinates of the mid-point R of PQ. Deduce the equation of the locus of R as t varies.

15. A and B are the points $(a, 0)$ and $(0, b)$ respectively and P is the mid-point of AB. Find the equation of the locus of P as a and b vary given that (a) $a + b = c$, where c is constant, (b) triangle OAB is of constant area k, (c) AB is of constant length l.

16. The line $y = mx$ intersects the curve $y = x^2 - 1$ at the points A and B. Find the equation of the locus of the mid-point of AB as m varies.

21.2 The circle

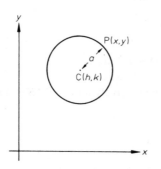

Let $P(x, y)$ be a point on the circle, centre (h, k) and radius a, then $PC^2 = a^2$

$$\therefore \quad (x - h)^2 + (y - k)^2 = a^2.$$

Thus the equation of the circle, centre (h, k), radius a, is

$$\boxed{(x - h)^2 + (y - k)^2 = a^2.}$$

In particular, the equation of the circle with centre the origin and radius a is

$$\boxed{x^2 + y^2 = a^2.}$$

Example 1   Find the equation of the circle with centre $(1, -2)$ and radius 3.

The equation is
$$(x - 1)^2 + (y + 2)^2 = 3^2$$
$$x^2 - 2x + 1 + y^2 + 4y + 4 = 9$$
i.e.
$$x^2 + y^2 - 2x + 4y - 4 = 0.$$

Given the equation of a circle, it can be expressed in the form $(x - h)^2 + (y - k)^2 = a^2$ in order to write down its centre and radius.

Example 2   Find the centre and radius of the circle $x^2 + y^2 + 6x - 2y - 6 = 0$.

[The aim is to rearrange the equation by completing squares of the form $(x - h)^2$ and $(y - k)^2$.]

The equation of the circle may be rearranged as

$$x^2 + 6x + y^2 - 2y = 6$$
$$x^2 + 6x + 9 + y^2 - 2y + 1 = 6 + 9 + 1$$

i.e.
$$(x+3)^2 + (y-1)^2 = 4^2.$$

Hence the circle has centre $(-3, 1)$ and radius 4.

We see that the equation $x^2 + y^2 + 2gx + 2fy + c = 0$ represents a circle for suitable values of g, f and c. Rearranging

$$x^2 + 2gx + y^2 + 2fy = -c$$
$$x^2 + 2gx + g^2 + y^2 + 2fy + f^2 = f^2 + g^2 - c$$

i.e.
$$(x+g)^2 + (y+f)^2 = f^2 + g^2 - c.$$

Thus the equation $x^2 + y^2 + 2gx + 2fy + c = 0$ represents a circle with centre $(-g, -f)$ and radius $\sqrt{(f^2 + g^2 - c)}$ provided that this is real.

Many problems in coordinate geometry concerning circles can be solved using their elementary properties.

Example 3 Find the equation of the tangent to the circle $x^2 + y^2 - 4x + 2y + 3 = 0$ at the point $(3, -2)$.

The centre of the circle is the point $(2, -1)$

$\therefore$ the gradient of the radius through $(3, -2) = \dfrac{-2 - (-1)}{3 - 2} = -1.$

Since any tangent to a circle is perpendicular to the radius through its point of contact, the gradient of the tangent through $(3, -2)$ is 1.

Hence its equation is $y + 2 = 1(x - 3)$, i.e. $y = x - 5$.

Example 4 Find the length of a tangent drawn from the point $A(3, -4)$ to the circle $x^2 + y^2 + 6x - 8y = 0$.

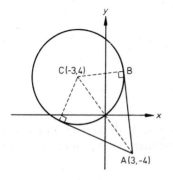

Rearranging the equation of the circle

$$x^2 + 6x + 9 + y^2 - 8y + 16 = 9 + 16$$

i.e.
$$(x+3)^2 + (y-4)^2 = 5^2$$

$\therefore$ the circle has centre $(-3, 4)$ and radius 5.

If C is the centre of the circle and B is the point of contact of a tangent through A, then

$$AC^2 = (3 - \{-3\})^2 + (-4 - 4)^2 = 6^2 + 8^2 = 100 \quad \text{and}$$
BC, a radius of the circle, is of length 5.

Hence, using Pythagoras' theorem,

$$AB^2 = AC^2 - BC^2 = 100 - 25 = 75 \qquad \therefore \quad AB = 5\sqrt{3}.$$

Thus the length of the tangents from A to the given circle is $5\sqrt{3}$.

Example 5   Find the equation of the circle whose diameter is the line joining the points $A(1, 3)$ and $B(-2, 5)$.

[One way of doing this is to write down the centre and the radius of the circle, then use these to obtain the equation. We give here an alternative method.]

If $P(x, y)$ is any point on the circle, then since AB is a diameter, the angle APB is a right angle.

$$\text{Gradient of } AP = \frac{y-3}{x-1}, \quad \text{gradient of } BP = \frac{y-5}{x+2}.$$

Since AP is perpendicular to BP, $\quad \dfrac{y-3}{x-1} \times \dfrac{y-5}{x+2} = -1$

$$\therefore \quad (x-1)(x+2) + (y-3)(y-5) = 0$$
$$x^2 + x - 2 + y^2 - 8y + 15 = 0.$$

Hence the equation of the circle is $\quad x^2 + y^2 + x - 8y + 13 = 0.$

Example 6   Find the equation of the circle which passes through the points $(-1, 0)$, $(1, 2)$ and $(-5, 4)$.

Let the equation of the circle be $x^2 + y^2 + 2gx + 2fy + c = 0$, then since the coordinates $(-1, 0)$, $(1, 2)$ and $(-5, 4)$ must satisfy this equation

$1 + 0 - 2g + c = 0$	$\therefore$	$-2g + c = -1$	(1)
$1 + 4 + 2g + 4f + c = 0$	$\therefore$	$2g + 4f + c = -5$	(2)
$25 + 16 - 10g + 8f + c = 0$	$\therefore$	$-10g + 8f + c = -41$	(3)

$$2 \times (2) - (3) \qquad\qquad 14g + c = 31$$
$$\text{Subtracting (1)} \qquad\qquad 16g = 32 \qquad \therefore \quad g = 2$$

Using (1) and (2) $c = 3$ and $f = -3$.

Hence the required equation is $\quad x^2 + y^2 + 4x - 6y + 3 = 0.$

Another way of finding the equation of a circle through three given points is to use the fact that if points A and B lie on a circle then the centre of the circle lies on the perpendicular bisector of AB.

Example 7   Find the equation of the circle that touches the line $y = x$ at the point $A(3, 3)$ and passes through the point $B(5, 9)$.

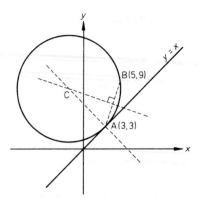

If C is the centre of the circle then C must lie on the perpendicular bisector of AB. Since the line $y = x$ is a tangent to the circle, C must also lie on the line through A perpendicular to $y = x$.

The gradient of $AB = \dfrac{9-3}{5-3} = 3$.

The coordinates of the mid-point of AB are $(4, 6)$

∴   the perpendicular bisector passes through $(4, 6)$ and has gradient $-\frac{1}{3}$.

Hence its equation is $y - 6 = -\frac{1}{3}(x - 4)$ i.e. $x + 3y = 22$ (1)

The line through A perpendicular to $y = x$ has gradient -1

∴   its equation is $y - 3 = -1(x - 3)$,   i.e. $x + y = 6$ (2)

Subtracting (2) from (1) $2y = 16$ ∴ $y = 8, x = -2$.

Hence the centre of the circle is the point $(-2, 8)$.

The radius of the circle $= AC = \sqrt{\{(-2-3)^2 + (8-3)^2\}} = \sqrt{50}$

Thus the equation of the circle is $(x + 2)^2 + (y - 8)^2 = 50$

$$\text{i.e.}\quad x^2 + y^2 + 4x - 16y + 18 = 0.$$

Exercise 21.2

1.   Find the equations of the circles with the following centres and radii
(a) $(3, 2)$, 4,   (b) $(-1, -2)$, 1,   (c) $(0, 0)$, 5,
(d) $(\frac{1}{2}, 0)$, $3/2$,   (e) $(4, -1)$, $\sqrt{3}$,   (f) $(-3, 5)$, $2\sqrt{5}$.

2.   Find the centres and radii of the following circles
(a) $x^2 + y^2 - 2x - 6y + 1 = 0$,   (b) $x^2 + y^2 = 4$,
(c) $x^2 + y^2 + 6x + 8y = 0$,   (d) $x^2 + y^2 - 4x + 2y + 4 = 0$,
(e) $4x^2 + 4y^2 - 5 = 0$,   (f) $2x^2 + 2y^2 - 6x + 10y + 7 = 0$.

3.   Find the range of values of k for which each of the following equations represents a circle with non-zero radius.
(a) $x^2 + y^2 = k$ (b) $x^2 + ky^2 - 2x - 8 = 0$,
(c) $kx^2 + y^2 + 4y + 9 = 0$,   (d) $2x^2 + 2y^2 + kxy - 9 = 0$,
(e) $x^2 + y^2 + 2x - 6y + k = 0$,   (f) $x^2 + y^2 + kx - 2y + 5 = 0$.

4. Find the equations of the tangents to the following circles at the given points
(a) $x^2 + y^2 = 5$, $(-2, 1)$,   (b) $x^2 + y^2 - 4x + 2y = 3$, $(0, -3)$,
(c) $x^2 + y^2 + 6y - 1 = 0$, $(3, -4)$,   (d) $2x^2 + 2y^2 + 9x - 4y + 4 = 0$, $(-2, 3)$.

5. Find the lengths of the tangents drawn from the following points to the given circles
(a) $(6, -1)$, $x^2 + y^2 = 12$,
(b) $(-1, 3)$, $x^2 + y^2 - 8x + 4y + 19 = 0$,
(c) $(4, -2)$, $x^2 + y^2 - 10y - 4 = 0$,
(d) $(3, -4)$, $x^2 + y^2 + x - 3y = 0$.

6. Find the equations of the circles which pass through the following sets of points
(a) $(0, 0)$, $(2, 0)$, $(0, 6)$,   (b) $(5, 6)$, $(5, -2)$, $(1, 6)$,
(c) $(5, 0)$, $(3, 2)$, $(-3, -4)$,   (d) $(2, 3)$, $(-2, 1)$, $(4, 7)$.

7. Find the greatest and the least distance of a point P from the origin as it moves round the circle
(a) $x^2 + y^2 - 24x - 10y + 48 = 0$,
(b) $x^2 + y^2 + 6x - 8y - 24 = 0$.

8. Find the equation of the circle which has the points $(-7, 3)$ and $(1, 9)$ at the ends of a diameter. Find also the equations of the tangents to this circle which are parallel (a) to the x-axis, (b) to the y-axis.

9. Find the equation of the locus of a point P which moves so that its distance from the point $A(1, 3)$ is twice its distance from the point $B(4, 6)$. Show that the locus is a circle giving its centre and radius.

10. Find the equation of the circle which passes through the point $A(6, 1)$ and touches the x-axis at the point $B(3, 0)$. If this circle cuts the y-axis at the points P and Q, find the area of the quadrilateral $ABPQ$.

11. Find the equation of the circle which touches the line $3x - 4y = 3$ at the point $(5, 3)$ and passes through the point $(-2, 4)$.

12. Show that the circle $x^2 + y^2 - 6x - 4y + 9 = 0$ touches the x-axis and that the circle $x^2 + y^2 - 2x - 6y + 9 = 0$ touches the y-axis. Find the coordinates of their points of intersection, showing that these lie on a straight line through the origin.

13. Find the equation of the circle C_1 which has as the ends of a diameter the points $P(1, 1)$ and $Q(6, 2)$. Show that the point $R(7, -3)$ lies outside this circle and find the equation of the circle C_2 which passes through P, Q and R. Hence show that the centre of C_2 lies on C_1.

21.3 Tangents and normals

As we have seen in previous chapters, the gradient of the tangent at any point on a curve in the x, y plane is given by dy/dx. The *normal* to a curve at any point is the line perpendicular to the tangent. Thus if the gradients of the tangent and the normal at a particular point are m_1 and m_2, then $m_1 m_2 = -1$.

Example 1   Find the equations of the tangent and the normal to the curve $x^2 - 3xy + y^2 = 5$ at the point $(1, 4)$.

Differentiating the equation with respect to x

$$2x - 3x\frac{dy}{dx} - 3y + 2y\frac{dy}{dx} = 0$$

$\therefore$ at the point $(1, 4)$ $2 - 3\dfrac{dy}{dx} - 12 + 8\dfrac{dy}{dx} = 0$

i.e. $5\dfrac{dy}{dx} - 10 = 0$ so that $\dfrac{dy}{dx} = 2.$

Hence the gradients of the tangent and normal at $(1, 4)$ are 2 and $-\frac{1}{2}$ respectively.

The equation of the tangent is $y - 4 = 2(x - 1)$

i.e. $y = 2x + 2.$

The equation of the normal is $y - 4 = -\frac{1}{2}(x - 1)$

i.e. $x + 2y = 9.$

Example 2   Find the equation of the normal to the curve with parametric equations $x = 3t + 5$, $y = 1 - t^2$, which is parallel to the line $3x + 4y = 7$.

Differentiating $\dfrac{dx}{dt} = 3$ $\dfrac{dy}{dt} = -2t$

$$\therefore \quad \frac{dy}{dx} = \frac{dy}{dt} \bigg/ \frac{dx}{dt} = -\frac{2t}{3}.$$

Hence the gradient of the normal at the point with parameter t is $3/2t$.

The gradient of the line $3x + 4y = 7$ is $-\frac{3}{4}$

$\therefore$ the normal to the curve is parallel to this line

when $\dfrac{3}{2t} = -\dfrac{3}{4}$,   i.e.   when $t = -2.$

Thus the required normal passes through the point $(-1, -3)$ on the curve.

$\therefore$ its equation is $3x+4y = 3(-1)+4(-3)$

i.e. $3x+4y+15 = 0$.

The next example shows a different way of obtaining the equation of a tangent to a curve.

Example 3 $P(2p, 2/p)$ and $Q(2q, 2/q)$ are two points on the curve $xy = 4$. Find the equation of the chord PQ. Deduce the equation of the tangent to the curve at P.

The gradient of $PQ = \dfrac{2/q-2/p}{2q-2p} = \dfrac{p-q}{pq(q-p)} = -\dfrac{1}{pq}$, assuming $p \neq q$.

$\therefore$ the equation of the chord PQ is

$$y - \frac{2}{p} = -\frac{1}{pq}(x-2p)$$
$$pqy - 2q = -x + 2p$$

i.e. $$x + pqy = 2(p+q).$$

As we consider points Q closer and closer to P, the chord PQ approaches the tangent at P

$\therefore$ letting $q \rightarrow p$, we obtain the equation of the tangent at P,

$$x + p^2y = 4p.$$

We now consider the equations of tangents to a curve through a given point not on the curve.

Example 4   Using the result of Example 3, find the equations of the tangents to the curve $xy = 4$ which pass through the point $(3, 1)$ and give the coordinates of their points of contact.

The tangent at the point $(2p, 2/p)$ has equation $x + p^2y = 4p$. If this tangent passes through $(3, 1)$ then $3 + p^2 = 4p$

i.e. $$p^2 - 4p + 3 = 0$$
$$(p-1)(p-3) = 0$$

$\therefore$ either $p = 1$ or $p = 3$.

Hence the required tangents have equations $x+y = 4$ and $x+9y = 12$, their points of contact being $(2, 2)$ and $(6, \frac{2}{3})$ respectively.

Example 5   Find the equations of the tangents to the circle $x^2 + y^2 = 5$ which pass through the point $(5, 0)$. Find also the equation of the chord of contact.

The equation of the line through $(5, 0)$ with gradient m is $y = m(x-5)$.

Substituting in the equation $x^2 + y^2 = 5$ to find the points of intersection

$$x^2 + m^2(x-5)^2 = 5$$
$$x^2 + m^2x^2 - 10m^2x + 25m^2 = 5$$
$$(1+m^2)x^2 - 10m^2x + 25m^2 - 5 = 0 \qquad (1)$$

The line will be a tangent to the circle if this quadratic equation in x has equal roots

i.e. if $(-10m^2)^2 - 4(1+m^2)(25m^2-5) = 0 \qquad [\,b^2 - 4ac = 0\,]$
$\therefore \qquad\qquad 100m^4 - 100m^4 - 80m^2 + 20 = 0$
$\therefore \qquad\qquad\qquad\qquad\qquad 1 - 4m^2 = 0$

$\therefore$ either $m = \frac{1}{2}$ or $m = -\frac{1}{2}$.

Hence the equations of the tangents through $(5, 0)$ are

$\qquad y = \frac{1}{2}(x-5)$ and $y = -\frac{1}{2}(x-5)$
i.e. $x - 2y = 5$ and $x + 2y = 5.$

When equation (1) has equal roots

$$x = \frac{10m^2}{2(1+m^2)} = \frac{10 \cdot \frac{1}{4}}{2(1+\frac{1}{4})} = 1 \qquad \left[x = \frac{-b}{2a} \right]$$

$\therefore$ the x-coordinate of both points of contact is 1.

Hence the equation of the chord of contact is $x = 1$.

Example 6 Show that the line $y = 3x - 4$ is a tangent to the curve $y = x^2 - x$. Find also the condition that the line $y = mx + c$ should be a tangent to the curve.

At the points of intersection between the curve $y = x^2 - x$ and the line $y = 3x - 4$,

$$x^2 - x = 3x - 4$$
$$x^2 - 4x + 4 = 0$$
$$(x-2)^2 = 0.$$

Since this equation has equal roots the line is a tangent to the curve.

At any points of intersection between the curve $y = x^2 - x$ and the line $y = mx + c$, $\qquad\qquad\qquad x^2 - x = mx + c$
$\therefore \quad x^2 - (m+1)x - c = 0$

The condition that this equation has equal roots is

$$(m+1)^2 - 4.1.(-c) = 0, \quad \text{i.e.} \quad (m+1)^2 + 4c = 0.$$

Hence the line $y = mx + c$ will be a tangent to the curve $y = x^2 - x$ if $(m+1)^2 + 4c = 0$.

Exercise 21.3

In questions 1 to 10 find the equations of the tangent and normal to the given curve at the stated point.

1. $y = 3 + 5x - x^2$; $(4, 7)$.

2. $y = x\sqrt{(x-1)}$; $(2, 2)$.

3. $y = x^2 - \dfrac{1}{x}$; $(-1, 2)$.

4. $y = \dfrac{2x-1}{x+2}$; $(-3, 7)$.

5. $3x^2 + y^2 = 39$; $(1, -6)$.

6. $x^2 + 5xy + 2y^2 = 8$; $(1, 1)$.

7. $y^2 = x^3 - 2$; $(3, 5)$.

8. $\dfrac{1}{x} - \dfrac{1}{y} = \dfrac{1}{6}$; $(2, 3)$.

9. $x = (2t+1)^2$, $y = t^2 - t$; $t = -1$.

10. $x = t(t^2 + 4)$, $y = t^2 + 4$; $t = 2$.

11. Show that the following lines are tangents to the given curves and find their points of contact
(a) $x + y = 4$; $y = 3x - x^2$,
(b) $2x + y = 5$; $x^2 + y^2 - 2x + 4y = 0$,
(c) $5x - 3y = 8$; $x = t + \dfrac{1}{t}$, $y = t - \dfrac{1}{t}$.

12. Find the equation of the normal to the curve $x = 3t - 2t^2$, $y = 2 + t^2$ which is parallel to the line $5x - 4y = 0$.

13. Find the equations of the tangents to the curve $y = x^3 - 10x + 5$ which are parallel to the line $2x - y = 5$.

14. Find the values of m for which the line $y = m(2x - 1)$ touches the curve $y = x^2 + 4x$.

15. Find the equations of the tangents to the curve $y = x^2 + 2x + 4$ which pass through the point $(2, 3)$ and the coordinates of the points of contact.

16. Find the equations of the tangents from the origin to the circle $x^2 + y^2 - 4x - 2y + 4 = 0$. Find also the length of the chord of contact.

17. Find the values of c for which the line $y = x + c$ is a tangent to the circle $x^2 + y^2 - 4x + 2y - 3 = 0$.

18. Find the values of c for which the line $2x - 3y = c$ is a tangent to the curve $x^2 + 2y^2 = 2$ and find the equation of the line joining the points of contact.

19. Find the equations of the tangents to the circle $x^2 + y^2 = 10$ which are parallel to the line $y = 3x$.

20. Find the equations of the tangents to the circle $x^2 + y^2 = 9$ which pass

through the point $(0, 5)$. Find also the acute angle between these tangents, giving your answer in degrees and minutes.

21. Show that the equation of the tangent to the curve $x = 4\cos\theta$, $y = 2\sin\theta$ at the point with parameter θ is $x\cos\theta + 2y\sin\theta = 4$. Hence find the equations of the tangents which pass through the point $(5, 0)$ and the coordinates of their points of contact.

22. Find the equation of the normal to the curve $x = 2t$, $y = t^2$ at the point with parameter t. If this normal meets the x- and y-axes at the points A and B respectively, find the equation of the locus of the mid-point of AB.

23. $P(p-1, p^2)$ and $Q(q-1, q^2)$ are two points on the curve $y = (x+1)^2$. Find the equation of the chord PQ and hence the equation of the tangent at P. If the tangent at P meets the line $x = -1$ at the point R, find the equation of the locus of the mid-point of PR.

24. $P(p^2, p^3)$ and $Q(q^2, q^3)$ are two points on the curve $y^2 = x^3$. Find the equation of the chord PQ and deduce the equation of the tangent at P. Given that the tangent at P passes through Q and is normal to the curve at Q, find the values of p and q.

21.4 The parabola

The locus of a point equidistant from a fixed point and a fixed line is called a *parabola*. The fixed point is the *focus* of the parabola and the fixed line is called the *directrix*.

The standard form of the equation of a parabola is obtained by letting the focus be the point $S(a, 0)$ and the directrix the line $x = -a$.

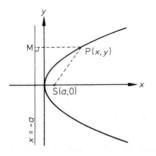

If $P(x, y)$ is any point on the parabola and M is the foot of the perpendicular from P to the directrix, then

$$SP^2 = PM^2$$

$$\therefore \quad (x-a)^2 + y^2 = (x+a)^2$$
$$\therefore \quad y^2 = 4ax$$

Thus the equation $y^2 = 4ax$ represents a parabola with focus $(a, 0)$, directrix $x = -a$.

The line of symmetry of the curve, in this case the x-axis, is called the *axis* of the parabola. The point where the curve cuts the axis, the origin $(0, 0)$, is referred to as the *vertex*.

Example 1   Find the equation of the tangent to the parabola $y^2 = 8x$ at the point $(2, 4)$.

$$y^2 = 8x$$

Differentiating with respect to x: $$2y\frac{dy}{dx} = 8$$

$$\therefore \quad \frac{dy}{dx} = \frac{8}{2y} = \frac{4}{y}.$$

$\therefore$ the gradient of the tangent at the point $(2, 4)$ is 1.

Hence its equation is $y - 4 = 1(x - 2)$, i.e. $y = x + 2$.

By the same method it can be shown that the tangent at (x_1, y_1) to the curve $y^2 = 4ax$ has equation $yy_1 = 2a(x + x_1)$.

It is usually more convenient to express the equation $y^2 = 4ax$ in the parametric form $x = at^2$, $y = 2at$. Substituting in the original equation, we find that the point $(at^2, 2at)$ lies on the parabola for all values of the parameter t.

Example 2   Find the equations of the tangent and the normal to the curve $y^2 = 4ax$ at the point $(at^2, 2at)$.

Differentiating the parametric equations

$$x = at^2 \qquad y = 2at$$

$$\frac{dx}{dt} = 2at \qquad \frac{dy}{dt} = 2a$$

$$\therefore \quad \frac{dy}{dx} = \frac{dy}{dt} \bigg/ \frac{dx}{dt} = \frac{2a}{2at} = \frac{1}{t}$$

$\therefore$ the gradient of the tangent at the point $(at^2, 2at)$ is $1/t$.

Hence its equation is $y - 2at = \dfrac{1}{t}(x - at^2)$, i.e.

$$ty - x = at^2.$$

The gradient of the normal at $(at^2, 2at)$ is $-t$.

Hence the equation of the normal is $y - 2at = -t(x - at^2)$,
i.e. $y + tx = 2at + at^3$.

A *chord* of a parabola is a straight line which joins any two points on the curve. A chord which passes through the focus is a *focal chord*. The focal chord parallel to the directrix is called the *latus rectum*.

Example 3   Find the equation of the chord PQ of the parabola $x = at^2$, $y = 2at$, where P and Q are the points with parameters p and q respectively. Given that the chord PQ passes through the focus of the parabola, find the equation of the locus of the mid-point M of PQ as p and q vary.

The coordinates of P and Q are $(ap^2, 2ap)$ and $(aq^2, 2aq)$

$\therefore$ the gradient of $PQ = \dfrac{2ap - 2aq}{ap^2 - aq^2} = \dfrac{2}{p+q}$.

Hence the equation of PQ is $y - 2ap = \dfrac{2}{p+q}(x - ap^2)$

$$(p+q)(y - 2ap) = 2(x - ap^2)$$
$$(p+q)y - 2ap^2 - 2apq = 2x - 2ap^2$$

i.e. $(p+q)y - 2x = 2apq.$

Given that the focus $(a, 0)$ lies on this line

$$-2a = 2apq \qquad \therefore \quad pq = -1$$

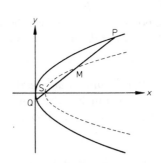

The coordinates of M are

$$(\tfrac{1}{2}\{ap^2 + aq^2\}, \tfrac{1}{2}\{2ap + 2aq\})$$

$\therefore$ the equations of the locus of M are

$$x = \tfrac{1}{2}a(p^2 + q^2), \quad y = a(p + q)$$

where $pq = -1$.

[To obtain the Cartesian equation of the locus, p and q must be eliminated from these three equations.]

Expressing x in terms of $p + q$ and pq,

$$x = \tfrac{1}{2}a\{(p+q)^2 - 2pq\}$$

$$\therefore \quad x = \tfrac{1}{2}a\left\{\dfrac{y^2}{a^2} - 2(-1)\right\} = \dfrac{y^2}{2a} + a.$$

Hence the equation of the locus of M is $y^2 = 2a(x - a)$.

Exercise 21.4

1. Sketch the following parabolas showing their foci and directrices
    (a) $y^2 = 4x$,   (b) $y^2 = 12x$,   (c) $y^2 = -8x$,
    (d) $y^2 = 4x + 1$,   (e) $y^2 = x$,   (f) $y = x^2$,
    (g) $x = 2t^2, y = 4t$,   (h) $x = 3t^2, y = -6t$.

2. Write down the equations of the parabolas with the following foci and directrices
    (a) $(2, 0)$; $x = -2$,   (b) $(5, 0)$; $x = -5$,
    (c) $(0, 1)$; $y = -1$,   (d) $(-3, 0)$; $x = 3$,
    (e) $(1, 1)$; $x = -1$,   (f) $(0, 2)$; $y = 0$.

3. Derive the equations of the parabolas with the following foci and directrices

(a) $(3,0)$; $x = 1$,   (b) $(3, 1)$; $x = -3$,
(c) $(-1, 1)$; $x = -3$,   (d) $(4, 2)$; $y = 3$.

4. Write down the parametric coordinates of a point on each of the following curves

(a) $y^2 = 8x$,   (b) $y^2 = 24x$,   (c) $y^2 = -16x$.

5. Find the equations of the tangents and normals to the following curves at the given points

(a) $y^2 = 16x$; $(1, 4)$,   (b) $y^2 = 6x$; $(6, -6)$,
(c) $x = t^2$, $y = 2t$; $t = 2$,   (d) $x = 3t^2$, $y = 6t$; $t = -1$.

6. The normal to the parabola $y^2 = 4ax$ at the point $P(at^2, 2at)$ meets the x-axis at A. Find the equation of the locus of the mid-point of AP as t varies.

7. The tangent to the parabola $y^2 = 4ax$ at the point $P(at^2, 2at)$ meets the x-axis at A and the y-axis at B. Find the equation of the locus of the mid-point of AB as t varies.

8. Show that, if the chord joining the points $P(ap^2, 2ap)$, $Q(aq^2, 2aq)$ on the parabola $y^2 = 4ax$ passes through $(a, 0)$, then $pq = -1$. Further, the tangent at P meets the line through Q parallel to the axis of the parabola at R. Prove that the line $x + a = 0$ bisects PR.   (O & C)

9. Find the condition that the line $y = mx + c$ should be a tangent to the parabola $y^2 = 4ax$. Use this result to find the equations of the tangents to the curve $y^2 = 4x$ which pass through the point $(-2, 1)$.

10. The tangents at $P(ap^2, 2ap)$ and $Q(aq^2, 2aq)$ to the parabola $y^2 = 4ax$ meet at a point R. Find the coordinates of R. If R lies on the line $2x + a = 0$, find the equation of the locus of the mid-point of PQ.

11. The points $P(ap^2, 2ap)$ and $Q(aq^2, 2aq)$ lie on the parabola $y^2 = 4ax$. Prove that if PQ is a focal chord then the tangents to the curve at P and Q intersect at right angles at a point on the directrix.

12. Find the gradient of the normal to the parabola $y^2 = 4ax$ at the point $P(ap^2, 2ap)$. Find the slope of the chord joining the point $P(ap^2, 2ap)$ to another point $Q(aq^2, 2aq)$. The normal at a point $P(ap^2, 2ap)$ meets the parabola again at a point $Q(aq^2, 2aq)$. By treating this line both as a normal and a chord, or otherwise, prove that $p^2 + pq + 2 = 0$.

The normal at a point $R(4a, 4a)$ meets the parabola again at a point S. The normal at S meets the parabola again at T. What are the coordinates of T? Find the length of RS, giving your answer in simplified surd form.   (SU)

13. The tangents at the points $P(ap^2, 2ap)$ and $Q(aq^2, 2aq)$ on the parabola $y^2 = 4ax$ intersect at the point R. Given that the tangent at P is perpendicular to the chord OQ, where O is the origin, find the equation of the locus of R as p varies.

14. Find the equation of the tangent to the parabola $y^2 = 4ax$ at the point $P(at^2, 2at)$. The line through O, parallel to this tangent, meets the parabola again at Q. Show that the line through P, parallel to the axis of the parabola, passes through the midpoint of OQ. Show also that, if the tangent and normal at P meet the x-axis at T and N respectively, the area of the triangle TPN is $2a^2t(1+t^2)$. (L)

15. The tangent to the parabola $y^2 = 4ax$ at the point $P(at^2, 2at)$ meets the x-axis at T. The straight line through P parallel to the axis of the parabola meets the directrix at Q. If S is the focus of the parabola, show that $PQTS$ is a rhombus. If M is the mid-point of PT and N is the mid-point of PM, find the equation of the locus of (i) M, (ii) N.   (AEB 1978)

21.5 Translations and change of origin

In the previous section the parabola $y^2 = 4ax$ was considered in some detail. We now look at ways of using our knowledge of this 'standard' parabola to determine the properties of a parabola whose vertex is not at the origin.

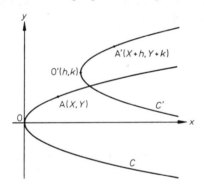

The sketch shows the curve C with equation $y^2 = 4ax$ and its image C' under a translation with column vector $\begin{pmatrix} h \\ k \end{pmatrix}$. The image of the origin O under this translation is the point $O'(h, k)$. The image of a typical point $A(X, Y)$ on the curve C is the point $A'(X+h, Y+k)$. Thus at a typical point A' on C'

$$x = X+h, \quad y = Y+k \quad \text{where} \quad Y^2 = 4aX.$$

Eliminating X and Y we find that the Cartesian equation of C' is

$$(y-k)^2 = 4a(x-h).$$

Hence the equation $(y-k)^2 = 4a(x-h)$ represents the image of the parabola $y^2 = 4ax$ under a translation in which the point $(0, 0)$ is mapped to the point (h, k).

Example 1   Show that the equation $y^2 = 4x - 8$ represents a parabola. Find its focus and directrix.

The equation may be written in the form $y^2 = 4(x-2)$.

Hence the equation represents the image of the curve $y^2 = 4x$ under a translation which maps the point $(0, 0)$ to the point $(2, 0)$. The curve $y^2 = 4x$ is a parabola with focus $(1, 0)$ and directrix $x = -1$

∴   the curve $y^2 = 4x - 8$ is a parabola with focus $(3, 0)$ and directrix $x = 1$.

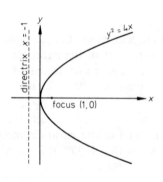

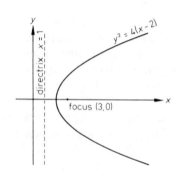

The result obtained for the parabola can be generalised. If the equation $f(x, y) = 0$ represents a curve in the x, y plane, then the equation $f(x-h, y-k) = 0$ represents the image of that curve under a translation with column vector $\begin{pmatrix} h \\ k \end{pmatrix}$.

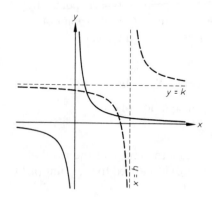

For instance, the curve $x^2 + y^2 = a^2$ is a circle, centre $(0, 0)$, radius a. The curve $(x-h)^2 + (y-k)^2 = a^2$ is a circle, centre (h, k), radius a.

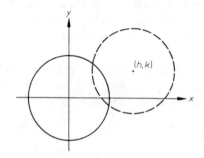

Similarly, the equation $xy = c^2$ represents a curve called a rectangular hyperbola with the lines $x = 0$, $y = 0$ as asymptotes. The curve $(x-h)(y-k) = c^2$ is a rectangular hyperbola with asymptotes $x = h$, $y = k$.

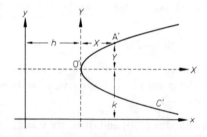

We now look briefly at a different approach to curves with equations of the form $f(x-h, y-k) = 0$. Consider again a typical point $A'(x, y)$ on the curve C' as defined earlier.

Relative to X- and Y-axes through a new origin $O'(h, k)$ the coordinates of A' are (X, Y), where $X = x - h$, $Y = y - k$. As before, since $Y^2 = 4aX$, the equation of C' must be $(y - k)^2 = 4a(x - h)$.

In general, if the point (x, y) has coordinates (X, Y) relative to a new origin at (h, k) then the equation $f(x - h, y - k) = 0$ may be written in the form $f(X, Y) = 0$.

Example 1 (alternative method)   Let the point (x, y) have coordinates (X, Y) with respect to a new origin at $(2, 0)$, then $X = x - 2$ and $Y = y$. The equation $y^2 = 4(x - 2)$ then becomes $Y^2 = 4X$. Relative to the new X- and Y-axes the equation $Y^2 = 4X$ represents a parabola with focus $(1, 0)$ and directrix $X = -1$. Hence, relative to the original x- and y-axes, the equation $y^2 = 4x - 8$ represents a parabola with focus $(3, 0)$ and directrix $x = 1$.

Exercise 21.5

1.   Find the images of the points $(1, 3)$, $(0, 4)$ and $(-2, 5)$ under a translation in which the point $(0, 0)$ is mapped to the point
(a) $(1, 0)$,   (b) $(0, -2)$,   (c) $(-1, 2)$.

2.   Find the coordinates of the image of the point $(0, 0)$ under a translation which maps the given curves as follows:
(a) $x^2 + y^2 = 4 \rightarrow (x - 1)^2 + (y - 2)^2 = 4$,
(b) $y^2 = 4x \rightarrow (y + 1)^2 = 4(x - 3)$,
(c) $x^2 + y^2 = 1 \rightarrow x^2 + y^2 - 2y = 0$,
(d) $xy = 4 \rightarrow xy + 3x + 2y + 6 = 4$.

3.   Show that each of the following equations represents a parabola and find its vertex, focus and directrix
(a) $y^2 = 4(x - 3)$,   (b) $(y - 1)^2 = 8x$,   (c) $y^2 = 4x + 20$,
(d) $y^2 + 4y = x$,   (e) $4y = x^2 + 4$,   (f) $y = x^2 - 2x$.

4.   Show that each of the following pairs of parametric equations represents a parabola and find its vertex, focus and directrix.
(a) $x = t^2 + 1$, $y = 2t + 3$,   (b) $x = 3t^2 - 2$, $y = 6t + 5$,
(c) $x = 1 - t^2$, $y = 2t$,   (d) $x = 4t - 3$, $y = 2t^2 - 1$.

5.   Sketch the following pairs of related curves
(a) $y = x^3$, $y = x^3 - 3x^2 + 3x + 1$,
(b) $xy = 1$, $xy = 2x + y - 1$,
(c) $y^2 = -12x$, $y^2 = 12(y - x)$.

Exercise 21.6   *(miscellaneous)*

1.   Find those points P on the curve with equation $y = x^2 - 2$ such that the normal to the curve at P passes through the point $(0, 0)$.   (AEB 1977)

2. Find the condition that the line $y = mx + c$ should be a tangent to the circle $x^2 + y^2 = a^2$.

3. A curve is defined by the parametric equations

$$x = \theta - \sin\theta, \quad y = 1 - \cos\theta, \quad 0 < \theta < 2\pi.$$

Show that $dy/dx = \cot\theta/2$, and find the equation of the tangent and of the normal to the curve at the point where $\theta = \pi/2$. (JMB)

4. Find the equations of the tangent and the normal to the curve $y^2(y-1) = x^2(x+3)$ at the point $(1, 2)$.

5. Find the centres and radii of the circles C_1 and C_2 whose equations are $x^2 + y^2 - 2x = 0$ and $x^2 + y^2 - 2y = 0$ respectively. Draw a figure to illustrate the circles. The line $y = mx$ through the origin O meets C_1 and C_2 again at P and Q. Find the coordinates of P and Q. Let A denote the second point of intersection of the two circles. Show that AP and AQ are perpendicular. (W)

6. Find the equation of the circle which passes through the points $A(2, 0)$, $B(10, 4)$ and $C(5, 9)$ and show that it touches the y-axis. If the tangents at A and B intersect at D find the coordinates of D, the length BD and the angle ADB correct to the nearest degree. (SU)

7. The circle S has the equation $x^2 + y^2 - 6x - 8y = 0$. (i) Find the coordinates of the centre, and the radius, of S. (ii) Prove, by calculation, that the point A, with coordinates $(7, 2)$, lies inside the circle S. (iii) Show that the chord of S which is bisected at A has equation $y - 2x + 12 = 0$. (iv) Find the equations of the two tangents to S which are parallel to the line $y = 2x$. (C)

8. Prove that the point $B(1, 0)$ is the mirror-image of the point $A(5, 6)$ in the line $2x + 3y = 15$. Find the equation of (a) the circle on AB as diameter, (b) the circle which passes through A and B and touches the x-axis. (L)

9. Find the equation of the tangent to the circle $x^2 + y^2 = a^2$ at the point $T(a\cos\theta, a\sin\theta)$. This tangent meets the line $x + a = 0$ at R. If RT is produced to P so that $RT = TP$, find the coordinates of P in terms of θ and find the coordinates of the points in which the locus of P meets the y-axis. (L)

10. A circle passes through the points A, B and C which have coordinates $(0, 3)$, $(\sqrt{3}, 0)$ and $(-\sqrt{3}, 0)$ respectively. Find: (i) the equation of the circle, (ii) the length of the minor arc BC, (iii) the equation of the circle on AB as diameter. A line $y = mx - 3$ of variable gradient m cuts the circle ABC in two points L and M. Find in cartesian form the equation of the locus of the mid-point of LM. (AEB 1976)

11. Find the equations of the two circles each of which touches both coordinate axes and passes through the point $(9, 2)$. Find (i) the coordinates of the second

point of intersection of these circles, (ii) the equation of the common chord of the two circles. (C)

12. The point $A(4, 1)$ lies on the line whose equation is $3x - 4y - 8 = 0$. A circle touches this line at A and passes through the point $B(5, 3)$. Find the equation of the circle, and show that it touches the y-axis. Find also the equation of the line parallel to AB on which the circle cuts off a chord equal in length to AB. (C)

13. Sketch for $0 \leqslant t \leqslant 2\pi$ the curve given parametrically by $x = a\cos^3 t$, $y = a\sin^3 t$. Show that $dy/dx = -\tan t$. Find and simplify the equation of the tangent at the point where $t = \alpha$. If this tangent meets the axes at A and B, show that the length of AB is independent of x. (L)

14. A curve is given parametrically by the equations $x = a(2 + t^2)$, $y = 2at$. Find the values of the parameter t at the points P and Q in which this curve is cut by the circle with centre $(3a, 0)$ and radius $5a$. Show that the tangents to the curve at P and Q meet on the circle, and that the normals to the curve at P and Q also meet on the circle. (L)

15. The parametric equations of a curve are $x = \cos 2t$, $y = 4\sin t$. Sketch the curve for $0 \leqslant t \leqslant \frac{1}{2}\pi$. Show that $dy/dx = -\csc t$ and find the equation of the tangent to the curve at the point $A(\cos 2T, 4\sin T)$. The tangent at A crosses the x-axis at the point M and the normal at A crosses the x-axis at the point N. If the area of the triangle AMN is $12\sin T$, find the value of T between 0 and $\frac{1}{2}\pi$. (AEB 1976)

16. A curve is defined by the parametric form $x = a(\cos 3\theta - 3\cos\theta)$, $y = a(\sin 3\theta + 3\sin\theta)$. Prove that $dy/dx = -\cot\theta$, and hence show that the normal to the curve at the point with parameter θ is given by $x\sin\theta - y\cos\theta + 4a\sin 2\theta = 0$. Prove that the distance between the points where the normal meets the coordinate axes is independent of θ. (O)

17. Find the equation of the tangent to the curve $y = \dfrac{5}{12}x^3 - \dfrac{13}{9}x$ at the point P at which $x = x_0$. Show that the x-coordinate of the point Q where this tangent meets the curve again is $-2x_0$, and find the values of x_0 for which the tangent at P is the normal at Q. (O)

18. Find the equations of the tangent and the normal to the parabola $y^2 = 4ax$ at the point P with parameter p. (i) Show that, if the tangent at P meets the directrix at L, then $PL = a(p^2 + 1)^{3/2}/p$. (ii) Show that, if the tangent at P is parallel to the normal at a point Q, then PQ passes through the focus of the parabola. (SU)

19. Find the equation of the normal to the parabola $y^2 = 4ax$ at the point $(at^2, 2at)$ and the coordinates of the point at which this normal cuts the x-axis. Show that the equation of the circle which touches this parabola at the points $(at^2, 2at)$ and $(at^2, -2at)$ is $(x - 2a - at^2)^2 + y^2 = 4a^2(1 + t^2)$. Find the values of t for which this circle passes through the point $(9a, 0)$. (L)

20. Prove that the chord joining the points $P(ap^2, 2ap)$ and $Q(aq^2, 2aq)$ on the parabola $y^2 = 4ax$ has the equation $(p+q)y = 2x + 2apq$. A variable chord PQ of the parabola is such that the lines OP and OQ are perpendicular, where O is the origin. (i) Prove that the chord PQ cuts the axis of x at a fixed point, and give the x-coordinate of this point. (ii) Find the equation of the locus of the mid-point of PQ. (C)

21. Prove that the tangent at the point $(at^2, 2at)$ on the parabola $y^2 = 4ax$ has the equation $ty = x + at^2$. Find, in their simplest form, the coordinates of T, the point of intersection of the tangents at the points $P(ap^2, 2ap)$ and $Q(aq^2, 2aq)$ on the parabola $y^2 = 4ax$. If PQ is of constant length d, show that T lies on the curve whose equation is

$$(y^2 - 4ax)(y^2 + 4a^2) = a^2 d^2.$$ (C)

22. Show that the equation of the normal to the parabola $y^2 = 4ax$ at the point $P(ap^2, 2ap)$ is $y + px = 2ap + ap^3$. Find the coordinates of R, the point of intersection of the normal at P and the normal at $Q(aq^2, 2aq)$. Given that the chord PQ passes through $S(a, 0)$, show that $pq = -1$ and find the equation of the locus of R. (AEB 1978)

23. Show that the equation of the normal to the parabola $y^2 = 4ax$ at the point $P(ap^2, 2ap)$ is $px + y = 2ap + ap^3$. The tangent at P meets the x-axis at A and the y-axis at B. The normal at P meets the x-axis at C and S is the point $(a, 0)$. Show that the areas of the triangles APS and SPC are equal. Show also that the locus of the mid-point of PS is a parabola through the mid-point of OS. If BS and OP meet at Q, show that the equation of the locus of Q is $2x^2 + y^2 = 2ax$. (AEB 1977)

24. Prove that the mid-points of chords of the parabola $y^2 = 4ax$ that are parallel to the line $y = mx$ lie on the line $y = 2a/m$. Hence, or otherwise, find the equation of the tangent to the parabola that is parallel to the line $y = mx$. (O)

25. Find the equation of the tangent to the parabola $y^2 = 4ax$ at the point $P(ap^2, 2ap)$. Show that the equation of the normal at P is $y = p(2a - x) + ap^3$. If the tangents at P and $Q(aq^2, 2aq)$ meet at T show that T is the point $(apq, ap + aq)$. The point N is the intersection of the normals at P and Q. Given that T lies on the line $x + 2a = 0$ show that N lies on the parabola with equation $y^2 = 4a(x - 4a)$. (L)

26. Prove that the line $y = mx + 15/4m$ is a tangent to the parabola $y^2 = 15x$ for all non-zero values of m. Using this result or otherwise find the equations of the common tangents to this parabola and the circle $x^2 + y^2 = 16$. (L)

27. Find the equation of the normal at the point $P(at^2, 2at)$ to the parabola $y^2 = 4ax$. The focus of the given parabola is the point $S, (a, 0)$. If PN is the normal at the point P and SN is parallel to the tangent at P, find the coordinates of the point N. Deduce that the locus of N for variable P is a parabola and find the coordinates of its vertex. (O & C)

28. Prove that the normal to the parabola $y^2 = 4ax$ at the point $(at^2, 2at)$ has equation $y + tx = 2at + at^3$. The normals at the points $P(ap^2, 2ap)$ and $Q(aq^2, 2aq)$ intersect at the point R. Find the coordinates of R in terms of $(p+q)$ and pq. If O is the vertex of the parabola and P and Q are variable points such that $P\hat{O}Q$ is a right angle, find the locus of R; verify that it is a parabola, and find the coordinates of its vertex.   (O & C)

29. A fixed point $P(ap^2, 2ap)$ is taken on the parabola $y^2 = 4ax$. Two points Q and R are chosen on the parabola so that the lines PQ and PR are perpendicular. Prove that the line QR passes through a fixed point F, independent of Q and R, and that PF is normal to the parabola at P.   (O)

30. A circle with centre at the point $P(h, k)$ touches the y-axis, and passes through the point $S(2, 0)$. Show that P lies on the curve $y^2 = 4x - 4$, and sketch this curve. Show that the straight line joining $P(h, k)$ to the point $Q(2 + h, 0)$ cuts the curve $y^2 = 4x - 4$ at right angles at P.   (L)

31. Find the equation of the normal to the parabola $y^2 = 4x$ at the point $P(p^2, 2p)$. If the normals at P and $Q(q^2, 2q)$ meet at $R(\alpha, \beta)$ show that $\alpha = 2 + p^2 + pq + q^2$, $\beta = -pq(p+q)$. The point of intersection of PQ with the x-axis divides PQ internally in the ratio $1 : \lambda$. Prove that $q = -\lambda p$. Given that $\lambda = 2$, find the Cartesian equation of the locus of R as p varies. Determine the coordinates of the point on this locus which is nearest the origin.   (JMB)

32. Find the equations of the tangents to the curve $27y^2 = 4x^3$ at the points $P(3p^2, 2p^3)$ and $Q(3q^2, 2q^3)$. Show that these tangents intersect at the point $R(\alpha, \beta)$ where $\alpha = p^2 + pq + q^2$, $\beta = pq(p+q)$. The points P and Q move along the curve in such a way that the tangents at P and Q are always perpendicular. Prove that R moves on the parabola $y^2 = x - 1$. Verify that this parabola touches the curve $27y^2 = 4x^3$ at the points $(3/2, \pm 1/\sqrt{2})$.   (JMB)

33. The straight line through the point $A(-a, 0)$ at an angle θ to the positive direction of the x-axis meets the circle $x^2 + y^2 = a^2$ at P, distinct from A. The circle on AP as diameter is denoted by C. (i) By finding the equation of C, or otherwise, show that if C touches the y-axis, then $\cos 2\theta = 2 - \sqrt{5}$. (ii) C meets the x-axis at M, distinct from A, and the tangents to C at A and M meet at Q. Find the coordinates of Q in terms of θ and show that, as θ varies, Q always lies on the curve $y^2 x + (x + a)^3 = 0$.   (JMB)

22 Inequalities

22.1 Basic inequalities

The most important rules for manipulating inequalities are as follows:
(1) Any number may be added to or subtracted from both sides of an inequality,
e.g.
$$x < y \Rightarrow x + 3 < y + 3.$$

(2) Both sides of an inequality may be multiplied or divided by the same *positive* number, e.g.
$$x < y \Rightarrow 3x < 3y.$$

(3) If both sides of an inequality are multiplied or divided by the same *negative* number, the inequality is reversed, e.g.
$$-2x < 6y \Rightarrow x > -3y.$$

[For further examples of the use of these rules see §1.4.]

In this chapter we will be considering two main types of inequality. This section deals briefly with basic inequalities which hold for all values of the variables involved. Later sections are concerned with finding the solution sets of inequalities which hold for only certain values of the variables. The distinction between these two types of inequality broadly corresponds to that made in earlier work between identities and equations.

Most basic inequalities are established using the fact that the square of a real number is never negative.

Example 1 Prove that $a^2 + b^2 \geqslant 2ab$ for all real values of a and b.

As any square is positive or zero, $(a - b)^2 \geqslant 0$
$$\therefore \quad a^2 - 2ab + b^2 \geqslant 0$$
Hence
$$a^2 + b^2 \geqslant 2ab.$$

The result of Example 1 can be used to derive further inequalities.

450

Example 2 Prove that $a^2 + b^2 + c^2 \geqslant ab + bc + ca$ for all real values of a, b and c.

$$a^2 + b^2 \geqslant 2ab, \quad b^2 + c^2 \geqslant 2bc, \quad c^2 + a^2 \geqslant 2ca.$$

Adding

$$2(a^2 + b^2 + c^2) \geqslant 2(ab + bc + ca).$$

Hence

$$a^2 + b^2 + c^2 \geqslant ab + bc + ca.$$

Example 3 Prove that $\dfrac{1}{a} + \dfrac{1}{b} \geqslant \dfrac{4}{a+b}$ for all positive values of a and b.

$$\frac{1}{a} + \frac{1}{b} - \frac{4}{a+b} = \frac{b(a+b) + a(a+b) - 4ab}{ab(a+b)}$$

$$= \frac{a^2 - 2ab + b^2}{ab(a+b)} = \frac{(a-b)^2}{ab(a+b)}$$

∴ provided that a and b are positive $\dfrac{1}{a} + \dfrac{1}{b} - \dfrac{4}{a+b} \geqslant 0$

i.e. $\dfrac{1}{a} + \dfrac{1}{b} \geqslant \dfrac{4}{a+b}$.

Alternative method

$$a^2 + b^2 \geqslant 2ab \Rightarrow (a+b)^2 \geqslant 4ab.$$

Dividing both sides by $ab(a+b)$, assumed positive

$$\frac{a+b}{ab} \geqslant \frac{4}{a+b}$$

∴ $\dfrac{1}{a} + \dfrac{1}{b} \geqslant \dfrac{4}{a+b}$.

Exercise 22.1

1. State whether each of the following statements is true or false. If you decide that a statement is false, show, by a numerical example, that this is so.
(a) $x^2 < 4 \Rightarrow x < 2$,
(b) $x^2 > 4 \Rightarrow x > 2$,
(c) $x < 4 \Rightarrow \dfrac{1}{x} > \dfrac{1}{4}$,
(d) $x > 4 \Rightarrow \dfrac{1}{x} < \dfrac{1}{4}$.

2. Given that a and b are positive, state the range of values of x for which the following statements are true.
(a) $ax \leqslant bx \Rightarrow a \leqslant b$,
(b) $x - a < x - b \Rightarrow a > b$,
(c) $\dfrac{x}{a} \geqslant \dfrac{x}{b} \Rightarrow a \geqslant b$,
(d) $a > b \Rightarrow \dfrac{1}{x+a} < \dfrac{1}{x+b}$.

3. Prove that for any real numbers p and q
(a) $p^2 + q^2 \geqslant 2pq$,
(b) $p^2 + 4q^2 \geqslant 4pq$,
(c) $(p+q)^2 \geqslant 4pq$,
(d) $(p+q)^2 \leqslant 2(p^2 + q^2)$.

4. Prove that if a and b are positive, then

(a) $\dfrac{a}{b} + \dfrac{b}{a} \geqslant 2$,   (b) $a^3 + b^3 \geqslant ab(a+b)$.

5. Prove that for any real numbers p, q, r, s
(a) $p^4 + q^4 \geqslant 2p^2q^2$,   (b) $p^4 + q^4 + r^4 + s^4 \geqslant 4pqrs$.

6. Prove that for any real numbers x, y, z
$3(xy + yz + zx) \leqslant (x+y+z)^2 \leqslant 3(x^2 + y^2 + z^2)$.

22.2 Graphical approach

We now consider the graphical approach to finding the set of values of x which satisfy an inequality of the form $f(x) < g(x)$. This method is particularly useful when the graphs of the functions $f(x)$ and $g(x)$ are fairly easy to sketch.

Example 1   Find the values of x for which $x^2 - 4x < 1$.

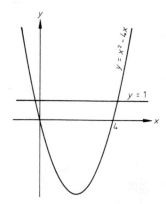

Consider the graphs of $y = x^2 - 4x$ and $y = 1$.
At their points of intersection

$$x^2 - 4x = 1$$

$$x^2 - 4x - 1 = 0$$

$$\therefore \quad x = \frac{4 \pm \sqrt{(4^2 - 4 \cdot 1\{-1\})}}{2}$$

$$= \frac{4 \pm \sqrt{20}}{2} = 2 \pm \sqrt{5}.$$

The inequality $x^2 - 4x < 1$ is satisfied when the curve $y = x^2 - 4x$ is below the line $y = 1$

i.e.   when $2 - \sqrt{5} < x < 2 + \sqrt{5}$.

Example 2   Find the values of x for which $0 \leqslant \dfrac{x-1}{x+2} \leqslant 2$.

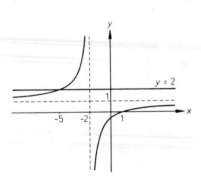

Consider the curve $y = \dfrac{x-1}{x+2}$.

It cuts the axes at the points $(1,0)$ and $(0, -\frac{1}{2})$. The lines $x = -2$ and $y = 1$ are asymptotes to the curve. [See §20.4.]

$$\frac{x-1}{x+2} = 2 \Rightarrow x-1 = 2x+4$$

$$\Rightarrow x = -5$$

$\therefore$ the curve cuts the line $y = 2$ when $x = -5$.

Hence, from the sketch, $0 \leqslant \dfrac{x-1}{x+2} \leqslant 2$ when $x \leqslant -5$ and when $x \geqslant 1$.

Example 3 Find the values of x for which $x-2 > \dfrac{3}{x}$.

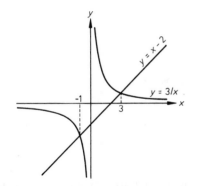

Consider the graphs $y = x-2$ and $y = \dfrac{3}{x}$.

At their points of intersection

$$x-2 = \frac{3}{x}$$

i.e.
$$x^2 - 2x - 3 = 0$$
$$(x+1)(x-3) = 0$$

$\therefore$ either $x = -1$ or $x = 3$.

The inequality $x-2 > \dfrac{3}{x}$ is satisfied when the line $y = x-2$ is above the curve $y = \dfrac{3}{x}$, i.e. when

$-1 < x < 0$ and when $x > 3$.

Exercise 22.2

With the aid of sketch graphs find the values of x which satisfy the following inequalities.

1. $x^2 - x - 6 < 0$.

2. $2x^2 + 7x + 3 \geqslant 0$.

3. $6x - x^2 \geqslant 5.$ 4. $x^2 - 2x > 5.$

5. $4x < 3 - x^2.$ 6. $x^3 - 3x^2 \leqslant 10x.$

7. $\dfrac{1}{x-3} > 1.$ 8. $\dfrac{x}{x+1} < 2.$

9. $0 < \dfrac{2x-4}{x-1} < 1.$ 10. $2 < \dfrac{x-7}{x-2} < 3.$

11. $3 - x > \dfrac{2}{x}.$ 12. $2x + 5 < \dfrac{3}{x}.$

13. $\dfrac{3}{x+2} < x.$ 14. $x - 2 > \dfrac{2}{x-1}.$

15. $\dfrac{12}{x-3} < x + 1.$ 16. $\dfrac{1}{x-1} < \dfrac{1}{x+1}.$

17. $\dfrac{x+1}{2x-3} < \dfrac{1}{x-3}.$ 18. $\dfrac{1}{x^2} < \dfrac{1}{x+2}.$

19. $\pi \sin x > 2x.$ 20. $2\pi \cos x < 3x.$

22.3 Analytical methods

If $(x - a)$ is a factor of $f(x)$ then $f(x)$ may change sign as x passes through the value a. This statement forms the basis of the analytical approach to inequalities.

Example 1   Find the values of x for which $x(2x - 5) > 3$.

$$x(2x - 5) > 3 \Leftrightarrow 2x^2 - 5x - 3 > 0$$
$$\Leftrightarrow (2x + 1)(x - 3) > 0.$$

[We now construct a table of signs for the function $(2x + 1)(x - 3)$, noting that sign changes may occur at $x = -\frac{1}{2}$ and $x = 3$.]

	$x < -\frac{1}{2}$	$-\frac{1}{2} < x < 3$	$x > 3$
$2x + 1$ $x - 3$	$-$ $-$	$+$ $-$	$+$ $+$
$(2x + 1)(x + 3)$	$+$	$-$	$+$

Hence $x(2x - 5) > 3$ when $x < -\frac{1}{2}$ or $x > 3$.

Example 2   Find the values of x for which $x(x^2 - 2) < x^2$.

$$x(x^2 - 2) < x^2 \Leftrightarrow \quad x^3 - x^2 - 2x < 0$$
$$\Leftrightarrow x(x+1)(x-2) < 0.$$

The function $f(x) = x(x+1)(x-2)$ changes sign at $x = -1$, $x = 0$ and $x = 2$.

	$x < -1$	$-1 < x < 0$	$0 < x < 2$	$x > 2$
$x+1$	$-$	$+$	$+$	$+$
x	$-$	$-$	$+$	$+$
$x-2$	$-$	$-$	$-$	$+$
$f(x)$	$-$	$+$	$-$	$+$

Hence the inequality holds for $x < -1$ and $0 < x < 2$.

Great care must be taken when dealing with inequalities involving fractions. In general, it is inadvisable to 'multiply through' by a variable denominator which may be positive or negative. The most reliable method is to collect all the terms on one side of the inequality.

Example 3   Find the values of x for which $\dfrac{x^2 - 12}{x} > -1$.

$$\frac{x^2 - 12}{x} > -1 \Leftrightarrow \frac{x^2 - 12}{x} + 1 > 0$$

$$\Leftrightarrow \frac{x^2 + x - 12}{x} > 0$$

$$\Leftrightarrow \frac{(x+4)(x-3)}{x} > 0.$$

The function $f(x) = (x+4)(x-3)/x$ changes sign as x passes through the values -4, 0 and 3.

	$x < -4$	$-4 < x < 0$	$0 < x < 3$	$x > 3$
$x+4$	$-$	$+$	$+$	$+$
x	$-$	$-$	$+$	$+$
$x-3$	$-$	$-$	$-$	$+$
$f(x)$	$-$	$+$	$-$	$+$

Hence the inequality holds for $-4 < x < 0$ and $x > 3$.

Example 4   For what values of x is $\dfrac{x}{x+8} \leqslant \dfrac{1}{x-1}$?

$$\frac{x}{x+8} \leqslant \frac{1}{x-1} \Leftrightarrow \frac{x}{x+8} - \frac{1}{x-1} \leqslant 0$$

$$\Leftrightarrow \frac{x(x-1)-(x+8)}{(x+8)(x-1)} \leqslant 0$$

$$\Leftrightarrow \frac{x^2-2x-8}{(x+8)(x-1)} \leqslant 0$$

$$\Leftrightarrow \frac{(x+2)(x-4)}{(x+8)(x-1)} \leqslant 0.$$

The function $f(x) = \dfrac{(x+2)(x-4)}{(x+8)(x-1)}$ passes through the value zero when $x = -2$ and when $x = 4$. The function is undefined at $x = -8$ and $x = 1$ and has different signs on either side of these values.

	$x < -8$	$-8 < x < -2$	$-2 < x < 1$	$1 < x < 4$	$x > 4$
$x+8$	$-$	$+$	$+$	$+$	$+$
$x+2$	$-$	$-$	$+$	$+$	$+$
$x-1$	$-$	$-$	$-$	$+$	$+$
$x-4$	$-$	$-$	$-$	$-$	$+$
$f(x)$	$+$	$-$	$+$	$-$	$+$

Hence $\dfrac{x}{x+8} \leqslant \dfrac{1}{x-1}$ when $-8 < x \leqslant -2$ and $1 < x \leqslant 4$.

Exercise 22.3

Find the values of x which satisfy the following inequalities.

1. $x^2 + 2x < 15$.

2. $3x^2 + 2 > 7x$.

3. $x^2(x-5) > 6x$.

4. $x(x^2+4) < 5x^2$.

5. $\dfrac{x^2+12}{x} > 7$.

6. $\dfrac{x^2+6}{x} > 5$.

7. $\dfrac{(x-1)(x+3)}{(x-2)} < 0$.

8. $\dfrac{(2x-3)}{(x+2)(x-5)} > 0$.

9. $\dfrac{6}{x-1} > 1$.

10. $\dfrac{2x-4}{x-1} < 1$.

11. $\dfrac{8}{x+2} > x.$

12. $\dfrac{6}{x-4} < x+1.$

13. $\dfrac{5-x}{x^2-3x+2} < 1.$

14. $\dfrac{x+6}{x(x+1)} < 6.$

15. $\dfrac{1}{x+1} < \dfrac{1}{x+4}.$

16. $\dfrac{1}{x+2} > \dfrac{1}{2x-3}.$

17. $\dfrac{x}{x-2} > \dfrac{1}{x}.$

18. $\dfrac{x+1}{2x-3} < \dfrac{1}{x-3}.$

22.4 Modulus inequalities

The modulus notation was introduced in §1.4. From the definition it follows that if $|f(x)| < a$, where a is a positive constant, then $-a < f(x) < a$.

Example 1   Find the values of x for which $|2x+1| < 3$.

$$|2x+1| < 3 \Leftrightarrow -3 < 2x+1 < 3$$
$$\Leftrightarrow -4 < 2x < 2$$
$$\Leftrightarrow -2 < x < 1.$$

Hence the inequality is satisfied when $-2 < x < 1$.

Some inequalities can be simplified by squaring both sides. However, it is important to remember that this method is valid only when both sides of the inequality are positive or zero for all values of x.

Example 2   Find the values of x for which $2|x-1| < |x+3|$.

Since both sides of the inequality are positive or zero for all values of x

$$2|x-1| < |x+3| \Leftrightarrow 4(x-1)^2 < (x+3)^2$$
$$\Leftrightarrow 4x^2-8x+4 < x^2+6x+9$$
$$\Leftrightarrow 3x^2-14x-5 < 0$$
$$\Leftrightarrow (3x+1)(x-5) < 0.$$

Hence the inequality holds if $-\tfrac{1}{3} < x < 5$.

In harder examples it may be possible to adapt the methods of Examples 1 and 2 or to use sketch graphs. [For work on graphs with equations of the form $y = |f(x)|$ see §§20.2, 20.3.]

Example 3   Find the values of x for which $|2x-3| > x$.

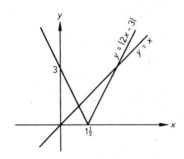

The graphs of $y = |2x-3|$ and $y = x$ have two points of intersection, where $2x-3 = x$, i.e. $x = 3$ and where $-(2x-3) = x$, i.e. $x = 1$. Hence, from the sketch, $|2x-3| > x$ when $x < 1$ and when $x > 3$.

Alternative method

Since $|2x-3|$ is never negative the inequality must hold when $x < 0$.

$$\text{When } x \geqslant 0, |2x-3| > x \Leftrightarrow \quad (2x-3)^2 > x^2$$
$$\Leftrightarrow 4x^2 - 12x + 9 > x^2$$
$$\Leftrightarrow 3x^2 - 12x + 9 > 0$$
$$\Leftrightarrow \quad x^2 - 4x + 3 > 0$$
$$\Leftrightarrow (x-1)(x-3) > 0$$
$$\Leftrightarrow \quad x < 1 \text{ or } x > 3.$$

Thus, considering all values of x, the inequality holds for $x < 1$ and $x > 3$.

Exercise 22.4

Find the values of x which satisfy the following inequalities.

1. $|x-2| > 1$.

2. $|x+3| \leqslant 5$.

3. $|3x+5| < 4$.

4. $|2x-1| \geqslant 11$.

5. $|x| \leqslant |x-1|$.

6. $2|x+2| > |x+3|$.

7. $3|x-2| \geqslant |x+6|$.

8. $5|2x-3| < 4|x-5|$.

9. $2|x-2| > x$.

10. $|3x+4| \leqslant x+2$.

11. $|2x+1| < 3x+2$.

12. $|x+1| > x-3$.

13. $|x^2-3x-2| < 2$.

14. $|x(x-5)| > 6$.

15. $\left|\dfrac{x}{x+4}\right| < 2$.

16. $\left|\dfrac{x^2-4}{x}\right| \leqslant 3$.

Exercise 22.5 (*miscellaneous*)

In questions 1 to 18 find the set of values of x which satisfy the given inequalities.

1. $x(x+6) \geqslant 7$.

2. $x(x-2) < 1$.

3. $(x^2-1)(2x+1) < 0.$

4. $x(x^2+10) < 7x^2.$

5. $x^4-3x^2-4 > 0.$

6. $x^4-5x^2+6 > 0.$

7. $0 < \dfrac{x}{2x-3} < 1.$

8. $2 < \dfrac{4x^2-1}{x^2} < 3.$

9. $\dfrac{5}{x+1} < x-3.$

10. $\dfrac{x}{x-2} > \dfrac{1}{x+1}.$

11. $\dfrac{4x-1}{x^2-2x-3} < 3.$

12. $\dfrac{5x^2+2x-11}{x^2+1} > 4.$

13. $|3-2x| \leqslant |x+4|.$

14. $|x^2+1| < |x^2-9|.$

15. $|3x-2| < x.$

16. $|5x-6| > x^2.$

17. $\left|\dfrac{x+1}{x-1}\right| < 1.$

18. $\left|\dfrac{x}{x-2}\right| < 2.$

19. Find the values of x between $-\pi/2$ and $+5\pi/2$ for which
(a) $\cos x > \tfrac{1}{2}$,   (b) $|\tan x| \leqslant 1$,   (c) $4\sin^2 x < 1.$

20. By considering $(a^2+b^2)^2$ or otherwise, prove that $a^4+b^4 \geqslant a^3b+ab^3$ for all real values of a and b.

21. Prove that the geometric mean of two positive real numbers p and q is less than or equal to their arithmetic mean.

22. Find the ranges of values of x between 0 and 2π for which $\sin 2x > \cos x$. (O)

23. For each of the following expressions, state, with reasons, for what set of values of x it is greater than -1:

(i) $\dfrac{1}{x}$,   (ii) $-(x-1)^2$,   (iii) $\dfrac{2x-1}{x-2}$.   (O&C)

24. Determine the range of values of x for which

(i) $\dfrac{6}{x+1} < x$,   (ii) $\dfrac{6}{|x|+1} < |x|$.   (O&C)

23 Work, energy and power

23.1 Work and kinetic energy

The *work done by a force* is the product of the force and the distance moved in the direction of the force by its point of application.

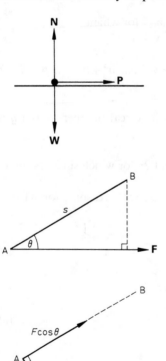

If a particle of weight **W** is pulled a distance s along a smooth horizontal plane by a horizontal force **P**, then the work done by this force is Ps. However, since the direction of motion is perpendicular to the directions of the weight **W** and the normal reaction **N**, these forces do no work.

Consider now a particle which moves a distance s from A to B under the action of a force **F**, where the angle between the displacement **AB** and the force **F** is θ.
The particle moves a distance $s \cos \theta$ in the direction of the force **F**. Hence the work done by the force is $Fs \cos \theta$.

Another way of obtaining the same result is to consider the force **F** as a sum of components along and perpendicular to **AB**. The component $F \sin \theta$ does no work but, since the particle moves a distance s in the direction of the component $F \cos \theta$, the work done is $F \cos \theta \times s$, i.e. $Fs \cos \theta$ as before.

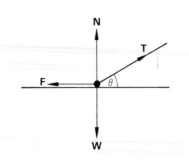

If a particle is pulled a distance s along a rough horizontal plane by a string inclined at an angle θ to the horizontal, then the work done by the tension **T** in the string is $Ts\cos\theta$. The weight **W** and the normal reaction **N** do no work. However, the displacement of the particle in the direction of the frictional force **F** which opposes motion is $-s$. Hence the work done by the force **F** is $-Fs$. The quantity Fs is called the work done *against* friction.

The SI unit of work is the joule (J). When the point of application of a force of 1 newton moves 1 metre, the work done is 1 joule, i.e. $1\,\text{J} = 1\,\text{N m}$.

To establish a relationship between the motion of a body and the work done on it, we consider a particle of mass m moving in a straight line under the action of a constant force F. We assume that the acceleration of the particle is a and that its initial velocity is u. If when the particle has moved a distance s its velocity is v, then, as shown in §12.2, $v^2 = u^2 + 2as$. By Newton's second law,

$$F = ma$$

$$\therefore \qquad Fs = mas \quad \text{and} \quad as = \tfrac{1}{2}(v^2 - u^2)$$

Hence $\qquad\qquad Fs = \tfrac{1}{2}mv^2 - \tfrac{1}{2}mu^2.$

The quantity $\tfrac{1}{2}mv^2$ is called the *kinetic energy* (K.E.) of a particle of mass m moving with velocity v and is measured in the same units as work, i.e. in joules.

Thus when a particle is moving in a straight line under the action of a constant force,

$$\text{work done by force} = \text{increase in kinetic energy.}$$

Example 1 A body of mass 4 kg is moving in a straight line under the action of a constant force. Given that its speed increases from $3\,\text{m s}^{-1}$ to $7\,\text{m s}^{-1}$, find the work done by the force.

Work done = increase in K.E. $= \tfrac{1}{2}.4.7^2 - \tfrac{1}{2}.4.3^2\,\text{J}$
$$= 98 - 18\,\text{J}.$$
Hence the work done by the force is 80 J.

Example 2 A car of mass 1200 kg travelling along a straight horizontal road accelerates uniformly from $10\,\text{m s}^{-1}$ to $15\,\text{m s}^{-1}$ in a distance of 200 m. If the thrust of the engine is 500 N, find the resistance to motion.

Let the resistance to motion be R N, then the total horizontal force acting on the car is $(500 - R)$ N.
$$\text{Work done} = \text{increase in K.E.}$$
$$\therefore \qquad (500 - R)200 = \tfrac{1}{2}.1200.15^2 - \tfrac{1}{2}.1200.10^2 = 600(225 - 100)$$
$$\therefore \qquad 500 - R = 3 \times 125 = 375.$$
Hence there is resistance to motion of 125 N.

When the motion of a particle in a straight line is opposed by a constant force, it may be convenient to express the relationship between work and kinetic energy in the form:

Work done again resistance = loss of K.E.

Exercise 23.1 [Take $g = 10$.]

1. A particle of weight W is pulled a distance s across a rough horizontal plane by a horizontal force P. Find the work done by the weight W, the normal reaction N, the force P and the frictional force F.

2. A body of weight W is lifted through a distance x by a force P directed vertically upwards. Find the work done by the weight W and the force P.

3. A particle of weight W slides a distance s down a rough plane inclined at an angle α to the horizontal. Find the work done by the weight W and the normal reaction N. Find also the work done against the frictional force F.

4. A particle of weight W is pulled a distance s up a rough plane inclined at an angle α to the horizontal by a string parallel to the plane. Find the work done by the tension T and the normal reaction N. Find also the work done against the frictional force F and against gravity.

5. A body of mass 10 kg moving under the action of a constant force accelerates from $2\,\mathrm{m\,s^{-1}}$ to $5\,\mathrm{m\,s^{-1}}$. Find the work done by the force. Given that the magnitude of the force is 7 N, find the distance moved by the body.

6. A railway truck of mass 2 tonnes moving on a straight horizontal track is brought to rest from a speed of $3\,\mathrm{m\,s^{-1}}$ by a force of P N. Find the work done by this force. Given that the truck travels 20 m before coming to rest, find the value of P.

7. A bullet of mass 0·04 kg travelling at $300\,\mathrm{m\,s^{-1}}$ hits a fixed wooden block and penetrates a distance of 4 cm. Find the average resistance of the wood.

8. A car of mass 1000 kg travelling along a straight horizontal road accelerates uniformly from $15\,\mathrm{m\,s^{-1}}$ to $25\,\mathrm{m\,s^{-1}}$ in a distance of 320 m. If the resistance to motion is 145 N, find the driving force of the engine.

9. A box of mass 16 kg is pulled from rest a distance of 5 m across a smooth horizontal floor by a cable inclined at $60°$ to the horizontal. Find the work done by the tension in the cable given that its magnitude is 25 N.

10. A particle of mass 6 kg sliding across a rough horizontal plane comes to rest in a distance of 8 m. Given that its initial velocity was $10\,\mathrm{m\,s^{-1}}$, find the work done against friction. Find also the coefficient of friction between the particle and the plane.

11. Find the kinetic energy gained by a body of mass 2 kg falling freely from rest through a distance of 10 m. If a vertical force of P N brings the body to rest in a further distance of 8 m, find the value of P.

12. A particle of mass 5 kg slides a distance 9 m down a rough plane inclined at an angle α to the horizontal, where $\sin \alpha = \frac{3}{5}$. The coefficient of friction is $\frac{1}{2}$. Find the work done by gravity and the work done against friction. Find also the velocity attained by the particle.

23.2 Conservation of mechanical energy

The *energy* of a body can be regarded as its capacity for doing work. For instance, the kinetic energy of a body is its capacity to do work by virtue of its motion. As we saw in the previous section, a moving body does work against forces opposing motion as it loses kinetic energy. There are many other forms of energy such as heat, light, sound, chemical and electrical energy. However, our interest will be confined to forms of *mechanical energy*, i.e. energy by virtue of motion or position.

A body is said to have *potential energy* if its position is such that when released it begins to move and to gain kinetic energy. For instance, a body of mass m released at a height h above the ground will fall freely under gravity until it hits the ground with velocity v.

$$\text{Work done by gravity} = \text{gain in K.E.}$$
$$\therefore \quad mgh = \tfrac{1}{2}mv^2.$$

Thus, initially the body is said to have *gravitational potential energy mgh* by virtue of its height h above the ground.

In any given situation a *zero level* for gravitational potential energy must be decided, and then the potential energy of a body can be calculated relative to this level. Bodies above the zero level will have positive potential energy and bodies below it negative potential energy.

As stated earlier, if a particle is moving in a straight line under the action of a resultant force F then,

$$\text{work done by } F = \text{gain in K.E.}$$

If the only force doing work is gravity this result becomes,

$$\text{work done by gravity} = \text{gain in K.E.}$$

However, by definition, when a mass m falls from a height h the work done by gravity, mgh, is equal to the loss of potential energy.

Hence,   loss of P.E. = gain in K.E.
so that   K.E. + P.E. = constant.

This result is a simple form of the *principle of conservation of mechanical energy*.

> If a particle is moving such that no external force other than gravity is doing work, then the total mechanical energy of the particle remains constant.

Example 1   A particle of mass m kg is released from rest at a point A on a smooth plane inclined at an angle α to the horizontal, where $\sin \alpha = 3/5$. Find the velocity acquired by the particle after travelling a distance of 6 m down the plane to a point B.

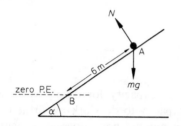

Let us assume that at B the particle has zero P.E. and velocity v m s^{-1}, then

at A: K.E. = 0, P.E. = $mg \times 6 \sin \alpha$
at B: K.E. = $\frac{1}{2}mv^2$, P.E. = 0.

Using the conservation of mechanical energy

$$\tfrac{1}{2}mv^2 + 0 = 0 + 6mg \sin \alpha$$

$$\therefore \quad v^2 = 2 \times 6 \times 9\cdot8 \times \frac{3}{5} = \frac{6^2 \times 7^2}{5^2}$$

$$\therefore \quad v = \frac{6 \times 7}{5} = 8\cdot4.$$

Hence the velocity acquired by the particle is $8\cdot4$ m s^{-1}.

[Note that although this problem could be solved using Newton's second law and one of the constant acceleration formulae, it is quicker to use an energy equation, since only velocity and distance moved are involved.]

It can be shown that the principle of conservation of mechanical energy also applies to the motion of a particle along a curved path, provided that no work is done by any external force other than gravity.

Example 2   A ring of mass m is threaded on a smooth circular wire fixed in a vertical plane. If the ring is projected from the lowest point A of the wire with velocity v and first comes to rest at a point B on the wire, express the vertical height h of B above A in terms of g and v.

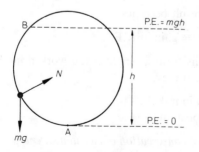

Since the wire is smooth the force exerted by the wire on the ring is always perpendicular to the direction of motion and therefore does no work. Hence mechanical energy is conserved throughout the motion, i.e.

$$\text{(K.E.} + \text{P.E.) at } B = \text{(K.E.} + \text{P.E.) at } A$$
$$0 + mgh = \tfrac{1}{2}mv^2 + 0.$$

Thus $h = v^2/2g.$

When using energy equations to solve problems it is helpful to distinguish between *conservative* and *non-conservative* forces. When work is done against a conservative force such as gravity, there is a gain in potential energy. However, when work is done against a non-conservative force, such as friction, energy is produced in various forms, including heat and sound, but there is no gain in potential energy. Thus the principle of conservation of mechanical energy can be applied to systems of particles provided that (i) no work is done by external forces other than gravity, (ii) no work is done against non-conservative internal forces such as friction.

Example 3   A particle *A* of mass 2 kg is connected to a particle *B* of mass 3 kg by means of a light inextensible string which passes over a smooth light pulley. The system is released from rest with both parts of the string taut and vertical. Find the speed attained by each particle when *A* has risen 2 m.

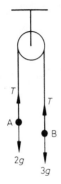

Let the initial P.E. of the system be zero and let the final speed of each particle be $v\,\mathrm{m\,s^{-1}}$.
Thus initially the system has zero mechanical energy.
Final P.E. of $A = 2g \times 2\,\mathrm{J} = 4g\,\mathrm{J}$.
Final P.E. of $B = 3g \times (-2)\,\mathrm{J} = -6g\,\mathrm{J}$.

Final K.E. of $A = \frac{1}{2} \times 2 \times v^2\,\mathrm{J} = v^2\,\mathrm{J}$.

Final K.E. of $B = \frac{1}{2} \times 3 \times v^2\,\mathrm{J} = \frac{3}{2}v^2\,\mathrm{J}$.

Thus the final mechanical energy of the system

$$= \left(v^2 + \frac{3}{2}v^2 + 4g - 6g\right)\mathrm{J} = \left(\frac{5}{2}v^2 - 2g\right)\mathrm{J}.$$

Using conservation of mechanical energy, $\dfrac{5}{2}v^2 - 2g = 0$.

$$\therefore \qquad v = \sqrt{\left(\frac{4g}{5}\right)} = \sqrt{\left(\frac{4}{5} \times \frac{49}{5}\right)} = \frac{2 \times 7}{5} = 2 \cdot 8.$$

Hence each particle attains a speed of $2 \cdot 8\,\mathrm{m\,s^{-1}}$.

When the total mechanical energy of a particle or system of particles is changed by the action of external forces an energy equation similar to the following may be used,

> initial (P.E. + K.E.) + work done = final (P.E. + K.E.).

If the mechanical energy of a system is reduced by work done against non-conservative forces a more convenient form may be,

---

initial (P.E. + K.E.) − work done against resistance = final (P.E. + K.E.).

Example 4   A small block of mass m slides down a rough plane inclined at an angle 30° to the horizontal. If the block acquires velocity v after travelling a distance x from rest, find an expression for the work done against friction.

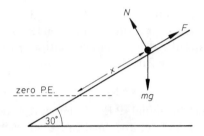

Let the final P.E. of the block be zero, then the initial P.E. is $mg \times x \sin 30°$, i.e. $\frac{1}{2}mgx$.
The initial K.E. is zero and the final K.E. is $\frac{1}{2}mv^2$.

Initial (P.E. + K.E.) − work done against friction = final (P.E. + K.E.).
Work done against friction
$\quad$ = initial (P.E. + K.E.) − final (P.E. + K.E.)
$\quad$ = $(\frac{1}{2}mgx + 0) - (0 + \frac{1}{2}mv^2)$
$\quad$ = $\frac{1}{2}m(gx - v^2)$.

Exercise 23.2

1.   A ball is thrown vertically downwards with a speed of $3\cdot5\,\mathrm{m\,s^{-1}}$. Find its speed when it has travelled a distance of 5 m.

2.   A particle is projected vertically upwards from a point A with a speed of $21\,\mathrm{m\,s^{-1}}$. Find its position relative to A when its speed is (a) $4\cdot2\,\mathrm{m\,s^{-1}}$, (b) $35\,\mathrm{m\,s^{-1}}$.

3.   A stone is thrown vertically upwards from a point A with velocity $v\,\mathrm{m\,s^{-1}}$. Find an expression for the greatest height above A reached by the stone.

4.   A small block is released from rest on a smooth plane inclined at 30° to the horizontal. Find the distance the particle has travelled down the plane when its speed is $6\cdot3\,\mathrm{m\,s^{-1}}$.

5.   A particle is projected with velocity $9\,\mathrm{m\,s^{-1}}$ up a line of greatest slope of a smooth plane inclined at an angle α to the horizontal, where $\sin\alpha = 4/7$. Find the speed of the particle when it has travelled 5 m up the plane.

6.   A tile slides from rest down a smooth roof inclined at an angle α to the horizontal, where $\sin\alpha = \frac{2}{3}$. It reaches the edge after travelling a distance of $4\cdot5\,\mathrm{m}$,

then falls to the ground 7 m below. Find the speed at which the tile is travelling as it hits the ground.

7.   A particle is attached to a fixed point by means of a light inextensible string of length 0·8 m. If when the particle is released from rest the string makes an angle of 60° with the downward vertical, find the speed of the particle as it passes through its lowest point.

8.   A force acts vertically upwards on a body of mass 4 kg. If the body rises vertically from rest through 5 m and acquires a speed of $3 \, \mathrm{m \, s^{-1}}$, find the work done by the force.

9.   Find the kinetic energy acquired by a body of mass 5 kg which falls freely from rest through a distance of 3 m. If the particle is then brought to rest by a vertical force of 70 N, find the further distance that it falls.

10.   A particle of mass m is connected to a particle of mass M by means of a light inextensible string which passes over a smooth light pulley. The system is released from rest with both parts of the string taut and vertical. Given that $M > m$, find an expression for the distance moved by each particle when the velocity attained is v.

11.   A particle of mass 3 kg standing on a smooth horizontal table is connected by means of a light inextensible string passing over a smooth pulley at the edge of the table to a particle of mass 2 kg hanging freely. Find the speed of the particles when they have travelled a distance of 25 cm from rest.

12.   A cyclist travelling initially at $2 \, \mathrm{m \, s^{-1}}$ free-wheels down a hill. At the bottom of the hill his speed is $10 \, \mathrm{m \, s^{-1}}$. If the total mass of the cyclist and his machine is 90 kg and the work done against resistance is 4500 J, find the difference in level between the top and bottom of the hill.

13.   In a fairground ride a car of mass 500 kg starts from rest at a point A, 30 m above the ground. After travelling 73 m along a track against a constant resistance of 600 N the car passes through a point B with a speed of $16 \, \mathrm{m \, s^{-1}}$. Find the height of B above the ground.

14.   A particle is moving at a speed of $7 \, \mathrm{m \, s^{-1}}$ when it begins to ascend a slope inclined at an angle α to the horizontal where $\sin \alpha = 5/13$. The coefficient of friction between the particle and the slope is 5/8. Find (a) the speed of the particle when it has travelled 1·95 m up the slope, (b) the total distance the particle travels up the slope before coming to rest.

15.   A block of mass 5 kg is projected with velocity $5·6 \, \mathrm{m \, s^{-1}}$ up a line of greatest slope of a rough plane inclined at an angle α to the horizontal where $\sin \alpha = 0·6$. If the block travels 2 m up the slope before coming to rest, find the work done against friction. Find also the coefficient of friction between the block and the plane.

23.3 Work done by a variable force

Consider a particle moving in a straight line under the action of a variable force **F** acting along the line. The work done by **F** in a small displacement δs is approximately equal to $F\delta s$. Thus an approximate value of the total work done in a displacement from $s = 0$ to $s = x$ is given by a sum of the form $\sum\limits_{s=0}^{s=x} F\delta s$.
Remembering that sums of this kind can be evaluated by integration, we write:

$$\text{work done} = \lim_{\delta s \to 0} \sum_{s=0}^{s=x} F ds = \int_0^x F ds.$$

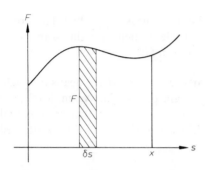

Geometrically this quantity is represented by the area under the force-displacement graph for the motion.

An important application of this theory is to the stretching of a spring or elastic string. By Hooke's law, when the extension in an elastic string is x, then the tension T is given by

$$T = \frac{\lambda x}{l}$$

where λ is the modulus of elasticity and l is the natural length of the string. Hence the work done against the tension when the extension is increased from x_1 to x_2 is

$$\int_{x_1}^{x_2} T dx = \int_{x_1}^{x_2} \frac{\lambda x}{l} dx = \left[\frac{\lambda x^2}{2l} \right]_{x_1}^{x_2} = \frac{\lambda}{2l}\left(x_2^2 - x_1^2 \right).$$

In particular, the work done when an elastic string is stretched from its natural length l to length $(l+x)$ is $\dfrac{\lambda x^2}{2l}$.

[A similar result can be obtained for the work done when a spring is compressed.]

Example 1   A particle of weight $10\,\text{N}$ is suspended from a fixed point by a light elastic string of natural length $1\cdot5\,\text{m}$. When the particle hangs in equilibrium the length of the string is $2\,\text{m}$. Find the modulus of elasticity of the string. Find also the work done in stretching the string from its natural length to a length of $2\,\text{m}$.

When the particle is in equilibrium the tension in the string is equal in magnitude to the weight of the particle.

Hence, using Hooke's law $T = \dfrac{\lambda x}{l}$, where $T = 10$, $l = 1\cdot5$ and $x = 0\cdot5$,

$$10 = \frac{\lambda \times 0\cdot5}{1\cdot5} \qquad \therefore \quad \lambda = 30$$

∴ the modulus of elasticity of the string is 30 N.
The work done in extending a string from its natural length by an amount x is $\dfrac{\lambda x^2}{2l}$

∴ the work done in extending this string by $0\cdot5$ m

$$= \frac{30 \times (0\cdot5)^2}{2 \times 1\cdot5} \, J = 2\cdot5 \, J.$$

Let us suppose that an elastic string is stretched and then released. As the tension acts to restore the string to its original length the work it does is equal to the work done against the tension in stretching the string. Thus, since no mechanical energy is lost, the tension in an elastic string is a conservative force. A spring or elastic string stretched from its natural length l to a length $(l+x)$ is said to have *elastic potential energy* of $\lambda x^2/2l$.

The principle of conservation of mechanical energy can be applied to systems involving springs and elastic strings, provided that no work is done by external forces (other than gravity).

Example 2   A particle of mass 2 kg is suspended from a fixed point by a spring of natural length 20 cm and modulus 10 N. Given that the extension in the spring when the particle is hanging in equilibrium is x m, find x. If the particle is released from rest with the spring vertical and unstretched, find the speed, $v\,\mathrm{m\,s^{-1}}$, of the particle when it passes through its equilibrium position.

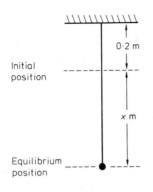

In the equilibrium position the tension in the spring is equal to the weight of the particle, i.e. $2g$ N.

Using Hooke's law, $T = \dfrac{\lambda x}{a}$

Substituting $T = 2g$, $\lambda = 10$, $a = 0\cdot2$

$$2g = \frac{10x}{0\cdot2}$$

∴ $x = 0\cdot04\,g = 0\cdot04 \times 9\cdot8 = 0\cdot392.$

Considering mechanical energy in the initial position,
let gravitational P.E. = 0,
energy stored in spring = 0,
kinetic energy = 0.

Considering mechanical energy in the equilibrium position,
gravitational P.E. $= 2g \times (-x)\text{J} = -0.08\,g^2\,\text{J}$

energy stored in spring $= \dfrac{\lambda x^2}{2a}\text{J} = \dfrac{10(0.04\,g)^2}{2+0.2}\text{J} = 0.04\,g^2\,\text{J}$

kinetic energy $= \frac{1}{2} \times 2 \times v^2\,\text{J} = v^2\,\text{J}$.

Using the principle of conservation of mechanical energy

$$-0.08\,g^2 + 0.04\,g^2 + v^2 = 0$$

$$\therefore \quad v^2 = 0.04\,g^2, \text{ i.e. } v = 0.2\,g = 1.96.$$

Hence the speed with which the particle passes through its equilibrium position
is $1.96\,\text{m s}^{-1}$.

[Note that in practice it is virtually impossible to construct a system in which
there is no loss of mechanical energy, since there are no truly frictionless pulleys,
smooth surfaces nor perfectly elastic strings. In this chapter we are dealing with
mathematical models of systems in which the work done against friction and
other non-conservative forces is negligible compared with the total energy
involved.]

Exercise 23.3

1.   Find the work done in stretching a spring of natural length 1 m and modulus
10 N to a length of (a) 1·2 m, (b) 2 m.

2.   Find the energy stored in an elastic string of natural length 4 m and modulus
2 N when its length is (a) 5 m, (b) 10 m.

3.   Find the work done in stretching a spring of natural length 1·5 m and
modulus 9 N from a length of 2 m to length 2·5 m.

4.   The work done in stretching an elastic string of natural length 2·5 m from 3 m
to 4 m in length is 6 J. Find the modulus of the string.

5.   When an elastic string of natural length 1·5 m is stretched to the length 1·8 m
the energy stored in it is 2·7 J. Find the energy stored in the string when its length
is 2 m.

6.   The work done in extending a spring from its natural length $6a$ to a length $7a$
is ka. Find the modulus of the spring in terms of k and the work done in
extending the length of the spring from $7a$ to $8a$.

7.   A particle of mass 8 kg is suspended from a fixed point by a spring of natural
length 0·5 m and modulus 140 N. If the particle is released from rest with the
spring vertical and unstretched, find the distance it falls before coming to rest
instantaneously.

8.   A particle of mass 5 kg on a smooth horizontal plane is attached to one end

of a string of natural length 1 m and modulus 20 N. The other end of the string is attached to a fixed point in the same plane. If the particle is released from rest with the string stretched to a length of 2 m, find its speed at the instant the string returns to its natural length.

9. A particle of mass m is attached by a light elastic string of natural length $4a$ and modulus $7mg$ to a fixed point. The particle is released from rest with the string taut and vertical, but unstretched. Find an expression for its speed when it has fallen a distance a.

10. A particle is suspended from a fixed point A by a light elastic string of natural length 2 m. When the particle hangs in equilibrium the length of the string is 2·5 m. Given that the particle is released from rest at A, find the distance it has fallen when it first comes to rest.

11. A particle of mass 7 kg is attached to one end of an elastic string of modulus $9g$ N and natural length 1·8 m. The other end of the string is fastened to a fixed point A. The point B is the position of the particle when it hangs in equilibrium. If the particle is released from rest at A, find its speed when it passes through B.

12. A particle is attached by a light elastic string of natural length 60 cm to a fixed point O. When the particle hangs in equilibrium its distance below O is 80 cm. If the particle is released from rest at a point 1 m vertically below O, find the distance of the particle from O when it next comes to rest. Find also the speed of the particle when it passes through its equilibrium position.

13. The ends of an elastic string of modulus 30 N and natural length 2 m are attached to fixed points A and B 2 m apart on the same horizontal level. A particle P is attached to the mid-point of the string then released at the mid-point of AB. If the particle falls 0·75 m before coming to rest instantaneously, find its weight.

14. A particle P of mass 2 kg is suspended from a point O by a light elastic string of length 1 m and modulus 98 N. Find the length of the string when P hangs in equilibrium. The particle is now pulled down and released from rest 1·6 m below O. Find (a) the speed of P as the string goes slack, (b) the depth below O at which P comes instantaneously to rest.

23.4 Power

Power is the rate at which a force does work. The SI unit of power is the watt (W). One watt is a rate of working of one joule per second, i.e. $1\,W = 1\,J\,s^{-1}$.
[For those who are familiar with the watt only as a unit of electrical power, it may be helpful to remember that a family saloon car has a maximum power output of approximately 40–50 kW. A sprinter may generate kinetic energy at an

average rate of about 1 kW for short periods, but most people produce mechanical energy at much lower rates.]

Example 1   Find the average power developed by a crane which lifts a load of 8 kN through a vertical distance of 15 m in 6 s.

Work done by the crane in 6 s is 8×15 kJ, i.e. 120 kJ
$\therefore$ average work done in 1 s is $120 \div 6$ kJ, i.e. 20 kJ.
Hence the average power of the crane is 20 kW.

Example 2   A pump raises water at a rate of 750 litres per minute through a vertical distance of 4 m and projects it into a reservoir at a speed of $6 \, \mathrm{m \, s^{-1}}$. Find the power developed by the pump.

Since the mass of 1 litre of water is 1 kg, the mass of water raised per second
$$= \frac{750}{60} \mathrm{kg} = \frac{25}{2} \mathrm{kg}.$$
Thus the P.E. given to the water each second

$$= \frac{25}{2} g \times 4 \mathrm{J} = 25 \times 9 \cdot 8 \times 2 \mathrm{J} = 490 \mathrm{J}.$$

The K.E. given to the water each second

$$= \frac{1}{2} \times \frac{25}{2} \times 6^2 \mathrm{J} = 25 \times 9 \mathrm{J} = 225 \mathrm{J}.$$

Hence the total power developed by the pump is

$(490 + 225)$ W, i.e. 715 W.

If the engine of a vehicle, such as a car or a train, produces a tractive force F newtons, then the power of the engine is the rate at which this force does work. When the vehicle is travelling at constant speed $v \, \mathrm{m \, s^{-1}}$, since the distance travelled each second is v m, the work done each second is Fv J. Hence the power of the engine is Fv watts.

A similar result is obtained for non-uniform motion by considering the work done when the vehicle moves a small distance δs in time δt. The rate of working in this interval is approximately $\frac{F \delta s}{\delta t}$, i.e. $F \frac{\delta s}{\delta t}$. However, as $\delta t \to 0, \frac{\delta s}{\delta t} \to \frac{ds}{dt}$. Thus the rate of working at the instant when the vehicle is moving with velocity $v \, \mathrm{m \, s^{-1}}$ must be Fv watts. Hence the power P watts, the tractive force F newtons and the velocity $v \, \mathrm{m \, s^{-1}}$ are connected by the formula

$$\boxed{P = Fv.}$$

Example 3   A car of mass 1200 kg is moving along a horizontal road against resistance to motion of 500 N. Find the power developed by the engine when the car is travelling at $15\,\mathrm{m\,s^{-1}}$ and accelerating at $0.25\,\mathrm{m\,s^{-2}}$. Find also the maximum speed reached by the car if the engine continues to work at the same rate.

If the pull of the engine is F newtons, then by Newton's second law,

$$F - 500 = 1200 \times 0.25$$
$$\therefore \qquad F = 300 + 500 = 800.$$

Hence the power of the engine

$$= 800 \times 15\ \mathrm{W} = 12\,000\ \mathrm{W} = 12\,\mathrm{kW}.$$

Let the maximum speed of the car be $v\,\mathrm{m\,s^{-1}}$. When maximum speed is reached the pull of the engine is equal in magnitude to the resistance to motion

$$\therefore \qquad 12\,000 = 500\,v, \text{ i.e. } v = 24.$$

Hence the maximum speed when the engine is working at the rate of $12\,\mathrm{kW}$ is $24\,\mathrm{m\,s^{-1}}$.

Example 4 A train of mass 150 tonnes is ascending an incline of 1 in 70 at a constant speed of $25\,\mathrm{m\,s^{-1}}$. If the power being exerted by the engine is $750\,\mathrm{kW}$, find the resistance to motion.

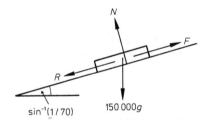

Let $F\,\mathrm{N}$ be the tractive force of the engine and $R\,\mathrm{N}$ the resistance to motion.
Using the formula $P = Fv$,
$$750\,000 = F \times 25$$
$$\therefore \quad F = 30\,000.$$

Applying Newton's second law up the incline

$$F - R - 150\,000\,g \times \frac{1}{70} = 0$$

$$\therefore \quad R = 30\,000 - \frac{150\,000 \times 9.8}{70} = 30\,000 - 21\,000 = 9000.$$

Hence the resistance to motion is $9\,\mathrm{kN}$.

Example 5 When a cyclist rides along a horizontal road at $v\,\mathrm{m\,s^{-1}}$ the resistance to motion is $(16 + \frac{1}{4}v^2)\,\mathrm{N}$. Find the rate at which the cyclist must work to maintain a steady speed of $6\,\mathrm{m\,s^{-1}}$.

When travelling at a steady speed the force exerted by the cyclist is equal to the resistance to motion,
$$\therefore \quad \text{his rate of work} = (16 + \tfrac{1}{4}v^2)v\ \mathrm{W}$$
$$= (16 + \tfrac{1}{4} \times 6^2)6\ \mathrm{W}$$
$$= (16 + 9)6\ \mathrm{W}$$

Hence to maintain a steady speed of $6\,\mathrm{m\,s^{-1}}$ the cyclist must work at a rate of $150\,\mathrm{W}$.

Exercise 23.4

1. A man of mass 80 kg climbs a staircase of height 5 m in 10 s. Find his average rate of work against gravity.

2. If the effective power developed by a crane is 15 kW, find the time taken to lift a load of weight 50 kN through a vertical distance of 12 m.

3. Find the minimum power needed to pull an object of weight 500 N a distance of 8 m up a smooth slope inclined at 30° to the horizontal in 5 seconds.

4. A lift of mass 1500 kg is being raised by an engine working at the rate of 24·5 kW. If the lift rises 50 m in 40 s, find the load it is carrying.

5. Water from a river, flowing at 4 m s^{-1}, completely fills a pipe of cross-sectional area 3 m^2. If, when the water is used to drive a generator, 30% of its kinetic energy is converted into electrical energy, find the power supplied by the generator.

6. A pump raises 40 kg of water a second through a vertical distance of 10 m and delivers it in a jet with speed 12 m s^{-1}. Find the effective power developed by the pump.

7. A pump raises water at the rate of 250 litres per minute through a vertical distance of 8 m and projects it into a lake at 2 m s^{-1}. Find the power developed by the pump.

8. If the engine of a train travelling at a steady speed of 25 m s^{-1} is working at a rate of 1800 kW, find the magnitude of the resistance to motion.

9. A motor-cyclist is riding along a horizontal road at a constant speed of 15 m s^{-1} against a resistance of 80 N. Find the power at which the motor-cycle engine is working. Given that the total mass of the rider and his machine is 200 kg, find the initial acceleration when the power is increased to 1·5 kW.

10. A car of mass 900 kg is travelling at 20 m s^{-1} along a horizontal road against a constant resistance of 600 N. Find the power developed by the engine if the car is (a) moving with constant speed, (b) accelerating at 0·15 m s^{-2}, (c) decelerating at 0·5 m s^{-2}.

11. A cable car of mass 1500 kg is pulled at a constant speed of 10 m s^{-1} up a smooth slope inclined at an angle α to the horizontal where $\sin \alpha = 1/7$. Find the power exerted on the car.

12. A car of mass 1000 kg is travelling down an incline of 1 in 28 against a constant resistance of 750 N. Find the maximum speed that can be reached by the car when the engine is working at the rate of 12 kW.

13. A lorry of mass 2400 kg is moving at a steady speed of 60 km/h against a constant frictional resistance of 2 kN. Find the power developed by the engine if the lorry is travelling (a) along a horizontal road, (b) up a hill of inclination $\sin^{-1}$ (1/21), (c) down the same hill.

14. A train of mass 50 tonnes is ascending an incline of 1 in 60 with the engine working at the rate of 200 kW. When the train is travelling at $10 \, \text{m s}^{-1}$, its acceleration is $0.1 \, \text{m s}^{-2}$. Find the frictional resistance to motion. Assuming that the resistance and the power of the engine remain constant, find the maximum speed the train can attain on this slope.

15. The constant non-gravitational resistance to the motion of a car of mass 1500 kg is 750 N. The engine of the car works at a constant rate of 20 kW. Find, in km/h, the maximum speed (a) on the level, and (b) directly up a road inclined at an angle $\arcsin (1/12)$ to the horizontal. Find also, in m/s^2, the acceleration of the car when it is travelling at 72 km/h on a level road. [Take g as $10 \, \text{m/s}^2$.]   (L)

16. The frictional resistance to the motion of a car of mass 1000 kg is $30v$ N where $v \, \text{m s}^{-1}$ is the speed of the car. The car ascends a hill of inclination $\arcsin \left(\dfrac{1}{10} \right)$, the power exerted by the engine being 12·8 kW. If the car is moving at a steady speed, prove (do not merely verify) that this steady speed is $10 \, \text{m s}^{-1}$. On reaching the top of the hill, find the immediate acceleration of the car, assuming the road becomes level.   (O&C)

17. The resistance to the motion of a car is $(160 + cV^2)$ newtons, where c is a constant and V is the speed of the car in m/s. The mass of the car is 900 kg and, when the engine is working at the rate of 19·2 kW, the maximum speed on a level road is 20 m/s. Find the value of c. The car ascends a slope inclined to the horizontal at an angle θ, where $\sin \theta = 1/30$. If the engine continues to work at the same rate, find the acceleration of the car when its speed is 10 m/s. [Take g as $10 \, \text{m/s}^2$.]   (L)

18. A car of mass 1·2 tonnes is travelling along a straight, horizontal road at a constant speed of 120 km/h against a resistance of 600 newtons. Calculate, in kW, the effective power being exerted. Given that the resistance is proportional to the square of the velocity, calculate also (i) the power required to go down a hill of 1 in 30 (along the slope) at a steady speed of 120 km/h, (ii) the acceleration of the car up this hill with the engine working at 20 kW at the instant when the speed is 80 km/h.   (AEB 1978)

Exercise 23.5 (miscellaneous)

1.   A particle is projected at a speed of $3.5 \, \text{m s}^{-1}$ across a rough horizontal plane. Given that it comes to rest after travelling a distance of 5 m, find the coefficient of friction.

2.   A particle released from rest on a smooth plane inclined at an angle α to the horizontal reaches a speed of $6\,\text{m s}^{-1}$ after travelling $3\,\text{m}$ down the plane. Find $\sin \alpha$.

3.   A body of mass $2\,\text{kg}$ is pulled from rest across a rough horizontal plane by a string inclined at an angle α to the horizontal where $\sin \alpha = 3/5$. The coefficient of friction is $1/3$. After travelling $7\cdot8\,\text{m}$ the body attains a speed of $5\cdot2\,\text{m s}^{-1}$. Assuming the tension in the string is constant, find its magnitude.

4.   A particle is attached to a fixed point by means of a light inextensible string of length $1\,\text{m}$. If the particle is projected horizontally through its lowest point with a speed of $2\cdot8\,\text{m s}^{-1}$, find the angle the string makes with the vertical when the particle first comes to rest.

5.   A particle of mass $5\,\text{kg}$ is projected with a speed of $2\cdot8\,\text{m s}^{-1}$ down a rough plane inclined at an angle α to the horizontal, where $\sin \alpha = 5/13$. If the particle comes to rest after travelling $2\cdot6\,\text{m}$, find the work done against friction. Find also the coefficient of friction between the particle and the plane.

6.   The work done in stretching an elastic string from $3\,\text{m}$ to $5\,\text{m}$ in length is $4\,\text{J}$. If the modulus of the string is $3\,\text{N}$, find its natural length.

7.   A particle of mass $2\,\text{kg}$ on a smooth plane inclined at $30°$ to the horizontal is attached by means of a light inextensible string passing over a smooth pulley at the top edge of the plane to a particle of mass $3\,\text{kg}$ which hangs freely. If the system is released from rest with both parts of the string taut, find the speed acquired by the particles when both have moved a distance of $1\,\text{m}$.

8.   A curved chute with a smooth surface is fixed so that a point A on the chute is $2\cdot5\,\text{m}$ above a horizontal plane. A particle is released from A and is travelling horizontally when it leaves the chute at a point B $1\cdot6\,\text{m}$ above the horizontal plane. Find the speed of the particle as it hits this plane and the horizontal distance it has travelled since leaving the chute.

9.   A particle P of mass m is attached to two elastic strings each of length $0\cdot6\,\text{m}$ and modulus kmg. The other ends of the strings are fixed to two points A and B $1\cdot2\,\text{m}$ apart on the same horizontal level. P is released from rest at M, the midpoint of AB and falls vertically $0\cdot8\,\text{m}$ before coming instantaneously to rest. Find k and also the speed of P as it passes through C, where $MC = 0\cdot45\,\text{m}$.

10. A light elastic string obeys Hooke's law, so that when it is stretched by an amount x the tension in it is given by $T = kx$ for some constant k. Prove that the elastic energy stored in the string when its extension is x is given by $E = \frac{1}{2}kx^2$.
   One end of such a string is attached to a fixed point A, and the other end is attached to a particle of mass m. The unstretched length of the string is a, and the particle hangs freely in equilibrium at a point B which is at a distance $\frac{5}{4}a$ below

A. Prove that $k = 4mg/a$. The particle is held at A, and released from rest. Find its velocity as it passes B, and the total distance it falls before coming instantaneously to rest. (C)

11. A bead of mass m slides on a smooth vertical rod and is attached to a light inextensible string which passes over a smooth peg at a distance a from the rod and has a particle of mass $2m$ fastened to its other end. The bead is held at the level of the peg and then released. Prove that it will descend $\dfrac{4}{3}a$ before coming to rest and find the tension in the string at this instant. (O)

12. A, B, C, D are fixed points situated at the mid-points of the sides of a square of side $2a$ with AC vertical (C above A) and BD horizontal. A weight W is held in the plane of the square by four springs, of modulus λ and natural length a, whose other ends are attached to A, B, C and D respectively. Show that in the equilibrium position the springs attached to B and D are inclined to the upward drawn vertical at an angle α where $2\cot\alpha - \cos\alpha = W/2\lambda$. Show that the amount of energy needed to raise the weight to the centre of the square must be at least $2a\lambda(\cot^2\alpha + \sin\alpha - 1)$. (W)

13. A small bead of mass m is threaded on a smooth wire which is shaped in the form of the curve $by = x^2 + b^2$ ($x \leqslant 0$). The positive y-axis is vertically upwards and the x-axis lies in a horizontal plane. The bead is released from rest at the point where $x = -2b$ and slides down the wire to its end at the point $(0, b)$, after which the bead moves freely under gravity until it strikes the horizontal plane at a point on the x-axis. Find expressions for (i) the speed of the bead on leaving the wire, (ii) the speed of the bead on striking the x-axis, (iii) the angle that the direction of motion of the bead makes with the x-axis when it meets the x-axis, (iv) the x-coordinate of the point where the bead meets the x-axis.

Suppose instead that it is required that the bead should hit a small target which is to be placed on the x-axis at the point $(c, 0)$. Find the x-coordinate of the point on the curve at which the bead has to be released from rest if it is to achieve this. (JMB)

14. An elastic string is such that when stretched from its natural length l to a length $l + x$ the tension in the string becomes kx, where k is a positive constant. Show that the work done when the length of the string is increased from $l + x_1$ to $l + x_2$ is $\frac{1}{2}k(x_2^2 - x_1^2)$.

A particle of mass m is attached to one end of the string, the other end of which is attached to a fixed point O in a smooth inclined plane of inclination α. The particle rests in equilibrium on the plane and the extension of the string is $\frac{1}{2}l$. Show That $kl = 2mg\sin\alpha$. The particle is then pulled a further distance $b - \frac{1}{2}l$ down the line of greatest slope, and released. By considerations of energy or otherwise, determine the distance up the plane, measured from the point of release, that the particle will travel, distinguishing the cases (i) $b \leqslant l$, (ii) $l < b \leqslant 2l$. Describe briefly in words what happens if $b > 2l$. (JMB)

15. A train of mass 3×10^5 kg travels along a straight level track. The resistance to motion is 1.5×10^4 N. Find the tractive force required to produce an acceleration of $0.1 \, \text{m s}^{-2}$, and the power in kW which is then developed by the engine when the speed of the train is $10 \, \text{m s}^{-1}$. Find also the maximum speed attainable on the same track when the engine is working at a rate of $360 \, \text{kW}$. (JMB)

16. A car of mass $1000 \, \text{kg}$ is travelling along a level road at a steady speed of $45 \, \text{km/h}$ with its engine working at $22\frac{1}{2} \, \text{kW}$. Calculate in newtons the total resistance (assumed constant) due to friction, etc. The engine is disconnected and simultaneously the brakes are applied, bringing the car to rest in 30 metres. Find the force, assumed constant, exerted by the brakes. (O & C)

17. A cable car in the Alps gains height at an average rate of $900 \, \text{m}$ in $2\frac{1}{2}$ minutes. Given that the cable car has mass $1300 \, \text{kg}$ and carries 30 passengers of average mass $90 \, \text{kg}$, determine the average power required if the efficiency of the machinery is 75%. [Take g to be $10 \, \text{m s}^{-2}$.] (C)

18. A pump raises water through a vertical height of $15 \, \text{m}$ and delivers it at a speed of $10 \, \text{m s}^{-1}$ through a circular pipe of internal diameter $12 \, \text{cm}$. Find the mass of water raised per second and the effective power of the pump, giving your answers to 3 significant figures.

19. A motor lorry weighing 5 metric tonnes can develop up to $15\,000$ watts. It is moving up a hill with slope 1 in 20 $\left(\theta = \sin^{-1} \dfrac{1}{20} \right)$ against a frictional force of $300 \, \text{N}$. This frictional force is independent of the speed. (i) Find the maximum steady speed in kilometres per hour at which the lorry can travel up this slope. (ii) Find the acceleration capable of being developed when the lorry travels up this slope at $10 \, \text{km h}^{-1}$. The lorry travels on a straight horizontal road at a speed of v kilometres per hour against a braking force of $10v$ newtons and other resistances of $300 \, \text{N}$. What is the maximum value of v? (W)

20. A locomotive of mass $15\,000 \, \text{kg}$, working at the rate of $220 \, \text{kW}$, pulls a train of mass $35\,000 \, \text{kg}$ up a straight track which rises $1 \, \text{m}$ vertically for every $50 \, \text{m}$ travelled along the track. When the speed is $10 \, \text{m/s}$ the acceleration is $0.23 \, \text{m/s}^2$. Find the frictional resistance at this speed. Given that the resistance is proportional to the speed of the train, find, in m/s, the greatest speed of the train up the slope if the rate of working is unchanged. [Take $g = 10 \, \text{m/s}^2$.] (L)

21. The engine of a car, of mass M kg, works at a constant rate of H kW. The non-gravitational resistance to the motion of the car is constant. The maximum speed on level ground is V m/s. Find, in terms of M, V, H, α and g, expressions for the accelerations of the car when it is travelling at speed $\frac{1}{2}V$ m/s (a) directly up a road of inclination α, (b) directly down this road. Given that the acceleration in case (b) is twice that in case (a), find $\sin \alpha$ in terms of M, V, H and g. Find also, in terms of V alone, the greatest steady speed which the car can maintain when travelling directly up the road. (L)

22. An engine is travelling at a steady speed of 10 m/s on a level track against a constant resistance of 40 000 N. Calculate, in kW, the power output of the engine. The engine is then coupled to a carriage by a towbar. The constant resistance to the motion of the carriage is 20 000 N. If the power output of the engine is now 900 kW, calculate, in m/s, the maximum speed of the train on a level track. State the tension, in newtons, in the towbar. The train then ascends an incline of inclination arcsin (1/50) to the horizontal with the same power output, 900 kW, against the same constant frictional forces. If the total mass of the train is 340 tonne, show that the acceleration of the train when it is travelling at 5 m/s is 13/85 m/s². [Take g as 10 m/s².]   (L)

23. A car of mass 2.5×10^3 kg travels at a constant speed of 20 m s^{-1} up a hill of inclination of sin^{-1} (1/20) to the horizontal, with the engine working at a power of 8×10^4 W. Given that the resistance is proportional to the cube of the speed show that the resistance at a speed of 24 m s^{-1} is 4.8×10^3 N, correct to 2 significant figures. Given that the engine is working at the same power, find the acceleration at an instant when the car is moving down a hill of inclination sin^{-1} (1/10) to the horizontal at a speed of 24 m s^{-1}.   (C)

24. A motor-cyclist, whose total mass with his machine is 200 kg, is ascending a hill of inclination θ to the horizontal, where sin $\theta = 1/8$. If the resistance forces total 150 N and the engine is working at 9 kW, calculate (i) the acceleration of the motor-cycle when the speed is 20 m/s, (ii) the maximum speed that can be attained by the motor-cycle up this hill. The motor-cyclist descends the same hill with a pillion rider whose mass is 80 kg. The resistive forces now total 170 N and the engine is switched off. Find the distance covered and the time taken as the speed of the motor-cycle increases from 10 m/s to 20 m/s. [Take the acceleration due to gravity to be 10 m/s².]   (AEB 1977)

25. A car has a maximum speed of 108 km/h when moving along a horizontal road with the engine working at 36 kW. Calculate, in newtons, the total resistance to the motion of the car. The car, which is of mass 800 kg, can move down a road inclined at an angle α to the horizontal at a maximum speed of 108 km/h with the engine working at 30 kW against the same total resistance. Calculate sin α. Given that the total resistance varies as the square of the speed, find, in kW, the rate at which the engine is working at the instant when the car is moving along a horizontal road with an acceleration of 0·5 m/s² at a speed of 54 km/h. [Take the acceleration due to gravity to be 10 m/s².]   (AEB 1978)

26. Two points A and B are at the same level, and they are connected by a road consisting of two straight sections AC and CB, each of uniform gradient. The point C is at a height h vertically above the line AB, and the inclinations to the horizontal of AC and CB are α and β respectively. A car of mass m starts from rest at A, accelerates uniformly to reach a speed $2U$ at C, and then moves with constant retardation, attaining a speed U at B. Show that the time taken to move from A to B is $h(\text{cosec}\,\alpha + \tfrac{2}{3}\text{cosec}\,\beta)/U$. The resistance to the motion of the car is a constant force R. Find the force exerted by the engine during the climb from A

to C, the maximum power developed by the engine and the total work done by the engine. Show that the described motion from C to B can be achieved by using the brakes, with the engine switched off, only if $R \leqslant \dfrac{m \sin \beta}{2h}(2gh + 3U^2)$.   (JMB)

24 Uniform motion in a circle

24.1 Angular velocity

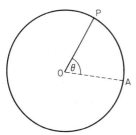

Consider a particle P moving round the circumference of a circle with centre O. If A is any fixed point on the circle and $\angle POA = \theta$, then the *angular velocity* of P is defined as the rate of change of θ and often denoted by ω.

Thus $\omega = \dot\theta$, i.e. $\omega = d\theta/dt$.

The value of ω may be positive or negative according to the direction of motion. Anti-clockwise rotations are usually taken to be positive. The *angular speed* of P is the magnitude of its angular velocity. The *angular acceleration* of P is the rate of change of the angular velocity, denoted by $\dot\omega$ or $\ddot\theta$.

The SI unit of angular velocity is one radian per second, i.e. 1 rad s^{-1}. Angular velocity can also be measured in revolutions per minute (rev min^{-1}).

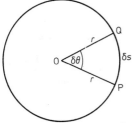

Suppose now that a particle is moving round the circumference of a circle radius r such that when the angular velocity of the particle is ω, its speed is v. If the particle moves from P to Q in time δt, let arc $PQ = \delta s$ and $\angle POQ = \delta\theta$.

Since $\delta s = r\,\delta\theta$ we have $\dfrac{\delta s}{\delta t} = r\dfrac{\delta\theta}{\delta t}$.

As $\delta t \to 0$, $\dfrac{\delta s}{\delta t} \to \dfrac{ds}{dt}$ and $\dfrac{\delta\theta}{\delta t} \to \dfrac{d\theta}{dt}$.

481

Hence
$$\frac{ds}{dt} = r\frac{d\theta}{dt}.$$

∴
$$\boxed{v = r\omega \quad \text{and} \quad \omega = \frac{v}{r}.}$$

Example 1   The tip of the second hand of a large clock is 60 cm from the centre of the clock face. Find the angular speed of the second hand and the speed at which the tip is travelling.

The second hand makes 1 revolution per minute, i.e. it moves through 2π rad in 60 s

∴   its angular speed $= \dfrac{2\pi}{60}\,\text{rad s}^{-1} = \dfrac{\pi}{30}\,\text{rad s}^{-1}$.

Hence the speed of the tip of the hand

$$= \frac{60}{100} \times \frac{\pi}{30}\,\text{m s}^{-1} = \frac{\pi}{50}\,\text{m s}^{-1} \approx 0.063\,\text{m s}^{-1}.$$

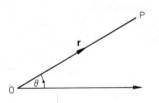

The definition of angular velocity can be extended to the motion in a plane of a point P with position vector $\mathbf{r}$. If the angle between $\mathbf{r}$ and some fixed direction is θ, then the angular velocity of P about O (i.e. the angular velocity of **OP**) is $\dot{\theta}$.

Example 2   At time t, the position vector of a point P with respect to the origin O is $2t\mathbf{i} + t^2\mathbf{j}$. Find the angular velocity of **OP** at time t.

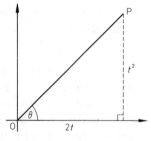

Let θ be the angle between **OP** and the direction of $\mathbf{i}$, then at time t,

$$\tan\theta = \frac{t^2}{2t} = \frac{t}{2}.$$

Differentiating with respect to t

$$\sec^2\theta\,\frac{d\theta}{dt} = \frac{1}{2}.$$

∴
$$(1+\tan^2\theta)\frac{d\theta}{dt} = \frac{1}{2}$$

$$\left(1 + \frac{t^2}{4}\right)\frac{d\theta}{dt} = \frac{1}{2}$$

i.e.
$$(4+t^2)\frac{d\theta}{dt} = 2.$$

Hence the angular velocity at time t is given by $\dfrac{d\theta}{dt} = \dfrac{2}{4+t^2}$.

Example 3 A particle P is moving anti-clockwise with constant angular speed ω round a circle with centre O and radius 5. Given that initially the position vector of P is $5\mathbf{i}$, find its position vector at time t. Deduce vector expressions for the velocity and acceleration of P at time t.

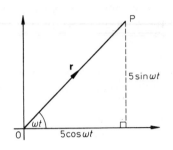

Since the angular speed is ω, at time t **OP** makes an angle ωt with the direction of **i**. Hence at time t the position vector of P is

$$\mathbf{r} = 5(\cos \omega t\mathbf{i} + \sin \omega t\mathbf{j}).$$

Differentiating with respect to t, the velocity of P is given by

$$\mathbf{v} = \dot{\mathbf{r}} = 5(-\omega \sin \omega t\mathbf{i} + \omega \cos \omega t\mathbf{j})$$

i.e. $$\mathbf{v} = 5\omega(-\sin \omega t\mathbf{i} + \cos \omega t\mathbf{j}).$$

Similarly the acceleration of P is given by

$$\mathbf{a} = \ddot{\mathbf{r}} = 5\omega(-\omega \cos \omega t\mathbf{i} - \omega \sin \omega t\mathbf{j})$$

i.e. $$\mathbf{a} = -5\omega^2(\cos \omega t\mathbf{i} + \sin \omega t\mathbf{j})$$

[We note that $v = 5\omega\sqrt{(\sin^2 \omega t + \cos^2 \omega t)} = 5\omega$, which is consistent with the result $v = r\omega$ obtained earlier in the section. We also see that $\mathbf{a} = -\omega^2\mathbf{r}$. Thus the acceleration of the particle is directed towards O and has magnitude $5\omega^2$.]

Exercise 24.1

1. A record of diameter 18 cm is rotating at 45 rev min^{-1}. Find its angular velocity in rad s^{-1} and the speed of a point on the rim of the record.

2. A point is travelling round a circle of radius 5 m at a speed of 20 m s^{-1}. Find the angular speed of the point about the centre of the circle.

3. Find the angular speed with which the earth is rotating about its axis. Hence, taking the radius of the earth to be 6400 km, find the speed of a point on the equator in m s^{-1}.

4. Find an approximate value for the angular speed at which the earth is moving about the sun. Assuming that the earth moves round the sun in an approximately circular orbit of radius 150 million kilometres, find the speed of the earth relative to the sun.

5. Given that the tip of the minute hand of a clock is moving at a speed of $0.001 \, \text{m s}^{-1}$, find the length of the hand to the nearest cm.

6. Find the angular velocity of a point P about the origin O at time $t > 0$, given that the position vector of P is

(a) $\mathbf{i} + t\mathbf{j}$,   (b) $2t\mathbf{i} - 3\mathbf{j}$,   (c) $\dfrac{1}{t}\mathbf{i} + t\mathbf{j}$.

7. A particle P moves anti-clockwise with constant angular speed ω round a circle with centre O and radius r. Write down the position vector of P at time t given that when $t = 0$ its position vector is (a) $r\mathbf{i}$, (b) $r\mathbf{j}$.

8. Use the results of question 7 to find the angular velocity of a particle P about the origin O and the speed at which the particle is travelling given that at time t its position vector is
(a) $3(\cos 2t\mathbf{i} + \sin 2t\mathbf{j})$,   (b) $5(-\sin t\mathbf{i} + \cos t\mathbf{j})$,
(c) $\cos 2\pi t\mathbf{i} - \sin 2\pi t\mathbf{j}$,   (d) $a(\sin kt\mathbf{i} - \cos kt\mathbf{j})$.

9. The points S and T are moving anti-clockwise round a circle with centre O (the origin) and radius a, with the same constant speed $a\omega$. When T is passing through the point with position vector $a\mathbf{j}$, S is passing through the point with position vector $a\mathbf{i}$. Find, at time t later, the position vectors of S and T. Find also, in vector form, the velocity of T relative to S.   (L)

24.2 Acceleration in circular motion

Consider a particle moving with angular velocity ω round the circumference of a circle centre O, radius r. Suppose that in time δt it moves from P to Q, where $\angle POQ = \delta\theta$. Let the speed of the particle be v at P and $v + \delta v$ at Q.

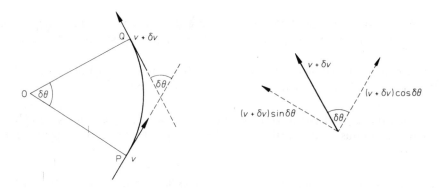

From the diagrams we see that the velocity of the particle at Q has components $(v + \delta v)\sin \delta\theta$ in the direction of $\vec{PO}$ and $(v + \delta v)\cos \delta\theta$ perpendicular to $\vec{PO}$.

Thus in the direction of $\vec{PO}$, the change in velocity in time δt

$$= (v + \delta v) \sin \delta\theta$$

$\therefore$ average acceleration $= \dfrac{(v + \delta v) \sin \delta\theta}{\delta t} = (v + \delta v) \dfrac{\sin \delta\theta}{\delta\theta} \cdot \dfrac{\delta\theta}{\delta t}.$

As $\delta t \to 0$, $\delta v \to 0$, $\dfrac{\sin \delta\theta}{\delta\theta} \to 1$ and $\dfrac{\delta\theta}{\delta t} \to \dfrac{d\theta}{dt}$

$\therefore$ acceleration along $\vec{PO} = v \dfrac{d\theta}{dt} = v\omega = \dfrac{v^2}{r}.$

Similarly, perpendicular to $\vec{PO}$, the change in velocity in time δt

$$= (v + \delta v) \cos \delta\theta - v$$

$\therefore$ average acceleration $= \dfrac{(v + \delta v) \cos \delta\theta - v}{\delta t}$

$$= \dfrac{v(\cos \delta\theta - 1)}{\delta t} + \dfrac{\delta v}{\delta t} \cos \delta\theta$$

$$= -\dfrac{v \cdot 2 \sin^2 \frac{1}{2}\delta\theta}{\delta t} + \dfrac{\delta v}{\delta t} \cos \delta\theta$$

$$= -v \dfrac{\delta\theta}{\delta t} \left(\dfrac{\sin \frac{1}{2}\delta\theta}{\frac{1}{2}\delta\theta} \right) \sin \tfrac{1}{2}\delta\theta + \dfrac{\delta v}{\delta t} \cos \delta\theta.$$

As $\delta t \to 0$, $\dfrac{\sin \frac{1}{2}\delta\theta}{\frac{1}{2}\delta\theta} \to 1$, $\dfrac{\delta\theta}{\delta t} \to \dfrac{d\theta}{dt}$, $\dfrac{\delta v}{\delta t} \to \dfrac{dv}{dt}$

but $\sin \tfrac{1}{2}\delta\theta \to 0$ and $\cos \delta\theta \to 1$

$\therefore$ acceleration perpendicular to $\vec{PO} = \dfrac{dv}{dt}.$

Hence the acceleration of a particle moving in a circle with variable speed v has, at any instant, components v^2/r towards the centre of the circle and dv/dt in the direction of motion.

If v is constant, then $dv/dt = 0$. Thus a particle moving in a circle with constant speed v has acceleration v^2/r towards the centre of the circle. In terms of the angular velocity ω the acceleration is $r\omega^2$.

[This last result can also be obtained by the vector methods used in Example 3 of the previous section.]

Using Newton's second law, we find that the resultant force acting on a particle of mass m, which moves in a circle radius r with constant speed v, is of magnitude mv^2/r and is directed towards the centre of the circle.

Example 1   One end of a light inextensible string of length 0·5 m is attached to a fixed point O on a smooth horizontal table and a particle of mass 2 kg is attached to its other end. If the particle moves in a horizontal circle, centre O, with a speed of 5 m s^{-1}, find the tension in the string and the reaction of the table on the particle.

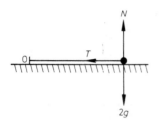

The acceleration of the particle is $5^2/0·5$ m s^{-2}, i.e. 50 m s^{-2} towards O

$\therefore$ using Newton's second law horizontally and vertically,

$$\uparrow N - 2g = 0$$
$$\leftarrow \qquad T = 2 \times 50 = 100.$$
$$\therefore \quad N = 2g = 19·6$$

Hence the tension in the string is 100 N and the reaction of the table is 19·6 N.

Example 2   A particle is placed at a point 20 cm from the centre O of a rough circular disc. When the disc rotates about a vertical axis through O at an angular speed of 3·5 rad s^{-1}, the particle is on the point of slipping. Find the coefficient of friction between the particle and the disc.

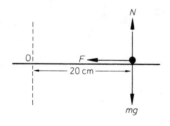

The acceleration of the particle is $0·2 \times (3·5)^2$ m s^{-2} towards O. Using Newton's second law horizontally and vertically,

$$\uparrow N - mg = 0 \quad \therefore \quad N = mg$$
$$\leftarrow \qquad F = m \times 0·2(3·5)^2 = m \times \frac{1}{5} \times \left(\frac{7}{2}\right)^2 = \frac{49m}{20}.$$

Since the particle is on the point of slipping, $F = \mu N$

$$\therefore \quad \frac{49m}{20} = \mu mg = 9·8 \,\mu m.$$

Hence $\mu = \dfrac{49}{20 \times 9·8} = \dfrac{1}{4}$

$\therefore$ the coefficient of friction between the particle and the disc is $\frac{1}{4}$.

Exercise 24.2

1.   Find the acceleration of a particle moving uniformly along a circular path
(a) radius 4 m at a speed of 6 m s^{-1},

(b) radius 25 m at a speed of 36 km/h,
(c) radius 80 cm with angular velocity 5 rad s^{-1},
(d) radius 10 m with angular velocity 24 rev/min.

2.   Find the magnitude of the resultant force acting on a particle of mass 5 kg which is moving in a circle of radius 2 m with a constant speed of 3 m s^{-1}.

3.   Find the magnitude of the resultant force acting on a particle of mass 1·5 kg which is moving in a circle of radius 50 cm with a constant angular velocity of 4 rad s^{-1}.

4.   One end of a light inextensible string of length 3 m is attached to a fixed point O on a smooth horizontal surface and a particle of mass 5 kg is attached to its other end. If the particle is moving in a horizontal circle, centre O, at a speed of 15 m s^{-1}, find the tension in the string and the reaction of the surface on the particle.

5.   One end of a light inextensible string of length 1·6 m is attached to a fixed point O on a smooth horizontal surface and a particle of mass 4 kg is attached to the other end. If the string will break when the tension in it exceeds 90 N, find the maximum speed at which the particle can move in a horizontal circle with centre O.

6.   A particle is placed at a point 0·5 m from the centre O of a rough circular disc. When the disc rotates about a vertical axis through O at an angular speed of 2·8 rad s^{-1}, the particle is on the point of slipping. Find the coefficient of friction between the particle and the disc.

7.   A small ring A is threaded on a horizontal rod PQ, the coefficient of friction between the rod and the ring being 2/7. When the rod rotates horizontally about P with a constant angular velocity of 2 rad s^{-1}, the ring does not slip along the rod. Find the maximum possible distance AP.

8.   A particle A is placed on a rough horizontal platform, the coefficient of friction between the particle and the platform being 3/8. The platform rotates about a vertical axis through a fixed point O on the platform. If $OA = 1·2$ m, find the angular velocity of the platform when the particle is on the point of slipping.

24.3 Further problems

Consider a particle P of mass m attached by means of a light inextensible string of length l to a fixed point O. The particle is set in motion so that it moves in a horizontal circle of radius r with constant angular velocity ω. Since the string traces out a cone in space, this system is called a *conical pendulum*.

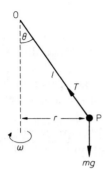

Applying Newton's second law horizontally and vertically,

$$\leftarrow T \sin \theta = mr\omega^2$$
$$\uparrow T \cos \theta - mg = 0.$$

Substituting $r = l \sin \theta$,

$$T \sin \theta = ml \sin \theta \omega^2$$

$\therefore$ the tension in the string is $ml\omega^2$.

Example 1   A particle of mass 2 kg is attached by a light inextensible string of length 1 m to a fixed point O. The particle is made to move in a horizontal circle whose centre is 0·8 m vertically below O. Find the tension in the string and the speed of the particle.

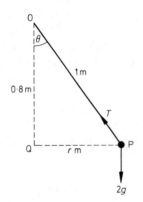

Let the speed of the particle be $v \, \text{m s}^{-1}$, then applying Newton's second law horizontally and vertically

$$\leftarrow T \sin \theta = \frac{2v^2}{r} \qquad (1)$$

$$\uparrow T \cos \theta - 2g = 0 \qquad (2)$$

In $\triangle OPQ$, by Pythagoras' theorem

$$r = \sqrt{\{1^2 - (0 \cdot 8)^2\}} = 0 \cdot 6$$

$\therefore$ $\sin \theta = 0 \cdot 6$ and $\cos \theta = 0 \cdot 8$

Using (2) $T = \dfrac{2 \times 9 \cdot 8}{0 \cdot 8} = 24 \cdot 5$

$\therefore$ the tension in the string is 24·5 N.

Using (1) $v^2 = 24 \cdot 5 \times 0 \cdot 6 \times \dfrac{0 \cdot 6}{2} = 7^2 \times (0 \cdot 3)^2$

$\therefore$ $v = 7 \times 0 \cdot 3 = 2 \cdot 1.$

Hence the speed of the particle is $2 \cdot 1 \, \text{m s}^{-1}$.

The methods used in Example 1 can also be applied to a particle sliding round the inner surface of a smooth sphere or cone in a horizontal circle. In such problems the acceleration towards the centre of the circle is produced by the horizontal component of the normal reaction of the surface on the particle.

Example 2   One end of a light inextensible string of length l is attached to a fixed point A and a particle P is attached to the other end. The ends of a second string of the same length are attached to P and to a fixed point B at a distance h ($<2l$) vertically below A. If the particle moves in a horizontal circle with uniform angular speed ω, find the least value of ω for which both strings are taut.

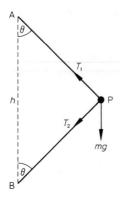

Let $\angle PAB = \theta$, then $h = 2l\cos\theta$ and the radius of the circle described by the particle is $l\sin\theta$.

Hence the acceleration of the particle

$$= \omega^2 l\sin\theta.$$

Applying Newton's second law horizontally and vertically

$$\uparrow T_1\cos\theta - T_2\cos\theta - mg = 0$$
$$\leftarrow T_1\sin\theta + T_2\sin\theta = m\omega^2 l\sin\theta$$

$$\therefore \quad T_1 + T_2 = m\omega^2 l$$

$$T_1 - T_2 = \frac{mg}{\cos\theta} = \frac{2mgl}{h}.$$

Subtracting

$$2T_2 = m\omega^2 l - \frac{2mgl}{h}$$

$$\therefore \quad T_2 = \frac{ml}{h}(h\omega^2 - 2g).$$

Since $T_1 > T_2$, both strings will be taut if $T_2 \geqslant 0$

i.e.   if $h\omega^2 - 2g \geqslant 0$.

Hence the least value of ω for which both strings are taut is $\sqrt{(2g/h)}$.

When a vehicle travels at constant speed round a bend, regarded as a circular arc, its acceleration is directed towards the centre of the arc. If the road surface is horizontal then this acceleration is produced by the frictional force between the wheels and the ground. However, if the road is banked, the horizontal components of the frictional force and of the normal reaction between the vehicle and the ground together provide the required acceleration. [We will ignore at this stage the possibility that the vehicle could overturn.]

Example 3   A car of mass m travels at constant speed v round a bend of radius r on a road banked at an angle α. The coefficient of friction between the car's tyres and the road surface is μ. Find an expression for v^2 given that (a) the car travels with no tendency to slip, (b) the car is about to slip outwards, (c) the car is about to slip inwards.

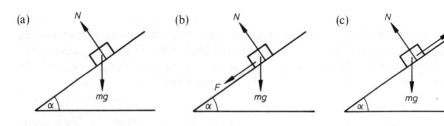

(a) In this case there is no frictional force acting.
Applying Newton's second law horizontally and vertically

$$\leftarrow N \sin \alpha = mv^2/r$$
$$\uparrow N \cos \alpha - mg = 0$$

$$\therefore \quad v^2 = \frac{rN \sin \alpha}{m} = \frac{r \sin \alpha}{m} \cdot \frac{mg}{\cos \alpha} = rg \tan \alpha.$$

Hence when there is no tendency to slip $v^2 = rg \tan \alpha$.

(b) The frictional force acts down the slope and takes its limiting value μN.
Applying Newton's second law horizontally and vertically

$$\leftarrow N \sin \alpha + \mu N \cos \alpha = mv^2/r$$
$$\uparrow N \cos \alpha - \mu N \sin \alpha - mg = 0$$

$$\therefore \quad v^2 = \frac{rN}{m}(\sin \alpha + \mu \cos \alpha) \quad \text{and} \quad N = \frac{mg}{\cos \alpha - \mu \sin \alpha}.$$

Hence when the car is about to slip outwards

$$v^2 = \frac{rg(\sin \alpha + \mu \cos \alpha)}{\cos \alpha - \mu \sin \alpha}, \quad \text{i.e.} \quad \frac{rg(\tan \alpha + \mu)}{1 - \mu \tan \alpha}.$$

(c) The frictional force acts up the slope and takes its limiting value μN.
Applying Newton's second law horizontally and vertically

$$\leftarrow N \sin \alpha - \mu N \cos \alpha = mv^2/r$$
$$\uparrow N \cos \alpha + \mu N \sin \alpha - mg = 0.$$

Hence when the car is about to slip inwards

$$v^2 = \frac{rg(\sin \alpha - \mu \cos \alpha)}{\cos \alpha + \mu \sin \alpha}, \quad \text{i.e.} \quad \frac{rg(\tan \alpha - \mu)}{1 + \mu \tan \alpha}.$$

[Note that if we substitute $\mu = \tan \lambda$, where λ is the angle of friction, the results in parts (b) and (c) may be written in the form $v^2 = rg \tan(\alpha + \lambda)$ and $v^2 = rg \tan(\alpha - \lambda)$ respectively.]

Exercise 24.3

1. A particle of mass 3 kg is attached by a light inextensible string of length 75 cm to a fixed point. The particle moves in a horizontal circle with constant speed such that the string makes an angle θ with the vertical, where $\tan \theta = 4/3$. Find the tension in the string and the speed of the particle.

2. A particle of mass 5 kg is attached by a light inextensible string of length 3·25 m to a fixed point. The particle moves in a horizontal circle with a constant angular velocity of 2·8 rad s^{-1}. Find the tension in the string and the radius of the circle.

3. A particle of mass m is moving with constant speed v in a horizontal circle of radius r around the smooth inner surface of a hollow cone of semi-vertical angle θ. Find the reaction between the particle and the cone and express v^2 in terms of r, θ and g.

4. A particle moving on the smooth inner surface of a sphere describes a horizontal circle with constant angular speed 2 rad s^{-1}. Find the depth of the circle below the centre of the sphere.

5. A particle of mass 2 kg is suspended from a fixed point by a light elastic string of natural length 1 m and modulus 19·6 N. The particle is moving in a horizontal circle with constant speed v m s^{-1}. Given that the length of the string is 2·25 m, find the value of v.

6. A light inextensible string of length $4a$ has one end fixed at a point A and the other end fixed at a point B which is at a distance $2a$ vertically below A. A smooth ring R of mass m is threaded on the string and moves in a horizontal circle with centre B such that the string is taut. Show that $AR = 5a/2$ and, assuming that the ring is free to move on the string, find its speed.

7. A light inextensible string of length $5a$ has one end fixed at a point A and the other end fixed at a point B which is vertically below A and at a distance $4a$ from it. A particle P of mass m is fastened to the midpoint of the string and moves with speed u, and with the parts AP and BP of the string both taut, in a horizontal circular path whose centre is the midpoint of AB. Find, in terms of m, u, a and g, the tensions in the two parts of the string, and show that the motion described can take place only if $8u^2 \geqslant 9ga$. (JMB)

8. A car is moving in a horizontal circle of radius 100 m on a track banked at an angle α, where $\tan \alpha = 5/16$. Find the speed of the car when there is no tendency for it to slip sideways.

9. A car travels at 90 km/h round a bend of radius 200 m on a road banked at an angle α. Find α, given that there is no tendency to skid sideways.

10. A car travels round a bend of radius 125 m on a road banked at an angle α where $\tan \alpha = 0·5$. The coefficient of friction between the car's tyres and the road is $\frac{1}{3}$. Find the maximum and minimum speeds at which the car can travel without slipping sideways.

11. A vehicle is approaching a bend of radius 50 m. The coefficient of friction between its wheels and the road is $\frac{1}{2}$. If the road surface is horizontal, find the

maximum speed at which the vehicle can negotiate the bend without skidding. If the vehicle is to drive round the bend at $20\,\mathrm{m\,s^{-1}}$ without skidding, find the least angle at which the road must be banked.

12.

The figure shows a particle P of mass $2\,m$ which is attached to a fixed point O by a light inextensible string of length l. A ring R of mass $3\,m$, which is attached to P by another light inextensible string of length l, is free to slide on a smooth vertical wire passing through O.

The plane OPR rotates about the wire with constant angular velocity ω, where $\omega^2 > 4g/l$. Show that $OR = 8g/\omega^2$. What happens if $\omega^2 \leqslant 4g/l$?   (C)

Exercise 24.4   (miscellaneous)

1.   A record is rotating at $33\frac{1}{3}\,\mathrm{rev/min}$. Find its angular speed in $\mathrm{rad\,s^{-1}}$. Find also the speed of a point on the record at a distance of $12\,\mathrm{cm}$ from the centre.

2.   A van, whose wheels are of diameter $0{\cdot}8\,\mathrm{m}$, is travelling at $72\,\mathrm{km/h}$. If the wheels do not slip on the road, find the number of revolutions made each second and hence the angular velocity of the wheels in $\mathrm{rad\,s^{-1}}$.

3.   A point P moves such that at time t its position vector is $2t\mathbf{i} + (3-t)\mathbf{j}$. Find an expression for the angular velocity of P about the origin O at time t.

4.   A particle A is moving in a horizontal circle of radius $3\,\mathrm{m}$ with constant speed $6\,\mathrm{m\,s^{-1}}$. A second particle B is moving in a concentric circle of radius $9\,\mathrm{m}$ with the same constant speed. At a given instant both particles are due east of the centre of the circles and are moving northwards. Find, in magnitude and direction, the velocity of A relative to B after a time $\frac{1}{4}\pi\,\mathrm{s}$.   (O & C)

5.   On an icy morning a car travelling at $42\,\mathrm{km/h}$ drives safely round a bend of radius $100\,\mathrm{m}$ but skids on a bend of radius $25\,\mathrm{m}$. Assuming that the road surface is horizontal, find the range of possible values of the coefficient of friction between the car's tyres and the road.

6.   One end of a light inextensible string of length $1{\cdot}2\,\mathrm{m}$ is attached to a fixed point A on a smooth horizontal surface. A particle of mass $2\,\mathrm{kg}$ is attached to the other end C and a particle of mass $1\,\mathrm{kg}$ is attached to the mid-point B. Each

particle is moving in a horizontal circle centre A, such that the string remains straight. If the particle at C is moving at a constant speed of $6\,\mathrm{m\,s^{-1}}$, find the tensions in BC and AB.

7. Show that when a particle moves in a circle of radius r with constant speed u its acceleration is of magnitude u^2/r and is directed towards the centre of the circle.

  A particle of mass m is attached to one end A of a light inelastic string of length a, the other end being attached to a fixed point O. A moves in a horizontal circle below O with constant speed $\sqrt{(3ag/2)}$. Find (i) the angle that OA makes with the vertical, (ii) the tension in the string, (iii) the time taken to complete one revolution.   (SU)

8. A particle P is attached by a light inextensible string of length l to a fixed point O. The particle is held with the string taut and OP at an acute angle α to the downward vertical, and is then projected horizontally at right angles to the string with speed u chosen so that it describes a circle in a horizontal plane. Show that $u^2 = gl\sin\alpha\tan\alpha$. The string will break when the tension exceeds twice the weight of P. Find the greatest possible value of α.   (JMB)

9. A particle moves in a circle, of radius a and centre O, at constant speed u. Prove that the acceleration of the particle is u^2/a towards O. The end A of a light rigid rod AB of length 1 metre is smoothly hinged at a fixed point. A ring is threaded on AB and the coefficient of friction between the ring and the rod is $\tfrac{1}{3}$. The rod is made to rotate horizontally at $2\,\mathrm{rad/s}$ about A. Show that the ring can remain at rest relative to the rod at any position between A and C, where $AC = 5/6$ metres. The ring is placed at C and a second identical ring is threaded on the rod and placed at the mid-point of AB. The rings are connected by a taut, light inextensible string. When the rod rotates horizontally at $\omega\,\mathrm{rad/s}$, the rings are on the point of sliding on the rod. Given that the mass of each ring is $0.2\,\mathrm{kg}$, calculate (i) the value of ω, (ii) the tension in the string. [Take the acceleration due to gravity to be $10\,\mathrm{m/s^2}$.]   (AEB 1977)

10. One end of a light inelastic string of length a is attached to a particle P of mass m, and the other end of the string is attached to a fixed point O. P moves with constant speed in a horizontal circle whose centre C is a fixed point vertically below O. If the angular velocity at which CP rotates is ω, show that $OC = g/\omega^2$. A second string, also of length a, is fixed to P and C, and the new system rotates about OC with uniform angular velocity Ω and with both strings taut. Find expression in terms of m, a, ω and Ω for the tensions in OP and PC, and deduce that $\Omega > \omega\sqrt{2}$.   (C)

11. A particle moves in a horizontal circle on the smooth inner surface of a fixed spherical bowl of radius a. If the depth of the circle below the centre of the sphere is h, find an expression for the speed of the particle.

12.

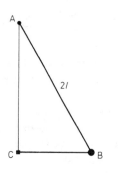

The ends of a light inextensible string ABC of length $3l$ are attached to fixed points A and C, C being vertically below A at a distance $\sqrt{3}l$ from A. At a distance $2l$ along the string from A a particle B of mass m is attached. When both portions of the string are taut, B is given a horizontal velocity u, and then continues to move in a circle with constant speed. Find the tensions in the two portions of the string and show that the motion is possible only if $u^2 \geqslant \frac{1}{3}gl\sqrt{3}$.

(JMB)

13.

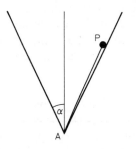

A circular cone of semi-vertical angle α is fixed with its axis vertical and its vertex, A, lowest, as shown in the diagram. A particle P of mass m moves on the inner surface of the cone, which is smooth. The particle is joined to A by a light inextensible string AP of length l. The particle moves in a horizontal circle with constant speed v and with the string taut. Find the reaction exerted on P by the cone. Find the tension in the string and show that the motion is possible only if $v^2 \geqslant gl\cos\alpha$.

(JMB)

14. A car moves in a horizontal circle of radius $500\,\text{m}$ on a track which is banked at an angle α to the horizontal. When the speed is $60\,\text{m/s}$ there is no sideways force on the wheels. Find the angle α. Find also the maximum speed round the track to avoid slipping if the coefficient of friction is $0{\cdot}5$.   (AEB 1978)

15. A light elastic string AB of natural length a and modulus of elasticity mg is joined to a light inextensible string BC of length a. The ends A and C of the string are fastened to two fixed points with A vertically above C and $AC = a$. A particle of mass m is fixed at B and rotates with speed v in a horizontal circle. Show that if AB makes an angle $\pi/6$ with the downward vertical then $v^2 = \frac{1}{2}\sqrt{3}ag$, and find the tension in BC.   (O)

16. A car moves with constant speed in a horizontal circle of radius r on a track which is banked at an angle α to the horizontal, where $\tan\alpha = \frac{3}{4}$. The coefficient of friction between the tyres and the track is $\frac{1}{2}$. Find, in terms of r and g, the range of speeds at which the car can negotiate this bend without the tyres slipping on the road surface. Show that the greatest possible speed is $\sqrt{11}$ times the least possible speed. [It may be assumed that the car will not overturn at these speeds.]   (L)

17. A small ring of mass m can slide on a smooth fixed vertical rod. One end of a light inextensible string of length $2a$ is fastened to the ring, and its other end is fastened to a point of the rod. A particle of mass $4m$ attached to the mid-point of the string is moving in a horizontal circle. Show that the angular velocity of the particle in this circle is greater than $\sqrt{(3g/2a)}$. Find the speed of the particle if the tension in the lower half of the string is $4mg$.   (L)

25 Elementary statistics

25.1 Graphical representation of data

Statistics may be described as a subject concerned with the analysis of data obtained from such sources as experiments and surveys. In elementary work this will involve making tables, charts and graphs to present the data as clearly as possible. In more advanced work various numerical measures of the properties of collections of data are studied. These may be used to compare sets of observations, to draw conclusions from information gathered and to make predictions about future events. An important aspect of the subject is measuring the significance of such conclusions and predictions.

We first consider three common ways of presenting information in graphical form. Below are two *pie charts* showing the results of an investigation into which brands of fabric softener housewives use. The charts both show the relative proportions of the different brands used, but the first is quite misleading as it gives no indication that the majority of housewives use no softener at all.

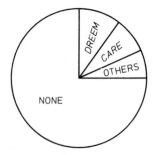

The same results can also be illustrated in the form of a *bar chart* or *block graph*. In both the given charts the vertical scale represents the number of housewives using the products. However, the first diagram exaggerates the lead of 'DREEM' in the market by starting the vertical axis at 60 instead of 0.

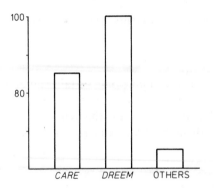
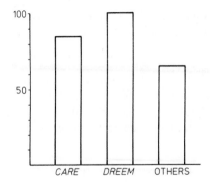

The following *line graphs* both represent the results of 5 successive surveys. This type of graph is used to illustrate a trend or pattern in results. However, in this case either graph on its own could be misleading because although sales of 'DREEM' are rising, its share of the market in fabric softeners is falling.

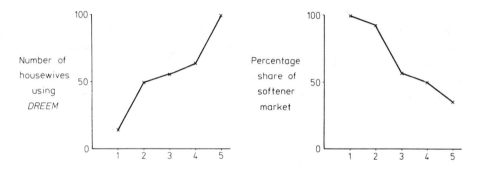

It is clear from these examples that care must be taken when making or interpreting graphs and charts, but it is also important to question the validity of the original data. In this survey a *sample* of 1000 housewives were questioned, but this is only a small proportion of the total *population* of housewives. The results could have been distorted because most of the people chosen belonged to the same income group or lived in the same area. The most reliable results are usually obtained by taking a *random sample*, in which each member of the population has an equal chance of being selected.

Another factor which may affect this type of investigation is the framing of the questions asked. For example, one person might give a different answer to each of the following four questions:

(1) Which product gives best value for money?
(2) Which product do you buy regularly?
(3) Which product did you buy this week?
(4) Which product do you prefer?

A survey based on only one of these questions could give misleading results.

Having considered some of the ways in which statistical information can be misunderstood or even misused, we now examine some more accurate ways of analysing and presenting results.

Example A   In a short test given to a set of 15 students the marks were as follows:

$$2 \quad 4 \quad 1 \quad 3 \quad 5 \quad 4 \quad 4 \quad 3 \quad 5 \quad 4 \quad 2 \quad 6 \quad 3 \quad 5 \quad 4$$

Since the test mark, x, of any particular student can take only certain values in the range 1 to 6, it is called a *discrete variable*. For a variable of this type it is helpful to make a table called a *frequency distribution* showing the frequency with which each value occurs. The corresponding diagram is called a *frequency chart*.

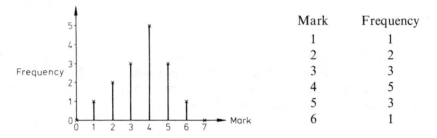

Mark	Frequency
1	1
2	2
3	3
4	5
5	3
6	1

When it is necessary to examine sums of frequencies, such as the number of students who scored 4 marks or less, a *cumulative frequency distribution* may be used. For a discrete variable x the cumulative frequency is the frequency of values *less than or equal to x*. As shown in the *cumulative frequency chart*, for a discrete variable, the cumulative frequency is a *step function*.

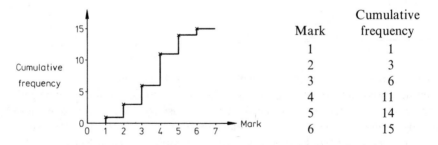

Mark	Cumulative frequency
1	1
2	3
3	6
4	11
5	14
6	15

Example B   The table below shows the numbers of children known to have contracted whooping-cough in 1965.

Age	No. of cases (to the nearest 100)
Under 1 year	1400
1–2 years	3300
3–4 years	3500
5–9 years	3900
10–14 years	500

Since the age of a child can take any real value between 0 and 15 years, it is called a *continuous variable*. Because the variable is continuous and because the table gives frequencies for age groups of different sizes, a simple frequency chart similar to the one used in Example A could be misleading. To allow for the various age ranges we draw a *histogram*, in which the *areas* (not the heights) of the columns are proportional to the frequencies.

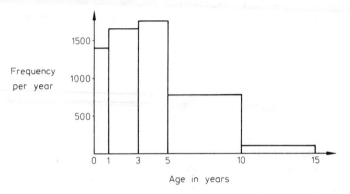

This means that the height of each column represents $\dfrac{\text{frequency}}{\text{column width}}$ (sometimes called *frequency density*). In this case the vertical scale may be taken to represent the number of cases of whooping-cough per one year age group.

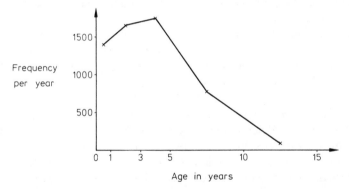

The *frequency polygon* shown here is obtained from the histogram by joining the mid-point of the tops of the columns. As several frequency polygons can be drawn in the same diagram, they are often used for comparing distributions.

For a continuous variable x, it is convenient to regard the cumulative frequency as the frequency of values *less than* x. In this example the cumulative frequency does not move in a series of jumps, but increases gradually throughout the range 0 to 15 years. Thus the values in the cumulative frequency table are used to draw a curve called the *cumulative frequency graph* or *ogive*.

Age	Cumulative frequency
Under 1 year	1400
Under 3 years	4700
Under 5 years	8200
Under 10 years	12 100
Under 15 years	12 600

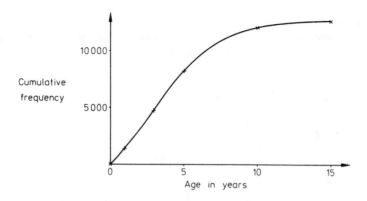

Example C The table below shows the number of staff meals issued per day in a college refectory over a period of 40 days.

56	50	63	56	58	53	52	71	59	52
45	55	42	62	62	60	57	48	63	68
55	65	50	59	54	54	46	51	73	64
53	50	61	51	58	52	57	67	53	69

In this example the variable is again discrete. However, a frequency table of the type used in Example A would have little meaning as there is a range of 31 from the lowest value 42 to the highest 73. Any pattern in the distribution of observations will be shown more clearly by grouping the numbers. The range could be divided into intervals 40–50, 50–60, 60–70 and 70–80. One difficulty about this choice arises when we come to observations such as 50 which appear in two classes. This could be avoided by choosing intervals such as 40–49, 50–59, ... [In certain circumstances, it may be better to regard 50 as $\frac{1}{2}$ in the 40–50 class and $\frac{1}{2}$ in the 50–60 class.] The other main difficulty is that this grouping produces only 4 classes and will not illustrate the distribution in any detail. From this point of view, the intervals 40–44, 45–49, ... would be better. However, the mid-points of intervals are often used in calculations, so it is convenient if these are round numbers. Thus a reasonable choice in this example is the grouping 38–42, 43–47, ... into intervals of length 5 units with mid-points 40, 45, ... The *class limits* are the lowest and highest observations which could occur in each class, namely 38 and 42, 43 and 47, ... However, the *class boundaries* or *end-points* are taken, in this case, to be 37·5, 42·5, ...

We may now draw up the frequency distribution for the grouped data using *tally marks* as shown below.

No. of meals		Frequency
38–42	I	1
43–47	I I	2
48–52	⊞⊞ I I I I	9
53–57	⊞⊞ ⊞⊞ I	11
58–62	⊞⊞ I I I	8
63–67	⊞⊞	5
68–72	I I I	3
73–77	I	1

In the histogram of this distribution, since the columns are of equal width, their heights as well as their areas will be proportional to the frequencies.

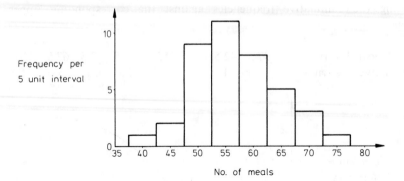

When constructing a frequency polygon it is possible to plot zero frequencies at each end of the diagram to produce a closed polygon.

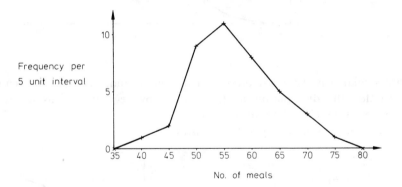

As in Example A the cumulative frequency chart constructed from the original data is a step diagram.

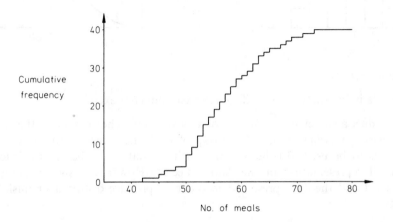

However, a cumulative frequency graph drawn using the grouped frequency distribution will provide a sufficiently good approximation for many purposes. To construct this graph we treat the number of meals as a continuous variable plotting the cumulative frequencies against the corresponding upper class boundaries.

No. of meals less than:	37·5	42·5	47·5	52·5	57·5	62·5	67·5	72·5	77·5
Cumulative frequency		0	1	3	12	23	31	36	39	40

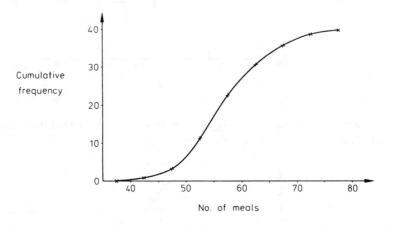

In a similar way we could produce a histogram and a cumulative frequency polygon for the distribution in Example A by regarding it as a grouped distribution with class boundaries 0·5, 1·5, 2·5, ... However, such artificially produced diagrams have very little true meaning.

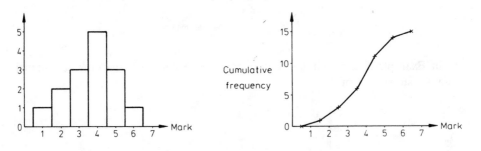

Exercise 25.1

[The data in questions 3 to 8 will be referred to in later exercises.]

1. Conduct a survey on a suitable topic of your own choice, such as the reading or television viewing habits of a group of students. Consider carefully (a) the questions to be used, (b) the way in which the data is to be recorded, (c) the graphical representation of the data collected. How could such a survey be conducted and the data presented in order to produce distorted or misleading results?

2.   The table below shows the sales in two successive years of a particular type of car.

	2-door super	4-door super	4-door de-luxe	estate
1st year	8100	10 800	18 000	3600
2nd year	15 000	21 000	25 000	11 000

(a) Draw two pie charts whose areas allow the total sales for the two years to be compared.
(b) Draw a comparative bar chart, i.e. one in which there are two bars for each model of the car.
(c) Comment briefly on the advantages of the two ways of representing the data.

3.   In an experiment the following 20 values of a variable x were recorded.

$$5 \quad 6 \quad 2 \quad 4 \quad 4 \quad 5 \quad 7 \quad 5 \quad 5 \quad 6$$
$$4 \quad 7 \quad 6 \quad 5 \quad 3 \quad 5 \quad 6 \quad 4 \quad 6 \quad 5$$

(a) Given that x is a discrete variable which can take only integral values, draw a frequency distribution chart and a cumulative frequency chart.
(b) Given that x is a continuous variable and that the above values are rounded to the nearest integer, draw a histogram and a cumulative frequency polygon.
(c) Given that x is a continuous variable and that the integral value n is recorded when $n \leqslant x < n+1$, draw a histogram and a cumulative frequency polygon.

4.   The table below gives the number of dresses sold each day in a small shop over a 5-week period.

	Monday	Tuesday	Wednesday	Friday	Saturday
Week 1	2	2	4	4	9
Week 2	0	4	7	3	7
Week 3	1	0	3	1	6
Week 4	3	3	2	2	8
Week 5	1	3	5	3	8

Draw a frequency distribution chart and a cumulative frequency chart for the distribution. Comment briefly on other ways of representing the data graphically in order to show any patterns in the distribution more clearly.

5.   In an investigation the following 50 values of a variable x were recorded:

$$6 \quad 11 \quad 12 \quad 8 \quad 9 \quad 8 \quad 9 \quad 4 \quad 2 \quad 7$$
$$9 \quad 7 \quad 4 \quad 11 \quad 7 \quad 5 \quad 8 \quad 7 \quad 6 \quad 4$$
$$9 \quad 6 \quad 3 \quad 4 \quad 6 \quad 7 \quad 5 \quad 6 \quad 6 \quad 9$$
$$4 \quad 9 \quad 7 \quad 1 \quad 4 \quad 7 \quad 7 \quad 4 \quad 10 \quad 9$$
$$5 \quad 8 \quad 10 \quad 7 \quad 4 \quad 9 \quad 5 \quad 7 \quad 9 \quad 11$$

Decide whether it would be more appropriate to represent the cumulative frequency distribution by a step diagram or an ogive given that (a) x is a discrete variable which takes only integral values, (b) x is a continuous variable with values rounded to the nearest integer. Draw the appropriate graph in each case.

6.   The length to the nearest minute of a man's journey to work over a period of 40 days are as follows:

57	56	33	56	49	56	44	56	45	58
62	57	42	68	49	44	59	63	50	58
53	58	50	56	60	59	45	51	65	53
50	38	65	43	51	63	56	61	67	61

Construct a suitable grouped frequency table, stating the class boundaries that you have used. Draw the corresponding histogram and cumulative frequency curve. Estimate from your graph the percentage of days on which the journey takes at least 45 minutes.

7.   In a particular year 80 children enter a certain large infants' school. The ages of the children on the day they first attend the school are given in the table below.

Age in years	$4\frac{1}{4}-$	$4\frac{1}{2}-$	$4\frac{3}{4}-$	$5-$	$5\frac{1}{4}-$	$5\frac{1}{2}-$	$6-$	$7-$	$7\frac{3}{4}-$
Frequency	0	10	35	18	7	3	5	2	0

Draw a histogram, a frequency polygon and a cumulative frequency polygon to represent this distribution. Can you account for the entries in the table for children of $5\frac{1}{2}$ years and over? Estimate the probability that a child in the catchment area of this school is under 5 years old when he first attends school, stating any assumptions you have made.

8.   The grouped frequency distribution of the test marks obtained by 150 students is given below.

Class limits	1–5	6–10	11–15	16–20	21–25
Frequency	0	4	12	22	30

Class limits	26–30	31–35	36–40	41–45	Over 45
Frequency	31	24	14	8	5

Draw a histogram to represent the data given that (a) the test was marked out of 50, (b) the test was marked out of 60, (c) the top mark was 54. Construct a cumulative frequency polygon in case (b). Give two reasons why this diagram represents an approximation to the true cumulative frequency distribution. Explain why a cumulative frequency curve may give a better approximation.

25.2 Mode, median and mean

The mode, median and mean of a distribution are all known as measures of *location* (or *central tendency*). For a collection of numerical data, they all indicate the general level or location of the observations by giving an 'average' or central value.

The *mode* is the observation which occurs most frequently. In some cases there are two or more modes.

The *median* is the middle observation when the observations are listed in order of magnitude. If the number of items is even, the average of the middle two is used.

The *mean* of a set of observations is the average or arithmetic mean. For a set of n observations $x_1, x_2, \ldots, x_n$, the mean $\bar{x} = \dfrac{1}{n}(x_1 + x_2 + \ldots + x_n)$. Using the Σ notation discussed in §16.5, this becomes $\bar{x} = \dfrac{1}{n}\sum\limits_{r=1}^{n} x_r$, or simply $\bar{x} = \dfrac{1}{n}\sum x$.

Of these measures of location, the mean is the one used most frequently, although it can be distorted by extreme values at the ends of a distribution. The median has the advantage that it is not affected in this way.

Example 1 Find the mode, median and mean of the following sets of observations (a) 6, 7, 7, 3, 8, 5, 3, 9; (b) 18, 13, 14, 13, 11, 12, 13, 12, 11.

(a) The distribution has two modes, 3 and 7.

Arranged in order of magnitude the observations are 3, 3, 5, 6, 7, 7, 8, 9. Thus the median is $\frac{1}{2}(6+7) = 6\cdot5$. The mean is $\dfrac{1}{8}(6+7+7+3+8+5+3+9) = 6$.

(b) The mode of this distribution is 13.

Arranged in order of magnitude the observations are 11, 11, 12, 12, 13, 13, 13, 14, 18. Thus the median is 13.

The mean is $\dfrac{1}{9}(2.11 + 2.12 + 3.13 + 14 + 18) = 13$.

[Note that arranging a set of observations in order of magnitude is called forming an *array*.]

We now return to Examples A, B and C discussed in the previous section.

Example A From both the frequency distribution and the frequency chart, it is clear that the mode is 4.

Since there are 15 marks, the median is the 8th mark in order of magnitude. Hence, using the cumulative frequency distribution or the corresponding chart, we find that the median is 4.

When calculating the mean, the total of the marks can be found using the frequency distribution.

Mark (x)	Frequency (f)	fx
1	1	1
2	2	4
3	3	9
4	5	20
5	3	15
6	1	6
	15	55

The formula for the mean can now be expressed as

$$\bar{x} = \frac{1}{n}\sum fx \quad \text{where} \quad n = \sum f.$$

Hence the mean mark $= \dfrac{55}{15} \approx 3{\cdot}67.$

Example B   For grouped data it is not possible to find a true mode, as the values of individual observations are not known. Instead the term *modal class* may be used to describe the interval corresponding to the highest column in the histogram. In this case the modal class is the 3–4 year age group. [Note that this is not the group with the highest number of cases of whooping-cough.]

The median is a value of the variable which has equal numbers of observations above and below it. For grouped data an approximate value of the median can be found either graphically or by calculation. In this example the median is the age below which 6300 whooping-cough cases lie. Thus the median can be estimated from the ogive by reading off the age corresponding to a cumulative frequency of 6300 cases, which is approximately 4 years.

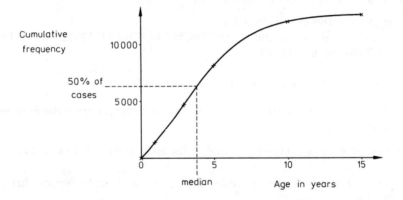

To estimate the median of a grouped frequency distribution by calculation, we assume that the observations in any class are evenly spaced, i.e. that the known points on the ogive can be joined with straight lines. This way of estimating intermediate values of a function is called *linear interpolation*. In this example there are 4700 cases under 3 years old and 8200 under 5 years old, so the median age lies in the 3–4 year class which contains 3500 cases. Since there are a total of 6300 cases below the median, 1600 of the cases in the 3–4 year class must lie below the median and 1900 above it. Thus we assume that the median divides this 2 year interval in the ratio $1600 : 1900$

$$\therefore \quad \text{the median} = 3 + \left(\frac{1600}{3500} \times 2\right) \text{years} \approx 3{\cdot}9 \text{ years.}$$

Finding the median by calculation is equivalent to using a cumulative frequency polygon rather than a curve. This calculated value can also be regarded as the value of the variable at which a vertical line divides the histogram into two equal areas.

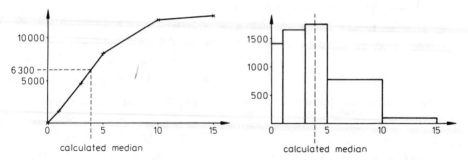

calculated median   calculated median

To find an approximate value for the mean of a grouped frequency distribution, we again assume that observations are evenly spaced. It then follows that the mean of the observations in any one class lies at the mid-point of the class interval. Hence if there are f observations in a class with mid-value x, then the sum of these observations is taken to be fx.

Class	Mid-value (x)	Frequency (f)	fx
Under 1 year	0·5	1400	700
1–2 years	2	3300	6600
3–4 years	4	3500	14 000
5–9 years	7·5	3900	29 250
10–14 years	12·5	500	6250
		$n = 12\,600$	56 800

$$\bar{x} = \frac{1}{n}\sum fx = \frac{56\,800}{12\,600} \approx 4\cdot 5$$

Hence the mean age is approximately 4·5 years.

Example C From the original data this distribution is found to have modes 50, 52 and 53. As these values occur only three times each, this information is probably of little value. The fact that the 53–57 group is the modal class may be more useful.

There are 40 observations, so the median is the average of the 20th and 21st, which is 56. An estimate of the median value can be obtained from the grouped frequency distribution, either graphically or by calculation, using the methods given in Example B. To estimate the median graphically we again use the ogive and obtain an approximate value of 56. To estimate the median by calculation we note, from the cumulative frequency table in §25.1, that 12 values of the variable are less than 52·5 and 23 values are less than 57·5. Since the median lies between the 20th and 21st value, it must be in the 5 unit interval from 52·5 to 57·5, with 8 values below it and 3 above. Assuming that these 11 observations are evenly spaced, the approximate value obtained for the median is $52\cdot5 + \left(\dfrac{8}{11} \times 5\right) \approx 56\cdot1$.

If it is necessary to find the mean from the original data without the aid of a calculator, it helps to use the deviations of the observations from an arbitrary origin. Taking 55 as origin the deviations are:

$$
\begin{array}{rrrrrrrrrr}
1 & -5 & 8 & 1 & 3 & -2 & -3 & 16 & 4 & -3 \\
-10 & 0 & -13 & 7 & 7 & 5 & 2 & -7 & 8 & 13 \\
0 & 10 & -5 & 4 & -1 & -1 & -9 & -4 & 18 & 9 \\
-2 & -5 & 6 & -4 & 3 & -3 & 2 & 12 & -2 & 14
\end{array}
$$

Cancelling equal and opposite numbers, the sum of the deviations reduces to

$$16+8+18+6+12+14 = 74$$

$$\therefore \quad \text{the mean} = 55 + \frac{74}{40} \approx 56 \cdot 9.$$

In this case a good approximation to the value of the mean can be obtained from the grouped data. This estimated value is found to be $56\frac{7}{8}$.

Exercise 25.2

1. Find the mode, median and mean of the following sets of integers
(a) 3, 4, 2, 3, 1, 3, 5, 7, 3, 4,
(b) 24, 38, 8, 13, 25, 3, 29,
(c) 21, 19, 20, 20, 22, 20, 21, 22, 21.

2. If an additional integer 20 is included in each of the lists given in question 1, find the new values of the mode, median and mean in each case.

In questions 3 to 8 use the data given in the corresponding questions in Exercise 25.1.

3. Find the mode, median and mean of the distribution, distinguishing where necessary between cases (a), (b) and (c).

4. Find the mode, median and mean of the distribution. Comment briefly on the usefulness of these measures in this context.

5. Use a frequency distribution table to find the mean of this distribution. Comment on the significance of your answer in cases (a) and (b).

6. (i) State the modal class of the grouped frequency distribution obtained in Exercise 25.1 and use this distribution to estimate the mean of the original data. Use the cumulative frequency curve to obtain a value for the median.
(ii) Form an array of the 40 items of data and write down the mode and the median of the distribution. Use a working origin of 50 to find the mean.
(iii) Comment on your answers to (i) and (ii).

7. (i) Calculate the approximate values of the median and mean of this distribution.

(ii) Assuming that children under $5\frac{1}{2}$ years of age enter a reception class, calculate the median age and the mean age of such children.
(ii) Discuss the significance of your answers.

8.   Use linear interpolation to find the median of this distribution. Estimate the value of the mean in each of the cases (a), (b) and (c).

9.   Construct a flow diagram for a procedure to find the mean of a set of numbers $x_1, x_2, \ldots, x_r$ which occur with frequencies $f_1, f_2, \ldots, f_r$ using a pocket calculator without built-in statistics functions. [Note that unless it is possible to calculate $\sum f$ and $\sum fx$ simultaneously, it will be necessary to input the numbers $f_1, f_2, \ldots, f_r$ twice.]

25.3 Measures of spread

Consider two sets of numbers (i) 1, 13, 15, 20, 21 and (ii) 12, 13, 14, 14, 17. Both have mean 14, but the first is more widely spread than the second. Thus, in addition to measures of location it is useful to have measures of *spread*, sometimes called *dispersion* or *variation*. The simplest of these is the *range*, which is the difference between the highest and the lowest items of data. In the above examples the ranges are (i) 20, (ii) 5. The main disadvantage of the range is that it is often determined by two freak values at the ends of a distribution.

Other measures of spread are obtained using quantities called *percentiles*. Theoretically, the percentiles divide a distribution into 100 equal parts. For instance, the 40th percentile is a value which has 40% of the observations in a distribution on or below it and 60% on or above it. Clearly the 50% percentile is the median. Other percentiles are found by similar methods to those used to find the median in the previous section. The interval between any two percentiles can be used as a measure of spread. Particularly useful in this way are the 25th percentile, referred to as the *first* or *lower quartile* Q_1, and the 75th percentile, the *third* or *upper quartile* Q_3. The *semi-interquartile range* is half the interval between the lower and upper quartiles, i.e. $\frac{1}{2}(Q_3 - Q_1)$. This is a more useful measure than the range because it is not affected by a few untypical values.

We return again to the examples discussed earlier in the chapter.

Example A

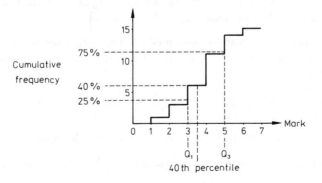

Using the cumulative frequency chart we find that the 40th percentile lies between 3 and 4. We take its value to be 3·5. The lower quartile is 3 and the upper quartile is 5

∴   the semi-interquartile range is $\frac{1}{2}(5-3)$, i.e. 1.

Example B

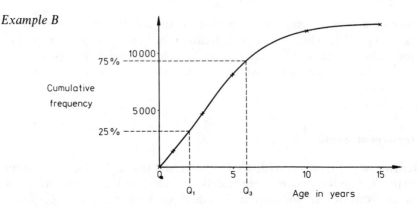

There are a total of 12 600 observations. Hence the lower quartile is the age below which 3150 of the cases occur. From the graph this is approximately 2·1 years. Similarly the upper quartile corresponds to a cumulative frequency of 9450 and is approximately 5·9 years. Hence the semi-interquartile range is $\frac{1}{2}(5\cdot9-2\cdot1)$ years, i.e. 1·9 years.

By calculation the lower quartile is estimated, using the cumulative frequency table, to be

$$1+\left(\frac{\{3150-1400\}}{3300}\times 2\right)\text{years}\approx 2\cdot1\text{ years.}$$

The upper quartile is $5+\left(\dfrac{\{9450-8200\}}{3900}\times 5\right)$ years $\approx 6\cdot6$ years.

Hence the calculated estimate of the semi-interquartile range is $\frac{1}{2}(6\cdot6-2\cdot1)$ years, i.e. 2·25 years.

[It appears likely that the whooping-cough cases are not spread evenly throughout the 5–9 year age group and that more cases occur in the lower half of the class. Thus it is probable that the estimate of 1·9 years obtained graphically is more accurate than the calculated value of 2·25 years.]

Further measures of spread are obtained by considering the set of *deviations* of the values of a variable x from the mean $\bar{x}$. For a set of n observations the sum of these deviations

$$=\sum(x-\bar{x})=\sum x-n\bar{x}=\sum x-n\left(\frac{1}{n}\sum x\right)=0.$$

Thus the mean deviation is always zero and cannot be used as a measure of spread. However, the *mean absolute deviation*, i.e. $\dfrac{1}{n}\sum|x-\bar{x}|$, is sometimes useful.

The main disadvantage when using this quantity is that moduli are difficult to manipulate algebraically.

The most frequently used measures of dispersion are the *variance*, which is the mean squared deviation, $\text{var}(x) = \frac{1}{n}\sum(x-\bar{x})^2$ and the *standard deviation*, which is the square root of the variance $\sqrt{\left\{\frac{1}{n}\sum(x-\bar{x})^2\right\}}$.

[Note that the variance is measured in square units, whereas the units of the standard deviation are the same as those of the original data.]

For examples (i) and (ii) considered at the beginning of the section, we have

(i) variance $= \frac{1}{5}\{(1-14)^2+(13-14)^2+(15-14)^2+(20-14)^2+(21-14)^2\}$

$$= \frac{1}{5}\{169+1+1+36+49\} = 51\cdot2$$

∴ the variance is 51·2 and the standard deviation is $\sqrt{(51\cdot2)} \approx 7\cdot2$.

(ii) variance $= \frac{1}{5}\{(-2)^2+(-1)^2+0^2+0^2+3^2\} = 2\cdot8$

∴ the variance is 2·8 and the standard deviation is $\sqrt{(2\cdot8)} \approx 1\cdot7$.

To avoid cumbersome calculations with a mean which is not a whole number, we now rearrange the formula for the variance, using the rules concerning Σ notation established in §16.5.

$$\text{var}(x) = \frac{1}{n}\sum(x-\bar{x})^2 = \frac{1}{n}\sum(x^2-2x\bar{x}+\bar{x}^2)$$

$$= \frac{1}{n}\sum x^2 - \frac{1}{n}\sum 2x\bar{x} + \frac{1}{n}\sum\bar{x}^2$$

$$= \frac{1}{n}\sum x^2 - 2\bar{x}\cdot\frac{1}{n}\sum x + \frac{1}{n}\cdot n\bar{x}^2$$

$$= \frac{1}{n}\sum x^2 - 2\bar{x}^2 + \bar{x}^2$$

Hence the variance, $\boxed{\text{var}(x) = \frac{1}{n}\sum x^2 - \bar{x}^2.}$

When working from a frequency table in which a value x has frequency f we may write:

$$\boxed{\text{var}(x) = \frac{1}{n}\sum f(x-\bar{x})^2 = \frac{1}{n}\sum fx^2 - \bar{x}^2.}$$

In Example A, assuming that a calculator with built-in statistical data programs is not available, the variance and standard deviation can be obtained in the following way.

Mark (x)	Frequency (f)	fx	fx^2
1	1	1	1
2	2	4	8
3	3	9	27
4	5	20	80
5	3	15	75
6	1	6	36
	15	55	227

$$n = \sum f = 15, \qquad \bar{x} = \frac{1}{n}\sum fx = \frac{55}{15}$$

$$\therefore \ \operatorname{var}(x) = \frac{1}{n}\sum fx^2 - \bar{x}^2 = \frac{1}{15} \times 227 - \left(\frac{55}{15}\right)^2.$$

Hence the variance is 1·69 and the standard deviation is 1·30 (both to 3 s.f.).

The variance of a grouped frequency distribution can be estimated using the formula $\operatorname{var}(x) = \frac{1}{n}\sum fx^2 - \bar{x}^2$ by taking the values of x to be the mid-values in each class.

In Example B the class intervals are quite long, so the estimated variance is likely to differ noticeably from the true value. However, in Example C the class intervals are shorter, so the grouped frequency distribution should give a better value for the variance.

From the original data the variance is found to be approximately 50·7 (measured in 'square meals') and the standard deviation is 7·12 meals. Using the grouped data the estimated variance is 57·1 and the standard deviation 7·56.

Exercise 25.3

1. Find the median, the quartiles and the semi-interquartile range for the following sets of integers.
(a) 8, 3, 5, 7, 11, 9, 7, 6, 7, 5, 8, 11,
(b) 28, 18, 32, 27, 42, 23, 49, 28, 33, 30,
(c) 2, −4, 10, 0, −5, 11, 6, −4, 11.

2. Find the mean, the variance and the standard deviation of the sets of integers given in question 1.

In questions 3 to 8 use the data given in the corresponding questions in Exercise 25.1.

3. Use the appropriate cumulative frequency diagram to find the following in cases (a), (b) and (c):
(i) the 40th and the 70th percentile,
(ii) the semi-interquartile range.

4. Find the variance and standard deviation of the distribution of sales.

5. (i) Use the appropriate cumulative frequency graph to estimate the lower quartile, median and upper quartile of this distribution in cases (a) and (b).
(ii) Use the frequency distribution table to find the standard deviation of the distribution.

6. Find the lower and upper quartiles of this distribution using (i) the original data, (ii) the grouped data and linear interpolation, (iii) an ogive based on the grouped data. Comment on any differences between your answers.

7. (i) Calculate approximate values, to the nearest month, of the 30th and the 65th percentile.
(ii) Estimate the variance of the distribution, stating the units in which your answer is measured.

8. The 150 test marks are to be graded as follows: the top 15% of students, grade I; the next 30%, grade II; the next 35%, grade III; the bottom 20%, grade IV. Use a cumulative frequency polygon to suggest suitable class limits for each grade.

9. Construct a flow diagram for a procedure to find the standard deviation of a set of numbers $x_1, x_2, \ldots, x_r$ which occur with frequencies $f_1, f_2, \ldots, f_r$ using a pocket calculator without built-in statistical functions. [Assume that $n = \sum f$ and $m = \frac{1}{n}\sum fx$ have already been calculated and may be read as data.]

25.4 Change of variable

Suppose we have a distribution of n values of a variable x with mean $\bar{x}$ and standard deviation σ_x. Consider a new variable y, such that $y = ax + b$, where a and b are constants. Let the distribution of the n values of y have mean $\bar{y}$ and standard deviation σ_y, then

$$\bar{y} = \frac{1}{n}\sum y = \frac{1}{n}\sum (ax+b) = \frac{1}{n}\sum ax + \frac{1}{n}\sum b$$

$$= a.\frac{1}{n}\sum x + \frac{1}{n}.nb = a\bar{x}+b$$

$$\text{var}(y) = \frac{1}{n}\sum(y-\bar{y})^2 = \frac{1}{n}\sum\{(ax+b)-(a\bar{x}+b)\}^2$$

$$= \frac{1}{n}\sum(ax-a\bar{x})^2 = a^2.\frac{1}{n}\sum(x-\bar{x})^2 = a^2\,\text{var}(x).$$

Hence,

$$\boxed{\text{if }\;y = ax+b,\quad\text{then}\quad \bar{y} = a\bar{x}+b\quad\text{and}\quad \text{var}(y) = a^2\,\text{var}(x).}$$

Example 1   During an industrial dispute a trade union negotiated a pay increase for its members of £8 a week plus 10%, so that a man earning £x per week received an increase of £$(8+x/10)$ per week. For the original wage distribution the mean was £120 per week and the standard deviation £20 per week. Find the new mean and standard deviation.

Let £y be the new weekly wage of a man originally earning £x per week, then

$$y = x+8 + \frac{x}{10} = \frac{11}{10}x+8$$

$$\therefore \qquad \bar{y} = \frac{11}{10}\bar{x}+8 = \left(\frac{11}{10}\times 120\right)+8 = 140.$$

$$\text{New s.d.} = \frac{11}{10}\times\text{original s.d.} = \frac{11}{10}\times 20 = 22.$$

Hence the new mean weekly wage is £140 and the standard deviation is £22.

A change of variable is sometimes used to simplify the calculation of the mean and standard deviation of a distribution. For instance, if a set of values of a variable x appear to have a mean approximately equal to x_0, then we can take this *assumed mean* as a working origin. If we let $y = x-x_0$, then $\bar{x} = x_0+\bar{y}$ and var (x) = var (y).

Example 2   The weights in kilogrammes of the players selected for a certain Rugby Union football team are 76, 79, 85, 86, 76, 75, 86, 105, 88, 104, 102, 103, 89, 92, 95. Find the mean and standard deviation of the distribution.

If x kg is the weight of a member of the team, let $y = x-90$, then the 15 values of y for the team are $-14, -11, -5, -4, -14, -15, -4, 15, -2, 14, 12, 13, -1, 2, 5$.

Cancelling equal and opposite values of y,

$$\bar{y} = \frac{1}{15}(-5-4) = -\frac{9}{15} = -0\cdot 6.$$

$$\sum y^2 = (-14)^2+(-11)^2+\ldots +2^2+5^2 = 1563$$

$$\therefore\quad \text{var}(y) = \frac{1}{15}\sum y^2 - \bar{y}^2 = \frac{1563}{15} - (0\cdot 6)^2 = 103\cdot 84$$

$$\therefore\quad \bar{x} = 90+\bar{y} = 89\cdot 4\quad\text{and}\quad \text{var}(x) = \text{var}(y) = 103\cdot 84$$

Hence the mean weight is 89·4 kg and the standard deviation is 10·2 kg.

When using a calculator a change of variable may be used to improve accuracy. For instance, if the values of a variable are given to 5 digits or more, squares of at least 9 digits will be produced in an accurate calculation of the variance. A change to a variable whose values have fewer digits would reduce the truncation and rounding errors generated by a calculator of limited capacity.

In certain situations a change of variable of the form $y = \dfrac{1}{c}(x - x_0)$ is useful.

If $y = \dfrac{1}{c}(x - x_0)$ then $x = x_0 + cy$

$\therefore$ $\bar{x} = x_0 + c\bar{y}$ and $\text{var}(x) = c^2 \text{var}(y)$.

Applying a transformation of this type to a distribution may be referred to as 'coding' the data. We illustrate the method using the grouped frequency distribution obtained in Example C, §25.1. We take an assumed mean of 55 as the value of x_0 and the class width 5 as the value of c, so that $y = \dfrac{1}{5}(x - 55)$.

Class	Mid-value (x)	y	Frequency (f)	fy	fy^2
38–42	40	-3	1	-3	9
43–47	45	-2	2	-4	8
48–52	50	-1	9	-9	9
53–57	55	0	11	0	0
58–62	60	1	8	8	8
63–67	65	2	5	10	20
68–72	70	3	3	9	27
73–77	75	4	1	4	16
			40	15	97

$$n = \sum f = 40, \quad \sum fy = 15, \quad \sum fy^2 = 97$$

$$\bar{y} = \frac{1}{n}\sum fy = \frac{15}{40} = \frac{3}{8}$$

$$\therefore \quad \bar{x} = 55 + 5 \cdot \frac{3}{8} \approx 56\cdot9$$

$$\text{var}(y) = \frac{1}{n}\sum fy^2 - \bar{y}^2 = \frac{97}{40} - \left(\frac{3}{8}\right)^2$$

$$\therefore \quad \text{s.d. of } x \text{ values} = 5\sigma_y = 5\sqrt{\left(\frac{97}{40} - \frac{9}{64}\right)} \approx 7\cdot56.$$

Hence the mean of the distribution is 56·9 and the standard deviation is 7·56.

Exercise 25.4

1. Find the mean, variance and standard deviation of the numbers

$$1, \quad 3, \quad 3, \quad 4, \quad 6, \quad 6, \quad 7, \quad 8, \quad 8, \quad 9.$$

Deduce the mean, variance and standard deviation of the following sets of numbers:
(a) 4, 6, 6, 7, 9, 9, 10, 11, 11, 12,
(b) 4, 12, 12, 16, 24, 24, 28, 32, 32, 36,
(c) 4·1, 4·3, 4·3, 4·4, 4·6, 4·6, 4·7, 4·8, 4·8, 4·9.

2. The mean and standard deviation of a set of mid-day temperatures for the month of June are $23°C$ and $5°C$ respectively. Find the mean and standard deviation for the same distribution measured in degrees Fahrenheit.

3. Use a frequency table and an assumed mean of 5 to calculate the mean, variance and standard deviation of the distribution of the discrete variable x given in question 3, Exercise 25.1. Deduce the mean, variance and standard deviation of the variable in case (c).

4. (i) Use an appropriate change of variable to estimate the standard deviation of the data given in question 6, Exercise 25.1 from the grouped frequency table.
(ii) With the aid of a calculator, find the standard deviation of the original distribution.
(iii) Comment on the difference between your answers.

5. Use an appropriate change of variable to find the mean and standard deviation of the distribution given in question 8, Exercise 25.1 assuming that the test is marked out of 60. The marks are to be scaled using the formula $z = ax + b$, where x is the original mark and z the new mark. If the mean and standard deviation of the scaled marks are to be approximately 50 and 15 respectively, find suitable values of a and b correct to 2 significant figures.

6. The variable x_1 represents the mark of a student in an examination. The distribution of marks for a certain group of students has mean $\bar{x}_1$ and standard deviation σ_1. Given that the marks are to be scaled so that the values of the new mark x_2 have mean $\bar{x}_2$ and standard deviation σ_2, prove that

$$x_2 = \bar{x}_2 + \frac{\sigma_2}{\sigma_1}(x_1 - \bar{x}_1).$$

Use this formula to standardise the marks of students A, B and C as given in the table below, taking $\bar{x}_2 = 50$ and $\sigma_2 = 15$. Hence find a combined mark for each student.

Subject	Mean	Standard deviation	A	B	C
Mathematics	60	20	90	96	70
English	45	10	52	46	70

Exercise 25.5   (*miscellaneous*)

1.   The following table gives an analysis by numbers of employees of the size of UK factories of less than 1000 employees manufacturing clothing and footwear.

Number of employees	11–19	20–24	25–99	100–199	200–499	500–999	Total
Number of factories	1500	800	2300	700	400	100	5800

Calculate as accurately as the data allow the mean and median of this distribution, showing your working. If 90% of the factories have less than N employees, estimate N.   (O & C)

2.   An inspection of 34 aircraft assemblies revealed a number of missing rivets as shown in the following table.

Number of rivets missing	0–2	3–5	6–8	9–11	12–14	15–17	18–20	21–23
Frequency	4	9	11	6	2	1	0	1

Draw a cumulative frequency curve. Use this curve to estimate the median and the quartiles of the distribution.   (O & C)

3.   The distribution of the times taken when a certain task was performed by each of a large number of people was such that its twentieth percentile was 25 minutes, its fortieth percentile was 50 minutes, its sixtieth percentile was 64 minutes and its eightieth percentile was 74 minutes. Use linear interpolation to estimate (i) the median of the distribution, (ii) the upper quartile of the distribution, (iii) the percentage of persons who performed the task in 40 minutes or less.   (JMB)

4.   The weights of a number of school children were measured and the mean weight was found to be 54·1 kg, with standard deviation 3·2 kg. Subsequently it was discovered the weighing machine was reading 0·9 kg under the true weight. How should the mean and standard deviation be revised? Give a clear indication of your reasoning.   (O & C)

5.   A frequency distribution has values $x_1, x_2, \ldots, x_n$ with frequencies $f_1, f_2, \ldots, f_n$ respectively. It has mean $\bar{x}$ and standard deviation σ. Find the mean and standard deviation of the distribution which has values $d_1, d_2, \ldots, d_n$ with frequencies $f_1, f_2, \ldots, f_n$ respectively, where $d_r = \dfrac{x_r - \bar{x}}{\sigma}$.   (AEB 1977)

6.   100 pupils were tested to determine their intelligence quotient (I.Q.), and the results were as follows:

I.Q.	45–	55–	65–	75–	85–	95–	105–	115–	125–134
No. of pupils	1	1	2	6	21	29	24	12	4

All I.Q.s are given to the nearest integer.

(i) Calculate the mean, and the standard deviation.

(ii) Draw a cumulative frequency graph, and estimate how many pupils have I.Q.s within 1 s.d. on either side of the mean.   (SU)

7. The haemoglobin levels were measured in a sample of 50 people and the results were as follows, each being correct to 1 place of decimals.

$$
\begin{array}{cccccccccc}
13{\cdot}5 & 15{\cdot}6 & 16{\cdot}3 & 12{\cdot}3 & 13{\cdot}1 & 14{\cdot}2 & 12{\cdot}4 & 11{\cdot}3 & 14{\cdot}0 & 14{\cdot}6 \\
13{\cdot}6 & 14{\cdot}8 & 12{\cdot}7 & 10{\cdot}9 & 11{\cdot}0 & 11{\cdot}4 & 15{\cdot}0 & 10{\cdot}1 & 15{\cdot}4 & 11{\cdot}3 \\
10{\cdot}7 & 14{\cdot}6 & 13{\cdot}5 & 15{\cdot}1 & 12{\cdot}1 & 12{\cdot}0 & 14{\cdot}2 & 11{\cdot}4 & 15{\cdot}0 & 13{\cdot}3 \\
13{\cdot}2 & 9{\cdot}1 & 16{\cdot}9 & 14{\cdot}2 & 15{\cdot}0 & 13{\cdot}6 & 14{\cdot}8 & 11{\cdot}4 & 14{\cdot}8 & 15{\cdot}7 \\
13{\cdot}5 & 13{\cdot}5 & 12{\cdot}9 & 13{\cdot}8 & 13{\cdot}7 & 16{\cdot}2 & 11{\cdot}6 & 13{\cdot}8 & 14{\cdot}2 & 10{\cdot}7
\end{array}
$$

(a) Group the data into eight classes, $9{\cdot}0$–$9{\cdot}9$, $10{\cdot}0$–$10{\cdot}9$, ..., $16{\cdot}0$–$16{\cdot}9$.

(b) What are the smallest and largest possible measurements which could be included in the class $9{\cdot}0$–$9{\cdot}9$?

(c) Draw a histogram of the grouped data and use it to estimate the median value of the sample, showing your working.

(d) Find the true median of the sample.   (SU)

8. A small firm has a total labour force of 267 of whom 250 are craftsmen and the rest managers. The distribution of the weekly pay of all employees is as follows, where, for example, $22{\cdot}5$– means that employees in this group earn at least £22.50, but less than £25.00.

Craftsmen Wages in £	No. of employees	Managers Wages in £	No. of employees
20.0–	29	50.0–	3
22.5–	33	52.5–	9
25.0–	35	55.0–	5
27.5–	40		
30.0–	51		
32.5–	37		
35.0–	25		

Display this information on a block diagram. Calculate (using a working zero of £31.25, or otherwise) the mean pay of (i) the craftsmen, (ii) the managers, (iii) all 267 employees. Comment on these means as they describe the situation in the firm and discuss other measures, such as the median or mode which might be used instead or as well.   (SU)

9. The following table gives the maximum temperature in 60 cities on a certain day. Readings are to the nearest degree.

```
72  61  43  54  54  48  59  55  61  50  55  48
30  66  41  55  48  57  61  48  46  61  30  50
66  73  54  48  66  61  45  57  48  70  68  43
52  50  46  64  46  50  50  50  48  37  45  53
64  50  39  32  66  68  41  70  48  73  39  43
```

Gather these measurements into classes at intervals of 5°, starting at 29·5°. From the information in this form construct a cumulative frequency table and draw a graph to display the cumulative frequency. Use this to find the median and semi-interquartile range. For the grouped distribution calculate the mean and standard deviation, explaining what effect the grouping is likely to have had on each.

(AEB 1978)

10. On 1 January 1975 the heights of ten boys, measured in centimetres, were 172·1, 168·6, 174·1, 174·6, 176·1, 167·6, 174·1, 170·1, 172·6, 171·1. Find the mean and standard deviation of these heights. If on 1 January 1976 each boy was 2·5 cm taller, calculate the mean height and variance on that date. If the heights had been measured in inches, find the variance of the heights of the ten boys stating the units in which your answer is given. [Take 2·54 cm = 1 inch.] (L)

11. For a finite population of size N, the mean of a measurement X is μ_X and the variance σ_X^2 is given by $\sigma_X^2 = \dfrac{1}{N} \sum (X - \mu_X)^2$. If $Y = aX + b$, where a and b are fixed constants, show that $(aN\sigma_X)^2 = N \sum Y^2 - (\sum Y)^2$.

For the finite population given in the table below, calculate the average earnings and the standard deviation of the earnings.

Earnings to nearest £	18–22	23–27	28–32	33–37	38–42
Number of workers	2	2	2	4	5

Earnings to nearest £	43–47	48–52	53–57	58–62	63–67
Number of workers	7	4	7	9	8

(L)

12. The set of positive integers $x_1, x_2, \ldots, x_9$ has a mean of 5 and $\sum_{i=1}^{9} (x_i - 5)^2 = 36$. A tenth integer (x_{10}) is added to the set, and the mean and variance of the set of ten integers are denoted by $\bar{x}$ and Var (x) respectively.
(a) Calculate the values of $\bar{x}$ and Var (x) if $x_{10} = 5$.
(b) Calculate the values of $\bar{x}$ and Var (x) if $x_{10} = 9$.
(c) If $0 < x_{10} < 5$ state two inequalities satisfied by $\bar{x}$.
(d) Without doing long calculations, suggest which values for x_{10} would result in Var (x) being greater than 4.

[In this question you should take Var $(x) = \dfrac{1}{10} \sum_{i=1}^{10} (x_i - \bar{x})^2$.] (L)

13. (i) A distribution, A, contains N_1 observations and has mean and standard deviation μ_1 and σ_1 respectively. A second distribution, B, of N_2 observations has mean and standard deviation μ_2 and σ_2 respectively. The standard deviation of the combined distribution of $N_1 + N_2$ observations is σ. Show that

$$(N_1 + N_2)\sigma^2 = N_1\sigma_1^2 + N_2\sigma_2^2 + \frac{N_1 N_2}{N_1 + N_2}(\mu_1 - \mu_2)^2.$$

(ii) Two teachers A and B each marked 100 exam papers with the following results

Mark	35	36	37	38	39	40	41	42	43	44	45
No. marked by A	4	10	14	18	12	12	10	8	4	5	3
No. marked by B	8	15	16	20	18	5	3	5	2	2	6

The mean mark for A is 39·23 with variance 6·4371. The mean mark for B is 38·51. Calculate the variance of B's marks. It is decided to add 0·72 to each of B's marks to make the mean for his papers the same as for A's. Calculate the mean and variance of the combined set of 200 papers (with B's marks adjusted).

(AEB 1975)

14. The following table gives the frequency distribution of 100 observed values of a random variable x.

x =	1	2	3	4	5	6
Frequency =	12	13	30	20	18	7

(a) Suppose x is the score obtained on a throw of a certain die.
   (i) Calculate the mean and the variance of the observed scores.
   (ii) Use the data to estimate the probability of throwing an even score with this die. Hence estimate the probability that six throws of the die will result in three even and three odd scores.
(b) Suppose, instead, that x is a continuous random variable and that in compiling the above frequency table the observed values were recorded to the nearest integer. Use linear interpolation to estimate the median of the observed values and the proportion of the observed values which exceeded 4·8.

(JMB)

Answers

Exercise 1.1

1. (a) True, (b) False, (c) True, (d) True,
   (e) False, (f) False, (g) True, (h) False.
2. $A = C$; $B \subset A$, $B \subset C$, $E \subset D$.
3. $\varnothing \subset X, \varnothing \subset Y, \varnothing \subset P, \varnothing \subset Q, X \subset Y$; $\varnothing \in Q, 0 \in X, 0 \in Y, X \in P$.
4. (a) $p \Rightarrow q$ (b) $p \Leftrightarrow q$ (c) $p \Rightarrow q$ (d) $p \neq q$.
5. (a) $X \subset Z$ (b) $X = Z$.
6. $A = \varnothing, B = \{2\}, C = \{2,3\}, D = \{1,2\}$.
7. (a) $(x,f), (x,g), (x,h)$, (b) $(x,p), (x,q)$, (c) $(p,x), (q,x)$, (d) (x,x).
8. $A = \{1,2,3\}, B = \{1,2\}$.
9. (a) $\{0,2\}$, (b) $\mathscr{E}$, (c) $\{2\}$, (d) $\varnothing$.
10. (a) True, (b) True only if $\mathscr{E} = \{\text{integers}\}$, (c) True, (d) False. A could be a subset of C.

Exercise 1.2

1. (a) $2, 0, -2$, (b) $2, 0, -2, 0\cdot\dot{7}, 7\cdot3$, (c) $\sqrt[3]{2}, \pi$ (d) $2, 0, -2, 0\cdot\dot{7}. \sqrt[3]{2}, 7\cdot3$, (e) π.
2. (a) $7, -1$, (b) $7, -1, 4/3, 3\cdot142$, (c) $\pi+2, \sqrt{\pi}, \sqrt{3}+2$,
   (d) $7, -1, 4/3, 3\cdot142, \sqrt{3}+2$, (e) $\pi+2, \sqrt{\pi}$.
3. (i) (a) $\{4\}$ (b) $\{\ \}$ (c) $\{1\}$ (d) $\{\ \}$
   (ii) (a) $\{4, -1\}$ (b) $\{\ \}$ (c) $\{1\}$ (d) $\{\ \}$
   (iii) (a) $\{4, -1\}$ (b) $\{\ \}$ (c) $\{1, \frac{1}{2}\}$ (d) $\{\ \}$
   (iv) (a) $\{4, -1\}$ (b) $\{-\sqrt{5}, \sqrt{5}\}$ (c) $\{1, \frac{1}{2}\}$ (d) $\{\ \}$.
4. $\dfrac{1}{2}, \dfrac{7}{16}, \dfrac{1}{25}, \dfrac{1}{625}, \dfrac{5}{64}, \dfrac{51}{75}$.
5. (a) $\dfrac{5}{9}$, (b) $\dfrac{1}{2}$, (c) $\dfrac{3}{11}$, (d) $\dfrac{1}{27}$, (e) $\dfrac{1}{7}$.

Exercise 1.3

1. $-5, -\frac{1}{4}, 2, -k, n$.

521

2. $\frac{1}{2}, -3, -1, 5/4, x.$
3. (a) not commutative, (b) not associative.
4. (a) commutative, (b) associative.
5. (a) commutative, (b) not associative.
6. (a) 4, (b) 9, (c) 2, (d) -2, (e) -3, (f) 0 or 2.

Exercise 1.4

2. (a) $x < 7$, (b) $x \geqslant 5$, (c) $x < -1$, (d) $x \leqslant 4$, (e) $x < -2$, (f) $x < -\frac{3}{4}$.
3. (a) $x < 1$, (b) $x \geqslant 8$, (c) $x < -3$, (d) $x \geqslant 7$, (e) $x \leqslant -5$ or $x \geqslant 5$,
    (f) $-6 < x < 6$.
4. (a) $-\frac{1}{2} < x < \frac{1}{2}$, (b) $-2 \leqslant x \leqslant 0$, (c) $1 < x < 2$, (d) $-\frac{1}{2} < x \leqslant 1$,
    (e) $-1 < x < 3$, (f) $-10 \leqslant x \leqslant 4$.
5. (a) False, (b) True, (c) False, (d) True, (e) False,
    (f) True, (g) False, (h) True, (i) False, (j) True.

Exercise 1.5

1. (a) 49, (b) 1, (c) 1/5, (d) 6, (e) 2, (f) 1/8, (g) $3\frac{1}{2}$, (h) $\frac{1}{3}$.
2. (a) 8, (b) $\frac{2}{3}$, (c) 16, (d) $\frac{1}{3}$, (e) $2\frac{7}{9}$, (f) $1\frac{1}{2}$, (g) 1, (h) 8.
3. (a) $3\sqrt{6}$, (b) $7\sqrt{2}$, (c) $6\sqrt{2}$, (d) $2\sqrt{17}$, (e) 18, (f) 2, (g) $\sqrt{3}$, (h) $\sqrt{2}$.
4. (a) $2\sqrt{5}$, (b) $5\sqrt{5}$, (c) $7\sqrt{2}$, (d) $-\sqrt{7}$.
5. (a) $\sqrt{18}$, (b) $\sqrt{125}$, (c) $\sqrt{x^2 y}$, (d) $\sqrt{4a^4 b}$.
6. (a) $2\sqrt{5}$, (b) $\dfrac{3\sqrt{7}+14}{7}$, (c) $\sqrt{15}-\sqrt{5}$, (d) $\dfrac{7+3\sqrt{5}}{4}$,

    (e) $\dfrac{17+13\sqrt{2}}{7}$, (f) $5+2\sqrt{6}$.
7. (a) $1\frac{1}{2}$, (b) $3\frac{1}{2}$, (c) -3.
8. (a) -5, (b) 1, (c) -1, (d) 2.
9. (a) 0,3, (b) 0,2, (c) 1,2, (d) $-1,2$, (e) 1,2, (f) $\frac{1}{2}, -\frac{1}{2}$.
10. (a) $\sqrt{(x^4-x)}$, (b) $\sqrt{(1+x^2)}$, (c) $x(x-2)^{1/2}$, (d) $2x(2x-1)^{-1/2}$.

Exercise 1.6

1. (a) $\log_2 16 = 4$, (b) $\log_3 (1/9) = -2$, (c) $\log_{16} 4 = \frac{1}{2}$.
2. (a) $64 = 4^3$, (b) $0 \cdot 2 = 5^{-1}$, (c) $27 = 9^{1 \cdot 5}$.
3. (a) 3, (b) -2, (c) 0, (d) 1, (e) $\frac{1}{2}$.
4. (a) 2, (b) $\frac{1}{2}$, (c) -1, (d) $1\frac{1}{2}$, (e) 0.
5. (a) 4, (b) $\frac{1}{3}$, (c) $-\frac{1}{2}$, (d) $\frac{1}{2}$, (e) $-\frac{2}{3}$, (f) -3, (g) $1\frac{1}{2}$, (h) -2.
6. (a) $2\log x + \log y$, (b) $\frac{1}{2}\log x + \frac{1}{2}\log y$, (c) $4\log x - 3\log y$.
7. (a) $\log A + 2\log B - 3\log C$, (b) $\log A + \frac{1}{2}\log B + 2\log C$,
    (c) $\frac{3}{2}\log A + \log B + \frac{1}{2}\log C$.
8. (a) $\log 4$, (b) $\log 80$, (c) $\log \frac{3}{4}$, (d) $\log \frac{1}{2}$.
9. (a) 5, (b) $1\frac{1}{2}$, (c) 2.
10. (a) $1 + 2p + q$, (b) $5p - q$, (c) $p/(p+q)$.
11. (a) $1 + 2t$, (b) $1/t$, (c) $\frac{1}{2}t$, (d) $-1/t$.

12. (a) 4,   (b) 1/5.
13. (a) 3,   (b) 1.
14. (a) $9, 1/9$,   (b) $2, 64$,   (c) $\frac{1}{2}, 64$,   (d) $5, 25$.
15. $x = 4, y = \frac{1}{2}$.

Exercise 1.7

1. (a) False,   (b) True,   (c) False,   (d) True.
2. (a) False,   (b) True,   (c) True,   (d) False.
3. (a) True,   (b) False,   (c) False,   (d) True.
4. (a) True,   (b) False,   (c) False,   (d) False.
5. (a) {irrational numbers},   (b) {non-negative numbers},
    (c) $\{x : x > 4\}$,   (d) $\{0\}$,   (e) $\mathbb{R}$,   (f) $\{x : x^2 \geqslant 1\}$.
6. (a) commutative,   (b) not associative.
8. (a) $x < -3$,   (b) $x \leqslant -8$,   (c) $\frac{1}{2} < x \leqslant 2$,   (d) $x > 3$ or $x < -3$,
    (e) $-7 < x < 3$,   (f) $x \leqslant 1$ or $x \geqslant 5$.
9. $2\sqrt{11}, 3\sqrt{5}, 4\sqrt{3}, 7, 5\sqrt{2}$.
10. (a) $1\frac{1}{2}$,   (b) 10,   (c) $\sqrt[3]{2}$.
11. (a) $3 + 2\sqrt{2}$,   (b) $17 - 12\sqrt{2}$,   (c) $1 - \sqrt{2}, -1 + \sqrt{2}$.
12. $x = 2, y = 0$.
13. (a) $-1\cdot89$,   (b) $0\cdot63$,   (c) $0\cdot26$ or $0\cdot37$.
14. (a) 16,   (b) 10,   (c) $\frac{1}{3}$,   (d) 2.
15. (a) -3,   (b) 2,   (c) 0.
16. (a) 4,   1/16,   (b) 1/9.

Exercise 2.1

1. (a) Yes,   (b) no,   (c) no.
2. (a) Yes,   (b) no,   (c) yes.
3. (a) 4, 10, 4,   (b) many-to-one,   (c) no,   (d) yes.
4. (a) 8, 38, 602,   (b) one-one,   (c) yes,   (d) no.
5. (a) 13, 343, 90301,   (b) one-one,   (c) yes,   (d) no.
6. (a) 7, 20, 352,   (b) many-to-one,   (c) yes,   (d) no.
7. $\mathbb{R}$ (a) yes,   (b) yes,   (c) yes.
8. $\{2\}$ (a) no,   (b) no,   (c) no.
9. $\mathbb{R}$ (a) yes,   (b) yes,   (c) yes.
10. $\{x \in \mathbb{R} : x \geqslant 0\}$ (a) no,   (b) no,   (c) no.
11. $\mathbb{Z}$ (a) yes,   (b) yes,   (c) yes.
12. $\mathbb{Z}$ (a) yes,   (b) yes,   (c) yes.
13. $\{x \in \mathbb{R} : x > 0\}$ (a) yes,   (b) no,   (c) no.
14. $\{x \in \mathbb{R} : x > 1\}$ (a) no,   (b) yes,   (c) no.
15. $\{1, 2, 4, 8, 16, \ldots\}$ (a) no,   (b) no,   (c) no.

Exercise 2.2

1. $y = 2 - x$.
2. $y = 2x + 1$.
3. $y = 4 - x^2$.
4. $y = x^3 - 2x^2$.
5. $y = 1/x^2$.
6. $y = 1/(1 - x)$.

7. Yes; $\mathbb{R},\mathbb{R}$; one-one.
8. Yes; $\mathbb{R},\mathbb{R}$; one-one.
9. Yes; $\mathbb{R},\mathbb{R}$; one-one.
10. Yes; $\mathbb{R}, \{y\in\mathbb{R}:y\geqslant 4\}$; many-to-one.
11. No.
12. Yes, $\mathbb{R}, \{y\in\mathbb{R}:y\leqslant 1\}$; many-to-one.
13. Yes; $\mathbb{R}-\{0\}, \mathbb{R}-\{0\}$; many-to-one.
14. Yes; $\mathbb{R}-\{1\}, \mathbb{R}-\{0\}$; one-one.
15. Yes; $\mathbb{R}-\{4\}, \mathbb{R}-\{0\}$; one-one.
16. No.
17. Yes; $\{x\in\mathbb{R}:x\geqslant 0\}, \{y\in\mathbb{R}:y\geqslant 0\}$; one-one.
18. Yes; $\mathbb{R}, \{y\in\mathbb{R}:y\geqslant 0\}$; many-to-one.

Exercise 2.3

1. (a) $(0,-1)$, $(1,0)$, (b) $(0,5)$, $(-2\frac{1}{2},0)$, (c) $(0,0)$, $(-3,0)$,
    (d) $(0,-2)$, $(-2,0)$, $(\frac{1}{2},0)$, (e) $(0,-1)$, (f) $(0,\frac{1}{2})$, $(3,0)$.
2. (a) $(-3,6)$, (b) $(3,1)$, (c) $(2,3)$, $(-5,115)$, (d) $(-2,-7)$, $(4,35)$, $(-1,-5)$,
    (e) $(-2,2)$, $(-1,-4)$, (f) $(7,-6)$, $(0,1)$.
3. (a) 15, (b) 5, (c) 13, (d) 10, (e) 17, (f) 25, (g) $\sqrt{13}$, (h) $3\sqrt{5}$.
4. (a) $(6,2\frac{1}{2})$, (b) $(-5,-6\frac{1}{2})$, (c) $(\frac{1}{2},8)$, (d) $(8,3)$, (e) $(0,-2\frac{1}{2})$,
    (f) $(-\frac{1}{2},8)$, (g) $(6,-2\frac{1}{2})$, (h) $(-\frac{1}{2},0)$.
5. (a) $(5,0)$, (b) $(0,-6)$, (c) $(1,-6)$, (d) $(2,1)$.
6. (a) $(9,-8)$, (b) $(0,1)$, (c) $(5,8)$, (d) $(2,-4)$.
7. (a) isosceles, (b) neither, (c) right-angled, (d) both.
9. 5.
10. 50.
11. $(9,4)$; 126.
12. $(-4,1)$; 80.
13. $S(7,0)$, $T(3,-6)$.
14. 2; $(10,5)$, 10.
15. 20.

Exercise 2.4

1. (a) $\frac{3}{4}$, (b) -7, (c) -2, (d) 3/7, (e) 5, (f) 1.
2. (a) $-4/3$, (b) 1/7, (c) $\frac{1}{2}$, (d) $-7/3$, (e) $-1/5$, (f) -1.
3. (a) No, (b) yes, (c) yes, (d) no.
6. (a) -1, (b) -5, (c) 3 or -4.
7. (a) $(4,0)$, (b) $(9,0)$, (c) $(-1,0)$.
8. $2\sqrt{10}$, $\sqrt{10}$, $\sqrt{10}$; 15 sq. units.
10. 0, -1; $7\frac{1}{2}$ sq. units, 10 sq. units.
11. 123 sq. units.

Exercise 2.5

1. (a) $y = 4x+11$, (b) $x+2y+4 = 0$, (c) $3x+y = 11$,

(d) $2x - 3y - 11 = 0$,   (e) $5x + 12y = 0$,   (f) $x - ty + t^2 = 0$.

2. (a) $y = x - 7$,   (b) $2x + y = 2$,   (c) $y = 7$,   (d) $3x - 4y = 0$,
    (e) $3x - 7y = 15$,   (f) $x = 6$,   (g) $(p+q)y = 2x + 2pq$,   (h) $x + pqy = p + q$.

3. (a) $5/2, 9/2$,   (b) $7/4, -13/4$,   (c) $-1/4, -19/8$,   (d) $1/2, -17/6$,
    (e) $0, -5/2$,   (f) $-5/9, 3$.

4. (a) $(1,1)$,   (b) $(3,2)$,   (c) $(0, -3)$,   (d) none,   (e) $(2,1)$,   (f) $(-\frac{2}{3}, \frac{1}{3})$.

5. (a) perpendicular,   (b) perpendicular,   (c) neither,
    (d) parallel,   (e) perpendicular,   (f) neither.

6. (a) $y = 2x + 5, x + 2y = 5$,   (b) $3x + 2y = 4, 3y - 2x = 6$,
    (c) $x = 3, y + 4 = 0$,   (d) $x - 3y = 1, 3x + y + 7 = 0$.

7. $(1, -2)$.   8. $(2, 0)$.

9. $y = 3x + 5$.   10. $y = x - 5$.

12. (a) $2y + x = 13, y = 2x + 9$,   (b) $(-1, 7)$.

13. $(2, -4)$.   14. $B(3, -1), D(1, 1)$.

15. $(-3, 7)$.   16. $(-1, 3), y = 7x + 10$.

Exercise 2.6

3. $y > -2$.

4. $x < 3$.

5. (a) $-1\frac{1}{2} < x < 6$,   (b) $0 < y < 3$,   (c) $0 < x + y < 9$.

6. (a) $-14/3 < x < 0$,   (b) $0 < y \leqslant 5$,   (c) $-14/3 < x + y < 17/4$.

Exercise 2.7

1. (a) $\{y \in \mathbb{R} : y \geqslant -4\}$; neither,
    (b) $\{y \in \mathbb{Q} : y = 1/n, n \in \mathbb{N}\}$; one-one,
    (c) $\mathbb{R}$; onto.

2. $-2\frac{1}{4} \leqslant y \leqslant 10$; $\{x \in \mathbb{R} : -3 \leqslant x \leqslant 0 \text{ or } 3 \leqslant x \leqslant 6\}$.

3. $A(-4, 0), B(0, 2)$; $2\sqrt{5}$.

6. $(10, -1), (2, 15), (-6, -1)$.

7. (a) $y = 2x + 1$,   (b) $9x + 5y - 2 = 0$,   (c) $4x - 3y = 11$,
    (d) $5x + 3y = 3$.

8. $A(2, -5), B(1, -2), C(4, 13)$.

9. $(3, \frac{2}{3})$.

10. $(6, -4)$; 50 sq. units.

11. $(5, 1), (-1, 7), (-1, -3)$.

12. (a) $\frac{1}{2}, 8\frac{1}{2}, -3$,   (b) 25 sq. units.

13. (a) $(\frac{1}{2}\{h + 3\}, \frac{1}{2}\{k + 10\})$,   (b) $(k - 10)/(h - 3)$.

14. (a) $(6, 2)$,   (b) $x + 7y = 45$.

16. (a) 6,   (b) $16x - 12y = 35$.

17. (a) $2x + 3y = 1$,   (b) $3\sqrt{13}$.

18. (a) $(2, 6)$,   (b) 20 sq. units.

19. $(0, 2), 5$.

20. $(2, 5), (5, 3), (6, 1)$; $19k/(2 + 3k)$.

Exercise 3.1

1. (a) Min, -9,   (b) Max, 1,   (c) Min, 7/4,   (d) Min, 4,   (e) Min, -12,
    (f) Min, 0,   (g) Max, 25/12,   (h) Max, $-25/3$.
2. (a) $\{y \in \mathbb{R} : y \geqslant -5\}$,   (b) $\{y \in \mathbb{R} : y \geqslant -9\}$,   (c) $\{y \in \mathbb{R} : y \geqslant \frac{1}{2}\}$,
    (d) $\{y \in \mathbb{R} : y \leqslant 12\frac{1}{4}\}$.
3. $(x-2)^2 + 4$.
6. (a) $0 < x < 2$,   (b) $1 \leqslant x \leqslant 5$,   (c) $x \leqslant -2$ or $x \geqslant 2$,
    (d) $x < 1$ or $x > 3$,   (e) $-1 < x < \frac{1}{3}$.   (f) $x \leqslant -3$ or $x \geqslant -\frac{1}{2}$.
7. $-7; c > 12$.
8. $k \geqslant 9/4$.
9. (a) $-3, 8$,   (b) $\frac{1}{2}, -5$,   (c) $1, 3$.
10. 8.

Exercise 3.2

1. (a) $\frac{1}{2}(3 \pm \sqrt{5})$,   (b) $\frac{1}{4}(-3 \pm \sqrt{17})$,   (c) $1 \pm \sqrt{2}$,   (d) no real roots,
    (e) $-3/2, 9/2$,   (f) $\frac{1}{2}(3 \pm \sqrt{3})$.
2. (a) $3, -2$,   (b) -1.
3. (a) $p < -5$ or $p > 0$,   (b) $-2 < p < 2$,   (c) $p < 1$ or $p > 9$,
    (d) all real $p \neq 0$.
4. (a) $a < -1$ or $a > 0$,   (b) all real $a \neq 0$,   (c) $0 < a < 3$,
    (d) $a > 1$.
5. $3, -1$.
7. 3.
8. $0 < k < 1$.
9. $k \leqslant -6$ or $2 \leqslant k < 3$.

Exercise 3.3

1. (a) $5/2, -3/2$,   (b) $1, -6$,   (c) $0, -5/3$,   (d) $-p, -p$,   (e) $7, 5$,
    (f) $\frac{3}{4}, -\frac{1}{4}$.
2. (a) $x^2 - 3x + 2 = 0$,   (b) $4x^2 + 2x + 3 = 0$,   (c) $x^2 - 4 = 0$,
    (d) $14x^2 - 6x - 1 = 0$,   (e) $x^2 - (p-q)x + p + q = 0$,
    (f) $abx^2 - a^2x + 1 = 0$.
3. 9.   4.   4, 6.
6. (i) 3,   (ii) -2,   (iii) 13,   (iv) 45.
7. (i) 4,   (ii) 24,   (iii) $2\sqrt{6}$,   (iv) $-4\sqrt{6}$.
8. (a) 17/4,   (b) 33/4,   (c) 25/8,   (d) $-1/4$,   (e) 17/16, (f) 25/32.
9. $3; x^2 - 5x + 3 = 0$.   10. $x^2 \pm 11x + 28 = 0$.
11. (a) $x^2 - 12x + 18 = 0$,   (b) $x^2 - 10x + 23 = 0$,
    (c) $x^2 - 16x + 56 = 0$.
12. (a) $x^2 + (2k-1)x + k^2 = 0$,   (b) $x^2 - x + k = 0$,
    (c) $x^2 + 3x + (k+2) = 0$.
13. (a) $cx^2 + bx + a = 0$,   (b) $a^2x^2 + 4ac - b^2 = 0$,
    (c) $b^2x^2 - b^2x + ac = 0$.
14. $2 + 2p - q, 2\frac{1}{2} + 2p - q$.   15. 0, 1.

Exercise 3.4

1. (a) $(1,2)$, $(4,17)$,   (b) $(2,-1)$ repeated (i.e. tangent),
    (c) $(1,-2)$, $(2,-1)$,   (d) none,   (e) $(2,-3)$, $(-5/4,-19/2)$,
    (f) $(-5,8/3)$.
2. $4,-3;\ -3,4.$ 3. $4,6;\ -4,-6.$
4. $3,-2.$ 5. $3,-1;\ -2,-2.$
6. $1,2$ (repeated).   7. $3,\frac{1}{2};\ 5/3,5/2.$
8. $5,1.$ 9. $1,-2;\ -2,1.$
10. $1,-1;\frac{1}{4},\frac{1}{2}.$ 11. $6,-5;\ -5,6.$
12. $-1,5/3;\ 5/2,-2/3.$
13. $2,1;\ -2,-1;\ \sqrt{3},2/\sqrt{3};\ -\sqrt{3},-2/\sqrt{3}.$
14. $5/2,-3/2;\ -3/2,5/2.$ 15. $-1,4$ (repeated).
16. $-2,0;\ -2,-2;\ 2/3,2/3;\ 2/3,-8/3.$
17. $\pm\sqrt{3},0;\ 2,-1;\ -2,1.$
18. $0,3$ (repeated); $1+2\sqrt{2},-1-\sqrt{2};\ 1-2\sqrt{2},-1+\sqrt{2}.$

Exercise 3.5

1. (a) $2x^3+3x^2+5x+14,$ (b) $6x^4-13x^3+16x^2-17x+3,$
    (c) $4x^4+11x^3-10x^2+5x-2,$ (d) $5x^5-7x^3+3x^2+2x-3,$
    (e) $9x^5-4x^3+18x^2-8,$ (f) $x^6-1.$
2. (a) $-16,-7,$ (b) $7,5,$ (c) $-5,3,$ (d) $-4,5.$
3. (a) $\dfrac{3}{(1-x)(2x-5)},$ (b) $\dfrac{x+2}{x^2-2x+4},$ (c) $x^2+x+1,$ (d) $\dfrac{2x+1}{2x-3}.$
4. (a) $19,$ (b) $112,$ (c) $15,$ (d) $-16,$ (e) $a^2,$ (f) $0.$
5. (a) $3x-5;\ 10,$ (b) $2x^2+x+4;\ 13,$ (c) $x^2-3x+9;\ -30,$
    (d) $x^4-2x^2+3x+1;0,$ (e) $x^2-x;5x-5,$ (f) $2x^2-x+7;\ -2x+7.$
6. (a) $(x-1)^2(2x+1),$ (b) $(x+1)(x-2)(3x+1),$
    (c) $(x+3)(x-3)(x^2+8),$ (d) $x(x^4+x^2+1),$
    (e) $(x+2)(2x-1)(2x-3),$ (f) $(x+1)(x-2)(2x+1)(2x-1).$
7. $2/3,7/3.$ 8. $3,-6.$ 9. $-5,7.$
10. $3,-5.$ 11. $3,-2.$ 12. $0,6,-6.$
13. $a=3:3(x-3)(x+1)^2;\ a=-3:-3(x+3)(x-1)^2.$
14. (a) $10,$ (b) $3.$
15. (a) $(2x-1)(2x-3)(2x-5),$ (b) $(2x+1)^2(3x-2),$
    (c) $(3x+2)(x^2-x+2).$

Exercise 3.6

1. $-1,2,7.$ 2. $1,-2,-3.$
3. $4,-1$ (repeated).   4. $-2,-2/3,-5/2.$
5. $\pm\frac{1}{2}.$ 6. $\pm3,\pm\sqrt{2}.$
7. $-2,1\pm\sqrt{3}.$ 8. $3.$
9. $-\frac{1}{2},\pm\sqrt{3}.$ 10. $2,-1,2\pm\sqrt{2}.$
11. $3.$ 12. $4.$
13. $1,5.$ 14. $-4.$

15. $-7, -6; 3, -2, -1.$ 16. $4; 2/3, -3/2.$
17. (a) $-3/2, 6,$ (b) $-5/2, 0,$ (c) $-4/3, 2/3,$ (d) $0, 6.$
18. $-8; a = -28, b = -32.$ 19. $\frac{1}{2}; p = -9, q = 12.$
20. (a) $6, -4,$ (b) $1/2, 3/2,$ (c) $0, 0,$ (d) $3/2, 1/2.$
21. 1 (repeated), $-3, -\frac{1}{3}.$ 22. $2, \frac{1}{2}.$

Exercise 3.7

1. (a) Min $(-1, -1)$,   (b) Min $(-\frac{1}{4}, -3\frac{1}{8})$,   (c) Max $(\frac{1}{2}, 4)$,   (d) Min $(3, 1)$.
2. $x > -1.$ 3. $5, -2, \frac{1}{2}(3 \pm \sqrt{5}).$
4. (a) $0, 3,$ (b) $4, -1.$
5. $x^2 + (2q - p - p^2)x + q(p + q + 1) = 0.$
6. $3x^2 + 10x - 9 = 0.$ 7. $p = -3, q = 19.$
8. (a) $p = 7, q = -3\frac{1}{2}$ or $p = -8, q = 4.$
9. (b) $k \leqslant 0$ or $k \geqslant 3,$ (c) $k \geqslant 3.$
11. $(1\frac{1}{2}, 8).$ 12. $(k + 2, 2k\{k + 2\}).$
13. $2, -1; 3/5, 2/5.$
14. $1, 2; -1, -2; 1/2, 5/2; -1/2, -5/2.$
15. $-13, 4; (x + 1), (x + 2), (2x - 1).$
16. $-2, 2, -8.$
17. $2x^2 + 5x - 3; x = -3$ or $\frac{1}{2}.$
18. $n(n - 1)(n + 1)(n + 2).$
20. $-2 \leqslant c \leqslant 2.$
21. (a) $1, 4, -1 \pm \sqrt{5}, 1 \pm \sqrt{5},$ (b) $\frac{1}{4}b^2(b^4 + 6a^2b^2 - 3a^4).$
22. $b = 1 - a; ax^2 + (1 - a)x + 1.$

Exercise 4.1

1. $44 \cdot 72 < z < 47 \cdot 52; z = 46 \pm 2.$
2. $92 \cdot 885 < z < 95 \cdot 625; z = 94 \pm 2.$
3. $0 \cdot 3883 < z < 0 \cdot 3900; z = 0 \cdot 389 \pm 0 \cdot 001.$
4. $4 \cdot 000 < z < 4 \cdot 243; z = 4 \cdot 1 \pm 0 \cdot 2.$
5. (a) $10 \cdot 65, 10 \cdot 73; 10 \cdot 7,$ (b) $10 \cdot 69, 10 \cdot 77; 11.$
6. (a) $1 \cdot 691, 1 \cdot 703; 1 \cdot 7,$ (b) $1 \cdot 690, 1 \cdot 702; 1 \cdot 7.$
7. (a) $20,$ (b) $0 \cdot 0720,$ (c) $14 \cdot 3,$ (d) $12 \cdot 7.$

Exercise 4.2

1. (a) $f^{-1} : x \to x/3,$ (b) $f^{-1} : x \to 2x,$ (c) $f^{-1} : x \to x + 1.$
   (d) $f^{-1} : x \to 1 - x,$ (e) $f^{-1} : x \to \frac{1}{2}(x - 3),$ (f) $f^{-1} : x \to \frac{1}{4}(x + 4).$
2. (a) $y = \frac{1}{2}x,$ (b) $y = 2/x,$ (c) $y = 2 - x,$ (d) $y = x + 2,$
   (e) $y = \log_2 x,$ (f) $y = 10^x + 2.$
4. (a) $1 \cdot 172,$ (b) $64 \cdot 82,$ (c) $19\,840,$ (d) $6 \cdot 463.$
5. (a) $1 \cdot 404,$ (b) $2 \cdot 096,$ (c) $0 \cdot 4307.$
6. (a) $3 \cdot 170,$ (b) $0 \cdot 3920,$ (c) $0 \cdot 5377.$
8. (a) $1 \cdot 292,$ (b) $1 \cdot 295,$ (c) $1 \cdot 613,$ (d) $4 \cdot 962.$
9. (a) $8 \cdot 432,$ (b) $1 \cdot 453.$
10. $1 \cdot 921.$

Exercise 4.3

1. (a) 1·2, (b) 3·36.
2. (a) REAL AND DISTINCT; 5, −3, (b) REAL AND EQUAL; −0·5,
   (c) NO REAL ROOTS, (d) REAL AND DISTINCT; 8, −0·6.

Exercise 4.4

1. (a) 12, −3, (b) 3,6, (c) 16,40.
2. (a) 10·52, 5·080, (b) 4·696, 20·84, (c) 18·10, 4·102,
   (d) 93·16, −5·533.
3. (a) 9·487, (b) 1·225, (c) 0·5435.
4. (a) 0·3832, (b) 3·889.
5. (a) 5·477, (b) 35·36.
6. (a) 1·556, (b) 9·595. (c) 9·595
7. (a) 255, (b) −4058, (c) 4·335.
8. (a) 3·922, (b) 2·900, (c) 6·423.

Exercise 4.5

1. $y = 0.30x + 0.60$.   2. $R = 88 + 2.3V$.
3. $E = 0.054L + 0.53$.   4. $\theta = 7.6 - 0.32T$.
5. 16.   6. $-2.7, 11$.
7. 1·5, 10.   8. 0·65, 1·4.
9. 2·5, 4·8.   10. 5·0, −2·0.

Exercise 4.6

1. $180 \cdot 12 \text{ cm}^2$, $187 \cdot 92 \text{ cm}^2$; $184 \pm 4 \text{ cm}^2$.
2. (a) 37 cm^2, (b) $37 \cdot 4 \text{ cm}^2$, (c) $37 \cdot 4 \text{ cm}^2$.
3. (a) $f^{-1}: x \to \frac{1}{2}(1-x)$, (b) $g^{-1}: x \to \log_5 x$, (c) $h^{-1}: x \to -1/x$.
4. 1·277.   5. 0, 1·262.
6. 3·78.   7. $z = 6p - 800$.
8. (a) $x = v^2, y = u^2$, (b) $x = u, y = v/u$, (c) $x = \lg v, y = \lg u$,
   (d) $x = u, y = \lg v$.
9. 0·25, 3.   10. 20, 0·50.
11. 0·411.
12. LET $S = 0$, LET $C = N$, $C = 0$?, LET $S = SX$; −1.
13. 5, 12, 13; 9, 12, 15.

Exercise 5.1

1. (a) 9, (b) 0, (c) 10, (d) 73.
2. (a) $\frac{1}{3}$, (b) $1\frac{1}{2}$, (c) $-\frac{1}{2}$, (d) 4.
3. (a) 3, (b) 1, (c) 2, (d) 1.
4. (a) 2, (b) 2.
5. As $x \to 1$ from above, $f(x) \to 0$; as $x \to 1$ from below, $f(x) \to 1$.

6. As $x \to 0$ from above, $f(x) \to \infty$; as $x \to 0$ from below, $f(x) \to -\infty$.
7. (a) $f(x) \to 0$, (b) $g(x) \to \infty$, (c) as $x \to k$ from above, $h(x) \to \infty$;
    as $x \to k$ from below, $h(x) \to -\infty$.
8. (a) 0, (b) 0, (c) 2, (d) 2.
9. (a) -5, (b) 3/2, (c) 1, (d) $\frac{1}{2}$.

Exercise 5.2

1. -1, -0.5, -0.1, -0.01, -0.001.
2. $1, 0$, -0.8, -0.98, -0.998.
3. $2 + h$; 2.
4. (a) 6, (b) -3, (c) 1, (d) 3.
5. $6x + 3h$; $6x$.
6. (a) $2x - 1$, (b) $-1/x^2$, (c) $3x^2$.
7. $8a + 4h$; $8a$.
8. (a) 3; 3, (b) $12 - 10a - 5h$; $12 - 10a$.

Exercise 5.3

1. $2x + \delta x$; $2x$.

2. (a) $-2x - \delta x$; $-2x$, (b) $2x + 2 + \delta x$; $2x + 2$, (c) $\dfrac{-2}{x(x + \delta x)}$; $-\dfrac{2}{x^2}$.

3. (a) $4x$, (b) $10x$, (c) $4x$.
4. (a) 1, (b) -1, (c) $1 + 2x$.
5. (a) 3, (b) $\frac{1}{2}$, (c) 0.
6. (a) $-3x^2$, (b) $6x^2$, (c) $4x^3$.

7. (a) $1 + \dfrac{1}{x^2}$, (b) $-\dfrac{2}{x^3}$, (c) $-\dfrac{3}{x^4}$.

Exercise 5.4

1. (a) $5x^4$, (b) $12x^{11}$, (c) $12x^3$, (d) 0.
2. (a) 2, (b) $6x - 4$, (c) $6x^2 - 42x^5$.

3. (a) $4 - 2x$, (b) $\dfrac{3}{2}x^2 + \dfrac{2}{3}x$, (c) $5 - 16x^3$.

4. (a) $4x + 3$, (b) $18x - 6$, (c) $3x^2 + 8x + 4$.
5. (a) $-2x^{-3}$, (b) $-4/x^5$, (c) $-9/x^{10}$, (d) $-6/x^4$.

6. (a) $2x + \dfrac{1}{x^2}$, (b) $4x^3 - \dfrac{8}{x^3}$, (c) $-\dfrac{1}{x^3} + \dfrac{1}{x^4}$.

7. (a) $1 - \dfrac{1}{x^2}$, (b) $\dfrac{15}{x^4} - \dfrac{4}{x^3}$, (c) $1 + \dfrac{2}{x^2}$.

8. (a) $3x^2(x^3 + 1), 0, -84$, (b) $9(x + 1)^2, 9, 9$.

9. (a) $2 - \dfrac{2}{x^3}, 1\frac{3}{4}, 4$, (b) $2x - \dfrac{2}{x^3}, 3\frac{3}{4}, 0$.

10. (a) 5, (b) 25, (c) 2, (d) $\frac{1}{4}$.

11. (a) $y = x - 4$, (b) $y = 11x + 7$, (c) $3x + y = 2$,
    (d) $8y + 121 = 0$, (e) $y = 4x - 3$, (f) $16y = 2x - 11$.

12. (a) $10t\,\mathrm{m\,s^{-1}}$, (b) $30\,\mathrm{m\,s^{-1}}$, $50\,\mathrm{m\,s^{-1}}$.

13. (a) $27 - 6t\,\mathrm{m\,s^{-1}}$, (b) $9\,\mathrm{m\,s^{-1}}$, $-3\,\mathrm{m\,s^{-1}}$.

14. 4π.   15. -2.

16. $3, -4$.   17. $2, \frac{1}{2}$.

18. (a) $\frac{1}{3}x^{-2/3}$, (b) $\frac{3}{2}x^{1/2}$, (c) $-\frac{1}{4}x^{-5/4}$, (d) $\frac{1}{5}x^{-4/5} = 1/(5\sqrt[5]{x^4})$,
    (e) $-\frac{1}{2}x^{-3/2} = -1/(2\sqrt{x^3})$, (f) $\frac{2}{3}x^{-2/3} = 2/(3\sqrt[3]{x^2})$.

19. (a) $\frac{1}{2}x^{-1/2} - \frac{3}{2}x^{-5/2}$, (b) $3x^{1/2} - \frac{1}{2}x^{-1/2}$, (c) $1 + \frac{1}{\sqrt{x}}$,
    (d) $\frac{1}{2\sqrt{x}} + \frac{1}{2\sqrt{x^3}}$, (e) $\frac{5}{2}x\sqrt{x} + \frac{3}{2}\sqrt{x}$, (f) $\frac{3}{2}\sqrt{x} + \frac{3}{2\sqrt{x}} - \frac{3}{2x\sqrt{x}} - \frac{3}{2x^2\sqrt{x}}$.

20. $4y = 5x + 4$.

Exercise 5.5

1. $(1, -4)$ min.   2. $(-\frac{3}{4}, 3\frac{1}{8})$ max.

3. $(-\frac{1}{2}, -2\frac{1}{4})$ min.   4. $(\frac{1}{2}, 2)$ min.

5. $(0,0)$ max, $(2, -4)$ min.   6. $(0,1)$ inflexion.

7. $(2, -8)$ inflexion.   8. $(-2,0)$ max, $(1, -27)$ min.

9. $(-1/3, -11/27)$ min, $(\frac{1}{2}, \frac{3}{4})$ max.

10. $(-1\frac{1}{4}, 15\frac{3}{16})$ max, $(\frac{1}{2}, -6\frac{1}{4})$ min.

11. $(-4, 256)$ max, $(0,0)$ min.   12. $(0,0)$ inflexion, $(1,1)$ max.

13. None.   14. None.

15. $(\frac{2}{3}, 3\frac{2}{3})$ max, $(2\frac{2}{3}, -32\frac{1}{3})$ min.

16. $(-1,4)$ min, $(0,6)$ max, $(2\frac{1}{2}, -17\frac{7}{16})$ min.

17. $(-1\frac{1}{2}, -12)$ max, $(1\frac{1}{2}, 12)$ min.   18. $(2,3)$ min.

19. $(0, -64)$ inflexion, $(3, 17)$ inflexion.

20. $(\frac{1}{4}, -\frac{1}{4})$ min.

21. $(0,21)$.   22. $-3, -9, 12$.

23. (i) $a^2 > 3b$, (ii) $a^2 = 3b$, (iii) $a^2 < 3b$.

24. $x = 0, y = x$.

Exercise 5.6

1. 6.   2. $30x - 2$.

3. $12x^2 - 12x + \dfrac{2}{x^3}$.   4. $-\dfrac{8}{x^3} - \dfrac{24}{x^4}$.

5. $-\dfrac{1}{4}x^{-3/2} + \dfrac{3}{4}x^{-5/2}$.   6. $\dfrac{3}{2}x^{-1/2} + \dfrac{1}{4}x^{-3/2} - \dfrac{9}{2}x^{-5/2}$.

7. $(-1, 2)$ max; $(1, -2)$ min.

8. $(-1, 0)$ min; $(0, 1)$ max; $(1, 0)$ min.

9. $(0,0)$ inflexion; $(1\frac{1}{2}, -1\frac{11}{16})$ min.

10. $(-\frac{1}{2}, -7)$ max; $(\frac{1}{2}, 1)$ min.

11. $(1,2)$ max; $(2,1)$ min.

12. $(-1,0)$ min; $(0,5)$ max; $(2,-27)$ min.
13. (a) $4, -17, 50$,   (b) $3, 0, 6$.
14. $(-1,-27)$ min; $(2,0)$ inflexion.
15. $(1,0)$, 3.
16. (a) $v = 18t - 9t^2$, $a = 18 - 18t$,   (b) $0\,\mathrm{m\,s^{-1}}, 18\,\mathrm{m\,s^{-2}}$,
    (c) $-27\,\mathrm{m\,s^{-1}}, -36\,\mathrm{m\,s^{-2}}$.
17. (a) $v = 24t^2 - 66t + 27$, $a = 48t - 66$,   (b) $27\,\mathrm{m\,s^{-1}}, -66\,\mathrm{m\,s^{-2}}$,
    (c) $45\,\mathrm{m\,s^{-1}}, 78\,\mathrm{m\,s^{-2}}$.
18. $2, -8, 5$.
19. (a) $\left(\dfrac{1}{\sqrt[3]{2}}, \dfrac{3}{2}\sqrt[3]{2}\right)$ min, $(-1,0)$ inflexion.

    (b) $(2,\frac{1}{4})$ max, $(3,\frac{2}{9})$ inflexion.

    (c) $\left(-\sqrt{3}, -\dfrac{2}{9}\sqrt{3}\right)$ min, $\left(\sqrt{3}, \dfrac{2}{9}\sqrt{3}\right)$ max, $\left(-\sqrt{6}, -\dfrac{5}{36}\sqrt{6}\right)$ inflexion,

$\left(\sqrt{6}, \dfrac{5}{36}\sqrt{6}\right)$ inflexion.
20. $(0,0)$.

Exercise 5.7

1. (a) 2,   (b) -4,   (c) 1.
2. (a) 0,   (b) 2,   (c) 0.
3. (b) 4.
4. (a) $3(x+1)^2$,   (b) $-(x+1)^{-2}$.
5. $\frac{1}{2}x^{-1/2}$.
6. (a) $6(3x+1)$,   (b) $1 - \dfrac{6}{x^2}$,   (c) $\dfrac{3}{2}\sqrt{x} - \dfrac{1}{\sqrt{x}}$.
7. $y = 4(x+1)$, $y = -3x$, $y = 12(x-3)$.
9. (a) $(0, -4)$ min,   (b) $\left(-\dfrac{2}{3}\sqrt{3}, \dfrac{16}{9}\sqrt{3}\right)$ max; $\left(\dfrac{2}{3}\sqrt{3}, -\dfrac{16}{9}\sqrt{3}\right)$ min,
    (c) $(-\sqrt{2}, -4)$ min; $(0,0)$ max; $(\sqrt{2}, -4)$ min,   (d) no turning points,
    (e) no turning points,   (f) $\left(2\sqrt{3}, \dfrac{1}{9}\sqrt{3}\right)$ max; $\left(-2\sqrt{3}, \dfrac{-1}{9}\sqrt{3}\right)$ min.
10. $(-2, -3\frac{3}{4})$ max; $(1,3)$ inflexion.
11. (a) $v = 7 + 10t - 6t^2\,\mathrm{m\,s^{-1}}$, $a = 10 - 12t\,\mathrm{m\,s^{-2}}$,
    (b) $7\,\mathrm{m\,s^{-1}}, 10\,\mathrm{m\,s^{-2}}$; $3\,\mathrm{m\,s^{-1}}, -14\,\mathrm{m\,s^{-2}}$.
14. $-b/3a$.

Exercise 6.1

3. (a) $\frac{1}{3}x^3 + c$,   (b) $-\dfrac{1}{x} + c$,   (c) $2x^4 + c$,   (d) $-\dfrac{3}{x^2} + c$.
4. $x^3 + 2x^2 + c$.
5. $\dfrac{1}{6}x^6 - x + c$.
6. $\dfrac{4}{3}x^3 + 2x^2 + x + c$.
7. $3x^2 + \dfrac{1}{12x^2} + c$.

8. $\frac{1}{3}x^3 + 2x - \frac{1}{x} + c.$

9. $-\frac{1}{x} + \frac{1}{3x^3} + c.$

10. $\frac{3}{4}x^{4/3} + \frac{3}{2}x^{2/3} + c.$

11. $\frac{2}{3}x^{3/2} - 2x^{1/2} + c.$

12. $\frac{2}{5}x^{5/2} + 2x^{-1/2} + c.$

13. (a) $t^3 + 7t^2 - 24t + c$, (b) $-\frac{4}{t} - \frac{2}{t^2} - \frac{1}{3t^3} + /c,$

   (c) $\frac{4}{5}t^{5/2} - \frac{10}{3}t^{3/2} + 4t^{1/2} + c.$

14. $x^3 - 4x^2 + 5x - 2.$

15. $-\frac{3}{4}.$

16. $y = x^3 - 4x - 1.$

17. $y = 6 + 4x - 2x^2.$

18. $2\sqrt{h^3} - \sqrt{h} - 4.$

19. $x^3 - 5x^2 + 8x - 3.$

20. $-5.$

Exercise 6.2

1. (a) $3x^2 + x + c$, (b) $3x^3 - 3x^2 + x + c$, (c) $\frac{4}{3}x^3 + \frac{1}{2}x^2 + c.$

2. (a) $3x + \frac{1}{x} + c$, (b) $\frac{1}{3}x^3 - 2x - \frac{1}{x} + c$, (c) $-\frac{3}{x} - \frac{1}{2x^2} + c.$

3. (a) $\frac{8}{3}x^{3/2} + c$, (b) $\frac{2}{3}x^{3/2} - 2x^{1/2} + c$, (c) $3x^{4/3} + 6x^{2/3} + c.$

4. (a) $t^3 + \frac{1}{t} + c$, (b) $\frac{1}{2}u^2 - \frac{2}{5}u^{5/2} + c$, (c) $s - \frac{1}{2s^2} + c.$

5. (a) 6, (b) 22, (c) $82\frac{2}{3}.$

6. (a) $4\frac{5}{6}$, (b) $2\frac{7}{24}$, (c) 54.

7. (a) -36, (b) 60, (c) $6\frac{1}{2}.$

8. (a) $1\frac{1}{3}$, (b) $2\frac{1}{4}$, (c) 8, (d) 38.

9. (a) $4\frac{1}{2}$, (b) $6\frac{3}{4}$, (c) $-5\frac{1}{3}$, (d) $-9\frac{3}{5}.$

10. (a) $4\frac{1}{2}$, (b) 36, (c) $1\frac{1}{3}$, (d) $1/12$, (e) $\frac{1}{3}$, (f) $\frac{3}{4}.$

11. $6\frac{2}{3}.$

12. $(0,0), (1,1); \frac{1}{3}.$

Exercise 6.3

1. (a) $21\frac{1}{3}$, (b) $13\frac{1}{2}$, (c) $3\frac{1}{12}$, (d) $1\frac{1}{3}.$

2. 36.

3. $10\frac{2}{3}.$

4. $20\frac{1}{4}.$

5. 1.

6. (a) $\frac{1}{2}$, (b) $1\frac{1}{3}$, (c) $10\frac{2}{3}.$

7. (a) 4, (b) $5\frac{5}{24}$, (c) $2\frac{2}{3}.$

Exercise 6.4

1. (a) 1220π, (b) $47\pi/10$, (c) 9π, (d) $405\pi/14$, (e) $377\pi/6$, (f) $57\pi/8.$

2. (a) $512\pi/15$, (b) $81\pi/35.$

3. (a) $15\pi/4$, (b) 36π.
4. (a) $32\pi/5$, (b) $7\pi/3$, (c) $\pi/2$, (d) $96\pi/5$.
5. (a) $202\pi/5$, (b) 20π.
6. (a) $9\pi/2$, (b) $53\pi/15$.
7. $\frac{1}{3}\pi r^2 h$.

8. $\frac{4}{3}\pi a^3$.

9. $32\pi/3$.

10. $8\pi/7$.

Exercise 6.5

1. $28\pi/15$.

2. $16\pi/15$.

3. (a) $24\,\text{m}$, (b) $21\,\text{m}$, (c) $14\frac{1}{3}\,\text{m}$, (d) $47\,\text{m}$, (e) $30\,\text{m}$, (f) $41\,\text{m}$.
4. $64\pi/15$.

5. 8π.

6. $\frac{1}{3}a^2 h\,\text{cm}^3$.

7. $960\sqrt{3}\,\text{cm}^3$.

Exercise 6.6

1. (a) $22\,\text{m}\,\text{s}^{-1}$, (b) $78\,\text{m}\,\text{s}^{-2}$.
2. (a) $3/4\,\text{m}\,\text{s}^{-1}$, (b) $4/5\,\text{m}\,\text{s}^{-2}$.
3. (a) 1, (b) $8\frac{2}{3}$, (c) -9, (d) 16.
4. $85\pi/6$.
5. 8.
6. (a) 93, (b) $1/64$, (c) $496/15$.

Exercise 6.7

1. (a) $y = 2x - x^2 + c$, (b) $y = x^{1/2} + c$, (c) $y = x - \dfrac{1}{2x^2} + c$.

2. (a) $\dfrac{2}{3}x^6 - \dfrac{1}{4}x^4 + c$, (b) $4x + \dfrac{3}{x} + c$, (c) $\dfrac{1}{2}x^2 - \dfrac{1}{2x^2} + c$.

3. (a) $4t - 6t^2 + 3t^3 + c$, (b) $x^5 + \dfrac{1}{15x^3} + c$, (c) $\dfrac{3}{2}u^{2/3} + c$.

5. $y = x^2 + \dfrac{2}{x} - 3$.

6. (a) $-\frac{11}{24}$, (b) $4\frac{1}{4}$.

7. 1.

8. $2\frac{2}{3}$.

9. $2\frac{1}{2}$.

10. $1\frac{1}{3}$.

11. $y = 2x - 2$; $1\frac{1}{3}$.

12. $4\frac{1}{2}$.

13. $\frac{1}{3}$.

14. 2.

15. (a) 24π, (b) 12π.
16. (a) $52\pi/5$, (b) 18π, (c) $76\pi/15$, (d) $10\pi/3$.
17. (a) $4\pi/3$, (b) $16\pi/15$.

19. $\pm\left(\dfrac{3}{2\sqrt{x}} - \dfrac{3}{2}\sqrt{x}\right)$; (a) 1, (b) 0; $27\pi/4$.

20. 3.

21. 10.

22. (a) $41\frac{23}{25}\,\text{m}$, (b) $101\,\text{m}$.

24. $\frac{2}{3}\pi a^2$.

Exercise 7.1

1. (a) $-\cos 15°$, (b) $-\sin 35°$, (c) $-\tan 50°$, (d) $\sin 68°$,
   (e) $\tan 85°$, (f) $\cos 42°$, (g) $-\sin 15°$, (h) $-\cos 57°$.
2. (a) $-\tan 56°$, (b) $\sec 23°$, (c) $\cot 68°$, (d) $-\sin 18°$,
   (e) $-\mathrm{cosec}\,80°$, (f) $-\cot 15°$, (g) $-\cos 6°$, (h) $-\sec 81°$.
3. (a) $-320°, -220°, 40°, 140°, 400°, 500°$,
   (b) $-350°, -10°, 10°, 350°, 370°, 710°$,
   (c) $-285°, -105°, 75°, 255°, 435°, 615°$,
   (d) $-180°, 180°, 540°$.
4. (a) $25°, 335°$, (b) $50°, 130°$, (c) $42·3°, 222·3°$,
   (d) $243·5°, 296·5°$.
5. (a) $\sqrt{3}/2$, (b) $\sqrt{3}$, (c) $-\sqrt{3}/2$, (d) -1, (e) $-1/\sqrt{2}$, (f) $\sqrt{3}/2$,
   (g) -2, (h) $2/\sqrt{3}$.
6. (a) 0, (b) $-\sqrt{3}$, (c) 0, (d) -2, (e) 0, (f) -1, (g) $-\sqrt{2}$, (h) -1.
7. (a) $\theta = 90°, 270°, \phi = 0°, 180°, 360°$,
   (b) $\theta = 0°, 180°, 360°, \phi = 90°, 270°$,
   (c) $\theta = 90°, \phi = 0°, 360°$,
   (d) $\theta = 90°, \phi = 180°$.
8. (a) $0° < \theta < 45°$, (b) $30° < \theta < 90°$, (c) $36° < \theta < 72°$,
   (d) $0° < \theta < 15°, 30° < \theta < 45°, 60° < \theta < 75°$.

Exercise 7.2

1. (a) $17·5°, 162·5°$, (b) $56·3°, 236·3°$, (c) $136·9°, 223·1°$,
   (d) $218·3°, 321·7°$.
2. (a) $-53·1°, 53·1°$, (b) $53·1°, 126·9°$, (c) $-31·8°, 148·2°$,
   (d) $-110°, 110°$.
3. (a) $45°, 135°$, (b) $150°, 330°$, (c) $45°, 225°$, (d) $120°, 240°$,
   (e) $180°$, (f) $30°, 150°$.
4. (a) $36·9°, 216·9°$, (b) $135°, 315°$, (c) $48·6°, 131·4°$
   (d) $48·2°, 311·8°$.
5. (a) $106·1°, 233·9°$, (b) $41·9°, 198·1°$, (c) $97·2°, 342·8°$,
   (d) $119·2°, 299·2°$, (e) $17·5°, 107·5°, 197·5°, 287·5°$,
   (f) $210·4°$, (g) $112·2°, 247·8°$, (h) $90°, 270°$
6. (a) $-75°, 15°, 45°$, (b) $-90°, -45°, 0°, 45°, 90°$,
   (c) $-81°, -45°, -9°, 27°, 63°$, (d) $-80°, -40°, 40°, 80°$.
7. (a) $60°, 120°, 240°, 300°$, (b) $60°, 90°, 270°, 300°$,
   (c) $0°, 78·5°, 180°, 281·5°, 360°$, (d) $270°$,
   (e) $0°, 30°, 150°, 180°, 360°$, (f) $56·3°, 90°, 236·3°, 270°$.
8. (a) $-135°, -26·6°, 45°, 153·4°$, (b) $-165·5°, -14·5°$,
   (c) $-60°, 0°, 60°$, (d) $-111·8°, -45°, 68·2°, 135°$.
9. (a) $x = 30°, y = 150°; x = 120°, y = 60°$,
   (b) $x = -45°, y = -165°; x = 75°, y = -45°; x = 135°, y = 15°$.

Exercise 7.3

2. (a) $\sin \theta$, (b) $\cos \theta$, (c) $\sin \theta$, (d) $-\cos \theta$, (e) $-\cos \theta$, (f) $-\cos \theta$.

3.　(a) $-\cot\theta$,　(b) $\tan\theta$,　(c) $\cot\theta$,　(d) $-\tan\theta$,　(e) $\tan\theta$,　(f) $-\cot\theta$.

4.　$3/5, 3/4, 5/4$.　　　　　　　　　5.　$8/17, -17/15, -15/8$.

6.　$-24/25, -7/25, -25/24$.　　　7.　$5/13, -12/13, 12/5$.

8.　(a) $\cos\theta$,　(b) $\tan\theta$,　(c) $\sin\theta$.

9.　(a) 1,　(b) $\sin^4\theta$,　(c) $\cos\theta+\sin\theta$,　(d) $1+\cos\theta$.

11.　(a) $x^2+y^2=1$,　(b) $(x-1)^2/4+(y-1)^2/9=1$,

　　　(c) $x^2-y^2=1$,　(d) $x^2+y^2=2$.

12.　$t=2/5, \theta=126\cdot9°$ or $t=-2/5, \theta=306\cdot9°$.

13.　$199\cdot5°, 340\cdot5°$.　　　　　　　　14.　$41\cdot4°, 120°, 240°, 318\cdot6°$.

15.　$60°, 300°$.　　　　　　　　　　16.　$51\cdot3°, 128\cdot7°$.

17.　$70\cdot5°, 120°, 240°, 289\cdot5°$.　　　18.　$19\cdot5°, 30°, 150°, 160\cdot5°$.

19.　$26\cdot6°,\ 116\cdot6°,\ 206\cdot6°,\ 296\cdot6°$.

20.　$18\cdot4°, 135°, 198\cdot4°, 315°$.　　　21.　$63\cdot4°, 135°, 243\cdot4°, 315°$.

22.　$45°, 76°, 225°, 256°$.

23.　$(\sin x+1)^2-2; 2, -2$.　　　　　24.　$\{y:y\geqslant 1\}$.

Exercise 7.4

1.　(a) $\frac14(\sqrt6+\sqrt2)$,　(b) $\frac14(\sqrt2-\sqrt6)$,　(c) $\sqrt3-2$,　(d) $2-\sqrt3$.

2.　(a) $\frac12$,　(b) $\frac12$,　(c) 0,　(d) 1.

3.　(a) $\sqrt3$,　(b) $1/\sqrt3$,　(c) $\frac12\sqrt6$,　(d) $\sqrt2$.

4.　$56/65, -33/65$.　　　　　　　5.　$-29/35, -8\sqrt6/35$.

6.　$225°$.　　　　　　　　　　　7.　$4\frac12$.

8.　(a) $2\sin A\cos B$,　(b) $2\cos A\cos B$,　(c) $\cos B$.

9.　$12°, 60°, 84°, 132°, 156°$.

10.　$-135°, 45°$.

11.　(a) $26\cdot6°, 108\cdot4°, 206\cdot6°, 288\cdot4°$,

　　　(b) $40\cdot9°, 220\cdot9°$,　(c) $0°, 180°, 360°$,　(d) $135°, 315°$.

12.　$\dfrac{\cot A\cot B-1}{\cot A+\cot B}$.

Exercise 7.5

1.　$24/25.\ -7/25, 44/125$.　　　2.　$1/9, -4\sqrt5, -79/81$.

3.　$4/5, 4/3, 5\frac12$.　　　　　　　4.　$-3/4, -\sqrt7/3, 2\sqrt2$.

5.　$-1/\sqrt5, 2/\sqrt5, -2/5\sqrt5$.

6.　(a) $1+1/\sqrt2$,　(b) $\sqrt3/2$,　(c) $1/4$,　(d) $1/\sqrt2$.

7.　(a) $0°, 60°, 180°$,　(b) $0°, 120°$,　(c) $0°, 60°, 120°, 180°$,

　　　(d) $0°, 45°, 135°, 180°$.

16.　(a) $120°, 240°$,　(b) $41\cdot8°, 138\cdot2°, 270°$,

　　　(c) $45°, 135°, 225°, 315°$,　(d) $0°, 104\cdot5°, 180°, 255\cdot5°, 360°$.

17.　(a) $135°, 161\cdot6°, 315°, 341\cdot6°$,

　　　(b) $23\cdot2°, 135°, 203\cdot2°, 315°$.

18.　(a) $-126\cdot9°, 43\cdot6°$,　(b) $-90°, 53\cdot1°$,　(c) $-143\cdot1°, 90°$,

　　　(d) $-36\cdot9°, 112\cdot6°$.

19.　$45°, 135°, 225°, 315°$.

20.　$4\cos^3\theta-3\cos\theta; 90°, 120°, 240°, 270°$.

Exercise 7.6

1. (a) $\sin 3x + \sin x$, (b) $\cos 8\theta + \cos 2\theta$, (c) $3(\cos A - \cos 5A)$,
   (d) $\frac{1}{2}(\sin 6t - \sin 2t)$, (e) $\cos 2A + \cos 2B$, (f) $\frac{1}{4}(1 + 2\sin 2x)$.
2. (a) $1 + \sqrt{3}$, (b) $5\sqrt{2}$, (c) $2 + \sqrt{3}$, (d) $3 + 2\sqrt{2}$.
3. (a) $0°$, $180°$, $360°$, (b) $60°$, $120°$, $240°$, $300°$,
   (c) $0°$, $60°$, $180°$, $240°$, $360°$,
   (d) $111.3°$, $168.7°$, $291.3°$, $348.7°$.
4. (a) $2\sin 4x \cos 3x$, (b) $2\cos 4A \cos A$, (c) $2\cos\theta \sin\alpha$,
   (d) $-2\sin 2x \sin x$.
5. (a) $\cot 5\theta$, (b) $-\tan 3x$, (c) $\cot 2A$, (d) $-\tan^2\theta$.
6. (a) $\sqrt{3}$, (b) $1/\sqrt{3}$.
7. (a) $0°$, $90°$, $180°$, (b) $0°$, $30°$, $90°$, $150°$, $180°$,
   (c) $30°$, $120°$, $150°$, (d) $22\frac{1}{2}°$, $45°$, $112\frac{1}{2}°$.
8. (a) $0°$, $90°$, $120°$, $180°$, (b) $15°$, $30°$, $75°$, $90°$, $150°$,
   (c) $0°$, $54°$ $126°$.

Exercise 7.7

1. (a) $\sqrt{2}\cos(\theta + 45°)$; $\sqrt{2}$, $\theta = -45°$, $315°$; $-\sqrt{2}$, $\theta = -225°$, $135°$,
   (b) $2\cos(\theta + 30°)$; 2, $\theta = -30°$, $330°$; -2, $\theta = -210°$, $150°$,
   (c) $13\cos(\theta - 67.4°)$; 13, $\theta = -292.6°$, $67.4°$; -13, $\theta = -112.6°$,
   $247.4°$,
   (d) $\sqrt{5}\cos(\theta - 26.6°)$; $\sqrt{5}$, $\theta = -333.4°$, $26.6°$; $-\sqrt{5}$, $\theta = -153.4°$,
   $206.6°$.
2. (a) $41.6°$, $244.7°$, (b) $26.2°$, $110.2°$, (c) $180°$, $313.6°$,
   (d) $98°$, $205.9°$.
3. $\sqrt{89}\sin(\theta - 58°)$; (a) $97.5°$, $198.5°$, (b) $90°$, $206°$.
4. (a) $10.9°$, $231°$, (b) $257.6°$, $349.8°$.

Exercise 7.8

1. (a) $70°$, $110°$, (b) $115°$, $155°$, $295°$, $335°$, (c) $145°$, $215°$,
   (d) $25°$, $85°$, $145°$, $205°$, $265°$, $325°$.
2. (a) 0.8, (b) -0.6, (c) -0.8, (d) -0.6, (e) -0.75, (f) 0.75.
3. (a) $56.3°$, $236.3°$, (b) $48.2°$, $120°$, $240°$, $311.8°$,
   (c) $0°$, $53.1°$, $180°$, $306.9°$, $360°$, (d) $23.6°$, $156.4°$.
4. (a) $2k^2 - 1$, (b) $(4k^2 - 1)\sqrt{(1 - k^2)}$, (c) $\sqrt{\left(\dfrac{1 + k}{2}\right)}$. (d) $\sqrt{\left(\dfrac{1 - k}{1 + k}\right)}$.
5. (a) $(x - 1)^2 + 4y^2 = 1$, (b) $xy = 1$, (c) $y = 2x^2$,
   (d) $y^2 = 4x(2 - x)(x - 1)^2$.
6. (a) $10.9°$, $100.9°$, (b) $11.8°$, $78.2°$,
   (c) $0°$, $45°$, $135°$, $180°$,
   (d) $0°$, $60°$, $90°$, $120°$, $180°$.
7. $56.3° \leqslant x \leqslant 236.3°$.   8. 8.
9. (a) max 1, min -1, (b) max 2, min 0, (c) max 2, min -2,
   (d) max 4, min 0.

11. (a) $79.7°$, $153.4°$,   (b) $\left(\dfrac{\lambda + 1}{\lambda - 1}\right) \tan \alpha$; $120°$, $300°$.

12. (a) $45°$, $120°$, $135°$, $225°$, $240°$, $315°$,
    (b) $5\cos(x - 36.9°)$; $103.3°$, $330.4°$,
    (c) $90°$, $270°$; $0°$, $180°$, $360°$.

13. $\dfrac{3\tan\theta - \tan^3\theta}{1 - 3\tan^2\theta}$; $0°$, $19.1°$, $160.9°$, $180°$.

14. $\dfrac{2t}{1 - t^2}$, $\dfrac{1 + t^2}{1 - t^2}$; $18°$.

15. (a) $25\sin(x - 73.7°)$; $110.6°$, $216.9°$,   (b) $60°$, $60°$.

16. $90°$, $323.1°$; $53.1°$, $180°$.   18. (b) $130.2°$, $342.4°$.

19. $36.9°$; $-36.9°$, $143.1°$; $0°$, $-73.7°$.

20. (a) $49.8°$, $130.2°$,   (b) $26.6°$, $108.4°$,   (c) $30°$, $90°$, $150°$,
    (d) $22\frac{1}{2}°$, $30°$, $67\frac{1}{2}°$, $112\frac{1}{2}°$, $150°$, $157\frac{1}{2}°$.

21. $36.9°$, $61.9°$.

22. (a) $30°$,   (b) $180°$,   (c) $-45°$,   (d) $30°$,   (e) $90°$,   (f) $90°$ or $-90°$,
    (g) $45°$,   (h) $45°$.

23. $k = 2$; $-\sqrt{2}$, $\frac{1}{2}(\sqrt{2} \pm \sqrt{6})$.

24. (a) $0°$, $48.6°$, $131.4°$, $180°$, $270°$,
    (b) $R = \sqrt{(2\lambda^2 - 2\lambda + 1)}$, $\tan\phi = (1 - \lambda)/\lambda$; $\min -\sqrt{(2\lambda^2 - 2\lambda + 1)}$.

Exercise 8.1

1. $\angle B = \angle A + \angle C$.

2.   (a) $x + y = 180°$,   (b) $z = 2y - 180°$.

3. $5 : 12$.

6. A is the incentre of $\triangle DEF$ when $\angle A$ is obtuse.

Exercise 8.2

1. $C = 70°$, $a = 4.61$, $b = 4.08$.

2. $B = 61°$, $a = 6.69$, $c = 8.70$.

3. $A = 33.6°$, $B = 95.2°$, $c = 7.04$.

4. $A = 26.5°$, $C = 108.5°$, $b = 3.80$.

5. $A = 40.8°$, $B = 60.6°$, $C = 78.6°$.

6. $A = 36.9°$, $B = 90°$, $C = 53.1°$.

7. $B = 67.3°$, $C = 45.3°$, $c = 7.71$.

8. $B = 35.6°$, $C = 23.4°$, $a = 10.2$.

9. $A = 103°$, $b = 2.57$, $c = 5.31$.

10. $A = 87.1°$, $C = 37.9°$, $a = 14.6$.

11. $A = 112.4°$, $B = 29.5°$, $C = 38°$.

12. $A = 60°$, $B = 90°$, $a = 5.20$.

13. 4.83, 1.86.   14. 15.9, 3.97.

15. $27.5\,\text{cm}^2$.   16. $30°$; $14\,\text{cm}$.

17. $4.07\,\text{cm}$, $11.5\,\text{cm}$.   18. $4.21\,\text{cm}$, $2.26\,\text{cm}$.

21. (a) $82.9°$, $45.1°$,   (b) $115.9°$, $24.1°$.

22. $\triangle = rs$; $r = \sqrt{3} = 1.73$.

Exercise 8.3

1. 13·4 km; 098°.
2. 14·1 km; S 23·9° E.
3. 19 min 15 sec.
4. 6·55 cm.
5. 13·5.
6. 20·2 m.
7. 1350 m.
8. 71·1 cm^2.
9. 10 cm; 36·9°.
10. 2/3, 7 cm.

Exercise 8.4

1. (a) 17·2°,   (b) 44·4°,   (c) 18·1°.
2. (a) 44·4°,   (b) 63°,   (c) 90°.
3. (a) 58°,   (b) 53·9°,   (c) 63·4°.
4. (a) 27°,   (b) 32·7°.
5. (a) 53°,   (b) 61·9°,   (c) 64·8°,   (d) 90°.
6. (a) 49·2°,   (b) 65·4°.
7. (a) $a\sqrt{2}$,   (b) 70·5°,   (c) 109·5°.
8. (a) 1/3,   (b) $2\sqrt{6}$.
9. (a) 22·5°,   (b) 39·7°,   (c) 112·9°,   (d) 117·8°.
10. (a) 73°,   (b) 83·3°,   (c) 70·6°,   (d) 96°.

Exercise 8.5

1. 76·7°, 474 cm^2.
2. 93·7 m, 56·5°.
3. (a) 27·6°,   (b) 53·4°.
4. 14·7°; 13·4° or 87·4°.
5. 060·5°, 38·2°.
6. 051·4°, 125·7°.
7. 20·4°, 43·2°.
8. 2, 26·6°, 18·4°.
10. 9·17 m, 15·6 m; 19·1°, 26·6°.
11. 6·6.

Exercise 8.6

1. 33·6°, 7·54 cm.
3. 62·6°, 91·1°, 153·7°; 7·70 cm.
5. (i) 2; $c = \frac{1}{2}(5\sqrt{3} \pm \sqrt{11})$,   (ii) none,   (iii) 1; $c = \sqrt{3} + 2\sqrt{2}$,
    (iv) 1; $c = 2\sqrt{2} - \sqrt{3}$.
9. $\tan^2\frac{1}{2}A = \dfrac{(s-b)(s-c)}{s(s-a)}$; area $= \sqrt{\{s(s-a)(s-b)(s-c)\}}$.
10. 124°, 28°, 28°.
11. S2θ° W.
12. 32·3 km h^{-1}.
15. (i) $\frac{1}{2}a\sqrt{10}$,   (ii) $a\sqrt{3}$,   (iii) 90°,   (iv) $\frac{1}{3}\sqrt{2}$.
16. (i) 90°,   (ii) 30°,   (iii) 45°.
17. 60°, 41·4°; 49·1°.
20. 60 cm; 688 cm^2.
21. $a\sqrt{6}$.
22. (i) $2a\sqrt{5}$,   (ii) $-1/5$,   (iii) $4/\sqrt{30}$.
23. (i) 836 m,   (ii) 042·9°,   (iii) 13·8°.

Exercise 9.1

1. (a) **PS**, (b) **AE**, (c) **AB**, (d) **0**, (e) **MR**, (f) **PS**.
2. (a) false, (b) true, (c) true, (d) true, (e) false, (f) true.
3. (a) true, (b) false, (c) true, (d) true, (e) false, (f) false.
4. $x+y$, $y+z$, $x+y+z$.
5. $a+d$, $d-b$, $d-b-c$, $a+d-c$.

Exercise 9.2

1. (a) 3 km E, (b) 10 km N, (c) 1 km N 30° E, (d) 1 km S.
2. (a) $2\sqrt{2}$ km NE, (b) $2\sqrt{2}$ km NE, (c) $2\sqrt{2}$ km SE, (d) $2\sqrt{2}$ km NW.
3. (a) 1 km S 30° E, (b) $\sqrt{3}$ km S, (c) $\sqrt{3}$ km S 60° E, (d) $\sqrt{3}$ km N 60° E.
4. $p+q$, $q-p$, $\frac{1}{2}(p+q)$, $\frac{1}{2}(q-p)$.
5. $-a$, $a-b$, $2b$, $2b-a$.   6. $2a+b$, $a+b$.
7. $2a-b$, $a+b$, $b-a$, $\frac{1}{2}b$.
8. (a) $2x+y$, (b) $u-2w$, (c) $3p+7q$.
16. **a** must be a positive multiple of **b**.

Exercise 9.3

1. (a) $2b$, (b) $-3a$, (c) $2b-6a$, (d) $b-a$, (e) a, (f) $2b-a$.
2. (a) z, (b) x, (c) $-x$, (d) z, (e) $-x+z$, (f) $y-z$.
3. (a) $-x+y$, (b) $x+z$, (c) $-x+y-z$, (d) $2x-y$,
     (e) $2x-y+2z$, (f) $-y+2z$.
4. (a) $5i+3j$, (b) $-i+6j$, (c) $-4i+2j$, (d) $2i-j$,
     (e) $3i+4j$, (f) $i+5j$.
5. (a) $-4i+5j+2k$, (b) $3j+2k$, (c) $i-j+k$, (d) $i+2j+3k$.
6. (a) 13, (b) $\sqrt{5}$, (c) 10, (d) 3, (e) 7, (f) $5\sqrt{3}$.
7. (a) $\frac{8}{9}i - \frac{4}{9}j - \frac{1}{9}k$, (b) $\frac{8}{9}i - \frac{4}{9}j - \frac{1}{9}k$, (c) $-\frac{8}{9}i + \frac{4}{9}j + \frac{1}{9}k$.
8. (a) 5, (b) 3, (c) 6.   9.   (a) $\sqrt{29}$, (b) $\sqrt{29}$, (c) 5.
10. (a) $\frac{1}{\sqrt{2}}i - \frac{1}{\sqrt{2}}j$, (b) $-\frac{7}{9}i + \frac{4}{9}j + \frac{4}{9}k$, (c) i.
11. ± 2.   12. $10 \leqslant |a+b| \leqslant 40$.
14. (a) $\frac{1}{2}, \frac{1}{2}$, (b) $-4, -2$.

Exercise 9.4

1. $b-a$, $b-2a$; $2b-2a$, $2b-3a$, $b-2a$.
2. $a-b$, $c-b$, $a-2b+c$; $2a-3b+2c$.
3. $\frac{1}{4}(a+b+c+d)$.
4. $2i+3j$, $5i-j$, $4i+4j$; $-i+5j$; $i+8j$, $(1,8)$.
5. (a) 6, $-\frac{1}{3}i+\frac{2}{3}j+\frac{2}{3}k$, $2i-3j+k$,
     (b) 22, $\frac{6}{11}i - \frac{9}{11}j - \frac{2}{11}k$, $-i+4j-2k$,

(c) $\sqrt{10}, \dfrac{3}{\sqrt{10}}\mathbf{j} - \dfrac{1}{\sqrt{10}}\mathbf{k}, \mathbf{i} + \dfrac{1}{2}\mathbf{j} + \dfrac{5}{2}\mathbf{k},$

(d) $3\sqrt{6}, -\dfrac{2}{\sqrt{6}}\mathbf{i} - \dfrac{1}{\sqrt{6}}\mathbf{j} - \dfrac{1}{\sqrt{6}}\mathbf{k}, 2\mathbf{i} + \dfrac{5}{2}\mathbf{j} - \dfrac{1}{2}\mathbf{k}.$

6. $2\mathbf{i} - \mathbf{j} + 5\mathbf{k}, 2\mathbf{j} - \mathbf{k}, -2\mathbf{i} + 4\mathbf{j} + 3\mathbf{k}$; 7, $(-1, 3, 1)$; $(0, 1, 9)$.

7. $\mathbf{XY} = 4\mathbf{i} + 2\mathbf{j} - 2\mathbf{k}, \mathbf{XZ} = -2\mathbf{i} - \mathbf{j} + \mathbf{k}.$

9. 50.

10. (a) 1,   (b) -1,   (c) -2,   (d) 4/3.

11. $\frac{1}{2}(\mathbf{p} + \mathbf{q}), \frac{1}{3}(\mathbf{p} + \mathbf{q}), -\frac{2}{3}\mathbf{p} + \frac{1}{3}\mathbf{q}$; $(1 - \frac{2}{3}k)\mathbf{p} + \frac{1}{3}k\mathbf{q}.$

12. $\dfrac{2}{9}(\mathbf{a} + \mathbf{b}); \dfrac{1}{2}, \dfrac{9}{4}; \dfrac{1}{2}(\mathbf{a} + \mathbf{b}).$

13. $\dfrac{1}{7}\mathbf{a} + \dfrac{4}{7}\mathbf{b}; \dfrac{1}{5}\mathbf{a} + \dfrac{4}{5}\mathbf{b}.$

Exercise 9.5

4. (ii) $\frac{1}{3}\mathbf{PQ} + \frac{2}{3}\mathbf{PS}, -\frac{2}{3}\mathbf{PQ} + \frac{2}{3}\mathbf{PS}, -\frac{1}{3}\mathbf{PQ} + \frac{1}{3}\mathbf{PS}.$

5. $\mathbf{a} + \mathbf{b} + \mathbf{c}, \mathbf{b} + \mathbf{c} - \mathbf{a}, \mathbf{a} + \mathbf{c} - \mathbf{b}, \mathbf{a} + \mathbf{b} - \mathbf{c}.$

6. (a) $\mathbf{b} - \mathbf{a}$,   (b) $3\mathbf{b} - 2\mathbf{a}$,   (c) $3\mathbf{b} - \mathbf{c}$,   (d) $\mathbf{a} - \dfrac{3}{2}\mathbf{b}$,   (e) $-\mathbf{a} - \dfrac{1}{2}\mathbf{b}$,

   (f) $\mathbf{a} + \dfrac{3}{2}\mathbf{b}.$

7. (a) $2\sqrt{6}, 5\sqrt{2}, 20$,   (b) $\frac{2}{3}\mathbf{i} + \frac{1}{3}\mathbf{j} - \frac{2}{3}\mathbf{k}.$

8. (a) $6\sqrt{3}$,   (b) $\dfrac{1}{9}\sqrt{3}\mathbf{i} - \dfrac{5}{9}\sqrt{3}\mathbf{j} + \dfrac{1}{9}\sqrt{3}\mathbf{k}$,   (c) $-2\mathbf{j} + 3\mathbf{k}.$

9. 1, 4.

11. (a) $-\dfrac{1}{\sqrt{2}}\mathbf{i} + \dfrac{1}{\sqrt{2}}\mathbf{j}$,   (b) $\dfrac{3}{2}\sqrt{29}.$

12. $\mathbf{a} + k(2\mathbf{b} - \mathbf{a})$; 3, $6\mathbf{b} - 2\mathbf{a}.$

13. $\frac{1}{3}\mathbf{a} + \frac{1}{3}\mathbf{b}.$

14. $(4, -7, 8)$: 54 sq. units.

15. $s + t = 1.$

16. $\frac{1}{3}\mathbf{p} + \frac{1}{3}\mathbf{q}; \frac{1}{3}\mathbf{a} + \frac{1}{3}\mathbf{b} + \frac{1}{3}\mathbf{c}.$

Exercise 10.3

1. (a) 13, 22·6°,   (b) 8·06, 60·3°,   (c) 9·85, 75·3°,
   (d) 14·5, 33°,   (e) 14·1, 37·3°,   (f) 8·86, 17·9°.

2. 133·5°.   3.   14·3 N at 81·4° to **G**.

4. 3·81 N, 15·7 N.   5.   9·43 N, N 12° W.

6. 13·0 N, S 37½° E.   7.   16·8 N, N 57·6° E.

8. (a) 5, N 53·1° E,   (b) 7, E,   (c) 6·71, S 63·4° E,   (d) 7·07, SW.

9. (a) 6,   (b) 17,   (c) 9·95.   10. 15·3 N.

11. $\sqrt{5}.$ 12. $4\sqrt{10}, 6\sqrt{5}.$

Exercise 10.4

1. (a) 15·3 N, 12·9 N, (b) 1·46 N, 6·85 N, (c) 7·05 N, 9·71 N.
2. (a) $-W$, $N\cos\alpha$, $T\sin(\alpha+\beta)$, (b) $-W\sin\alpha$, 0, $T\cos\beta$,
   (c) $W\cos\alpha$, $-N$, $-T\sin\beta$.
3. 11·5 N, N 15·5° W.   4. 29·3 N, N 15·8° W.
5. 344·7°.   6. 15·6, 76·6°.
7. 15·2 N, 66·8°.   8. 4·03 N.
9. (a) 48 N, 40 N, 46·2 N, (b) 14 N, 30 N, 19·2 N, (c) 134·2 N,
   (d) 20·6 N, 157·4°.

Exercise 10.5

1. 31·4 N, S 84·6° W.   2. 12·9, 6·39.
3. $5\sqrt{3}$.   ·4. 6, 8.
5. (a) $20\sqrt{3}$, (b) 20, (c) $10\sqrt{3}$.
6. $3\frac{1}{2}$ N, $12\frac{1}{2}$ N.   7. 1·87 N, 4·41 N.
8. 37·1 N, 19·7 N.
9. (a) $5W$, $5\sqrt{3}W$, (b) $10W/\sqrt{3}$, $10W/\sqrt{3}$.
10. 4·14 N, 43°.   11. $8W$. $6W$.
12. $10\sqrt{3}$ N, $20\sqrt{3}$ N at 30° to vertical.
13. (a) $\sqrt{5}W$, $2\sqrt{2}W$, (b) 63·4°, 45°.
14. $2\sqrt{2}W$, $\sqrt{5}W$, 26·6°.
15. (a) $5\cdot80W$, $4\cdot24W$, $5\cdot02W$, (b) $7\cdot07W$, $5\cdot39W$, $5\cdot39W$.

Exercise 10.6

1. 69·6°.   2. $\sqrt{10}$, 51·3°.
3. $0 \leqslant \theta \leqslant 36\cdot9°$.   4. $\sqrt{65}$, N 60·3° E.
6. $2\sqrt{5}$ N, $4\sqrt{2}$ N. 7. $5\sqrt{3}$ N.
8. 154·9°.   9. 6·32 N, S 71·6° W.
10. 15·4 N, S 37·6° W.   11. $2\sqrt{5}W$.
12. 12·9 N, 45·2 N.   13. 46·8°, 11·5 N.
15. (a) 25/8 N, (b) 2·62 N.   16. $W/\sqrt{3}$.
19. 16·6 N, 6·50 m.

Exercise 11.1

1. 13, $25\mathbf{i}+6\mathbf{j}$.   2. $-2\mathbf{i}+4\mathbf{j}$.
3. $24\mathbf{i}+3\mathbf{j}+12\mathbf{k}$; $8\mathbf{i}+\mathbf{j}+4\mathbf{k}$, 11.
4. 85 m.   5. $5\mathbf{i}+8\mathbf{j}$; 7.
6. $-3\mathbf{i}+6\mathbf{j}$.
7. (a) 13, N 67·4° E, (b) $2\sqrt{10}$, N 71·6° E, (c) $4\sqrt{5}$, S 63·4° E,
   (d) $\sqrt{97}$, S 24° W.
8. (a) 13·6 m s^{-1}, N 45·2° E, (b) 59·7 km/h, S 74·4° E,
   (c) 16·4 m s^{-1}, N 6·6° W, (d) 98·0 km/h, S 49° E.
9. (a) 4·18 m s^{-1}, (b) 30 m.
10. 14·6 km/h, N 87° E.

Exercise 11.2

1. 9·43 km/h from S 58° W.
2. 41·6 km/h, S 70·1° E.
3. 15, 126·9°.
4. $i - 4j, 4i + 5j$.
5. 33·1 km/h, S 65° W.
6. $20\sqrt{3}$ km/h, N 60° W.
7. 10·7 km/h from N 37·5° W.
8. 90 km/h.
9. N 73·7° W; 2 min.
10. $2·25\,m\,s^{-1}$; at 63·3° to the bank.
11. N 47·3° W; 1 h 30 min.
12. 76·8 km/h; S 64·8° W.

Exercise 11.3

1. 151·8°, 1 h 47 min.
2. $4\sqrt{3}$ km/h, N 11·4° E, 6 km/h.
3. 28·8 km/h on S 33·7° W; 8·32 km.
4. $5·57\,m\,s^{-1}$ on 321·1°, 115 s.
5. 25 km/h on S 36·9° E; 60 m.
6. 576 km/h, N 51·3° E; 31·2 km; 10.04, N 38·7° W.
7. $500\sqrt{2}$ km/h, SE; 25·9 km.
8. (ii) $3i + 4j$.
9. 3, 500 s; $1000\sqrt{10}$ m.
10. N 73·1° E; 2·9 km.

Exercise 11.4

1. 203°; 157°; 10·4 knots E.
2. $6\sqrt{3}$ knots, 120°.
3. (i) due north, 100 s, 583 m,   (ii) N 36·9° W, 125 s, 500 m.
4. 95·2 s; 92·3 s; 125 s.
5. 032·8°, 237·2°; 64·4 min.
6. $65\,km\,h^{-1}$ from N 32·6° E; N 5·8° E, S 54·2° W.
7. 97 min, 69 min, 137 min.
8. 14.36, N 45·2° W; 13.42.
9. N; 26·8 min.
10. (a) $3\,m\,s^{-1}$,   (b) $i + 3j$; $\sqrt{181}\,m\,s^{-1}$.
11. (i) $50\,km\,h^{-1}$ towards N 73·7° W,   (ii) 112 m,   (iii) 31·36 m, 107·52 m.
12. 42 km/h, S 8·2° E; $13.06\frac{1}{4}$, 8·66 km; 13.25.
13. N 41·4° E, 45 min; N 48·2° E, 7·45 km.
14. 18·0 km/h; 13.12; 4·16 km/h.

Exercise 12.1

1. (a) $9t^2 - 5, 18t$,   (b) $4t + 3, 4$.
2. (a) $6t^2 + 7, 2t^3 + 7t$,   (b) $20 - 10t, 5 + 20t - 5t^2$,
    (c) $8t^3 - 12t^2 + 4, 2t^4 - 4t^3 + 4t - 2$.
3. 0, 3; 4.
4. 24; 32 m.
5. 36 m, $9\,m\,s^{-1}$.
6. 1 s, 9 s; $16\,m\,s^{-1}$.
7. 29, 52 m.
8. $-12\,m\,s^{-1}$, 16 m; 48 m.
9. (a) $2\frac{1}{2}$,   (b) 5 m.
10. (a) $-15\,m\,s^{-1}, -4\,m\,s^{-2}$,   (b) 0 s, 5 s,   (c) 72 m.
11. (a) $9t - 3t^2, \frac{9}{2}t^2 - t^3$,   (b) 27 m.

12. (a) $0\,\mathrm{s}, 6\,\mathrm{s}$,   (b) $32\,\mathrm{m}$,   (c) $64\,\mathrm{m}$,   (d) $36\,\mathrm{m\,s^{-1}}$.
13. (a) $1\,\mathrm{s}, 4\,\mathrm{s}$,   (b) $10\,\mathrm{m}$,   (c) $18\tfrac{1}{2}\,\mathrm{m}$,   (d) $7\,\mathrm{m\,s^{-1}}$.
14. (a) $0\,\mathrm{s}, 2\tfrac{1}{4}\,\mathrm{s}, 6\,\mathrm{s}$,   (b) $60\tfrac{3}{4}\,\mathrm{m}$,   (c) $171\tfrac{1}{2}\,\mathrm{m}$,   (d) $90\,\mathrm{m\,s^{-1}}$.
15. $5\tfrac{1}{3}, 10, 14\tfrac{2}{3}; 6\,\mathrm{m\,s^{-1}}$.
16. (a) $27\,\mathrm{m}$,   (b) $16\,\mathrm{m\,s^{-1}}$,   (c) $12\,\mathrm{m\,s^{-2}}$.

Exercise 12.2

1. 11, 24.
2. 12, 2.
3. 16, -3.
4. $-7, -4$.
5. $\tfrac{1}{2}$, 10.
6. 0, 24.
7. 10, 10 or 15, -10.
8. 1, 60.
9. $-2\tfrac{1}{2}, 2$.
10. $0, 1\tfrac{1}{2}$.
11. $\tfrac{1}{4}\,\mathrm{km}$.
12. 1/48.
13. $31\tfrac{1}{4}\,\mathrm{m}, 5\,\mathrm{s}$.
14. $14\,\mathrm{m\,s^{-1}}, 8\,\mathrm{m}$.
15. $5\,\mathrm{m\,s^{-1}}, 10\,\mathrm{s}$.
16. $34\,\mathrm{m}, 20\,\mathrm{m\,s^{-1}}$.
17. $2\tfrac{1}{2}\,\mathrm{m\,s^{-2}}, 1\,\mathrm{s}$.
18. $40\,\mathrm{m\,s^{-1}}, 80\,\mathrm{m}$.
19. $1750\,\mathrm{m}; 48, 100, 32, 45$.
20. $30, 28{\cdot}8\,\mathrm{km}$.
21. 12 min.
22. 2, 10·05.
23. 20.
24. 1 min, 900 m.
25. $3\,\mathrm{s}, 15\,\mathrm{m}$.
26. 20 m.

Exercise 12.3

1. $1\tfrac{1}{4}\,\mathrm{s}; 2\tfrac{1}{2}\,\mathrm{s}, 53{\cdot}0\,\mathrm{m}$.
2. $60\,\mathrm{m}; 35{\cdot}8\,\mathrm{m\,s^{-1}}$ at $33{\cdot}2°$ to horizontal.
3. $50\,\mathrm{m}; y = 2x - x^2/50$.
4. $5, 2{\cdot}5; 30{\cdot}1$ at $54{\cdot}2°$ to the horizontal.
5. $20\,\mathrm{m\,s^{-1}}, 19{\cdot}6\,\mathrm{m\,s^{-1}}; 44{\cdot}1\,\mathrm{m}$.
6. $11{\cdot}2\,\mathrm{m\,s^{-2}}; 3{\cdot}2\,\mathrm{m}$.   8. $4{\cdot}5\,\mathrm{km}; 26{\cdot}6°$.
9. $y = \dfrac{5}{3}x - \dfrac{49x^2}{90u^2}; 2{\cdot}1, 13{\cdot}5\,\mathrm{m}$.   10. $14, 36{\cdot}9°$.
11. $28; 15° < \alpha < 75°$.   12. $6u^2/g; 2u/g, 4u/g$.
13. $u/(g \sin \alpha)$.   14. $2V \sin \theta (U + V \cos \theta)/g$.
15. $12V^2/13g; 2V/(g\sqrt{13})$.
16. $\sqrt{(V^2 - 2gh)}, 2\sqrt{(V^2 \sin^2 \theta - 2gh)/g}; 60°$.
18. $63{\cdot}4°, 26{\cdot}6°; 4:5$.   19. $81\tfrac{2}{3}; 69{\cdot}3\,\mathrm{m\,s^{-1}}$.
20. $2u_2/g$.

Exercise 12.4

1. (a) $3\mathbf{i} + 6t^2\mathbf{k}, 12t\mathbf{k}$,   (b) $2t\mathbf{i} + 4t^3\mathbf{j} - 6\mathbf{k}, 2\mathbf{i} + 12t^2\mathbf{j}$.
2. $-3\,\mathrm{m\,s^{-1}}, 13\,\mathrm{m\,s^{-1}}$.
3. $3(t-2)\mathbf{i} + 2(t^2 - 1)\mathbf{j} + t(1 - t^2)\mathbf{k}$.
4. $(3t^2 + 1)\mathbf{i} + \mathbf{j} - 4t\mathbf{k}, (t^3 + t)\mathbf{i} + (t + 3)\mathbf{j} - 2t^2\mathbf{k}$.
5. $3\mathbf{i} + (2 - 10t)\mathbf{j}, 3t\mathbf{i} + (2t - 5t^2)\mathbf{j}$.
6. $31\tfrac{1}{4}\,\mathrm{m}, 200\,\mathrm{m}, 5\,\mathrm{s}$.

7. $10\mathbf{i}+(10\sqrt{3}-10t)\mathbf{j},\ 10t\mathbf{i}+(10\sqrt{3}t-5t^2)\mathbf{j}.$
8. $48\,\text{m},\ 24\,\text{m};\ 32{\cdot}0\,\text{m}\,\text{s}^{-1}.$

Exercise 12.5

1. $25\tfrac{11}{27}\text{m},\ 20\,\text{m}\,\text{s}^{-1};\ t<2\tfrac{1}{6}.$
2. $343\tfrac{3}{4}\text{m};\ 5\,\text{s};\ 49\,\text{m}\,\text{s}^{-2};\ 108\,\text{m}\,\text{s}^{-1}.$
3. (a) 7,   (b) $2t(35+3t-2t^2)$,   (c) $0<t<5.$
6. (i) $b+\tfrac{1}{2}ft^2-ut$,   (ii) $b+(V-u)t-V^2/2f$;
    $b-u^2/2f$; car overtakes bus after time $(2fb-V^2)/2f\,(u-V).$
7. (a) U/g,   (b) $2U/g.$
8. $18{\cdot}3\,\text{m},\ 61{\cdot}6\,\text{m},\ 3{\cdot}83\,\text{s}.$
9. (i) $40\,\text{m}$,   (ii) $10\,\text{m}$,   (iii) $20\,\text{m}\,\text{s}^{-1}$ at $45°$ to horizontal,
    (iv) $2\sqrt{2}\,\text{s};\ (20+5\sqrt{2},\,8\tfrac{3}{4}),\ (20-5\sqrt{2},\,8\tfrac{3}{4}).$
10. $40,\ 45°;\ 15°.$
11. $D^2\tan^2\theta-20D\tan\theta+20h+D^2=0.$
15. $\dfrac{3}{5}\mathbf{i}-\dfrac{4}{5}\mathbf{k};\ 2\,\text{s},\ 2\mathbf{j}+16\mathbf{k}.$
16. $5,\ 25.$

Exercise 13.1

1. (a) $2x^2-1$,   (b) $4x^2+4x$,   (c) x^4-2x^2,   (d) $4x+3.$
2. (a) $2x^2-2$,   (b) $4x-4$,   (c) $4x^2-4x+1$,   (d) $x^4-1.$
3. $x+2.$ 4. $4-x^2.$
5. (a) $\sqrt{u}$,   (b) $u^5/32$,   (c) $-1/u$,   (d) $\tfrac{1}{2}u^8(u+1).$
6. $6(x+3)^5.$ 7. $8(2x-1)^3.$
8. $-21(5-3x)^6.$ 9. $-6(3x-2)^{-3}.$
10. $15x^2(x^3-1)^4.$ 11. $8x(4-x^2)^{-5}.$
12. $\dfrac{3}{2}(1+3x)^{-1/2}.$ 13. $-2(6x+1)^{-4/3}.$
14. $15x(3x^2-1)^{3/2}.$ 15. $-3\left(1+\dfrac{1}{x^2}\right)\left(x-\dfrac{1}{x}\right)^{-4}.$
16. $\dfrac{x}{\sqrt{(1-x^2)^3}}.$ 17. $\dfrac{-1}{\sqrt{x}(1+\sqrt{x})^3}.$
18. $-12;\ 12x+y+11=0.$ 19. $3x+8y=22.$
20. $1\,\text{m}\,\text{s}^{-1},\ -2\,\text{m}\,\text{s}^{-2}.$ 21. $1-4/(t+1)^2,\ 3\,\text{m}\,\text{s}^{-1}.$
22. (a) $(6x+1)(x+1)^4$,   (b) $2/(x+1)^2$,   (c) $\tfrac{1}{2}(x+2)(x+1)^{-3/2}.$
23. (a) $\dfrac{4(2x-3)^3}{\sqrt{\{(2x-3)^4-1\}}}$,   (b) $\dfrac{-4x}{\{4+\sqrt{(4x^2+1)}\}^2\sqrt{(4x^2+1)}}.$
24. $(1,256)\,\text{max},\ (5,0)\,\text{min}.$ 25. $(1,1)\,\text{max}.$

Exercise 13.2

1. $3/y.$ 2. $-x/y.$
3. $-4x/y.$ 4. $2x/3y^2.$

5. $1/(y-1)$.

6. $(x^2-1)/2y$.

7. $\dfrac{2-2x}{1+2y}$.

8. $\dfrac{1}{(x+y)^2}-1$.

9. 2.

10. $\frac{1}{2}, -\frac{1}{2}$.

11. $3x+4y+5=0$.

12. $4y=5x-9$.

13. $-5/3, -1$.

14. $(3,9), (3,-1)$.

Exercise 13.3

1. $5x^4+12x^3-1$.

2. $3x^2-7$.

3. $12x+15x^2-75x^4$.

4. $4x^3+4/x^3$.

5. $x(5x+6)(x+3)^2$.

6. $(18x-1)(3x-1)^4$.

7. $(x+1)(7x+9)(x+2)^4$.

8. $-2(3+7x)(2x+3)^2(1-x)^3$.

9. $\dfrac{x(2+3x^2)}{\sqrt{(1+x^2)}}$.

10. $\dfrac{2-9x^2-5x^3}{2\sqrt{(1-x^3)}}$.

11. $\dfrac{4}{(x+2)^2}$.

12. $\dfrac{2x(x+1)}{(2x+1)^2}$.

13. $\dfrac{11}{(2-3x)^2}$.

14. $\dfrac{x(2-3x-x^3)}{(x^3+1)^2}$.

15. $\dfrac{6x}{(x^2+1)^2}$.

16. $\dfrac{1+x^2}{(1-x^2)^2}$.

17. $\dfrac{4(x+1)}{(x+2)^3}$.

18. $\dfrac{-x(2+x)}{(x-1)^4}$.

19. $-\dfrac{8x(2x+1)}{(4x+3)^4}$.

20. $\dfrac{1}{(1-x^2)^{3/2}}$.

21. $\dfrac{x^2(3-4x^2)}{(1-2x^2)^{3/2}}$.

22. $\dfrac{x-4}{(x+1)^2\sqrt{(x^2+4)}}$.

23. $\dfrac{2(x-1)}{(2-x)^3}$.

24. $\dfrac{-1}{\sqrt{\{2x(x-1)^3\}}}$.

25. $\dfrac{x^2-1}{2x\sqrt{(x+x^3)}}$.

26. (a) $\dfrac{2x-y}{x-2y}$, (b) $-\dfrac{2xy}{x^2+3y^2}$, (c) $\dfrac{2x-6y+3}{6x-2y+2}$.

27. $-5/12$.

28. $4y=3x+14$.

29. (a) $\dfrac{2(3x^2-1)}{(x^2+1)^3}$, (b) $\dfrac{2}{(x-1)^3}$, (c) $\dfrac{1}{(x^2+1)^{3/2}}$.

30. (a) $\dfrac{2x(x-1)(3x^2-3x+2)}{(2x-1)^2}$, (b) $\dfrac{x^2+3x-2}{2(x+1)^2\sqrt{(x-1)}}$,

(c) $\dfrac{3(x-1)}{2\sqrt{\{(x+1)^3(x^3+1)\}}}$.

32. $-3, -12$.

33. $1/12, -1/32$.

Exercise 13.4

1. Min. 0 when $x = 2\frac{1}{2}$.   2. Min. -27 when $x = -1$.
3. Min. 0 when $x = 0$ and 3; max. 16 when $x = 1$.
4. Min. -1 when $x = -1$; max. $\frac{1}{2}$ when $x = 2$.
5. Min. -3 when $x = -1$; max. 1 when $x = 1$.
6. Min. -1 when $x = 0$.   7. $17, -7$.
8. $24\frac{4}{9}, 12$.   9. $0 < y \leqslant 4$; $(-1, 3)$, $(1, 3)$.
10. $-2\sqrt{3} \leqslant y \leqslant 2\sqrt{3}$; $(-3, -3)$, $(0, 0)$, $(3, 3)$.
11. 8.   12. 3.
13. 40 m.   14. 108 cm^2.
15. $V = 2\pi h(R^2 - h^2)$.   16. $\frac{1}{3}a$.
17. $6 - \sqrt{3}$.   18. $\sqrt{\left(\dfrac{S^3 p^2}{8\pi(1 + p)^3}\right)}$; 2.

Exercise 13.5

1. 60 cm^2/s.   2. 4/25 cm/s.
3. 4 cm/s.   4. -6.
5. 2000.   6. -50.
7. $(s^4 - 1)/s^3$ m s^{-2}.   8. $-2/(2s + 1)^3$ m s^{-2}.
9. 4 cm^2/s.
10. (a) 6π cm^3/s, (b) 2π cm^2/s.
11. 0·01 m/s.   12. $\frac{1}{3}$ cm/s.

Exercise 13.6

1. 0·3.   2. 0·01.
3. $\pi/50$.   4. $1·2\pi$ cm^2.
5. (a) $7·5\pi$ cm^3, (b) $2·7\pi$ cm^3.
6. 2%.   7. 6%, 4%.
8. $2\frac{1}{2}$%.   9. 1% decrease.
10. 3; (a) 8·03, (b) 7·94.
11. (a) 24·24, (b) 23·52.
12. (a) 2·001, (b) 0·198, (c) 4·996, (d) 1·02.

Exercise 13.7

1. (a) $\dfrac{6(x-1)}{(5 + 2x - x^2)^4}$, (b) $\dfrac{3x^2}{\sqrt{(2x^3 + 5)}}$.
3. $2x + y = 5$.   4. $\pm\frac{3}{4}$.
5. $5x - 8y + 22 = 0$.
6. (a) $2(9x^2 + 12x + 1)(x^2 + 1)^3$, (b) $\dfrac{5x^2 - 2x - 2}{\sqrt{(2x - 1)}}$,
    (c) $\dfrac{5x - 7}{(x + 1)^3}$, (d) $-\dfrac{1}{\sqrt{\{(x + 3)^3 (x + 5)\}}}$.

7. $\dfrac{4x}{\sqrt{(1+4x^2)}}$, $\dfrac{4}{\sqrt{(1+4x^2)^3}}$,

8. $\dfrac{x(2x^2-y^2)}{y(x^2-2y^2)}$.   9.   1, 6.

11. $(0,0)$ max, $(2,-108)$ min, $(5,0)$ inflexion.

12. $(1,\frac{1}{2})$ max, $(-1,-\frac{1}{2})$ min; $(0,0)$ inflexion, $(\sqrt{3},\frac{1}{4}\sqrt{3})$ inflexion, $(-\sqrt{3},-\frac{1}{4}\sqrt{3})$ inflexion.

13. (i) $g'(x+g(a))$.   (ii) $g'(x).g'(a+g(x))$,   (iii) $2xg'(x^2)$.

15. $20/(\pi+2)$.   16. (a) $25\,\text{cm}^2$,   (b) $8\sqrt{5}\,\text{cm}$.

17. -2.

18. (a) $4\pi\,\text{cm}^3/\text{s}$,   (b) $17\pi/5\,\text{cm}^2/\text{s}$.

19. (a) $4\cdot001$,   (b) $0\cdot2008$,   (c) $2\cdot001$.

20. (a) $72\cdot01$,   (b) $68\cdot98$.   21. $1\frac{1}{2}\%$.

22. $0\cdot08$.   23. $64R^3/81$.

24. $0\cdot5\,\text{cm/s}$.   25. $16\frac{1}{2}\,\text{s}$, $a/27\,\text{cm/s}$.

Exercise 14.1

1.   (a) $180°$,   (b) $45°$,   (c) $30°$,   (d) $240°$.

2.   (a) $\pi/2$,   (b) $\pi/3$,   (c) $5\pi/6$,   (d) $7\pi/4$.

3.   (a) -1,   (b) $-\frac{1}{2}$,   (c) 1,   (d) $2/\sqrt{3}$.

4.   (a) $\pi/4$,   (b) $-\pi/3$,   (c) $\pi/6$,   (d) π.

5. $15\,\text{cm}$, $75\,\text{cm}^2$.   6. $8-2\pi\,\text{cm}^2$.

7. $\frac{1}{2}r^2(2\theta+\sin 2\theta)$, $2r(1+\theta+\cos\theta)$.

8.   (a) $1\cdot2$,   (b) $0\cdot4$,   (c) $2\pi/3$,   (d) $\pi/6$.

9.   6.

10. (a) $\pi/6, 5\pi/6$,   (b) $\pi/3, 2\pi/3, 4\pi/3, 5\pi/3$,   (c) $\pi/4, 5\pi/4$.

11. (a) $\pi/12, 5\pi/12, 3\pi/4$,   (b) $\pi/18, \pi/6, 5\pi/18, \pi/2, 13\pi/18, 5\pi/6, 17\pi/18$.

12. $-\pi<x<-\pi/2, \pi/4<x<\pi/2, 3\pi/4<x<3\pi/2$.

13. $13\cdot7\,\text{cm}^2$, $29\cdot7\,\text{cm}$.   14. $98\cdot8\,\text{cm}^2$.

15. $0\cdot403$, $13\cdot0\,\text{cm}^2$.

Exercise 14.2

1. $2n\pi\pm2\pi/5$.   2. $n\pi+\pi/3$.

3. $n\pi+(-1)^n\pi/8$.   4. $(2n+1)\pi$.

5. $n\pi-\pi/4$.   6. $n\pi+(-1)^n\pi/6$.

7. $(2n+1)\pi/2$.   8. $(2n+1)\pi/4$.

9. $n\pi, 2n\pi\pm\pi/3$.   10. $2n\pi/3$.

11. $2n\pi\pm\pi/3$.   12. $n\pi+\pi/4$.

13. $n\pi/3$.   14. $(2n+1)\pi/10$.

15. $(2n+1)\pi/8$.   16. $(4n+1)\pi/16, (4n+1)\pi/4$.

17. $360n° + 122\cdot3°, 360n° + 20\cdot8°$.

18. $360n° + 121\cdot9°, 360n° + 1\cdot9°$.

Exercise 14.3

1.   (a) $0\cdot0209$,   (b) $0\cdot0122$,   (c) $0\cdot000262$.

2. (a) 2,   (b) 3/2,   (c) 4/9.
3. (a) $\dfrac{\sqrt{3}}{2} + \dfrac{1}{2}\theta - \dfrac{\sqrt{3}}{4}\theta^2$,   (b) $\dfrac{\sqrt{2}}{2} + \dfrac{\sqrt{2}}{2}\theta - \dfrac{\sqrt{2}}{4}\theta^2$,   (c) $1 - \dfrac{5}{2}\theta^2$.
4. (a) 1/2,   (b) 3,   (c) 5/4.
5. (a) 2,   (b) 2/3.

Exercise 14.4

1. (a) $2\cos 2x$,   (b) $-12\sin 4x$,   (c) $2\sec^2 \tfrac{1}{2}x$.
2. (a) $-4\cos^3 x \sin x$,   (b) $\cos x/2\sqrt{(\sin x)}$,   (c) $5\sin 10x$.
3. (a) $3(x + \sin x)^2(1 + \cos x)$,   (b) $2x\cos(x^2)$,   (c) $\dfrac{\pi}{30}\sec^2(6x°)$.
4. (a) $x^3(4\cos x - x\sin x)$,   (b) $8x\tan x + (4x^2 + 1)\sec^2 x$,
    (c) $x(2\sin 3x + 3x\cos 3x)$.
5. (a) $-\operatorname{cosec}(x + 1)\cot(x + 1)$,   (b) $4\operatorname{cosec}^2(1 - 2x)$,
    (c) $3\sec(3x - 4)\tan(3x - 4)$.
6. (a) $-\sin x \sec^2(\cos x)$,   (b) $\dfrac{1}{2\sqrt{x}}\sec(1 + \sqrt{x})\tan(1 + \sqrt{x})$,

    (c) $\dfrac{1}{x^2}\operatorname{cosec}^2\left(\dfrac{1}{x}\right)$.

7. (a) $\cos x \cos 2x - 2\sin x \sin 2x$,   (b) $3\cos^2 x \cos 4x$,   (c) $\cos x$.
8. (a) $\dfrac{x\cos x - 2\sin x}{x^3}$,   (b) $\dfrac{\sin x + \cos x}{(\cos x - \sin x)^2}$,   (c) $-\dfrac{3}{1 + \sin 3x}$.
9. $(\pi/2, 5)$ max; $(3\pi/2, -3)$ min.
10. $(\pi/3, \pi/3 - \sqrt{3})$ min; $(5\pi/3, 5\pi/3 + \sqrt{3})$ max.
11. $(3\pi/4, -1)$ min; $(7\pi/4, -1)$ min.
12. $(0,0)$ inflexion; $(\pi/3, 3\sqrt{3}/16)$ max; $(2\pi/3, -3\sqrt{3}/16)$ min;
    $(\pi, 0)$ inflexion; $(4\pi/3, 3\sqrt{3}/16)$ max; $(5\pi/3, -3\sqrt{3}/16)$ min;
    $(2\pi, 0)$ inflexion.
13. (a) $\pi/3$,   (b) 10 m,   (c) $30\,\mathrm{m\,s^{-1}}$, at O.
14. $4500\,k\pi\,\mathrm{cm^3/s}$.
15. (a) $2\cdot5\,\mathrm{cm^2/s}$,   (b) $\tfrac{1}{2}\sqrt{3} + 1\,\mathrm{cm/s}$.
16. (a) $0\cdot860$,   (b) $1\cdot0105$,   (c) $0\cdot870$.
17. $\tfrac{1}{3}$; $20 < P \leqslant 10(1 + \sqrt{10})$.
18. (a) $8\,\mathrm{m\,s^{-2}}$,   (b) $3\sqrt{3}\,\mathrm{m\,s^{-1}}$.

Exercise 14.5

1. (a) $-\tfrac{1}{3}\cos 3x + c$,   (b) $2\sin\tfrac{1}{2}x + c$,   (c) $2\tan 2x + c$.
2. (a) $\tfrac{1}{2}x + \tfrac{1}{4}\sin 2x + c$,   (b) $-\cot x - x + c$,   (c) $\tfrac{1}{3}\sin^3 x + c$.
3. (a) $-\dfrac{1}{4}\cos^4 x + c$,   (b) $\dfrac{1}{4}\cos 2x - \dfrac{1}{8}\cos 4x + c$,

    (c) $\dfrac{1}{6}\sin 3x + \dfrac{1}{2}\sin x + c$.

4. (a) $\tfrac{1}{2}$,   (b) $(\pi - 2)/24$,   (c) $\tfrac{1}{3}$.

5. (a) $\sqrt{3}-\pi/3$, (b) $3/10$, (c) $1/4$.
6. (a) 1, (b) $1-\pi/2$, (c) $2/3$.
7. $\frac{1}{3}\tan^3 x - \tan x + x + c$.
8. $-2\cos^2 x +$ constant.
9. $\dfrac{3}{8} + \dfrac{1}{2}\cos 2x + \dfrac{1}{8}\cos 4x$; $(3\pi+8)/32$.
10. 1.
11. $11\pi^2$.
12. $4/3\pi$.

Exercise 14.6

1. $\dfrac{1}{\sqrt{(4-x^2)}}$.

2. $\dfrac{3}{1+9x^2}$.

3. $-\dfrac{1}{\sqrt{(-x-x^2)}}$.

4. $\dfrac{x}{1+x^2} + \tan^{-1} x$.

5. -1.

6. $1 - \dfrac{x\sin^{-1} x}{\sqrt{(1-x^2)}}$.

7. $-\dfrac{1}{x\sqrt{(x^2-1)}}$.

8. $\dfrac{2-8x\tan^{-1} 2x}{(1+4x^2)^2}$.

9. $\dfrac{1}{1+x^2}$.

10. $\sin^{-1}\dfrac{x}{5} + c$.

11. $\dfrac{1}{3}\tan^{-1}\dfrac{x}{3} + c$.

12. $\dfrac{1}{2}\sin^{-1} 2x + c$.

13. $\dfrac{1}{3}\tan^{-1} 3x + c$.

14. $\dfrac{1}{3}\sin^{-1}\dfrac{3x}{4} + c$.

15. $\dfrac{1}{6}\tan^{-1}\dfrac{2x}{3} + c$.

16. $\pi/6$.

17. $\pi/8$.

18. $\pi/9$.

19. $\pi/15$.

20. $\pi/72$.

21. $\pi/8$.

Exercise 14.7

1. $(2+\sqrt{3})\pi x/6$, $(5\pi-6\sqrt{3})x^2/24$.
4. (a) $0, \pi/2, \pi, 3\pi/2, 2\pi$, (b) $0, 2\pi/3, \pi, 4\pi/3, 2\pi$.
5. (a) $(4n+1)\pi/6$, (b) $(2n+1)\pi/6$, $n\pi\pm\pi/3$.
6. $360n° + 12\cdot4°$, $360n° - 79\cdot8°$.
7. $\frac{1}{2}\cos\alpha$.
9. $\cos x$, $-\sin x$; $-\operatorname{cosec}^2 x$, $-\operatorname{cosec} x\cot x$.
10. (a) $2\sin(4x-10)$, (b) $4x^3(\tan 4x + x\sec^2 4x)$,
    (c) $\dfrac{2\sec x}{(\sec x - \tan x)^2}$, (d) $\dfrac{\cos x}{(\cos 2x)^{3/2}}$.
11. (a) $(\pi/6, 2\sqrt{3})$ max; $(5\pi/6, -2\sqrt{3})$ min; $(3\pi/2,0)$ inflexion,
    (b) $(0,0)$ inflexion; $(\pi/2,4)$ max; $(\pi,0)$ inflexion;
    $(3\pi/2, -4)$ min; $(2\pi,0)$ inflexion.
12. (a) $\frac{1}{2}\sin(2x-1)+c$, (b) $\frac{1}{2}x-\frac{1}{2}\sin x+c$,
    (c) $\frac{1}{2}\tan^2 x+c$, (d) $-\dfrac{1}{10}\cos 5x - \dfrac{1}{6}\cos 3x+c$.
13. (a) $21/128$, (b) $(3\pi+4)/48$.
14. $x\sin x+\cos x+c$; $2\cos x+2x\sin x-x^2\cos x+c$.
15. $\pi-\sqrt{3}$.   16. $\frac{3}{4}\pi^2$.

17. $\tan x + \sec x + c$; $4(\sqrt{3}-1)/\pi$.

18. (a) $\dfrac{(\sin^{-1}x)\sqrt{(1-x^2)}-x}{(\sin^{-1}x)^2\sqrt{(1-x^2)}}$,   (b) $\dfrac{\sec x \tan x}{1+\sec^2 x}$,   (c) $-\dfrac{x}{\sqrt{(1-x^2)}}$.

19. (a) $\sin^{-1}\dfrac{x}{3}+c$,   (b) $\dfrac{1}{15}\tan^{-1}\dfrac{3x}{5}+c$.

20. (a) $\pi/3\sqrt{3}$,   (b) $\pi/12$.

21. (ii) $\dfrac{2+p^2}{2(1+p^2)}$.

24. $\dfrac{1}{x^2+2ax+a^2+1}$; (a) $\tan^{-1}(x+2)+c$,   (b) $\dfrac{1}{2}\tan^{-1}\left(\dfrac{x+1}{2}\right)+c$.

Exercise 15.1

1. (a) 5040, (b) 600, (c) 1680, (d) 120, (e) 4, (f) 6720.
2. (a) 5!, (b) 10!/6!, (c) $n!/(n-3)!$.
3. 720.   4. 40 320.
5. 20 160.   6. 151 200.
7. 13 800.   8. (a) 240, (b) 600.
9. (a) 120, (b) 432.   10. (a) 1000, (b) 720, (c) 990.
11. (a) 720, (b) 20 160, (c) 60, (d) 50 400.
12. 60, 30.   13. 10 080.
14. 720.   15. 120, 72.
16. 6720; (a) 720, (b) 2400.   17. (a) 48, (b) 180.
18. 96.
19. (a) 168, (b) 2016, (c) 20; 2520.
20. 9.

Exercise 15.2

1. (a) 35, (b) 5, (c) 6, (d) 1.
3. (a) 210, (b) 1140, (c) 1287, (d) 12.
4. 1260.
5. (a) 462, (b) 5775, (c) 10 395.
6. 15 840.   7. 38, 16.
8. (a) 2277, (b) 2541.   9. 238.
10. 79, 24.   11. (a) 90, (b) 7560.
12. (a) 15 015, (b) 37 037.

Exercise 15.3

1. (a) $\{x\}$, (b) $\{x, y, p, q, r\}$, (c) $\{x, y, z, p, q, r\}$,
    (d) $\{x, p\}$, (e) $\{x\}$, (f) $\{x, y, p, q, r\}$.
2. (a) $\{x\in\mathbb{R}: -2 < x < 7\}$, $\{x\in\mathbb{R}: 0 < x < 4\}$,
    (b) $\{x\in\mathbb{R}: x \leqslant 9\}$, $\{x\in\mathbb{R}: 2 \leqslant x \leqslant 5\}$,
    (c) $\{x\in\mathbb{R}: |x| < 10\}$, $\{x\in\mathbb{R}: -4 < x < 3\}$,
    (d) $\{x\in\mathbb{R}: -3 < x \leqslant 6\}$, $\varnothing$.

3. (a) $\varnothing$, (b) $\{4\}$, (c) $\{3, 4, 5\}$, (d) $\{1, 2\}$, (e) $\{1, 2, 3\}$, (f) $\{4, 5\}$.

4. (a) X', (b) $X \cap Y$, (c) $X' \cap Y'$, (d) $X \cup Y'$, (e) $X' \cup Y'$,
   (f) $(X \cup Y) \cap (X' \cup Y')$ or $(X \cap Y') \cup (X' \cap Y)$.

9. (a) A, (b) $\varnothing$, (c) A, (d) A.

10. 27.

11. $n(A) + n(B) + n(C) - n(A \cap B) - n(B \cap C) - n(C \cap A) + n(A \cap B \cap C)$.

12. $n(M \cap F) \geqslant 3, n(M \cap P) \geqslant 10, n(M \cap F \cap P) \leqslant 12$.

14. (a) A, (b) B, (c) $A \cap C$, (d) $B \cup C$.

Exercise 15.4

1. $\dfrac{1}{6}, \dfrac{1}{2}, \dfrac{2}{3}$.

2. $\dfrac{1}{2}, \dfrac{1}{13}, \dfrac{3}{13}$.

3. $\dfrac{1}{8}, \dfrac{1}{2}, \dfrac{3}{4}$.

4. $\dfrac{1}{21}, \dfrac{2}{7}, \dfrac{10}{21}$.

5. (a) $\dfrac{2}{3}$, (b) 0, (c) $\dfrac{5}{6}$, (d) $\dfrac{1}{6}$.

6. (a) $\dfrac{7}{13}$, (b) $\dfrac{1}{26}$, (c) $\dfrac{1}{13}$, (d) $\dfrac{5}{13}$.

7. (a) 0·8, (b) 0·07, (c) 0·1, (d) 0·43.

8. (a) $\dfrac{1}{4}$, (b) $\dfrac{5}{9}$, (c) $\dfrac{7}{36}$.

9. (a) $\dfrac{1}{14}$, (b) $\dfrac{3}{28}$, (c) $\dfrac{3}{7}$, (d) $\dfrac{4}{7}$.

10. 0·32.

11. (a) $\dfrac{1}{9}$, (b) $\dfrac{2}{3}$.

12. (a) $\dfrac{1}{16}$, (b) $\dfrac{3}{8}$, (c) $\dfrac{5}{8}$, (d) $\dfrac{7}{8}$.

13. (a) $\dfrac{1}{3}$, (b) $\dfrac{2}{9}$, (c) $\dfrac{8}{27}$.

14. (a) $\dfrac{3}{10}$, (b) $\dfrac{3}{10}$, (c) $\dfrac{3}{10}$, (d) $\dfrac{1}{5}$.

15. (a) $\dfrac{3}{20}$, (b) $\dfrac{11}{20}$.

16. $\dfrac{11!}{3!2!}$; (a) $\dfrac{2}{11}$, (b) $\dfrac{1}{77}$.

17. $\dfrac{15}{28}$.

18. (a) $\dfrac{1}{2}$, (b) $\dfrac{1}{4}$, (c) $\dfrac{1}{2}$.

19. (a) 0·204, (b) 0·052, (c) 0·398.

20. $\dfrac{21}{55}, \dfrac{27}{55}, \dfrac{27}{220}, \dfrac{1}{220}$.

21. (a) 0·108, (b) 0·374.

22. (a) $\dfrac{1}{42}$, (b) $\dfrac{1}{7}$, (c) $\dfrac{2}{3}$, (d) $\dfrac{1}{2}$.

Exercise 15.5

1. $\dfrac{1}{2}$.

2. 0·33.

3. $\dfrac{1}{12}$.

4. (i) (a) A and B, A and C, (b) A and B;

   (ii) (a) 0, (b) $\dfrac{3}{4}$, (c) $\dfrac{1}{4}$, (d) $\dfrac{1}{2}$, (e) $\dfrac{1}{4}$, (f) 1.

5. (i) (a) none,   (b) *A* and *C*;

    (ii) (a) $\frac{1}{9}$,   (b) 1,   (c) $\frac{7}{36}$,   (d) $\frac{5}{6}$,   (e) 0,   (f) $\frac{8}{9}$.

6. $0.6 \leqslant P(A') \leqslant 0.9$.

7. $\frac{1}{2}, \frac{1}{52}, \frac{3}{13}, 0, \frac{1}{52}, \frac{3}{26}$; *A* and *C*.

8. $\frac{1}{3}, \frac{1}{4}, \frac{1}{6}, \frac{1}{12}, \frac{1}{12}, \frac{1}{6}$; *A* and *B*.

9. (a) $\frac{3}{20}$,   (b) $\frac{7}{10}$,   (c) $\frac{1}{10}$,   (d) $\frac{17}{20}$.

10. $\frac{5}{6}, \frac{2}{5}$.

11. $\frac{9}{10}, \frac{9}{10}$.

12. (a) $\frac{9}{40}$,   (b) $\frac{1}{28}$.

13. (a) $\frac{1}{16}$,   (b) $\frac{7}{16}$.

14. (a) $\frac{1}{27}$,   (b) $\frac{1}{8}$,   (c) $\frac{91}{216}$.

15. (a) $\frac{1}{36}$,   (b) $\frac{1}{36}$,   (c) $\frac{1}{2}$.

16. (a) $\frac{5}{9}$,   (b) 0,   (c) $\frac{20}{81}$.

17. 0·302.

18. (a) $\frac{1}{16}$,   (b) $\frac{9}{64}$,   (c) $\frac{1}{4}$,   (d) $\frac{1}{16}$.

19. 0·590.

20. (a) $\frac{5}{32}$,   (b) $\frac{5}{16}$,   (c) $\frac{13}{16}$.

21. (a) $\frac{1}{64}$,   (b) $\frac{2}{9}$,   (c) $\frac{5}{36}$.

Exercise 15.6

1. $\frac{1}{4}, \frac{1}{4}, \frac{4}{7}$.

2. $\frac{1}{7}, \frac{1}{7}, \frac{1}{28}, \frac{1}{28}$.

3. $\frac{1}{4}, \frac{4}{7}, \frac{1}{4}, \frac{4}{7}$.

4. $\frac{1}{7}, \frac{3}{28}, \frac{3}{7}$.

5. $\frac{3}{14}, \frac{15}{28}, \frac{3}{14}$.

6. $\frac{2}{13}, \frac{1}{2}, \frac{1}{4}$.

7. $\frac{1}{2}, \frac{1}{4}, \frac{1}{13}, \frac{1}{26}$.

8. $0, 0, \frac{2}{3}, \frac{1}{2}$.

9. $0, \frac{1}{2}, \frac{1}{4}$.

10. $\frac{3}{26}, \frac{11}{26}, \frac{33}{52}$.

11. (a) $\frac{4}{81}$,   (b) $\frac{1}{15}$.

12. (a) $\frac{27}{5000}$,   (b) $\frac{1}{140}$.

13. $\frac{2}{3}$.

14. $\frac{3}{4}$.

15. $\frac{1}{5}$.

16. $\frac{7}{12}$.

17. (a) 0·7, (b) 0·6, (c) 0·75.

18. (a) $\dfrac{25}{72}$, (b) $\dfrac{19}{36}$.

19. (a) $\dfrac{18}{35}$, (b) $\dfrac{5}{12}$.

20. $\dfrac{1}{2}, \dfrac{1}{9}$.

Exercise 15.7

1. (a) 1260, (b) 8232.
2. 30 240; (a) 840, (b) 10 080, (c) 13 440.
3. 63.   4. 20.
5. 280, 36.   6. (a) 243, (b) 21.
7. $K \cap R = \{squares\}$; $K \cap P = \{rhombuses\}$.
8. (a) true, (b) false, (c) false, (d) false.

9. 1200.

10. (a) $\dfrac{1}{7}$, (b) $\dfrac{3}{35}$.

11. (a) (i) $\dfrac{5}{18}$, (ii) $\dfrac{1}{3}$, (b) $\dfrac{12}{35}$, (c) 294.

12. (i) $\dfrac{7}{20}$, (ii) $\dfrac{4}{13}$.

13. (a) $\dfrac{6}{77}$, (b) 0, (c) $\dfrac{2}{15}$, (d) $\dfrac{2}{33}$.

14. $\dfrac{1}{2}, \dfrac{1}{2}, \dfrac{1}{2}, \dfrac{1}{4}, \dfrac{1}{4}$.

15. (i) $\dfrac{5}{12}$, (ii) $\dfrac{8}{15}$, (iii) $\dfrac{1}{12}$.

16. $\dfrac{5}{8}$; (a) $\dfrac{55}{64}$, (b) $\dfrac{25\pi}{256}$.

17. $4\cdot12 \times 10^{-6}$.

18. (a) $\dfrac{1}{22}$, (b) $\dfrac{41}{55}$, (c) $\dfrac{3}{11}$, (d) $\dfrac{3}{44}$.

19. (a) $\dfrac{1}{3}, \dfrac{1}{2}$, (b) (i) $\dfrac{63}{64}$, (ii) $\dfrac{5}{16}$, (iii) $\dfrac{5}{72}$.

20. (a) $\dfrac{81}{256}$, (b) $\dfrac{1}{256}$, (c) $\dfrac{3}{128}$, (d) $\dfrac{3}{32}$.

21. (i) 0·3, (ii) 0·3, (iii) 0·5.

22. (a) $P(9) = \dfrac{25}{216}$, $P(10) = \dfrac{27}{216}$,

(b) P(at least one six) $= 1 - \left(\dfrac{5}{6}\right)^4 = 0\cdot518$;

P(at least one double six) $= 1 - \left(\dfrac{35}{36}\right)^{24} = 0\cdot491$.

(c) P(at least one six) $= 1 - \left(\dfrac{5}{6}\right)^6 = 0\cdot665$;

P(at least two sixes) $= 1 - \left(\dfrac{5}{6}\right)^{12} - 2\left(\dfrac{5}{6}\right)^{11} = 0\cdot619$.

23. (a) (i) $\dfrac{1}{4}$, (ii) $\dfrac{1}{4}$, (iii) $\dfrac{1}{17}$, (iv) $\dfrac{15}{34}$; (b) $\dfrac{13}{30}$;

(c) (i) $\dfrac{95}{253}$, (ii) $\dfrac{285}{506}$.

24. (a) $\dfrac{33}{16\,660}$, (b) $\dfrac{6}{4165}$; $\dfrac{11}{1081}$.

26. (a) $\dfrac{1}{221}$, (b) $\dfrac{32}{221}$, (c) $\dfrac{13}{17}$; $\dfrac{25}{33}$.

27. (a) (i) $\dfrac{3}{8}$, (ii) $\dfrac{1}{2}$, (iii) $\dfrac{1}{6}$; (b) 0·14.

28. (a) 0·01, (b) 0·87; (a) 0·9606, (b) 0·0847.

29. (a) $\dfrac{1}{36}$, (b) $\dfrac{5}{12}$; $\dfrac{73}{648}$, $\dfrac{25}{81}$.

30. (i) $\dfrac{2}{3}$, (ii) $\dfrac{1}{6}$, (iii) $\dfrac{5}{18}$, (iv) $\dfrac{1}{4}$, (v) $\dfrac{3}{5}$.

31. (i) $\dfrac{31}{120}$, (ii) $\dfrac{91}{240}$, (iii) 1 ; $\dfrac{2}{7}$.

32. (i) $P(A \cap B) = P(A) . P(B)$, (ii) $P(A \cap C) = 0$; $\dfrac{13}{25}, \dfrac{1}{2}, \dfrac{9}{35}$; $\dfrac{3}{10}$.

33. (i) 0·22, (ii) 0·30, (iii) 0·44.

Exercise 16.1

1. 25, 30; $5n$.
2. 16, 19; $3n+1$.
3. $\dfrac{1}{5}, \dfrac{1}{6}; \dfrac{1}{n}$.
4. $\dfrac{5}{6}, \dfrac{6}{7}; \dfrac{n}{n+1}$.
5. 16, 32; 2^{n-1}.
6. 35, 48; n^2-1.
7. $\dfrac{4}{3}, \dfrac{8}{3}; \dfrac{1}{3} . 2^{n-3}$.
8. $\dfrac{1}{30}, \dfrac{1}{42}; \dfrac{1}{n(n+1)}$.
9. 5, -6; $(-1)^{n-1}n$.
10. 4, 17; $10+(-1)^{n+1}n$.
11. 11, 21, 35; $14n-7$ $(n \geqslant 2)$.
12. $-4, -46, -130; -21n^2+21n-4$.
13. 2, 2, 4; 2^{n-1} $(n \geqslant 2)$.
14. 1, $-\dfrac{1}{2}, -\dfrac{1}{6}; -\dfrac{1}{n(n-1)}$ $(n \geqslant 2)$.
15. 1, 8, 27; n^3.
16. 1, 4, 9; n^2.
17. 88 200.
18. 22 100, 20 825.

Exercise 16.2

1. (a) 31, $4n+3$, (b) $-17, 25-7n$, (c) 37, $2n-9$,
   (d) $10\frac{1}{2}, \frac{1}{2}(n+5)$.
2. (a) 1325, (b) 1188, (c) 88, (d) $193\frac{1}{2}$.
3. (a) 904, (b) 1105, (c) -80, (d) 152.
4. 1512.
5. 9072.
7. 2600, $6\frac{1}{4}$.
8. 7, 4, 1; -333.
9. 607, 825.
10. 8.
11. 4, 16.
12. $-17, 3$.
14. 6, 11, 16.
15. (a) 15, (b) -12, (c) $\dfrac{5}{27}$, (d) lg 9.

16. 8, 11, 14.   17. 3, 10, 17.
18. 4, 6.   19. 20, 190.
20. 7.

Exercise 16.3

1. (a) $64, 2^{n-2}$, (b) $\frac{2}{3}, 2 \cdot 3^{5-n}$, (c) $\frac{25}{32}, 200(-4)^{1-n}$, (d) $-\frac{81}{16}, -\left(\frac{3}{2}\right)^{n-3}$.

2. (a) $9, 127\frac{3}{4}$, (b) $9, 42\frac{3}{4}$, (c) $6, 1249 \cdot 92$, (d) $7, 14\frac{15}{32}$.

3. (a) $111 \cdot 1111$, (b) $\frac{182}{243}$, (c) $1-(-2)^n$, (d) $a^p(a^{3k}-1)/(a^3-1)$.

4. $\frac{3}{4}$.   5. $-5, 2\frac{1}{2}$.

6. (a) 9, (b) $\frac{1}{9}$, (c) 10^{15}.   7. 13.

8. $-53\frac{1}{3}, -385\frac{5}{6}$.   9. $-40, -8$.
10. $182; \frac{1}{2}, -1$.   11. 2.
12. 12.   13. 2, 6.
14. $(3^n-1)/24; 8$.   15. 10.

Exercise 16.4

1. (a) 9, (b) $\frac{2}{3}$, (c) $11\frac{1}{9}$, (d) 27.
2. (a) 6/11, (b) 8/111, (c) 31/54.
3. $25 \cdot 6$.   4. $-\frac{2}{3}, 81$.
5. $6, \frac{1}{4}, 8$.

6. (a) $\frac{1}{1-3x}, |x| < \frac{1}{3}$, (b) $\frac{2}{2+x}, |x| < 2$,

   (c) $\frac{2}{1+2x}, |x| < \frac{1}{2}$, (d) $\frac{3x}{3-x}, |x| < 3$.

7. $x < 0, -3$.   8. $243\left\{1-\left(\frac{5}{9}\right)^n\right\}; 243; 8$.

Exercise 16.5

1. (a) $4+14+36+76 = 130$,   (b) $50+29+6 = 85$,
   (c) $11+15+19+23+27+31 = 126$,
   (d) $120+60+40+30+24+20 = 294$,
   (e) $\sin\frac{\pi}{3} + \sin\frac{2\pi}{3} + \sin\pi + \sin\frac{4\pi}{3} + \sin\frac{5\pi}{3} + \sin 2\pi = 0$,
   (f) $3-5+9-17+33-65 = -42$.

2. (a) $\sum_{1}^{12}(2r+3)$,   (b) $\sum_{1}^{16}r^3$,   (c) $\sum_{1}^{40}(-1)^r(r+1)$,

   (d) $\sum_{1}^{7}360(-\frac{1}{2})^{r-1}$,   (e) $\sum_{1}^{10}\frac{2r-1}{2r}$,   (f) $\sum_{1}^{12}\frac{r^2}{3r+1}$.

3. (a) 940,   (b) 16,   (c) -1275,   (d) 855,   (e) 2,   (f) 1/90.

Exercise 16.6

1. (a) $x^4 + 4x^3y + 6x^2y^2 + 4xy^3 + y^4$,
   (b) $a^7 - 7a^6b + 21a^5b^2 - 35a^4b^3 + 35a^3b^4 - 21a^2b^5 + 7ab^6 - b^7$,
   (c) $64 + 192p^2 + 240p^4 + 160p^6 + 60p^8 + 12p^{10} + p^{12}$,
   (d) $32h^5 - 80h^4k + 80h^3k^2 - 40h^2k^3 + 10hk^4 - k^5$,
   (e) $x^3 + 3x + \dfrac{3}{x} + \dfrac{1}{x^3}$,
   (f) $z^8 - 4z^6 + 7z^4 - 7z^2 + \dfrac{35}{8} - \dfrac{7}{4z^2} + \dfrac{7}{16z^4} - \dfrac{1}{16z^6} + \dfrac{1}{256z^8}$.

2. (a) $210x^4$, (b) $16\,128x^2$, (c) $1760a^3b^9$, (d) $-945p^4q^6$, (e) -20,
   (f) $36/x^3$.

3. $64x^5 + \dfrac{160}{x} + \dfrac{20}{x^7}$.   4.   14.

5. 15.   6.   2.
7. $8, -\frac{1}{2}$.   8.   1·13.
9. 30·43168.
10. $1 + 30x + 420x^2 + 3640x^3$; 1·03042.
11. (a) $1 + 7x + 21x^2 + 35x^3$, (b) $1 + 7x + 14x^2 - 7x^3$.
12. $1 + 16x + 136x^2 + 784x^3$.   13. $1 + 9x + 24x^2$.
14. (a) $-16\,464$, (b) -19, (c) -1760, (d) 26.

Exercise 16.7

1. $1 - 2x + 3x^2$; $(-1)^r(r+1)x^r$.
2. $1 + x + x^2$; x^r.
3. $1 - 6x + 24x^2$; $(-1)^r(r+1)(r+2)2^{r-1}x^r$.
4. $1 + \dfrac{1}{2}x - \dfrac{1}{8}x^2$; $|x| < 1$.   5. $1 - x - x^2$; $|x| < \frac{1}{3}$.

6. $1 - \dfrac{1}{4}x + \dfrac{3}{32}x^2$; $|x| < 2$.   7. $\dfrac{1}{3} - \dfrac{1}{9}x + \dfrac{1}{27}x^2$; $|x| < 3$.

8. $2 - \dfrac{1}{4}x - \dfrac{1}{64}x^2$; $|x| < 4$.   9. $27 - 18x + 2x^2$; $|x| < \dfrac{9}{4}$.

10. $1 - x^2 + \dfrac{3}{2}x^4 - \dfrac{5}{2}x^6$.   11. $1 - \dfrac{1}{2}x - \dfrac{1}{8}x^2 - \dfrac{1}{16}x^3$; 0·9487.

12. $2 + \dfrac{9}{2}x^2 + \dfrac{17}{6}x^3$.

13. $\dfrac{1}{4} + \dfrac{1}{4}x + \dfrac{3}{16}x^2 + \dfrac{1}{8}x^3 + \dfrac{5}{64}x^4$; 0·309.

14. $1 + x - \frac{1}{2}x^2 + \frac{1}{2}x^3$; 1·732.   15. $1 - 7x + 28x^2 - 112x^3$; $|x| < \frac{1}{4}$.

16. (a) $1 - x + 2x^2 - 3x^3$, (b) $1 + \dfrac{1}{2}x + \dfrac{7}{8}x^2 - \dfrac{7}{16}x^3$.

Exercise 16.8

1. $n(n-1)$; 15th.

2. (a) $3, 4\frac{1}{2}, 4\frac{1}{6}, 4 + \dfrac{1}{n(n-1)}$,   (b) $2, 4, 18, n.n!$.

3. (a) 250,   (b) 510,   (c) $n(n+1)$.

4. b^2.   5.   11, 20.

6. $-2\frac{1}{2}, \frac{1}{2}$; 205.   7. $2\frac{1}{2}$.

8. 29 mm.   9. (i) $1, \dfrac{2}{3}$, (ii) $\dfrac{53}{165}$.

10. (i) $1 + \dfrac{1}{4} + \dfrac{1}{16} + \dfrac{1}{64}$, (ii) all values of r; $1+r^2$ for $r \neq 0$, 0 for $r = 0$.

12. (a) 1 200 000,   (b) 72.   13. $-2 < x < 2$; 6.

14. $1, -1+\frac{1}{2}\sqrt{2}, -1-\frac{1}{2}\sqrt{2}$; (i) $\dfrac{8}{7} + \dfrac{2}{7}\sqrt{2}$, (ii) $-35 + \dfrac{45}{2}\sqrt{2}$.

15. $\dfrac{4}{7}$.   16. (i) $p^2 + q^2$, (ii) $\dfrac{p^2}{p^2+q^2}$.

18. (i) $-\frac{1}{3} < x < 1, \frac{1}{2}$, (ii) 149.

19. (a) $1 + 12x + 54x^2$,   (b) $1 - 3x - 9x^2$.

20. (a) 28/243,   (b) -307.   21. 14.

22. 255, 1.   23. $-\dfrac{3}{2}, -8$; 405.

24. $\dfrac{5}{2}x + \dfrac{15}{4}x^2 + \dfrac{65}{8}x^3$.

25. $1 + x - x^2 + \dfrac{5}{3}x^3 - \dfrac{10}{3}x^4$; $|x| < \frac{1}{3}$; 0·98990.

26. $1 - 6x + 24x^2 - 80x^3$, $|x| < \frac{1}{2}$; 0·000008500.

27. (i) $1 + \dfrac{1}{2}x - \dfrac{1}{8}x^2$, (ii) $1 + \dfrac{1}{2}x + \dfrac{3}{8}x^2$; $1 + x + \dfrac{1}{2}x^2$; 3·315.

28. (a) $u^3 + 3u, u^5 + 5u^3 + 5u$,

    (b) $1 + kx + \dfrac{1}{2}(3k^2 - 1)x^2 + \dfrac{k}{2}(5k^2 - 3)x^3$.

30. 0·015625, 0·140625, 0·421875, 0·421875.

Exercise 17.1

1. 48 N.   2.   0·2 m s^{-2}.
3. 12 kg.   4.   40 N.
5. $2\mathbf{i} - 5\mathbf{j}$.   6. $12\mathbf{i} + 4\mathbf{j}$.
7. 6·4 m s^{-2} at 17·4° to the 9 N force.
8. 84·9°.   9.   6·25 m s^{-2}, S 53·1° E.
10. 2·05 m s^{-2}, N 61·6° W.   11. 1·8 s.
12. 15 N.   13. 10 kg.
14. 10 m s^{-1}.   15. 8, 540 m.
16. 20 m s^{-1}; $6\frac{2}{3}$, 10.

Exercise 17.2

1. (a) 98 N,   (b) 14·7 N,   (c) 2·45 N.

2. $6.2 \, \text{m s}^{-2}$.   3. $62.4 \, \text{N}$.
4. (a) $440 \, \text{N}$,   (b) $392 \, \text{N}$,   (c) $360 \, \text{N}$.
5. (a) decelerating at $1.2 \, \text{m s}^{-2}$,   (b) accelerating at $0.8 \, \text{m s}^{-2}$,
    (c) moving with uniform speed.
6. $33.1 \, \text{N}, 280 \, \text{N}$.   7. $1.71 \, \text{m s}^{-2}, 87 \, \text{N}$.
8. $5 \, \text{N}, 11.5°$.   9. $8.49 \, \text{m s}^{-2}, 24.5 \, \text{N}$.
10. $2.44 \, \text{s}$.   11. (a) $8.28 \, \text{N}$,   (b) $14.3 \, \text{N}$.
12. $2.12 \, \text{m s}^{-2}, 55.4 \, \text{N}$.

Exercise 17.3

1. (a) $3 \, \text{m s}^{-2}$,   (b) $350 \, \text{N}$.
2. (a) $240 \, \text{N}, 80 \, \text{N}$,   (b) $2640 \, \text{N}, 880 \, \text{N}$.
3. $6000 \, \text{N}, 1200 \, \text{N}, 600 \, \text{N}; 0.25 \, \text{m s}^{-2}, 7200 \, \text{N}, 3600 \, \text{N}$.
4. $4.2 \, \text{m s}^{-2}, 28 \, \text{N}$.   5. $4.9 \, \text{m s}^{-2}, 44.1 \, \text{N}$.
6. $\dfrac{M-m}{M+m}g, \dfrac{2Mm}{M+m}g$.   7. $24mg/5$.
8. $2.45 \, \text{m s}^{-2}, 8.82 \, \text{N}$.
9. $0.392 \, \text{m}; mg/\sqrt{2} \, \text{N} = 6.93m \, \text{N}$ at $45°$ to vertical.
10. $3.92 \, \text{m s}^{-2}, 17.64 \, \text{N}; 30.6 \, \text{N}$ at $30°$ to vertical.
11. $m \sin \alpha > M$.   12. $0.5 \, \text{s}$.
13. $3.75 \, \text{m}$.
14. $0.98 \, \text{m s}^{-2}, 34.3 \, \text{N}; 1.4 \, \text{m s}^{-1}; 1\frac{1}{6} \, \text{m}$.

Exercise 17.4

1. $7.84 \, \text{m s}^{-2}$, upwards; $17.64 \, \text{N}$.
2. $2.8 \, \text{m s}^{-2}$, downwards; $28 \, \text{N}$.
3. $6.2 \, \text{m s}^{-2}$ upwards, $5.8 \, \text{m s}^{-2}$ downwards; $32 \, \text{N}, 16 \, \text{N}$.
4. $1.96 \, \text{m s}^{-2}$ down, $5.88 \, \text{m s}^{-2}$ up; $31.4 \, \text{N}, 15.7 \, \text{N}$.
5. $1.51 \, \text{m s}^{-2}$ down, $2.26 \, \text{m s}^{-2}$ down; $41.5 \, \text{N}, 15.1 \, \text{N}$.
6. $1.89 \, \text{m s}^{-2}$.   7. $49 \sin \alpha/(4 + \sin^2 \alpha) \, \text{m s}^{-1}$.
8. $0.8 \, \text{m s}^{-2}$ vertically, $3.2 \, \text{m s}^{-2}$ down plane.

Exercise 17.5

1. $3.70 \, \text{kg}$.   2. $24.1, 242.5°$.
3. $1250, 40 \, \text{s}$.   4. $10\mathbf{i} + 20\mathbf{j}, 92\mathbf{i} + 10\mathbf{j}$.
5. $9.16 \, \text{N}$.   6.   (a) $791 \, \text{N}$,   (b) $651 \, \text{N}$.
7. $2 \, \text{m s}^{-1}; 1980 \, \text{N}, 1960 \, \text{N}, 1920 \, \text{N}$.
8. (a) $784 \, \text{N}, 0°$,   (b) $809 \, \text{N}, 14.3°$.
9. $\sqrt{(2gx \sin \alpha)}$.   10. $12.5 \, \text{m}$.
11. (a) 2,   (b) $56.4°$ 12. $1071 \, \text{N}, 315 \, \text{N}$.
13. $3.15 \, \text{m}$.   14. (a) $2\frac{1}{3} \, \text{m s}^{-1}$,   (b) $1\frac{13}{15} \, \text{m s}^{-1}$.
15. $1.70 \, \text{m s}^{-2}, 83.4 \, \text{N}$ at $25°$ to the vertical.

16. 47·04 N, 23·52 N; 2·06 m.

17. 30°; $mg\sqrt{3}$ at 30° to vertical; $\dfrac{3}{2}mg$, $\dfrac{1}{4}g$, $\dfrac{3\sqrt{3}}{2}mg$ at 30° to vertical.

18. $\dfrac{2}{3}mg$.   19. 48 N, 24 N.

20. 525 N, 11·5°.

Exercise 18.1

1. (a) 10 N,   (b) 0,   (c) 5 N.
2. (a) 5 N down plane,   (b) 0,   (c) 5 N up plane.
3. (a) 3,   (b) $12\sqrt{2/5} = 3\cdot39$,   (c) $4\sqrt{2} = 5\cdot66$.
4. $\tan 25° = 0\cdot466$.   5. (a) 3·98,   (b) 2·45.
6. $0\cdot733\,\mathrm{m\,s^{-2}}$.   7. $5\cdot75\,\mathrm{m\,s^{-2}}$.
8. $3\cdot20\,\mathrm{m\,s^{-2}}$.   9. 0·147.
10. $1\cdot34\,\mathrm{m\,s^{-2}}$.

Exercise 18.2

1. 9·8 N, $\frac{1}{4}$.   2. 1·5 s.
3. (a) $6\cdot21\,\mathrm{m\,s^{-2}}$,   (b) $2\cdot79\,\mathrm{m\,s^{-2}}$.   4. 7·38.
5. (a) $\tan\alpha$,   (b) $(g\sin\alpha - a)/g\cos\alpha$.
6. $23\cdot04\,\mathrm{m\,s^{-2}}$, 61·0 N.   7. 36·3 N, 0·48.
8. $\mu < 1$, 5/13.   9. $4\cdot2\,\mathrm{m\,s^{-2}}$, 22·4 N.
10. $\frac{2}{3}$; 34·6 N at 45° to horizontal.
11. $1\cdot94\,\mathrm{m\,s^{-2}}$ up the plane, 39·3 N.
12. (a) $1\cdot86\,\mathrm{m\,s^{-2}}$ up the plane,   (b) $1\cdot25\,\mathrm{m\,s^{-2}}$ down the plane.

Exercise 18.3

1. 14°.   2. (a) 12·5 N,   (b) 5·47 N.

3. (a) $\mu \geqslant \frac{1}{3}$,   (b) $\mu \geqslant \frac{1}{5}\sqrt{3}$,   (c) $\mu \geqslant \frac{1}{7}\sqrt{3}$.

5. $\dfrac{1}{6}\sqrt{3}$.   6. $2\cdot87 \leqslant P \leqslant 4\cdot79$.

7. (a) 16,   (b) $24\frac{8}{13}$.   8. $M \geqslant 3m$.

Exercise 18.4

1. (a) 6,   (b) 5·66,   (c) 9·74.
2. 12·5 N, 36·9°.
3. 1·94 N at 14° to horizontal.
4. $W\sin\lambda$ at λ to horizontal.
5. (a) 1·20 N at 18·4° to plane, i.e. at 11·6° above horizontal,
   (b) 4·49 N at 18·4° to plane, i.e. at 48·4° above horizontal.
6. (a) $W\sin(\lambda - \alpha)$ at λ to plane.   (b) $W\sin(\lambda + \alpha)$ at λ to plane.

Exercise 18.5

1. 5 N.   2. 9 N.
3. 2·4 N.   4. 1·2 m.
5. 26·6°, $a(3 + \sqrt{5})$.   6. 6·2 m s^{-2}.
7. 1·8 m.   8. 2·5W/8.
9. 4·8 N in PA, 2 N in PB; 0·9 m.
10. 20.
11. 18 N; 4·5 cm, 7·5 cm, 12 cm.
12. 1·25 m, 1·75 m.

Exercise 18.6

1. (a) 140 N,   (b) 135 N.   2. $\frac{1}{2}W$.
3. (a) 15·7,   (b) 7·84; 3·92 m s^{-2}.
4. 65, $\frac{1}{2}$.   5. (i) $\dfrac{\mu W}{1+2\mu^2}$, (ii) $\dfrac{3\mu W}{\sqrt{(1+4\mu^2)}}$.

8. 99·0, 33·0 N.
9. (a) 4·5 m s^{-2}; 42 N, 24 N,   (b) 2 m s^{-2}; 42 N, 24 N.
11. (a) $\mu < \dfrac{1}{8}$,   (b) $\mu < \dfrac{1}{8}$.   12. (ii) $\dfrac{1}{8}g(2-\sqrt{3})$, $\dfrac{1}{8}Mg(6+\sqrt{3})$.
14. 14°, 2·60 m s^{-2}.
15. (a) 0·09 N at 1·3° to horizontal,
    (b) 1·45 N at 21·3° to horizontal.
16. $\lambda(2\sec\theta - 1)$, $W + \lambda(2 - \cos\theta)$, $\lambda(2\tan\theta - \sin\theta)$.
17. (i) $5mg$, $4\sqrt{2}mg$, (ii) 53·1°, 45°.

Exercise 19.1

1. $\begin{pmatrix} 3 & 12 & 0 & 9 \\ -6 & 0 & 6 & -3 \end{pmatrix}$; $\begin{pmatrix} 0 & 1 & 3 & -1 \\ -3 & -5 & -1 & 2 \end{pmatrix}$; $\begin{pmatrix} 1 & 3 & -3 & 4 \\ 1 & 5 & 3 & -3 \end{pmatrix}$;

$\begin{pmatrix} 1 & 5 & 3 & 2 \\ -5 & -5 & 1 & 1 \end{pmatrix}$; $\begin{pmatrix} 2 & 11 & 9 & 3 \\ -13 & -15 & 1 & 4 \end{pmatrix}$.

2. $\mathbf{A}^2 = \begin{pmatrix} 3 & 5 \\ -5 & 8 \end{pmatrix}$, $\mathbf{AB} = \begin{pmatrix} 2 & 2 & 8 \\ -1 & 13 & 3 \end{pmatrix}$, $\mathbf{BC} = \begin{pmatrix} 3 & -3 & 2 \\ 4 & 16 & -2 \end{pmatrix}$,

$\mathbf{BD} = \begin{pmatrix} -10 & -1 \\ 2 & 2 \end{pmatrix}$, $\mathbf{C}^2 = \begin{pmatrix} 5 & 1 & 3 \\ -2 & 16 & -5 \\ -4 & 2 & -3 \end{pmatrix}$, $\mathbf{CD} = \begin{pmatrix} 5 & 10 \\ 11 & -1 \\ -1 & -7 \end{pmatrix}$,

$\mathbf{DA} = \begin{pmatrix} 6 & -11 \\ 4 & 2 \\ -7 & 0 \end{pmatrix}$, $\mathbf{DB} = \begin{pmatrix} 1 & -17 & -5 \\ 2 & -2 & 6 \\ -3 & 7 & -7 \end{pmatrix}$.

3. $\mathbf{A} + \mathbf{C} = \mathbf{C} + \mathbf{A} = \begin{pmatrix} 1 & 2 & 0 \end{pmatrix}$, $\mathbf{B} + \mathbf{D} = \mathbf{D} + \mathbf{B} = \begin{pmatrix} 8 \\ 0 \\ 3 \end{pmatrix}$, $\mathbf{AB} = (0)$.

$$\mathbf{AD} = (17), \mathbf{CB} = (7), \mathbf{CD} = (-16), \mathbf{BA} = \begin{pmatrix} 5 & -10 & 15 \\ 1 & -2 & 3 \\ -1 & 2 & -3 \end{pmatrix},$$

$$\mathbf{DA} = \begin{pmatrix} 3 & -6 & 9 \\ -1 & 2 & -3 \\ 4 & -8 & 12 \end{pmatrix}, \quad \mathbf{BC} = \begin{pmatrix} 0 & 20 & -15 \\ 0 & 4 & -3 \\ 0 & -4 & 3 \end{pmatrix},$$

$$\mathbf{DC} = \begin{pmatrix} 0 & 12 & -9 \\ 0 & -4 & 3 \\ 0 & 16 & -12 \end{pmatrix}.$$

4. $\begin{pmatrix} -4 & 2 \\ 10 & 9 \end{pmatrix}, \begin{pmatrix} -7 & -4 \\ 7 & 12 \end{pmatrix}, \begin{pmatrix} 9 & -1 \\ 1 & 10 \end{pmatrix}, \begin{pmatrix} 9 & -1 \\ 1 & 10 \end{pmatrix}, \begin{pmatrix} -8 & -2 \\ 8 & 15 \end{pmatrix}, \begin{pmatrix} -5 & 4 \\ 11 & 12 \end{pmatrix}.$

5. $\begin{pmatrix} 16 & 22 \\ 27 & 17 \end{pmatrix}, \begin{pmatrix} 40 & 6 \\ 7 & -7 \end{pmatrix}, \begin{pmatrix} -2 & 4 \\ -13 & 26 \end{pmatrix}, \begin{pmatrix} 18 & -4 \\ -27 & 6 \end{pmatrix},$

$\begin{pmatrix} 2 & -4 \\ 13 & -26 \end{pmatrix}, \begin{pmatrix} -12 & 8 \\ 18 & -12 \end{pmatrix}.$

6. $1, -3.$ 8. $5.$

Exercise 19.2

1. (a) $\begin{pmatrix} 3 & -4 \\ -2 & 3 \end{pmatrix}$,   (b) $\begin{pmatrix} 3 & -2 \\ 7 & -5 \end{pmatrix}$,   (c) No inverse,   (d) $\begin{pmatrix} \frac{1}{3} & -\frac{1}{3} \\ \frac{1}{3} & \frac{2}{3} \end{pmatrix}$,

(e) $\begin{pmatrix} -2 & -1 \\ 3 & 2 \end{pmatrix}$,   (f) No inverse,   (g) $\begin{pmatrix} \frac{1}{3} & 0 \\ 0 & -\frac{1}{2} \end{pmatrix}$,   (h) $\begin{pmatrix} -3/7 & 5/7 \\ 2/7 & -1/7 \end{pmatrix}$.

2. (a) $x = 0, y = -1$,   (b) $p = -1, q = -2$,   (c) $a = 4, b = 7$.
3. (a) $(-1,2)$,   (b) None,   (c) $(-1,1)$.
4. (a) $\mathbf{A}^{-1}\mathbf{B}$,   (b) $\mathbf{BA}^{-1}$,   (c) $\mathbf{B}$,   (d) $\mathbf{A}^{-1}\mathbf{BA}$.
5. (a) $\begin{pmatrix} 0 & -1 \\ 1 & 4 \end{pmatrix}$,   (b) $\begin{pmatrix} 3 & 1 \\ 2 & 1 \end{pmatrix}$,   (c) $\begin{pmatrix} 3 & -2 \\ 0 & 2 \end{pmatrix}$.

6. (a) $\dfrac{1}{k}, \dfrac{k+1}{k}$; $k \neq 0$; parallel,

(b) $\dfrac{2}{k+2}, \dfrac{-1}{k+2}$; $k \neq \pm 2$; $k = 2$, coincident; $k = -2$, parallel.

Exercise 19.3

1. $(-1,2), (5,14), (-5,3).$ 2. $(4,-2), (1,-1), (10,-5).$
3. (a) $\begin{pmatrix} -8 \\ 2 \end{pmatrix}$,   (b) $\begin{pmatrix} 4 \\ -6 \end{pmatrix}$,   (c) $\begin{pmatrix} 0 \\ 0 \end{pmatrix}$,   (d) $\begin{pmatrix} 1 \\ 1 \end{pmatrix}$.
4. (a) $y = 2x$,   (b) $y = x + 6$,   (c) $y = 11 - 4x$,   (d) $2x + 3y = 6$.
5. (a) $(-1,1), (-3,0), (1,-4)$,   (b) $(1,1), (0,3), (-4,-1)$,
    (c) $(1,4), (-3,9), (-7,-7)$,   (d) $(1,-1), (0,3), (-4,7)$.

6. $\begin{pmatrix} 4 & -1 \\ -5 & 7 \end{pmatrix}$.

7. $\begin{pmatrix} 3 & -2 \\ -1 & 0 \end{pmatrix}$.

8. (a) $y = 2x$, (b) $y = -2x$, (c) $y = -x$, (d) $y = x+2$.

9. $\begin{pmatrix} 3 \\ -2 \end{pmatrix}$; $(7, -1), (0,0), (1,1), (-2, -2)$.

10. $\begin{pmatrix} 4 & -3 \\ 3 & -2 \end{pmatrix}$; $(3,2), (1,1), (-2,1)$.

11. $(1,2), (0,0), (2, -3)$.

Exercise 19.4

1. (a) $\begin{pmatrix} -1 & 0 \\ 0 & 1 \end{pmatrix}$, (b) $\begin{pmatrix} 2 & 0 \\ 0 & 2 \end{pmatrix}$, (c) $\begin{pmatrix} 1 & 0 \\ 0 & 3 \end{pmatrix}$, (d) $\dfrac{1}{\sqrt{2}}\begin{pmatrix} 1 & 1 \\ -1 & 1 \end{pmatrix}$,

(e) $\begin{pmatrix} 0 & 1 \\ 1 & 0 \end{pmatrix}$, (f) $\begin{pmatrix} 1 & 0 \\ 2 & 1 \end{pmatrix}$.

2. (a) Enlargement, $\times 3$, (b) Clockwise rotation, 90°,
(c) Shear, y-axis fixed, (d) Identity transformation,
(e) Stretch parallel to x-axis, $\times 2$,
(f) All points mapped onto origin,
(g) All points mapped onto the line $y = x$,
(h) Two-way stretch: $\times 2$ along x-axis, $\times 3$ along y-axis.

3. (a) Rotation, 53·1° anti-clockwise,
(b) Reflection in $y = \frac{1}{3}x$,
(c) Reflection in $y = (\sqrt{2}-1)x$ [$= x \tan 22\frac{1}{2}°$],
(d) Rotation, 45° clockwise and enlargement $\times \sqrt{2}$,
(e) Reflection in $y = 5x$,
(f) Rotation, 36·9° anti-clockwise and enlargement $\times 5$.

4. $\begin{pmatrix} 5 & 1 \\ 3 & 7 \end{pmatrix}$; 16.

5. $\frac{1}{2}|x_1y_2 - x_2y_1|$.

6. $y = -2x$; $y = \frac{1}{2}x$.

7. (a) $y = 3x$; $y = \frac{1}{2}x$, (b) $y = 3x$; $y = -\dfrac{3}{2}x$.

8. Reflection in $y = x/\sqrt{3}$; $y = x/\sqrt{3}$, $y = -\sqrt{3}x$; $y = \sqrt{3}x$.

9. Shear parallel to $y = x$; $y = x$; $y = \frac{1}{2}x$.

10. (a) $k\begin{pmatrix} 1 \\ -2 \end{pmatrix}$, (b) $k\begin{pmatrix} 6 \\ 1 \end{pmatrix}$.

11. (a) $y = x$, $y = 5x$, (b) $y = x$, $y = -\dfrac{3}{2}x$.

12. $y = \frac{1}{2}x$, $y = -2x$; one way stretch parallel to $y = \frac{1}{2}x$, factor 6.

Exercise 19.5

1. (a) **A**: stretch parallel to x-axis, $\times 2$.
   B: enlargement, scale factor $\frac{1}{2}$,
   BA and **AB**: stretch parallel to y-axis, $\times \frac{1}{2}$.

(b) **A**: reflection in the line $y = x$,
    B: reflection in the y-axis,
    BA: anti-clockwise rotation through $90°$,
    AB: clockwise rotation through $90°$.
(c) **A**: anti-clockwise rotation through $30°$,
    B: anti-clockwise rotation through $60°$,
    BA and **AB**: anti-clockwise rotation through $90°$.
(d) **A**: rotation through $180°$,
    B: reflection in the line $y = \frac{1}{3}x$,
    BA and **AB**: reflection in the line $y = -3x$.

2. (a) **A**: anti-clockwise rotation through $45°$,
    $\textbf{A}^2$: anti-clockwise rotation through $90°$,
    $\textbf{A}^{-1}$: clockwise rotation through $45°$.
  (b) **A**: enlargement, scale factor 3,
    $\textbf{A}^2$: enlargement, scale factor 9,
    $\textbf{A}^{-1}$: enlargement, scale factor $\frac{1}{3}$.
  (c) **A**: reflection in the line $y = -x$,
    $\textbf{A}^2$: identity transformation,
    $\textbf{A}^{-1}$: reflection in the line $y = -x$.

4. (a) $(13, 7)$, (b) $(2, -1)$, (c) $(-5, -4)$, (d) $(9\frac{1}{2}, -15\frac{1}{2})$.

5. $\mathbf{R} = \begin{pmatrix} \cos\theta & -\sin\theta \\ \sin\theta & \cos\theta \end{pmatrix}$, $\mathbf{S} = \begin{pmatrix} 1 & 1 \\ 0 & 1 \end{pmatrix}$, $\mathbf{T} = \begin{pmatrix} 1 & 0 \\ 0 & -1 \end{pmatrix}$.
  (a) reflection in $y = x\tan\theta$, (b) reflection in x-axis,
  (c) shear fixing $y = -x\tan\theta$.

6. (a) reflection in the line $y = 1$,
  (b) $180°$ rotation about $(1, 1)$,
  (c) enlargement, $\times 3$, centre $(\frac{1}{2}, \frac{1}{2})$,
  (d) glide-reflection along $y = x - 1$, $(1, 0) \rightarrow (2, 1)$.

7. $(1, 2)$, $x + y = 3$, $y = x + 1$, enlargement, scale factor 2 about $(1, 2)$
  with reflection in the line $x + y = 3$.

8. (a) $\begin{pmatrix} x' \\ y' \end{pmatrix} = \begin{pmatrix} 0 & 1 \\ -1 & 0 \end{pmatrix}\begin{pmatrix} x \\ y \end{pmatrix} + \begin{pmatrix} -3 \\ -1 \end{pmatrix}$, (b) $\begin{pmatrix} x' \\ y' \end{pmatrix} = \begin{pmatrix} 0 & -1 \\ -1 & 0 \end{pmatrix}\begin{pmatrix} x \\ y \end{pmatrix} + \begin{pmatrix} 2 \\ 2 \end{pmatrix}$,

  (c) $\begin{pmatrix} x' \\ y' \end{pmatrix} = \begin{pmatrix} 2 & 0 \\ 0 & 2 \end{pmatrix}\begin{pmatrix} x \\ y \end{pmatrix} + \begin{pmatrix} 1 \\ 1 \end{pmatrix}$, (d) $\begin{pmatrix} x' \\ y' \end{pmatrix} = \begin{pmatrix} 1 & 0 \\ 0 & -1 \end{pmatrix}\begin{pmatrix} x \\ y \end{pmatrix} + \begin{pmatrix} 2 \\ 4 \end{pmatrix}$.

Exercise 19.6

1. $\mathbf{A} + \mathbf{D} = \begin{pmatrix} 3 & 2 \\ -2 & 1 \\ 3 & -3 \end{pmatrix}$; $\mathbf{E} + \mathbf{G} = \begin{pmatrix} -1 \\ 3 \end{pmatrix}$; $\mathbf{AE} = \begin{pmatrix} 7 \\ -2 \\ 0 \end{pmatrix}$; $\mathbf{AG} = \begin{pmatrix} 1 \\ 4 \\ -5 \end{pmatrix}$;

  $\mathbf{BA} = \begin{pmatrix} 0 & 7 \\ 4 & -3 \\ -1 & -11 \end{pmatrix}$; $\mathbf{B}^2 = \begin{pmatrix} 7 & -1 & -4 \\ -9 & 4 & 9 \\ -12 & 3 & 10 \end{pmatrix}$; $\mathbf{BC} = \begin{pmatrix} 6 \\ -1 \\ -10 \end{pmatrix}$;

  $\mathbf{BD} = \begin{pmatrix} 3 & 0 \\ 3 & -5 \\ -4 & 0 \end{pmatrix}$; $\mathbf{CF} = \begin{pmatrix} 3 & 0 & 3 \\ -1 & 0 & -1 \\ 0 & 0 & 0 \end{pmatrix}$; $\mathbf{CH} = \begin{pmatrix} -3 & 0 \\ 1 & 0 \\ 0 & 0 \end{pmatrix}$;

$$\mathbf{DE} = \begin{pmatrix} 0 \\ 2 \\ -3 \end{pmatrix}; \quad \mathbf{DG} = \begin{pmatrix} -5 \\ 1 \\ -4 \end{pmatrix}; \quad \mathbf{EF} = \begin{pmatrix} 1 & 0 & 1 \\ 2 & 0 & 2 \end{pmatrix}; \quad \mathbf{EH} = \begin{pmatrix} -1 & 0 \\ -2 & 0 \end{pmatrix};$$

$$\mathbf{FA} = (3 \quad 2); \quad \mathbf{FB} = (-1 \quad 1 \quad 1); \quad \mathbf{FC} = (3); \quad \mathbf{FD} = (3 \quad -3);$$

$$\mathbf{GF} = \begin{pmatrix} -2 & 0 & -2 \\ 1 & 0 & 1 \end{pmatrix}; \quad \mathbf{GH} = \begin{pmatrix} 2 & 0 \\ -1 & 0 \end{pmatrix}; \quad \mathbf{HE} = (-1); \quad \mathbf{HG} = (2).$$

2. 5, 7.

3. $\begin{pmatrix} 5 & 0 \\ 0 & 5 \end{pmatrix}; \dfrac{1}{5}(\mathbf{X} - 2\mathbf{I}).$

4. (a) $2(\mathbf{BA} - \mathbf{AB})$, (b) $2\mathbf{B}(\mathbf{AB} + \mathbf{BA})$.

5. (a) $m = p, n = q$, (b) $n = p$, (c) $m = q$, (d) $m = n$, (e) $n = p = q$.

6. $a^2 + bc = 1, a + d = 0$.

8. (a) $3k \neq 2h$, (b) $\dfrac{3}{2} = \dfrac{h}{k} = \dfrac{p}{q}$, (c) $\dfrac{3}{2} = \dfrac{h}{k} \neq \dfrac{p}{q}$.

9. True only if $\det \mathbf{A} \neq 0$.

10. (a) enlargement, scale factor 2 and reflection in $y = x$,
(b) rotation through $90°$ followed by stretch parallel to x-axis $\times 2$ or stretch parallel to y-axis $\times 2$ followed by rotation through $90°$,
(c) projection onto x-axis,
(d) shear fixing the line $x + y = 0$, $(0, 1) \to (1, 0)$.

11. $2x + 3y = 1; y = -2x$.

12. A: reflection in $y = x \tan 22\frac{1}{2}° = x(\sqrt{2} - 1)$,
B: anti-clockwise rotation through $45°$,
AB: reflection in x-axis, BA: reflection in $y = x$.

13. (a) $y = 2x$, (b) $y = \frac{1}{3}x$, (c) $y = \left(\dfrac{9 + 7m}{7 + m}\right)x; \pm 3$.

14. (i) (a) $(2, -1)$, (b) $y = 9x$, (ii) (a) $\{(2k, 3k)\}$, (b) $(1, 2)$.

15. $\begin{pmatrix} 1 & 0 \\ 0 & 1 \end{pmatrix}, -1; \left\{\begin{pmatrix} 3k \\ 2k \end{pmatrix}\right\}, \left\{\begin{pmatrix} 2k \\ -3k \end{pmatrix}\right\};$ reflection in the line $y = \frac{2}{3}x$.

16. Reflection in $y = x \tan 22\frac{1}{2}° = x(\sqrt{2} - 1)$ and enlargement $\times \sqrt{2}$;
$y = x(\sqrt{2} - 1), y = -x(1 + \sqrt{2})$.

17. $\begin{pmatrix} 1 & 7 \\ 5 & 11 \end{pmatrix}; \dfrac{1}{2}\begin{pmatrix} 1 & 7 \\ 5 & 11 \end{pmatrix};$ T_1 is T_2 and an enlargement $\times 2$.

18. $\dfrac{1}{5}\begin{pmatrix} -3 & 4 \\ 4 & 3 \end{pmatrix}; y^2 = x$.

19. $\begin{pmatrix} \cos 2(\phi - \theta) & -\sin 2(\phi - \theta) \\ \sin 2(\phi - \theta) & \cos 2(\phi - \theta) \end{pmatrix};$ a rotation through $2(\phi - \theta)$.

20. (a) $\begin{pmatrix} \cos 2\theta & \sin 2\theta \\ \sin 2\theta & -\cos 2\theta \end{pmatrix}$, (b) $\begin{pmatrix} \cos \theta & -\sin \theta \\ \sin \theta & \cos \theta \end{pmatrix}; \begin{pmatrix} 0 & 1 \\ 1 & 0 \end{pmatrix}, \dfrac{1}{2}\begin{pmatrix} -1 & \sqrt{3} \\ \sqrt{3} & 1 \end{pmatrix};$
$\dfrac{1}{2}\begin{pmatrix} \sqrt{3} & -1 \\ 1 & \sqrt{3} \end{pmatrix},$ rotation through $\dfrac{\pi}{6}; \begin{pmatrix} \sqrt{3} & -1 \\ 1 & \sqrt{3} \end{pmatrix}; (-1 - \sqrt{3}, -1 + \sqrt{3}).$

21. (a) $6:1; \dfrac{1}{6}\begin{pmatrix} 1 & -2 \\ 1 & 4 \end{pmatrix}$, $(1/3, -1/6); x+y = 0, x+2y = 0,$

    (b) $\begin{vmatrix} 4 & 2 \\ -2 & -1 \end{vmatrix} = 0; 2y = 1-4x; x+2y = 0.$

22. (i) anti-clockwise rotation through $90°$ about $(1, 1)$ and enlargement $\times 2$,
    (ii) reflection in the line $x+2y-2 = 0$.

Exercise 20.1

1. (a) even, (b) odd, (c) odd, (d) even, (e) neither, (f) odd.
3. (a) even, (b) odd, (c) neither, (d) neither, (e) odd, (f) even.
4. (a) odd, (b) even, (c) odd, (d) even, (e) neither, (f) even.

Exercise 20.2

5. $(0, -1); (0, -1).$
6. $(0, 1); (0, 1).$
7. $(1, -1); (1, -1).$
8. $(-2, 0);$ none.
9. $(2, 1); (2, 1).$
10. $(\frac{1}{2}, 2\frac{1}{4}); (\frac{1}{2}, \frac{4}{9}).$
11. $(-2, 4), (0, 0); (-2, \frac{1}{4}).$
12. $(-2, 16), (2, -16); (-2, \frac{1}{16}), (2, -\frac{1}{16}).$
15. $(1, -4); (1, 4).$
16. $(3, 0); (3, 0).$

Exercise 20.4

1. $x = -2, y = 0.$
2. $x = 4, y = 0.$
3. $x = 1\frac{1}{2}, y = 0.$
4. $x = -2, y = 1.$
5. $x = -2, y = 1.$
6. $x = 1\frac{1}{2}, y = 1.$
7. $x = -2, y = 2.$
8. $x = 3, y = 3.$
9. $x = \dfrac{2}{5}, y = -\dfrac{2}{5}.$
10. $y = \dfrac{x}{x+1}.$
11. $y = \dfrac{1-x}{1+x}.$

Exercise 20.5

2. (a) $x+4y = 12,$ (b) $y = x^2+2x+2,$ (c) $8x = (y+4)^2,$
   (d) $y = x+2$ $(x \geqslant -1),$ (e) $y = 1/(x-2),$ (f) $y = x^2+2x+2$ $(x \geqslant -1).$
3. (a) $4x^2+9y^2 = 36,$ (b) $25x^2-y^2 = 25,$
   (c) $4x^2+y^2-8x-2y+1 = 0,$ (d) $5x^2-6xy+2y^2 = 1,$
   (e) $2x^2+2xy+y^2 = 1.$
4. (a) $(0, 0), (9, 27),$ (b) $(-3, -3), (1\frac{1}{2}, 6),$ (c) $(1, -4), (16, 16),$
   (d) $(7, 9), (-\frac{7}{3}, \frac{25}{9}),$ (e) $(1, 0), (\frac{1}{2}, \frac{1}{2}),$ (f) $(-3, 4), (4, -3).$

Exercise 20.6

1. $t.$
2. $2/t.$

3. $1/4t^2$.

4. $1/6(1-2t)$.

5. $-\frac{3}{4}\cot t$.

6. $-\tan t$.

7. $-\frac{1}{2}t^3, \frac{3}{4}t^5$.

8. $\dfrac{1}{t} - t^2, \ -\dfrac{1}{12t^3} - \dfrac{1}{6}$.

9. $\dfrac{2t+1}{3t^2}, \ -\dfrac{2(t+1)}{9t^5}$.

10. $\dfrac{t}{t+1}, \ \dfrac{1}{2(t+1)^3}$.

11. $-2\sin t, \ -1$.

12. $-2\cot 2t, \ -4\operatorname{cosec}^3 2t$.

13. 9.

14. $(3t^2+1)y = 2tx + (t^2+1)^2$.

15. $y = x\operatorname{cosec}\theta - \cot\theta$.

16. $\dfrac{t-3}{15-9t}$; $(16, 9)$.

17. $(2,4), (10,0)$; $(4, 12\sqrt{3}-18), (4, -12\sqrt{3}-18)$.

Exercise 20.7

1. (a) neither, (b) odd, (c) even, (d) even, (e) even, (f) neither.

2. (a) $a = c = 0$, (b) $b = d = 0$, (c) $a = 0, b > 0, c^2 < 4bd$,
   (or $a = b = c = 0, d \geqslant 0$), (d) $a = 0, b < 0, c^2 < 4bd$ (or $a = b = c = 0, d \leqslant 0$).

3. $f(x) = k/x$.

6. $(-1, 0)$; $1, -3$; $1/\sqrt[3]{2}$, minimum.

7. $f(x)$: max $(0,0)$, min $(\pm\sqrt{2}, 1)$,

   $g(x)$: max $(\pm\sqrt{2}, 1)$, inflexion $\left(\pm\sqrt{\dfrac{10}{3}}, \dfrac{21}{25}\right)$.

9. (a) $(1, C), (2, E), (4, A), (5, B)$, (c) $y = \left|\dfrac{1}{x+1}\right|$ is a possibility.

12. $a > 4$: one; $0 < a < 4$: none; $-4 < a < 0$: two; $a < -4$: one.

13. (a) $9x^2 - 4y^2 = 144$, (b) $x^2 - y^2 = 1$,
    (c) $10x^2 - 2xy + 5y^2 = 49$, (d) $4x^2 - 6\sqrt{3}xy + 9y^2 = 9$.

14. $\dfrac{3t^2 - 4}{2t}$; $\left(\dfrac{8}{3}, \pm\dfrac{16}{9}\sqrt{3}\right)$.

15. $\dfrac{t-1}{t+1}$, $(4,0)$; $x + y = 2$.

16. $(-\frac{1}{2}, -6), (3,1)$; $2, y = 2x - 5 \ (x \neq 0)$.

17. $x = \dfrac{\sqrt{3}}{2}\sin(\theta+45°), y = \dfrac{1}{2}\cos(\theta+45°)$; $4x^2 + 12y^2 = 3$.

18. (a) $\dfrac{12t-14}{9+18t}, \ \dfrac{40}{81(1-2t)(1+2t)^3}$.

    (b) $-\cot\left(\dfrac{3t}{2}\right), \ -\dfrac{3}{8}\operatorname{cosec}^3\left(\dfrac{3t}{2}\right)\sec\left(\dfrac{t}{2}\right)$.

19. $(1 - \sin\theta - 2\sin^2\theta)/\cos\theta$, $(\frac{1}{2}, \pm\frac{3}{4}\sqrt{3})$, $(1,0)$; $1\cdot3\pi$.

Exercise 21.1

1. $x^2 + y^2 + 6x - 8y = 0$.

2. $x + y = 4$.

3. $x^2 = 4$.

4. $y = 4x + 13$.

5. $(x-y)(x+y-2) = 0$.

6. $x^2 = 4(y-1)$.

7. $xy = 9$.

8. $x^2 + y^2 - 2x - 4y + 1 = 0$.

9. $y^2 = 12x$.

10. $x^2 + y^2 - 12x - 2y + 21 = 0$.

11. $x^2 + y^2 + 2x - 2y = 0$.

12. $3x + y + 3 = 0$; $(-3, 6)$.

13. $y^2 = 2x$.

14. $(\frac{1}{2}t + 1, \frac{1}{4}t^2 - 2)$; $y = 2x^2 - 4x$.

15. (a) $x + y = \frac{1}{2}c$, (b) $xy = \frac{1}{2}k$, (c) $x^2 + y^2 = \frac{1}{4}l^2$.

16. $y = 2x^2$.

Exercise 21.2

1. (a) $x^2 + y^2 - 6x - 4y - 3 = 0$, (b) $x^2 + y^2 + 2x + 4y + 4 = 0$,
   (c) $x^2 + y^2 = 25$, (d) $x^2 + y^2 - x - 2 = 0$,
   (e) $x^2 + y^2 - 8x + 2y + 14 = 0$, (f) $x^2 + y^2 + 6x - 10y + 14 = 0$.

2. (a) $(1, 3)$, 3, (b) $(0, 0)$, 2, (c) $(-3, -4)$, 5,
   (d) $(2, -1)$, 1, (e) $(0, 0)$, $\frac{1}{2}\sqrt{5}$, (f) $(1\frac{1}{2}, -2\frac{1}{2})$, $\sqrt{5}$.

3. (a) $k > 0$, (b) $k = 1$, (c) No values of k,
   (d) $k = 0$, (e) $k < 10$, (f) $k > 4$ or $k < -4$.

4. (a) $y = 2x + 5$, (b) $x + y + 3 = 0$, (c) $y = 3x - 13$,
   (d) $x + 8y = 22$.

5. (a) 5, (b) 7, (c) 6, (d) $2\sqrt{10}$.

6. (a) $x^2 + y^2 - 2x - 6y = 0$, (b) $x^2 + y^2 - 6x - 4y - 7 = 0$,
   (c) $x^2 + y^2 - 2x + 4y - 15 = 0$, (d) $x^2 + y^2 + 6x - 16y + 23 = 0$.

7. (a) 24, 2, (b) 12, 2.

8. $x^2 + y^2 + 6x - 12y + 20 = 0$; (a) $y = 1$, $y = 11$, (b) $x = 2$, $x = -8$.

9. $x^2 + y^2 - 10x - 14y + 66 = 0$; $(5, 7)$, $2\sqrt{2}$.

10. $x^2 + y^2 - 6x - 10y + 9 = 0$; 27 sq. units.

11. $x^2 + y^2 - 4x - 14y + 28 = 0$.

12. $(1, 2)$, $(9/5, 18/5)$.

13. $x^2 + y^2 - 7x - 3y + 8 = 0$; $x^2 + y^2 - 8x + 2y + 4 = 0$.

Exercise 21.3

1. $y + 3x = 19$; $3y = x + 17$.

2. $y = 2x - 2$; $2y + x = 6$.

3. $x + y = 1$; $y = x + 3$.

4. $y = 5x + 22$; $5y + x = 32$.

5. $2y = x - 13$; $2x + y + 4 = 0$.

6. $9y + 7x = 16$; $7y = 9x - 2$.

7. $10y = 27x - 31$; $27y + 10x = 165$.

8. $4y = 9x - 6$; $9y + 4x = 35$.

9. $4y = 3x + 5$; $3y + 4x = 10$.

10. $4y = x + 16$; $y + 4x = 72$.

11. (a) $(2, 2)$, (b) $(3, -1)$, (c) $(2\frac{1}{2}, 1\frac{1}{2})$.

12. $5x - 4y + 34 = 0$.

13. $2x - y = 11$, $2x - y + 21 = 0$.

14. 1, 4.

15. $y = 3$ at $(-1, 3)$, $y = 12x - 21$ at $(5, 39)$.

16. $y = 0$, $3y = 4x$; $4/\sqrt{5}$.

17. 1, -7.

18. $\pm\sqrt{17}$, $3x + 4y = 0$.

19. $y = 3x \pm 10$.

20. $3y = 15 \pm 4x$; $73 \cdot 7°$.

21. $2x \pm 3y = 10$; $(16/5, 6/5)$, $(16/5, -6/5)$.

22. $x + ty = 2t + t^3$; $x^2 = 2y^2(y - 1)$.

23. $y = (p + q)x + p + q - pq$, $y = 2px + 2p - p^2$; $y = 0$.

24. $(p + q)y = (p^2 + pq + q^2)x - p^2q^2$; $2y = 3px - p^3$;
    $p = \frac{2}{3}\sqrt{2}$, $q = -\frac{1}{3}\sqrt{2}$ or $p = -\frac{2}{3}\sqrt{2}$, $q = \frac{1}{3}\sqrt{2}$.

Exercise 21.4

1. (a) $(1,0)$; $x = -1$, (b) $(3,0)$; $x = -3$, (c) $(-2,0)$; $x = 2$,
    (d) $(\frac{3}{4},0)$; $x = -5/4$, (e) $(\frac{1}{4},0)$; $x = -\frac{1}{4}$, (f) $(0,\frac{1}{4})$; $y = -\frac{1}{4}$,
    (g) $(2,0)$; $x = -2$, (h) $(3,0)$; $x = -3$.
2. (a) $y^2 = 8x$, (b) $y^2 = 20x$, (c) $x^2 = 4y$,
    (d) $y^2 + 12x = 0$, (e) $(y-1)^2 = 4x$, (f) $x^2 = 4(y-1)$.
3. (a) $y^2 = 4(x-2)$, (b) $(y-1)^2 = 12x$, (c) $(y-1)^2 = 4(x+2)$,
    (d) $x^2 - 8x + 2y + 11 = 0$.
4. (a) $(2t^2, 4t)$, (b) $(6t^2, 12t)$, (c) $(-4t^2, -8t)$.
5. (a) $y = 2x + 2$, $x + 2y = 9$, (b) $x + 2y + 6 = 0$, $y = 2x - 18$,
    (c) $2y = x + 4$, $2x + y = 12$, (d) $x + y + 3 = 0$, $y = x - 9$.
6. $y^2 = a(x-a)$.   7. $2y^2 + ax = 0$.
9. $mc = a$; $x + y + 1 = 0$, $2y = x + 4$.
10. $(apq, a(p+q))$; $y^2 = a(2x-a)$.
12. $-p$; $\dfrac{2}{p+q}$; $\left(\dfrac{121a}{9}, \dfrac{22a}{3}\right)$; $5\sqrt{5a}$.
13. $x + 2a = 0$.   14. $ty = x + at^2$.
15. (i) $x = 0$, (ii) $2y^2 = 9ax$.

Exercise 21.5

1. (a) $(2,3)$, $(1,4)$, $(-1,5)$, (b) $(1,1)$, $(0,2)$, $(-2,3)$,
    (c) $(0,5)$, $(-1,6)$, $(-3,7)$.
2. (a) $(1,2)$, (b) $(3,-1)$, (c) $(0,1)$, (d) $(-2,-3)$.
3. (a) $(3,0)$, $(4,0)$, $x = 2$, (b) $(0,1)$, $(2,1)$, $x = -2$,
    (c) $(-5,0)$, $(-4,0)$, $x = -6$, (d) $(-4,-2)$, $(-3\frac{3}{4}, -2)$, $x = -4\frac{1}{4}$,
    (e) $(0,1)$, $(0,2)$, $y = 0$, (f) $(1,-1)$, $(1,-\frac{3}{4})$, $y = -1\frac{1}{4}$.
4. (a) $(1,3)$, $(2,3)$, $x = 0$, (b) $(-2,5)$, $(1,5)$, $x = -5$,
    (c) $(1,0)$, $(0,0)$, $x = 2$, (d) $(-3,-1)$, $(-3,1)$, $y = -3$.

Exercise 21.6

1. $(0,-2)$, $\left(\sqrt{\dfrac{3}{2}}, -\dfrac{1}{2}\right)$, $\left(-\sqrt{\dfrac{3}{2}}, -\dfrac{1}{2}\right)$. 2. $c^2 = a^2(1+m^2)$.
3. $y = x + 2 - \frac{1}{2}\pi$, $x + y = \frac{1}{2}\pi$.   4. $8y = 9x + 7$, $8x + 9y = 26$.
5. $(1,0)$, 1; $(0,1)$, 1; $\left(\dfrac{2}{1+m^2}, \dfrac{2m}{1+m^2}\right)$, $\left(\dfrac{2m}{1+m^2}, \dfrac{2m^2}{1+m^2}\right)$.
6. $x^2 + y^2 - 10x - 8y + 16 = 0$; $(10,-6)$, 10, $53°$.
7. (i) $(3,4)$, 5, (iv) $y = 2x - 2 + 5\sqrt{5}$, $y = 2x - 2 - 5\sqrt{5}$.
8. (a) $x^2 + y^2 - 6x - 6y + 5 = 0$, (b) $3x^2 + 3y^2 - 6x - 26y + 3 = 0$.
9. $x \cos\theta + y \sin\theta = a$; $(a(2\cos\theta + 1), a(1 - \cos\theta - 2\cos^2\theta)/\sin\theta)$;
    $(0, 2a/\sqrt{3})$, $(0, -2a/\sqrt{3})$.
10. (i) $x^2 + y^2 - 2y - 3 = 0$, (ii) $4\pi/3$,
    (iii) $x^2 + y^2 - \sqrt{3}x - 3y = 0$; $x^2 + y^2 + 2y - 3 = 0$.
11. $x^2 + y^2 - 10x - 10y + 25 = 0$; $x^2 + y^2 - 34x - 34y + 289 = 0$;
    (i) $(2,9)$, (ii) $x + y = 11$.

12. $x^2 + y^2 - 5x - 6y + 9 = 0$; $y = 2x + 3$.

13. $y \cos \alpha + x \sin \alpha = a \sin \alpha \cos \alpha$.   14. ± 2.

15. $y \sin T + x = 1 + 2 \sin^2 T$; $\frac{1}{4}\pi$.

17. $y = \left(\frac{5}{4}x_0^2 - \frac{13}{9}\right)x - \frac{5}{6}x_0^3$; $\pm\frac{1}{3}\sqrt{5}$, $\pm\frac{2}{3}\sqrt{2}$.

18. $py = x + ap^2$, $y + px = 2ap + ap^3$.

19. $y + tx = 2at + at^3$, $(2a + at^2, 0)$; $\pm\sqrt{3}$, $\pm\sqrt{15}$.

20. (i) $4a$,   (ii) $y^2 = 2a(x - 4a)$.   21. $(apq, a(p + q))$.

22. $(a(2 + p^2 + pq + q^2), -apq(p + q))$; $y^2 = a(x - 3a)$.

24. $y = mx + a/m$.   25. $py = x + ap^2$.

26. $y = \frac{3}{4}x + 5$, $y = -\frac{3}{4}x - 5$.   27. $(a(1 + t^2), at)$; $(a, 0)$.

28. $(a\{2 + (p + q)^2 - pq\}, -apq(p + q))$; $y^2 = 16a(x - 6a)$; $(6a, 0)$.

31. $y + px = 2p + p^3$; $27y^2 = 4(x - 2)^3$, $(2, 0)$.

32. $y = px - p^3$, $y = qx - q^3$.   33. (ii) $(-a \sin^2 \theta, -a \cos^3 \theta/\sin \theta)$.

Exercise 22.1

1. (a) True,   (b) False,   (c) False,   (d) True.

2. (a) $x > 0$,   (b) all x,   (c) $x < 0$,   (d) $x < -a$ or $x > -b$.

Exercise 22.2

1. $-2 < x < 3$.
2. $x \leqslant -3$ or $x \geqslant -\frac{1}{2}$.
3. $1 \leqslant x \leqslant 5$.
4. $x < 1 - \sqrt{6}$ or $x > 1 + \sqrt{6}$.
5. $-2 - \sqrt{7} < x < -2 + \sqrt{7}$.
6. $x \leqslant -2$ or $0 \leqslant x \leqslant 5$.
7. $3 < x < 4$.
8. $x < -2$ or $x > -1$.
9. $2 < x < 3$.
10. $-3 < x < -\frac{1}{2}$.
11. $x < 0$ or $1 < x < 2$.
12. $x < -3$ or $0 < x < \frac{1}{2}$.
13. $-3 < x < -2$ or $x > 1$.
14. $0 < x < 1$ or $x > 3$.
15. $-3 < x < 3$ or $x > 5$.
16. $-1 < x < 1$.
17. $0 < x < 1\frac{1}{2}$ or $3 < x < 4$.
18. $-2 < x < -1$ or $x > 2$.
19. $x < -\frac{1}{2}\pi$ or $0 < x < \frac{1}{2}\pi$.
20. $x > \frac{1}{3}\pi$.

Exercise 22.3

1. $-5 < x < 3$.
2. $x < \frac{1}{3}$ or $x > 2$.
3. $-1 < x < 0$ or $x > 6$.
4. $x < 0$ or $1 < x < 4$.
5. $0 < x < 3$ or $x > 4$.
6. $0 < x < 2$ or $x > 3$.
7. $x < -3$ or $1 < x < 2$.
8. $-2 < x < 1\frac{1}{2}$ or $x > 5$.
9. $1 < x < 7$.
10. $1 < x < 3$.
11. $x < -4$ or $-2 < x < 2$.
12. $-2 < x < 4$ or $x > 5$.
13. $x < -1$, $1 < x < 2$, $x > 3$.
14. $x < -1\frac{1}{2}$, $-1 < x < 0$, $x > \frac{2}{3}$.
15. $-4 < x < -1$.
16. $-2 < x < 1\frac{1}{2}$ or $x > 5$.
17. $x < 0$ or $x > 2$.
18. $0 < x < 1\frac{1}{2}$ or $3 < x < 4$.

Exercise 22.4

1. $x < 1$ or $x > 3$.
2. $-8 \leqslant x \leqslant 2$.
3. $-3 < x < -\frac{1}{3}$.
4. $x \leqslant -5$ or $x \geqslant 6$.
5. $x \leqslant \frac{1}{2}$.
6. $x < -7/3$ or $x > -1$.
7. $x \leqslant 0$ or $x \geqslant 6$.
8. $-5/6 < x < 5/2$.
9. $x < 4/3$ or $x > 4$.
10. $-3/2 \leqslant x \leqslant -1$.
11. $x > -3/5$.
12. All real x.
13. $-1 < x < 0$ or $3 < x < 4$.
14. $x < -1, 2 < x < 3, x > 6$.
15. $x < -8$ or $x > -8/3$.
16. $-4 \leqslant x \leqslant -1$ or $1 \leqslant x \leqslant 4$.

Exercise 22.5

1. $x \leqslant -7$ or $x \geqslant 1$.
2. $1 - \sqrt{2} < x < 1 + \sqrt{2}$.
3. $x < -1$ or $-\frac{1}{2} < x < 1$.
4. $x < 0$ or $2 < x < 5$.
5. $x < -2$ or $x > 2$.
6. $x < -\sqrt{3}, -\sqrt{2} < x < \sqrt{2}, x > \sqrt{3}$.
7. $x < 0$ or $x > 3$.
8. $-1 < x < -1/\sqrt{2}$ or $1/\sqrt{2} < x < 1$.
9. $-2 < x < -1$ or $x > 4$.
10. $x < -1$ or $x > 2$.
11. $x < -1, -\frac{2}{3} < x < 3, x > 4$.
12. $x < -5$ or $x > 3$.
13. $-\frac{1}{3} \leqslant x \leqslant 7$.
14. $-2 < x < 2$.
15. $\frac{1}{2} < x < 1$.
16. $-6 < x < 1$ or $2 < x < 3$.
17. $x < 0$.
18. $x < 4/3$ or $x > 4$.
19. (a) $-\dfrac{\pi}{3} < x < \dfrac{\pi}{3}, \dfrac{5\pi}{3} < x < \dfrac{7\pi}{3}$,

    (b) $-\dfrac{\pi}{4} \leqslant x \leqslant \dfrac{\pi}{4}, \dfrac{3\pi}{4} \leqslant x \leqslant \dfrac{5\pi}{4}, \dfrac{7\pi}{4} \leqslant x \leqslant \dfrac{9\pi}{4}$,

    (c) $-\dfrac{\pi}{6} < x < \dfrac{\pi}{6}, \dfrac{5\pi}{6} < x < \dfrac{7\pi}{6}, \dfrac{11\pi}{6} < x < \dfrac{13\pi}{6}$.

22. $\dfrac{\pi}{6} < x < \dfrac{\pi}{2}$ or $\dfrac{5\pi}{6} < x < \dfrac{3\pi}{2}$.

23. (i) $x < -1$ or $x > 0$, (ii) $0 < x < 2$, (iii) $x < 1$ or $x > 2$.

24. $-3 < x < -1$ or $x > 2$, (ii) $x < -2$ or $x > 2$.

Exercise 23.1

1. $0, 0, Ps, -Fs$.
2. $-Wx, Px$.
3. $Ws \sin \alpha, 0; Fs$.
4. $Ts, 0; Fs, Ws \sin \alpha$.
5. $105\,\text{J}, 15\,\text{m}$.
6. $-9000\,\text{J}, 450\,\text{N}$.
7. $45\,\text{kN}$.
8. $770\,\text{N}$.
9. $62 \cdot 5\,\text{J}$.
10. $300\,\text{J}; 5/8$.
11. $200\,\text{J}; 45$.
12. $270\,\text{J}, 180\,\text{J}; 6\,\text{m s}^{-1}$.

Exercise 23.2

1. $10 \cdot 5\,\text{m s}^{-1}$.
2. (a) $21 \cdot 6\,\text{m}$ above A, (b) $40\,\text{m}$ below A.
3. $v^2/2g$.
4. $4 \cdot 05\,\text{m}$.

5. $5\,\mathrm{m\,s^{-1}}$.
6. $14\,\mathrm{m\,s^{-1}}$.
7. $2{\cdot}8\,\mathrm{m\,s^{-1}}$.
8. $214\,\mathrm{J}$.
9. $147\,\mathrm{J}$; $7\,\mathrm{m}$.
10. $(M+m)v^2/2(M-m)g$.
11. $1{\cdot}4\,\mathrm{m\,s^{-1}}$.
12. $10\,\mathrm{m}$.
13. $8\,\mathrm{m}$.
14. (a) $3{\cdot}5\,\mathrm{m\,s^{-1}}$, (b) $2{\cdot}6\,\mathrm{m}$.
15. $19{\cdot}6\,\mathrm{J}$; $\frac{1}{4}$.

Exercise 23.3

1. (a) $0{\cdot}2\,\mathrm{J}$, (b) $5\,\mathrm{J}$.
2. (a) $0{\cdot}25\,\mathrm{J}$, (b) $9\,\mathrm{J}$.
3. $2{\cdot}25\,\mathrm{J}$.
4. $15\,\mathrm{N}$.
5. $7{\cdot}5\,\mathrm{J}$.
6. $12k$, $3ka$.
7. $0{\cdot}56\,\mathrm{m}$.
8. $2\,\mathrm{m\,s^{-1}}$.
9. $\frac{1}{2}\sqrt{(ag)}$.
10. $4\,\mathrm{m}$.
11. $7\,\mathrm{m\,s^{-1}}$.
12. $60\,\mathrm{cm}$; $1{\cdot}4\,\mathrm{m\,s^{-1}}$.
13. $2{\cdot}5\,\mathrm{N}$.
14. $1{\cdot}2\,\mathrm{m}$; (a) $\dfrac{7}{5}\sqrt{3}$, (b) $0{\cdot}7\,\mathrm{m}$.

Exercise 23.4

1. $392\,\mathrm{W}$.
2. $40\,\mathrm{s}$.
3. $400\,\mathrm{W}$.
4. $500\,\mathrm{kg}$.
5. $28{\cdot}8\,\mathrm{kW}$.
6. $6{\cdot}8\,\mathrm{kW}$.
7. $335\,\mathrm{W}$.
8. $72\,\mathrm{kN}$.
9. $1{\cdot}2\,\mathrm{kW}$; $0{\cdot}1\,\mathrm{m\,s^{-2}}$.
10. (a) $12\,\mathrm{kW}$, (b) $14{\cdot}7\,\mathrm{kW}$, (c) $3\,\mathrm{kW}$.
11. $21\,\mathrm{kW}$.
12. $30\,\mathrm{m\,s^{-1}}$.
13. (a) $33\frac{1}{3}\,\mathrm{kW}$, (b) $52\,\mathrm{kW}$, (c) $14\frac{2}{3}\,\mathrm{kW}$.
14. $6\frac{5}{6}\,\mathrm{kN}$, $13\frac{1}{3}\,\mathrm{m\,s^{-1}}$.
15. (a) $96\,\mathrm{km/h}$, (b) $36\,\mathrm{km/h}$; $1/6\,\mathrm{m/s^2}$.
16. $0{\cdot}98\,\mathrm{m\,s^{-2}}$.
17. 2, $1{\cdot}4\,\mathrm{m/s^2}$.
18. $20\,\mathrm{kW}$; (i) $6{\cdot}93\,\mathrm{kW}$, (ii) $0{\cdot}201\,\mathrm{m\,s^{-2}}$.

Exercise 23.5

1. $1/8$.
2. $30/49$.
3. $10\,\mathrm{N}$.
4. $53{\cdot}1°$.
5. $68{\cdot}6\,\mathrm{J}$, $7/12$.
6. $2{\cdot}4\,\mathrm{m}$.
7. $2{\cdot}8\,\mathrm{m\,s^{-1}}$.
8. $7\,\mathrm{m\,s^{-1}}$, $2{\cdot}4\,\mathrm{m}$.
9. 3, $2{\cdot}57\,\mathrm{m\,s^{-1}}$.
10. $\dfrac{3}{2}\sqrt{(ga)}$, $2a$.
11. $30\,mg/19$.
13. (i) $2\sqrt{(2gb)}$, (ii) $\sqrt{(10gb)}$, (iii) $26{\cdot}6°$, (iv) $4b$; $-\frac{1}{2}c$.
14. (i) $2b-l$, (ii) b^2/l.
15. $4{\cdot}5 \times 10^4\,\mathrm{N}$, $450\,\mathrm{kW}$; $24\,\mathrm{m\,s^{-1}}$.
16. $1800\,\mathrm{N}$, $804\frac{1}{6}\,\mathrm{N}$.
17. $320\,\mathrm{kW}$.
18. $113\,\mathrm{kg}$, $22{\cdot}3\,\mathrm{kW}$.
19. (i) $19\frac{7}{11}\,\mathrm{km\,h^{-1}}$, (ii) $0{\cdot}53\,\mathrm{m\,s^{-2}}$; 60.

20. 500 N, 20 m/s.
21. (a) $(1000H - MVg \sin \alpha)/MV$, (b) $(1000H + MVg \sin \alpha)/MV$; $1000H/3MVg$, $3V/4$.
22. 400 kW, 15 m/s, 20 000 N.   23. $0.395 \, \mathrm{m \, s^{-2}}$.
24. (i) 0·25 m/s,   (ii) 22·5 m/s; $233\frac{1}{3}$ m, $15\frac{5}{9}$ s.
25. 1200 N; 1/40; 10·5 kW.
26. $R + mg \sin \alpha + 2mU^2 \sin \alpha/h$, $2U(R + mg \sin \alpha + 2mU^2 \sin \alpha/h$, $mgh + 2mU^2 + Rh \operatorname{cosec} \alpha$.

Exercise 24.1

1. $\dfrac{3\pi}{2} \mathrm{rad \, s^{-1}}, 0.42 \, \mathrm{m \, s^{-1}}$.   2. $4 \, \mathrm{rad \, s^{-1}}$.
3. $7.3 \times 10^{-5} \mathrm{rad \, s^{-1}}, 470 \, \mathrm{m \, s^{-1}}$.   4. $2 \times 10^{-7} \mathrm{rad \, s^{-1}}, 30 \, \mathrm{km/s}$.
5. 57 cm.
6. (a) $\dfrac{1}{1+t^2}$,   (b) $\dfrac{6}{9+4t^2}$,   (c) $\dfrac{2t}{1+t^4}$.
7. (a) $r(\cos \omega t \mathbf{i} + \sin \omega t \mathbf{j})$,   (b) $r(-\sin \omega t \mathbf{i} + \cos \omega t \mathbf{j})$.
8. (a) $2 \, \mathrm{rad \, s^{-1}}, 6 \, \mathrm{m \, s^{-1}}$,   (b) $1 \, \mathrm{rad \, s^{-1}}, 5 \, \mathrm{m \, s^{-1}}$,
   (c) $-2\pi \, \mathrm{rad \, s^{-1}}, 2\pi \, \mathrm{m \, s^{-1}}$,   (d) $k \, \mathrm{rad \, s^{-1}}, ak \, \mathrm{m \, s^{-1}}$.
9. $a(\cos \omega t \mathbf{i} + \sin \omega t \mathbf{j})$, $a(-\sin \omega t \mathbf{i} + \cos \omega t \mathbf{j})$,
   $a\omega(\sin \omega t - \cos \omega t)\mathbf{i} - a\omega(\cos \omega t + \sin \omega t)\mathbf{j}$.

Exercise 24.2

1. (a) $9 \, \mathrm{m \, s^{-2}}$,   (b) $4 \, \mathrm{m \, s^{-2}}$,   (c) $20 \, \mathrm{m \, s^{-2}}$,   (d) $63.2 \, \mathrm{m \, s^{-2}}$.
2. $22\frac{1}{2}$ N.   3.   12 N.
4. 375 N, 49 N.   5. $6 \, \mathrm{m \, s^{-1}}$.
6. 0·4.   7.   0·7 m.
8. $1.75 \, \mathrm{rad \, s^{-1}}$.

Exercise 24.3

1. 49 N, $2.8 \, \mathrm{m \, s^{-1}}$.   2.   127·4 N, 3 m.
3. $mg \operatorname{cosec} \theta$; $rg \cot \theta$.   4.   2·45 m.
5. $3.15 \, \mathrm{m \, s^{-1}}$.   6. $\sqrt{(3ag)}$.
7. $5m(8u^2 + 9ag)/72a$ in *AP*, $5m(8u^2 - 9ag)/72a$ in *BP*.
8. $17\frac{1}{2} \mathrm{m \, s^{-1}}$.   9.   17·7°.
10. $35 \, \mathrm{m \, s^{-1}}, 5\sqrt{7} \, \mathrm{m \, s^{-1}}$.   11. $7\sqrt{5} \, \mathrm{m \, s^{-1}}$; 12·7°.

Exercise 24.4

1. $10\pi/9 \, \mathrm{rad \, s^{-1}}, 0.42 \, \mathrm{m \, s^{-1}}$.   2. $25/\pi$, 50.
3. $-6/(5t^2 - 6t + 9)$.   4. $6 \, \mathrm{m \, s^{-1}}$ S 30° W.
5. $5/36 \leqslant \mu < 5/9$.   6.   60 N, 75 N.
7. (i) 60°,   (ii) $2mg$,   (iii) $\pi\sqrt{(2a/g)}$ s.
8. 60°.   9.   (i) $\sqrt{5}$,   (ii) 1/6 N.

10. $\frac{1}{2}ma(2\omega^2 + \Omega^2)$, $\frac{1}{2}ma(\Omega^2 - 2\omega^2)$.
11. $\sqrt{\{g(a^2 - h^2)/h\}}$.　　　　　　　12. $\frac{2}{3}mg\sqrt{3}$, $m(3u^2 - gl\sqrt{3})/3l$.
13. $m(v^2\cos\alpha + gl\sin^2\alpha)/l\sin\alpha$, $m(v^2 - gl\cos\alpha)/l$.
14. $36\cdot3°$; $97\cdot8\,\mathrm{m\,s^{-1}}$.　　　　　　15. $mg(\sqrt{3} - 1)$.
16. $\sqrt{(2rg/11)}$ to $\sqrt{(2rg)}$.　　　　　17. $\sqrt{(45ag/8)}$.

Exercise 25.1

5. (a) step diagram,　(b) ogive.
6. One possibility is $32\cdot5$, $37\cdot5$, $42\cdot5$, $\ldots$, $67\cdot5$, $72\cdot5$; 88%.
7. 9/14, excluding the 10 late entrants (possibly transfers from schools in other
    areas).

Exercise 25.2

1. (a) 3, 3, $3\frac{1}{2}$.　(b) no mode, 24, 20,　(c) 20 and 21, 21, $20\frac{2}{3}$.
2. (a) 3, 3, 5,　(b) no mode, 22, 20,　(c) 20, $20\frac{1}{2}$, $20\frac{3}{5}$.
3. (a) 5, 5, 5,　(b) 5, $5\frac{1}{14}$, 5,　(c) 5, $5\frac{4}{7}$, $5\frac{1}{2}$.
4. 3, 3, $3\cdot64$.　　　　　　　　　　　　5. $6\cdot84$.
6. (i) $52\cdot5 \leqslant x < 57\cdot5$, $54\frac{3}{8}$, 55,　(ii) 56, 56, $54\cdot175$.
7. (i) $4\cdot96$, $5\cdot14$,　(ii) $4\cdot93$, $4\cdot95$.
8. $26\cdot6$; (a) 27,　(b) $27\cdot2$,　(c) $27\cdot1$.

Exercise 25.3

1. (a) 7; $5\frac{1}{2}$, $8\frac{1}{2}$; $1\frac{1}{2}$,　(b) 29; 27, 33; 3,　(c) 2; -4, 10; 7.
2. (a) $7\cdot25$, $5\cdot19$, $2\cdot28$,　(b) 31, $71\cdot8$, $8\cdot47$,　(c) 3, $39\cdot8$, $6\cdot31$.
3. (a) (i) 5, 6, (ii) 1;　(b) (i) $4\cdot8$, $5\cdot7$, (ii) $0\cdot8$;
    (c) (i) $5\cdot3$, $6\cdot2$, (ii) $0\cdot8$.
4. $6\cdot31$, $2\cdot51$.
5. (i) (a) 5, 7, 9, (b) $4\cdot9$, $6\cdot9$, $8\cdot8$;　(ii) $2\cdot45$.
6. (i) $49\cdot5$, $59\cdot5$,　(ii) $48\cdot9$, $60\cdot3$,　(iii) 49, 60.
7. (i) 4 yrs 10 months, 5 yrs 1 month,　(ii) $46\cdot5$ months^2.
8. I: $38-$,　II: $28-37$,　III: $19-27$,　IV: $0-18$.

Exercise 25.4

1. $5\cdot5$, $6\cdot25$, $2\cdot5$; (a) $8\cdot5$, $6\cdot25$, $2\cdot5$,　(b) 22, 100, 10,
    (c) $4\cdot55$, $0\cdot0625$, $0\cdot25$.
2. $73\cdot4°\mathrm{F}$, $9°\mathrm{F}$.　　　　　　3. 5, $1\cdot5$, $1\cdot22$; $5\cdot5$, $1\cdot5$, $1\cdot22$.
4. (i) $7\cdot84$,　(ii) $7\cdot98$.　　　　　5. $27\cdot2$, $9\cdot69$; $1\cdot5$, $9\cdot2$.
6. Mathematics: $72\frac{1}{2}$, 77, $57\frac{1}{2}$; English: $60\frac{1}{2}$, $51\frac{1}{2}$, $87\frac{1}{2}$;
    Combined: 133, $128\frac{1}{2}$, 145.

Exercise 25.5

1. $86\cdot6$, $44\cdot1$; $188\cdot1$.　　　　　　2. $6\cdot5$; $4\cdot2$, $9\cdot0$.
3. (i) 57 min,　(ii) $71\cdot5$ min,　(iii) 32%.
4. mean 55 kg, s.d. $3\cdot2$ kg.　　　　　5. 0, 1.

6. (i) 100·7, 14·5,   (ii) 72.
7. (b) 8·95, 9·95,   (c) 13·53,   (d) 13·55.
8. (i) £28·87,   (ii) £54·04,   (iii) £30·47.
9. 52°, 7·5°; 53·2°, 10·7°.
10. 172·1 cm, 2·608 cm; 174·6 cm, 6·8 cm^2; 1·054 in^2.
11. £49, £12·85.
12. (a) 5, 3·6,   (b) 5·4, 5·04,   (c) $4·5 < \bar{x} < 5·0$,
    (d) $x_{10} \geqslant 8$ or $x_{10} \leqslant 2$.
13. (ii) 6·8299; 39·23, 6·6335.
14. (a) (i) 3·4, 2, (ii) 0·4, 0·276;   (b) $3\frac{1}{3}$, 19·6%.

Index